Lecture Notes in Chemistry

Volume 98

The Lecture Notes in Chemistry

The series Lecture Notes in Chemistry (LNC) reports new developments in chemistry and molecular science-quickly and informally, but with a high quality and the explicit aim to summarize and communicate current knowledge for teaching and training purposes. Books published in this series are conceived as bridging material between advanced graduate textbooks and the forefront of research. They will serve the following purposes:

- provide an accessible introduction to the field to postgraduate students and nonspecialist researchers from related areas,
- provide a source of advanced teaching material for specialized seminars, courses and schools, and
- be readily accessible in print and online.

The series covers all established fields of chemistry such as analytical chemistry, organic chemistry, inorganic chemistry, physical chemistry including electrochemistry, theoretical and computational chemistry, industrial chemistry, and catalysis. It is also a particularly suitable forum for volumes addressing the interfaces of chemistry with other disciplines, such as biology, medicine, physics, engineering, materials science including polymer and nanoscience, or earth and environmental science.

Both authored and edited volumes will be considered for publication. Edited volumes should however consist of a very limited number of contributions only. Proceedings will not be considered for LNC.

The year 2010 marks the relaunch of LNC.

More information about this series at http://www.springer.com/series/632

Fabien Gatti · Benjamin Lasorne
Hans-Dieter Meyer · André Nauts

Applications of Quantum Dynamics in Chemistry

Springer

Fabien Gatti
Institut des Sciences Moléculaires d'Orsay (ISMO)
University of Paris-Sud
Orsay, Essonne
France

Hans-Dieter Meyer
Institute of Physical Chemistry, Theoretical Chemistry Group
University of Heidelberg
Heidelberg
Germany

Benjamin Lasorne
Institute Charles Gerhardt, CNRS5253,
Université de Montpellier
Montpellier
France

André Nauts
Faculté des Sciences, Ecole de Physique, and Institute of Condensed Matter and Nanosciences
Université Catholique de Louvain
Louvain-la-Neuve
Belgium

and

Laboratoire de Chimie Physique (UMR8000)
Université de Paris-Sud
Orsay, Essonne
France

Additional material to this book can be downloaded from http://extras.springer.com

ISSN 0342-4901 ISSN 2192-6603 (electronic)
Lecture Notes in Chemistry
ISBN 978-3-319-53921-8 ISBN 978-3-319-53923-2 (eBook)
DOI 10.1007/978-3-319-53923-2

Library of Congress Control Number: 2017932629

Printed on acid-free paper

This Springer imprint is published by Springer Nature
The registered company is Springer International Publishing AG
The registered company address is: Gewerbestrasse 11, 6330 Cham, Switzerland

Foreword

Quantum physics is certainly one of the most astonishing intellectual constructions ever achieved. Its Rosetta stone consisted in some unexplained regularities in the fluorescence light of simple elements, such as hydrogen or helium atoms. The next building block was the radiation emitted by an oven in thermal equilibrium, the famous blackbody problem. From these mere facts and some strokes of genius, quantum theory was built. Quite rapidly the intuition of the founding fathers became a well-established theory, still full of surprises and paradoxical conclusions, but entirely consistent and with an unprecedented range of validity. Quantum theory applies to microscopic objects like nuclei, atoms, molecules, as well as to macroscopic systems such as superconductors and superfluids, and even astrophysical objects. Quantum physics is simultaneously a framework where uncertainty is included at the core of the theory, and a method of calculation that agrees with experimental facts with an outstanding precision, at the level of 10^{-12} for some observables, like the gyromagnetic ratio of the electron.

The present book explains how to use the quantum formalism in order to address the dynamics of molecular systems, and it provides a nice illustration of the diversity of the quantum world. I had the chance to have one of its four authors, Fabien Gatti, in my quantum mechanics class about twenty years ago, and it is a real pleasure to see how some simple ideas that we had been discussing at that time have flourished in such a fruitful manner. The book is organized in a remarkably progressive way, starting with simple systems like the hydrogen molecule, and then moving towards notably more complex molecular edifices. This smooth progression is quite valuable for readers with a physics background like me, who were taught by A. Schawlow, 1981 physics Nobel Prize winner, that "a diatomic molecule is a molecule with one atom too many"! This statement illustrates the fear that potential readers could have had in front of the subject if they were not properly guided. To tell the whole truth, Schawlow himself knew a lot about "non-monoatomic molecules" and he performed outstanding experiments in molecular spectroscopy.

When reading through "Application of Quantum Dynamics in Chemistry", one realizes the wealth of phenomena that can occur in the dynamics of two- or several-atom molecules. With well-conceived lab sessions, the authors guide the reader through key molecular processes such as photo-dissociation, simple chemical reactions and coherent control. The key notion is the concept of molecular wave-packets, created and probed using ultra-short pulses of light. This is at the basis of the rapidly developing field of femto-chemistry initiated by the late Ahmed Zewail, 1999 chemistry Nobel Prize winner, whose major discoveries were acknowledged in both the chemistry and the physics communities. With femto-chemistry it is now possible to manipulate chemical reactions, using proper light pulses that influence the breaking of particular molecular bonds.

An appealing aspect of this book is that it will ring many bells in the mind of physicists, in addition of course to the chemistry community towards which it is targeted at first. Fundamental notions like quantum coherence revealed by interferences between various quantum paths, atomic or molecular states dressed with laser light, are very dear to a physicist's heart. These physicists will thus find in many instances a renewed playground for such notions, which illustrates once more the universality of quantum concepts. For instance, the authors discuss the Berry phase accumulated when travelling along a path around in a conical intersection; the essence of this phenomenon is very similar to what is found for the properties of graphene in condensed matter physics, or in quantum gas physics, when atoms move in an optical lattice with Dirac points in the Brillouin zone. In a different, but related perspective, the Multi-Configuration Time-Dependent Hartree (MCTDH) package that is intensively exploited here to address the dynamics of molecular edifices can be used to approach other problems emerging in quantum many-body physics. This is notably the case for the non-linear Schrödinger equation that governs the evolution of light beams in non-linear materials, or of interacting Bose–Einstein condensates, two research domains that have been also very active worldwide over the last two decades.

Overall I am convinced that F. Gatti, B. Lasorne, H.-D. Meyer and A. Nauts have produced a very useful text, with a clear orientation towards solving practical problems. All the necessary background is presented in a pedagogical manner, with numerous illustrations. Molecular spectra now replace the hydrogen fluorescence lines deciphered by the founding fathers, but the desire to understand complex phenomena from simple microscopic modeling is still present. I am convinced that the material developed in the book will constitute a perfect guide for the reader who wishes to start a fruitful journey in the rapidly developing field of Quantum Molecular Dynamics.

Paris, France
November 2016

Jean Dalibard

The original version of the book was revised: Second affiliation for book author has been included. The erratum to the book is available at https://doi.org/10.1007/978-3-319-53923-2_14

Acknowledgements

We want to express our gratitude to all those who, by their assistance and encouragement, have made this book possible and, in particular, have helped to improve the text by their constructive comments and criticism.

First and foremost, it is a particular pleasure for us to warmly thank Jean Dalibard for his inspiring foreword in which he emphasizes the importance of interdisciplinarity and the universality of quantum concepts.

The material of this book is largely based on our research but also on lectures and seminars we have given to a variety of audiences during the past 10 to 15 years so that it has been largely tested through teaching.

For instance, in 2016 in Rolla, at the Missouri University of Science and Technology (USA) in the group of Richard Dawes, in Beijing at the Peking University (China) in the group of Jian Liu and in Dalian (China) in the groups of Zhigang Sun and Dong-Hui Zhang in the context of the French-Chinese network of theoretical chemistry (GDRI RFCCT).

Fabien Gatti also tested the content of the book through teaching in the context of the French Network of Theoretical Chemistry (RFCT) and, together with Benjamin Lasorne and André Nauts, in the context of the CNRS thematic summer schools (Aussois 2009 and Gif-sur-Yvette 2013). Benjamin Lasorne also gave lectures on related topics during an intensive course of the European Master in Theoretical Chemistry and Computational Modeling organized in Porto in 2010.

We are particularly grateful to the Ecole de Physique of the Université catholique de Louvain (Louvain-la-Neuve) for repeatedly inviting Fabien Gatti over the last 15 years. It initiated an intensive master-level teaching collaboration with André Nauts, combining an introduction to the fundamentals of molecular dynamics with hands-on lab-sessions in MCTDH example simulations. The idea of the present book largely arose out of this longstanding collaboration.

Many people contributed to improving this book and some of its chapters. We would like to thank them here for their help and for their critical reading of parts of the manuscript.

David Lauvergnat read preliminary versions of Chaps. 3 (*Molecular Hamiltonian Operators*) and 6 (*The Kinetic Energy Operator in Curvilinear Coordinates*). We thank him for his insightful comments and remarks.

A large part of Chap. 8 (*Introduction to Numerical Methods*) is extracted from the MCTDH lecture notes of Hans-Dieter Meyer written by Daniel Pelaez-Ruiz. The authors sincerely thank him for authorizing them to use his script.

Section 10.2 (*Indirect Photo Dynamics*) is based mainly on Steve Ndengué's PhD work. The authors gratefully thank him for his contribution and his help. Very fruitful discussions with Rémy Jost on the photodissociation of ozone are also gratefully acknowledged.

Section 12.3 (*Importance of Interferences Around Conical Intersections*) is partly based on the work of Loïc Joubert-Doriol with Artur Ismaylov. The authors thank Loïc Joubert-Dorriol for having read this section and for his constructive help.

Chapter 13 (*Control of Molecular Processes*) is based mainly on Matthieu Sala's PhD work. The authors gratefully thank him for his contribution and his help. They also thank Stéphane Guérin, Matthieu Sala's PhD supervisor, for having read and corrected this chapter.

We also thank Roberto Marquardt who gave Fabien Gatti his lectures notes on molecular quantum dynamics. Very helpful and lively discussions with him on quantum dynamics are also gratefully acknowledged.

Lorenz Cederbaum, Horst Koeppel, Xavier Chapuisat, Claude Leforestier, Christophe Iung, Yves Justum and David Lauvergnat are the initiators of many ideas and methods presented in this book. We very warmly thank them for the enlightening, fruitful and in-depth exchanges we had over the years.

Finally, we would like to thank all the students, as well as our colleagues, whose questions, constructive comments and criticism greatly helped to improve our teaching, our lecture notes and, in the end, the present book.

We thank the French National Center for Scientific Research (CNRS), the French Agence Nationale de la recherche (ANR), the German Deutsche Forschungsgemeinschaft (DFG), the universities of Montpellier, Paris-Saclay, Paris-Sud, Louvain-la-Neuve and Heidelberg for their continuous support to our work.

We are grateful to our collaborators from Springer Verlag for their constructive help and their efficiency.

Contents

Chapter 1
Introduction

The main goal of this book is to illustrate how the concept of a *wavepacket* becomes central in quantum mechanics when turning to concrete applications, for instance in molecular physics and chemistry. In other words, *the Schrödinger equation in its time-dependent form provides the central framework* here. This book is not a textbook: the foundations of quantum mechanics have been detailed in many seminal references (see for instance [1, 2]). More recently, D. Tannor has given a comprehensive description of the time-dependent perspective in quantum mechanics [3].

Here, our approach is different and complementary. Indeed, the most original aspect of the present book is to propose lab-sessions using the Heidelberg Multi-Configuration Time-Dependent Hartree (MCTDH) package. The latter is freely available and can be easily installed. MCTDH can be seen as an algorithm to solve the time-dependent Schrödinger equation (i.e. to propagate wavepackets) for multidimensional dynamical systems consisting of distinguishable particles [4–9]. The present book and the lab-sessions have two levels: one more dedicated to Master's students, typically for advanced courses on quantum mechanics for physicists or on theoretical chemistry for chemists. The teachers and the students can then use MCTDH as a black box to visualize the time evolution of quantum systems and observe pure quantum effects, on which special emphasis will be placed. At a higher level, the book may come in useful for thematic schools for Ph.D. students and postdocs in different fields of quantum physics or computational chemistry. Here, we offer the possibility to have a look at the input files, the users can even change themselves the characteristics of the wavepackets they propagate. The lab-sessions should also be helpful for any scientist who wishes to learn how to use the MCTDH package.

The processes presented as illustrations correspond to realistic situations involving several degrees of freedom. Most of the examples have been chosen so that the students can compare their results with data that have been measured experimentally (photoabsorption spectra, cross sections, etc.). The systems studied in this book are all *molecular systems*, i.e. *we apply quantum physics (more precisely quantum*

F. Gatti et al., *Applications of Quantum Dynamics in Chemistry*,
Lecture Notes in Chemistry 98, DOI 10.1007/978-3-319-53923-2_1

dynamics) to chemistry. However, *the concepts developed in the present book can be easily transferred to applications in physics and even in biology*. Generally speaking, quantum dynamics is a very diverse field ranging from entangled photons to biologically relevant response to laser light, cold atoms and molecules, etc. In this context it is worth noting that MCTDH has been extended to treat the dynamics of fermions (MCTDHF) or bosons (MCTDHB) and is commonly used to study Bose-Einstein condensates [10–13] and the dynamics of electrons in solid-state physics, atomic physics or in the context of attophysics [14–20].

The authors of the present book are all researchers in the field of "molecular quantum dynamics", an emerging field at the border between quantum physics and chemistry. We do not intend to give a general introduction to this field here and we refer the readers to a previous book edited in 2014 in the series "physical chemistry in action" by Springer [21]. We just remind that, from the point of view of a physicist, a molecule can be viewed as a quantum-mechanical aggregate composed of electrons and nuclei. However, chemistry is rarely taught within a full quantum-mechanical perspective. In particular, an elementary chemical process is generally not described as the evolution of a quantum system. However, very early within the advent of quantum mechanics, it became clear that the wave aspects of electrons could not be neglected as well as the quantization of the electronic states. This led to the field known as *quantum chemistry* with the concepts of molecular orbitals and potential energy surfaces. On the other hand, the motion of the nuclei, which is crucial in chemistry since there is no chemical process if the nuclei in molecules do not change their relative arrangement (the *reaction coordinate* describes a collective motion of the nuclei), is often conceived classically.

However, there is growing evidence that a significant number of chemical reactions are impacted by strong quantum-mechanical effects involving both the electrons *and the nuclei*. Let us simply consider two examples. Chemical reaction rates, when light particles such as protons, hydrogen atoms, and hydride ions are exchanged, can be greatly enhanced by quantum tunnelling, namely by the fact that particles can tunnel through a barrier that they classically cannot surmount [22–25]. Another quantum effect is the involvement of *quantum resonances* in reactivity. Resonant states are metastable states, the nuclei being temporarily trapped during a reactive collision. The presence of resonances can change a chemical reaction decisively, in particular at low temperatures [26–28].

Perhaps even more important is the fact that such quantum effects can be used to create a radically new chemistry at a higher level of efficiency and selectivity. For instance, lasers are sources of coherent light. After absorption of the latter by a molecular system, a coherent superposition of quantum eigenstates can be produced, i.e., a *molecular wavepacket*. Indeed, we know that, in quantum mechanics, a system can be in a *coherent superposition* of different quantum states. The different components of the quantum superposition can interfere, yielding new properties that can be measured and that have no classical counterpart. The exact definitions of a wavepacket and quantum coherence will be given in Chap. 2. We just mention here that experimentalists can now produce vibrational wavepackets in a systematic way. In addition, time-resolved pump-probe laser techniques allow them to study chemical processes on the femtosecond time scale (10^{-15} s), i.e., on the scale of a typical period of molecular vibrations [29–31]. This was at the origin of the development of femtochemistry that earned Ahmed Zewail the 1999 Nobel Prize in Chemistry. This technique allows one to follow the motion of the nuclei in real time: when chemical bonds break, form, or change geometrically. Moreover, the quantum coherence itself can be exploited to change and guide the reactivity [32, 33]. The quantum coherence can be preserved during a time that is sufficient to drastically modify the reactivity of complex systems even when they are embedded in an environment (in general, since a system is never isolated, it interacts with its environment that dissipates quantum coherence). In particular, recent experiments provided observation of long-lived electronic quantum coherence, after excitation by laser pulses, for energy transfer processes in light-harvesting complexes of biological systems such as photosynthetic systems at ambient temperature and in the condensed phase [34].

Since the advent of femtochemistry, remarkable and decisive progress has been achieved on the experimental front with the possibility to align molecules with lasers [35, 36] and study electron motion using tools from the new field of attophysics [37–40]. In other words, it is now feasible to create rotational, vibrational, and electronic wavepackets and to control all the different aspects of a chemical elementary act. In particular, we are close to what some already call the field of *attochemistry* where, at each step of a molecular process, the *coupled motion of electrons and nuclei* could be controlled on its natural time scale [41]. This will clearly lead, on the long term, to a major breakthrough: a new chemistry working at an elementary microscopic level and based on the systematic use of quantum phenomena. It is thus not surprising that molecular quantum dynamics and the description of molecular systems in terms of wavepackets has become an enormously active field of research.

As we will show in the present book, for a theoretician, a wavepacket has a broader significance than a quantum state created in a molecule after absorption of the light produced by a laser. Most of the wavepackets that we will encounter in the different chapters can be considered as "artificial" wavepackets, in the sense that their utility is often independent of whether these wavepackets can be created or not experimentally. From their mathematical properties, wavepackets can be exploited to obtain indirectly observable data including all the quantum effects involved in the process. We will see, for instance in the applications and in the corresponding lab-sessions, that the propagation of wavepackets can provide access to photoabsorption spectra of molecules or to cross sections of collisions.

The book is divided into two parts: Part I, Concepts and Methods; Part II, Applications. Some parts of the book are labelled as advanced topics, indicated by a asterisk in the table of contents. They can be skipped for teaching at the Master's level. In Part I, Chap. 2 is a very brief reminder of the main concepts of quantum mechanics that are essential to the understanding of the book: eigenstate, wavepacket, coherent superposition, quantum decoherence, etc. Chapter 3 presents the molecular Hamiltonian operator along with the Born-Oppenheimer separation, i.e. the separation of the electronic and nuclear motions. The cases where the Born-Oppenheimer approximation fails are described in detail by the introduction of the diabatic representation and of conical intersections. Chapter 6 is dedicated to the derivation of the nuclear kinetic energy operator in any set of coordinates and we provide examples of operators for several molecular systems. In Chap. 5, we address the major issue of the choice of the set of nuclear coordinates used to describe molecular processes. Again several examples are given. The last chapter of Part I (Chap. 8) contains a rather thorough description of the numerical methods used to solve the Schrödinger equation for the nuclei. Special emphasis is placed on the MCTDH algorithm.

Part II focuses on illustrations of the theoretical background described in the first Part. At the end of each chapter, we propose a lab-session using the MCTDH package that allows one to visualize the evolution of wavepackets for realistic systems. The input files and a version of the code are provided as supporting material. As they are conceived, the lab-sessions are rather adapted to Ph.D. students and postdocs. However, teachers can easily simplify the lab-sessions (by removing some parts of the text) so that Master's students can use MCTDH as a "black box". It is worth noting that Sect. 9.6 explains how to install the code.

For each chapter of application, there is a preliminary part explaining the physical context with several figures. We have tried to present a wide variety of processes from infrared spectroscopy (Chap. 9) to photodissociation processes (Chap. 10); from non-adiabatic (or non-Born-Oppenheimer) couplings (Chap. 12) to reactive collisions (Chap. 11). To recall that wavepackets are not only "artificial" tools for theoreticians, and that they can also be produced and measured in experiments, we conclude the book with applications in the field of coherent control by laser pulses (Chap. 13).

References

1. Cohen-Tannoudji C, Diu B, Laloe F (1992) Quantum mechanics. Wiley-VCH
2. Messiah A (1962) Quantum mechanics, vol 1. Wiley, New York
3. Tannor DJ (2007) Introduction to quantum dynamics: a time-dependent perspective. University Science Books, Sausalito, Cal
4. Meyer H-D, Manthe U, Cederbaum LS (1990) The multi-configurational time-dependent Hartree approach. Chem Phys Lett 165:73
5. Manthe U, Meyer H-D, Cederbaum LS (1992) Wave-packet dynamics within the multiconfiguration Hartree framework: general aspects and application to NOCl. J Chem Phys 97:3199
6. Beck MH, Jäckle A, Worth GA, Meyer H-D (2000) The multi-configuration time-dependent Hartree (MCTDH) method: a highly efficient algorithm for propagating wave packets. Phys Rep 324:1
7. Meyer H-D, Worth GA (2003) Quantum molecular dynamics: propagating wavepackets and density operators using the multiconfiguration time-dependent Hartree (MCTDH) method. Theor Chem Acc 109:251
8. Meyer H-D, Gatti F, Worth GA (eds) (2009) Multidimensional quantum dynamics: MCTDH theory and applications. Wiley-VCH, Weinheim
9. Meyer H-D (2012) Studying molecular quantum dynamics with the multiconfiguration time-dependent Hartree method. WIREs: Comput Mol Sci**2**:351
10. Sakmann K, Streltsov AI, Alon OE, Cederbaum LS (2005) Exact ground state of finite Bose-Einstein condensates on a ring. Phys Rev A 72(3):033613
11. Lode AUJ, Streltsov AI, Alon OE, Meyer H-D, Cederbaum LS (2009) Exact decay and tunneling dynamics of interacting few boson systems. J Phys B 42:044018

12. Lode AUJ, Streltsov AI, Alon OE, Meyer H-D, Cederbaum LS (2010) Corrigendum: exact decay and tunneling dynamics of interacting few boson systems. J Phys B 43:029802
13. Sakmann K, Streltsov AI, Alon OE, Cederbaum LS (2010) Quantum dynamics of attractive versus repulsive Josephson junctions: Bose-Hubbard and full-Hamiltonian results. Phys Rev A 82:013620
14. Zanghellini J, Kitzler M, Fabian C, Brabec T, Scrinzi A (2003) An MCTDHF approach to multi-electron dynamics in laser fields. Laser Phys 13:1064
15. Caillat J, Zanghellini J, Scrinzi A (2004) Parallelization of the MCTDHF code. AURORA Tech Rep 19:1
16. Caillat J, Zanghellini J, Kitzler M, Koch O, Kreuzer W, Scrinzi A (2005) Correlated multielectron systems in strong laser fields—an MCTDHF approach. Phys Rev A 71:012712
17. Kato T, Kono H (2004) Time-dependent multiconfiguration theory for electronic dynamics of molecules in an intense laser field. Chem Phys Lett 392:533
18. Nest M, Klamroth T, Saalfrank P (2005) The multiconfiguration time-dependent Hartree-Fock method for quantum chemical calculations. J Chem Phys 122:124102
19. Hochstuhl D, Bauch S, Bonitz M (2010) Multiconfigurational time-dependent Hartree-Fock calculations for photoionization of one-dimensional helium. J Phys: Conf Ser 220:012019
20. Haxton DJ, Lawler KV, McCurdy CW (2011) Multiconfiguration time-dependent Hartree-Fock treatment of electronic and nuclear dynamics in diatomic molecules. Phys Rev A 83:063416
21. Gatti F (ed) (2014) Molecular quantum dynamics. Springer, Heidelberg
22. McMahon RJ (2003) Chemical reactions involving quantum tunneling. Science 299:833
23. Wu T, Werner H-J, Manthe U (2004) First-principles theory for the $H + CH_4 \rightarrow H_2 + CH_3$ reaction. Science 306:2227
24. Wassermann TN, Luckhaus D, Coussan S, Suhn MA (2006) Proton tunneling estimates for malonaldehyde vibrations from supersonic jet and matrix quenching experiments. Phys Chem Chem Phys 8:2344
25. Giese K, Petković M, Naundorf H, Kühn O (2006) Multidimensional quantum dynamics and infrared spectroscopy of hydrogen bonds. Phys Rep 430:211
26. Qiu M, Ren Z, Che L, Dai D, Harich SA, Wang X, Yang X, Xu C, Xie D, Gustafsson M, Skoedje RT, Sun Z, Zhang D-H (2006) Observation of Feshbach resonances in the $F + H_2 \rightarrow HF + H$ reaction. Science 311:1440
27. Althorpe SC (2010) Setting the trap for reactive resonances. Science 327:1460
28. Dong W, Xiao C, Wang T, Dai D, Yang X, Zhang DH (2010) Transition-state spectroscopy of partial wave resonances in the F+HD. Science 327:1501
29. Chergui M (ed) (1996) Femtochemistry. World Scientific, Singapore
30. Zewail AH (1994) Femtochemistry—ultrafast dynamics of the chemical bond. World Scientific, Singapore
31. Ihee H, Lobastov V, Gomez U, Goodson B, Srinivasan R, Ruan C-Y, Zewail AH (2001) Direct imaging of transient molecular structures with ultrafast diffraction. Science 291:385
32. Shapiro M, Brumer P (2003) Principles of the quantum control of molecular processes. Wiley, New York
33. Fleming GR, Ratner MA (2008) Grand challenges in basic energy sciences. Phys Today 61:28
34. Engel GS, Calhoun TR, Read EL, Ahn T-K, Mancal T, Cheng Y-C, Blankenship RE, Fleming GR (2007) Evidence for wavelike energy transfer through quantum coherence in photosynthetic systems. Nature 446:782
35. Holmegaard L, Hansen JL, Kalhøj L, Kragh SL, Stapelfeldt H, Filsinger F, Küpper J, Meijer G, Dimitrovski D, Martiny C, Madsen LB (2010) Photoelectron angular distributions from strong-field ionization of oriented molecules. Nat Phys 6:428
36. Madsen CB, Madsen LB, Viftrup SS, Johansson MP, Poulsen TB, Holmegaard L, Kumarappan V, Jorgensen KA, Stapelfeldt H (2009) Manipulating the torsion of molecules by strong laser pulses. Phys Rev Lett 102:073007
37. Drescher M, Hentschel M, Kienberger R, Uiberacker M, Scrinzi A, Westerwalbesloh T, Kleineberg U, Heinzmann U, Krausz F (2002) Time-resolved atomic inner-shell spectroscopy. Nature 419:803

38. Goulielmakis E, Loh Z-H, Wirth A, Santra R, Rohringer N, Yakovlev VS, Zherebtsov S, Pfeifero T, Azzeer AM, Kling MF, Leone SR, Krausz F (2010) Real-time observation of valence electron motion. Nature 466:739
39. Kling MF, Siedschlag C, Verhoef AJ, Khan JI, Schultze M, Uphues T, Ni Y, Uiberacker M, Drescher M, Krausz F, Vrakking MJJ (2006) Control of electron localization in molecular dissociation. Science 312:246
40. Niikura H, Légaré F, Hasbani R, Bandrauk AD, Ivanov MY, Villeneuve DM, Corkum PB (2002) Sub-laser-cycle electron pulse for probing molecular dynamics. Nature 417:917
41. Kuleff AI, Cederbaum LS (2012) Ultrafact correlation-driven electron dynamics. J Phys B 47:124002

Part I
Concepts and Methods

Chapter 2
Quantum Mechanical Background

Quantum mechanics is certainly one of the most successful theories in science. It has deeply influenced many areas of pure and applied physics and pervades many branches of science, from physics, matter sciences, computer science to chemistry and even to molecular biology. However, quantum mechanics has to face several conceptual difficulties of which most relate to the process of quantum measurement and its randomness so that, almost one century after its birth, a complete consensus has still not been reached concerning the interpretation of the theory and its foundations.

Quantum mechanics is also known to be *counter-intuitive* and to lead to representations of physical phenomena very different from our daily experience, such as *superposition, entanglement* and *non-locality*. To pinpoint this microscopic "quantum strangeness", the fathers of the quantum theory, especially Einstein and Bohr, resorted to "*thought experiments*" involving the manipulations of isolated particles. These experiments, which were believed to remain virtual, have now been performed (see "Exploring the Quantum" by Haroche and Raimond [1]) and may help shed light on the conceptual difficulties mentioned above. In their turn, they may also raise other intriguing issues such as, for instance, the connection between quantum and classical physics. Indeed, macroscopic systems, i.e. systems directly accessible to our senses, never display non-locality and other strange features of quantum mechanics such as *state superposition* (Schrödinger's cat) or *quantum interference*.

A nice, lively and non-technical overview of the major interpretations and strangeness of quantum mechanics can be found in "Beyond measure" by Baggot [2]. More in-depth and technical discussions are given in "Do we really understand Quantum Mechanics?" by Laloë [3].

It is perhaps also worth mentioning that, by casting doubts on fundamental concepts such as space, material objects, and causality quantum mechanics demands serious *reconsideration* of most of traditional philosophy and has become a central issue in the realm of the *philosophy of science*. A huge amount of literature exists on the subject and we will only mention here two examples: "On Physics and Philosophy" by d'Espagnat [4] and "Making Sense of Quantum Mechanics" by Bricmont [5].

F. Gatti et al., *Applications of Quantum Dynamics in Chemistry*,
Lecture Notes in Chemistry 98, DOI 10.1007/978-3-319-53923-2_2

Nevertheless, *in spite of all these conceptual difficulties, the extraordinary feature of quantum mechanics is that, although we do not understand it nor know how to interpret it, we can apply it* and, by means of the rules of calculation it inspires, compute *properties of matter with unparalleled accuracy*. However, in the present book, we will adopt what can be viewed as a *pragmatic approach* in which quantum mechanics is regarded as an operational theory designed to predict the outcomes of measurements on physical systems under well-defined conditions.

The purpose of the present chapter is not to provide a general and detailed introduction to quantum mechanics, which is available in most textbooks, e.g. Refs. [6–15]. Our purpose here is merely to provide a *quantum mechanical background* with emphasis on the concepts and rules needed for the various topics covered in the book. (for a lively and pedagogical further reading see "Lectures on Quantum Mechanics" by Basdevant [16].)

2.1 General Principles

2.1.1 Wavefunctions

In quantum mechanics, the *state of a system*, at a given time t, parametrized by a set of coordinates $\boldsymbol{R}$, is completely determined by a complex *wavefunction*, $\Psi(\boldsymbol{R}, t)$ [10]. In our case, the system will often be a *molecular system* in the wider sense of the term, and the coordinates $\boldsymbol{R}$, a set of *generalized or curvilinear coordinates well-adapted* to the description and the evolution in time of the molecular and chemical processes under consideration. According to the standard interpretation of quantum mechanics, the square of the modulus of $\Psi(\boldsymbol{R}, t)$,

$$|\Psi(\boldsymbol{R}, t)|^2 = \Psi^\star(\boldsymbol{R}, t)\Psi(\boldsymbol{R}, t)\,, \tag{2.1}$$

has the meaning of a *probability density*. The *probability* of finding the system in the volume $d\boldsymbol{R}$ around the point $\boldsymbol{R}$ at time t, is given by

$$d\mathcal{P}(\boldsymbol{R}, t) = |\Psi(\boldsymbol{R}, t)|^2 d\boldsymbol{R}\,. \tag{2.2}$$

The wavefunction $\Psi(\boldsymbol{R}, t)$ is also called the *probability amplitude* of finding the system at point $\boldsymbol{R}$. It is square integrable and, in view of the probabilistic interpretation, must be normalized to unity[1]:

$$\int |\Psi(\boldsymbol{R}, t)|^2 d\boldsymbol{R} = 1\,. \tag{2.3}$$

[1]In the following, $\int$ denotes the integral over the complete domain of definition.

2.1.2 Superposition Principle

Another property of wavefunctions, regarded as a fundamental principle of quantum mechanics, is the *superposition principle*: it means that, if $\Psi_1(\boldsymbol{R}, t)$ and $\Psi_2(\boldsymbol{R}, t)$ describe possible states for the system, any linear combination

$$\Psi(\boldsymbol{R}, t) = c_1\Psi_1(\boldsymbol{R}, t) + c_2\Psi_2(\boldsymbol{R}, t), \tag{2.4}$$

where c_1 and c_2 are arbitrary complex coefficients, also represents a possible state. The additivity of probability *amplitudes* is at the origin of *interference phenomena* in quantum mechanics (see Eq. (2.38) below). Moreover, from a more theoretical perspective, this additivity property hints to the fact that the set of all the possible wavefunctions of a given system has the properties of a *linear vector space* $\mathcal{E}$ (see below).

The non-classical aspect of the superposition principle is illustrated by Schrödinger's famous cat, which can be alive and dead simultaneously. In other words, *Schrödinger's cat* can be in a *coherent superposition* of both a dead state and an alive state. These two states can interfere to create new behaviors that cannot be observed for a cat that is either alive or dead. For instance, let us consider an assembly of, let us say, one thousand Schrödinger's cats. This assembly is not constituted of cats that are either dead or alive but of one thousand cats where the dead and alive aspects interact to create a completely different behavior. Schrödinger's cat can be seen as a *paradox* only because the cat is a large-scale system and creating a coherent superposition for a cat is not realistic.

2.1.3 Measurements of Physical Quantities

One of the most intriguing features of quantum mechanics is, as indicated by its name, the hypothesis of *quantization*. This notion implies that, under particular conditions, physical observables measured experimentally can only take certain discrete values. More precisely, in quantum mechanics, physical quantities measured experimentally can only take a *restricted set of values* that can be discrete or continuous or a mixture of both.

This is the reason why, whereas in classical mechanics the physical observables are represented by functions of time such as position, in quantum mechanics, to each physical quantity $\mathcal{A}$, we associate an observable $\hat{A}$, which is a *linear Hermitian operator* acting in the space $\mathcal{E}$ of *wavefunctions* called a *Hilbert space*.

More explicitly, let us define in the space $\mathcal{E}$ a *Hermitian scalar product*[2] of two wavefunctions $\Psi(\boldsymbol{R})$ and $\Phi(\boldsymbol{R})$ as follows

[2]Equation (2.5) is also known as the *overlap between the wavefunctions* $\Psi(\boldsymbol{R})$ and $\Phi(\boldsymbol{R})$.

$$\int \Phi^\star(\boldsymbol{R})\Psi(\boldsymbol{R})d\boldsymbol{R} = \langle\Phi|\Psi\rangle\,, \tag{2.5}$$

where the so-called Dirac *bracket notation* is introduced. This bracket notation is very elegant and allows, among other things, to avoid to explicitly specify the coordinates, $\boldsymbol{R}$, their conjugate momenta $\boldsymbol{P}$, or any other set of coordinates, used to describe the system. Combined with the superposition principle, this leads naturally to considering $|\Psi\rangle$, called a "*ket*", as a *vector* of space $\mathcal{E}$, and $\langle\Phi|$ called a "bra" as a linear form[3] that acts on the ket $|\Psi\rangle$ to yield the "*bracket*" $\langle\Phi|\Psi\rangle$, which, being a scalar product, is generally a complex number.

$|\Psi\rangle$ and $|\Phi\rangle$ are quantum *state vectors* corresponding to the "$\boldsymbol{R}$ representation" wavefunctions $\Psi(\boldsymbol{R})$ and $\Phi(\boldsymbol{R})$. The *normalization condition* of Eq. (2.3) now reads

$$\langle\Psi|\Psi\rangle = 1\,, \tag{2.6}$$

and the *orthogonality condition*

$$\langle\Phi|\Psi\rangle = \int \Phi^\star(\boldsymbol{R})\Psi(\boldsymbol{R})d\boldsymbol{R} = 0 \tag{2.7}$$

may be regarded as an extension of the geometrical notion of orthogonality to state vectors and wavefunctions.

An *operator* $\hat{A}$ transforms any given vector $|\Psi\rangle$ into another vector $|\Psi'\rangle = |\hat{A}\Psi\rangle = \hat{A}|\Psi\rangle$. When acting on a vector $\alpha|\Psi\rangle + \beta|\Phi\rangle$, where α and β are complex numbers, such as

$$\hat{A}(\alpha|\Psi\rangle + \beta|\Psi\rangle) = \alpha\hat{A}|\Psi\rangle + \beta\hat{A}|\Phi\rangle\,. \tag{2.8}$$

the operator is called a *linear operator*. We will work only with linear operators. *Matrix elements* of an operator $\hat{A}$ are written as

$$\langle\Psi|\hat{A}|\Phi\rangle = \langle\Psi|\hat{A}\Phi\rangle = \int \Psi^\star(\boldsymbol{R})\hat{A}\Phi(\boldsymbol{R})d\boldsymbol{R}\,. \tag{2.9}$$

---

[3] In a finite Hermitian space $\mathbb{C}^n$, the "kets" are the column matrices $|u\rangle = \begin{bmatrix} u_1 \\ u_2 \\ \vdots \\ u_n \end{bmatrix}$ and the "bra" are the row matrices $\langle v| = \left[v_1^\star v_2^\star \dots v_n^\star\right]$ so that $\langle v|u\rangle = \sum_{i=1}^n v_i^\star u_i$ is a matrix product.

The *product* of two operators $\hat{A}$ and $\hat{B}$ is defined as $(\hat{A}\hat{B})|\Psi\rangle = \hat{A}(\hat{B}|\Psi\rangle)$. It is interesting to note that multiplying a ket by a bra, $|\Psi\rangle\langle\Phi|$ is an operator.[4] Indeed, applied to a vector $|\varphi\rangle$, it yields another vector: $|\Psi\rangle\langle\Phi|\varphi\rangle = \langle\Phi|\varphi\rangle|\Psi\rangle$, where $\langle\Phi|\varphi\rangle$ is a complex number. In particular, $\hat{\Pi}_\Psi = |\Psi\rangle\langle\Psi|$ is the *projection* operator onto $|\Psi\rangle$, since

$$\hat{\Pi}_\Psi^2 = |\Psi\rangle\langle\Psi|\Psi\rangle\langle\Psi| = |\Psi\rangle\langle\Psi| = \hat{\Pi}_\Psi \tag{2.10}$$

holds, which shows that $\hat{\Pi}_\Psi$ is a projector, and since

$$\hat{\Pi}_\Psi|\Psi\rangle = |\Psi\rangle\langle\Psi|\Psi\rangle = |\Psi\rangle \tag{2.11}$$

holds, which shows that the projector projects onto $|\Psi\rangle$. Note that a normalized $|\Psi\rangle$ is assumed.

The *commutator* of two operators $\hat{A}$ and $\hat{B}$ is defined as

$$\left[\hat{A}, \hat{B}\right] = \hat{A}\hat{B} - \hat{B}\hat{A}\,. \tag{2.12}$$

When $\left[\hat{A}, \hat{B}\right] = 0$, which implies $\hat{A}\hat{B} = \hat{B}\hat{A}$, the two operators are said *to commute*.

For each linear operator $\hat{A}$, there exists an *adjoint operator*,, $\hat{A}^\dagger$, defined as follows:

$$\langle\Psi|\hat{A}^\dagger|\Phi\rangle = \langle\hat{A}\Psi|\Phi\rangle = \langle\Phi|\hat{A}|\Psi\rangle^\star \tag{2.13}$$

or, in integral form, with wavefunctions

$$\begin{aligned}\langle\Psi|\hat{A}^\dagger|\Phi\rangle &= \int \Psi^\star(\boldsymbol{R})\hat{A}^\dagger\Phi(\boldsymbol{R})d\boldsymbol{R} = \int (\hat{A}\Psi(\boldsymbol{R}))^\star\Phi(\boldsymbol{R})d\boldsymbol{R} \\ &= \left(\int \Phi^\star(\boldsymbol{R})\hat{A}\Psi(\boldsymbol{R})d\boldsymbol{R}\right)^\star = \langle\Phi|\hat{A}|\Psi\rangle^\star\,.\end{aligned} \tag{2.14}$$

An operator $\hat{A}$ is *Hermitian* if

$$\hat{A}^\dagger = \hat{A}\,. \tag{2.15}$$

If $\hat{A}$ is Hermitian, its *expectation value*, i.e. $\langle\Psi|\hat{A}|\Psi\rangle$ for a given state vector $|\Psi\rangle$, is real. Indeed, in view of Eq. (2.14) and since $\hat{A}^\dagger = \hat{A}$

$$\langle\Psi|\hat{A}|\Psi\rangle^\star = \langle\Psi|\hat{A}^\dagger|\Psi\rangle = \langle\Psi|\hat{A}|\Psi\rangle\,. \tag{2.16}$$

---

[4]In a finite Hermitian space $\mathbb{C}^n$, $|u\rangle\langle v| = \begin{bmatrix} u_1 \\ u_2 \\ \vdots \\ u_n \end{bmatrix} \left[v_1^\star v_2^\star \dots v_n^\star\right] = \begin{bmatrix} u_1v_1^\star & u_1v_2^\star & \dots & u_1v_n^\star \\ u_2v_1^\star & u_2v_2^\star & \dots & u_2v_n^\star \\ \vdots & \vdots & \vdots & \vdots \\ u_nv_1^\star & u_nv_2^\star & \dots & u_nv_n^\star \end{bmatrix}$, which is a $(n \times n)$ matrix, i.e. an operator in $\mathcal{C}^n$.

A *nonzero* vector $|\varphi_k\rangle$ is said to be an *eigenvector* of an operator $\hat{A}$ if

$$\hat{A}|\varphi_k\rangle = a_k|\varphi_k\rangle\,, \tag{2.17}$$

and a_k is the *eigenvalue* associated with this eigenvector. The set $\{a_k\}$ is called the *spectrum* of $\hat{A}$. For simplicity, we assume the spectrum to be discrete and non-degenerate (i.e. there are no two or more equal eigenvalues).

If $\hat{A}$ is Hermitian, the eigenvalues a_k are real and the eigenvectors corresponding to different eigenvalues are orthogonal. Thus, taking eigenvectors normalized to unity, we have

$$\langle\varphi_k|\varphi_l\rangle = \delta_{kl}\,, \tag{2.18}$$

where δ_{kl} is the *Kronecker delta*. In addition, the set $\{\varphi_k\}$ of normalized eigenvectors forms a complete set of orthonormal basis vectors (*spectral theorem*), i.e. any state vector $|\Psi\rangle$ can be expanded as follows

$$|\Psi\rangle = \sum_k c_k|\varphi_k\rangle\,, \tag{2.19}$$

or, in terms of wavefunctions,

$$\Psi(\boldsymbol{R}) = \sum_k c_k\varphi_k(\boldsymbol{R})\,, \tag{2.20}$$

where the c_k are complex coefficients. In fact,

$$c_k = \langle\varphi_k|\Psi\rangle = \int \varphi_k^\star(\boldsymbol{R})\Psi(\boldsymbol{R})d\boldsymbol{R}\,, \tag{2.21}$$

and, since $\Psi(\boldsymbol{R})$ is normalized to unity, i.e. $\langle\Psi|\Psi\rangle = 1$,

$$\sum_k |c_k|^2 = 1\,. \tag{2.22}$$

If two observables *commute*, there exists a basis of *eigenvectors common* to the two operators.

To summarize: to each *physical quantity* $\mathcal{A}$ corresponds an *observable* $\hat{A}$ that is a Hermitian linear operator acting in the space of wavefunctions $\mathcal{E}$, and characterized by its spectrum, i.e. the set $\{a_k\}$ of its eigenvalues and the set $\{\varphi_k(\boldsymbol{R})\}$ of the corresponding normalized eigenfunctions, which constitute an orthonormal basis set of $\mathcal{E}$ (spectral theorem).

We are now ready to state[5] the *measurement principles* of quantum mechanics:

(a) In the measurement of a physical quantity $\mathcal{A}$, the only possible results of the measurement are the eigenvalues a_k of the corresponding observable $\hat{A}$ (it is also known as the *quantization principle*).

(b) Owing to the *spectral theorem*, the wavefunction *before* the measurement can be expressed in terms of the normalized eigenfunctions of $\hat{A}$:

$$\Psi(\boldsymbol{R}) = \sum_k c_k \varphi_k(\boldsymbol{R}) . \tag{2.23}$$

When the measurement occurs, the *probability* of finding the value a_k as *result* is

$$\mathcal{P}(a_k) = |c_k|^2 = |\langle \varphi_k | \Psi \rangle|^2 . \tag{2.24}$$

The above equation is also known as *Born's probability rule.*

(c) Immediately after the measurement of the physical quantity $\mathcal{A}$ has been performed and *has given the result* a_k, the new state wavefunction of the system is the (normalized) eigenfunction $\varphi_k(\boldsymbol{R})$. This "instantaneous" change from $\Psi(\boldsymbol{R})$ to $\varphi_k(\boldsymbol{R})$ is known as the *reduction of the wavepacket* or *wavefunction collapse*. According to the standard interpretation of quantum mechanics, this wavefunction collapse is due to the interaction between the system and the measuring apparatus.[6]

As already briefly mentioned, for a given wavefunction $\Psi(\boldsymbol{R})$, the *expectation value* of any physical quantity $\mathcal{A}$ is given by

$$\int \Psi^\star(\boldsymbol{R}) \hat{A} \Psi(\boldsymbol{R}) d\boldsymbol{R} = \langle \Psi | \hat{A} | \Psi \rangle . \tag{2.25}$$

Introducing Eq. (2.23) into Eq. (2.25) yields

$$\begin{aligned}
\int \Psi^\star(\boldsymbol{R}) \hat{A} \Psi(\boldsymbol{R}) d\boldsymbol{R} &= \sum_k \sum_l c_k^\star c_l \int \varphi_k^\star(\boldsymbol{R}) \hat{A} \varphi_l(\boldsymbol{R}) d\boldsymbol{R} \\
&= \sum_k \sum_l c_k^\star c_l \int \varphi_k^\star(\boldsymbol{R}) a_l \varphi_l(\boldsymbol{R}) d\boldsymbol{R} \\
&= \sum_k \sum_l c_k^\star c_l a_l \int \varphi_k^\star(\boldsymbol{R}) \varphi_l(\boldsymbol{R}) d\boldsymbol{R} \\
&= \sum_k \sum_l c_k^\star c_l a_l \delta_{kl} ,
\end{aligned} \tag{2.26}$$

[5] We do it mainly in terms of wavefunctions since the "$\boldsymbol{R}$ representation" will be predominantly used in the present book.

[6] The mechanism and meaning of the wavepacket collapse is a difficult and much debated topic in quantum mechanics. For an introduction see, for instance Chap. 8 of Ref. [3].

so that

$$\langle\Psi|\hat{A}|\Psi\rangle = \sum_l |c_l|^2 a_l = \sum_l \mathcal{P}(a_l)a_l\,, \tag{2.27}$$

which is the usual definition of the *expectation value* in standard probability theory. As pointed out in Chap. 9 of Ref. [3], in the majority of experiments, what is really observed is the sum over a very large number of particles of one individual microscopic observable (sum of atomic dipoles for instance), which is accurately described by the *average value of this observable*, i.e. the *expectation value* (see Eq. (2.27)).

2.1.4 Time Evolution Principle and Schrödinger Equation

As long as the system does not undergo any observation, its time evolution is postulated to be given by the *time-dependent Schrödinger equation* (TDSE):

$$\hat{H}(t)\Psi(\boldsymbol{R},t) = i\hbar\frac{\partial\Psi(\boldsymbol{R},t)}{\partial t}\,, \tag{2.28}$$

with $\hbar = \frac{h}{2\pi}$, the *reduced Planck constant*. The operator $\hat{H}(t)$ is the *energy observable* or *Hamiltonian operator*, which may or may not depend on time.

If the *Hamiltonian operator is time-independent*, i.e. if *the system is isolated*, one may solve the eigenvalue equation

$$\hat{H}\Psi_n(\boldsymbol{R}) = E_n\Psi_n(\boldsymbol{R})\,, \tag{2.29}$$

which is known as the *time-independent Schrödinger equation* (TISE).

Equation (2.28) can then be formally solved as:

$$\Psi(\boldsymbol{R},t) = e^{-i\hat{H}t/\hbar}\Psi(\boldsymbol{R},t=0)\,, \tag{2.30}$$

where $e^{-i\hat{H}t/\hbar}$ is known as the *evolution operator*.

Then, two cases must be distinguished:

(i) At a time of reference, say $t = 0$, *the wavefunction is equal to one of the eigenfunctions* given by Eq. (2.29):

$$\Psi(\boldsymbol{R},t=0) = \Psi_n(\boldsymbol{R})\,. \tag{2.31}$$

Inserting Eq. (2.31) into Eq. (2.30) yields

$$\Psi(\boldsymbol{R},t) = e^{-iE_nt/\hbar}\Psi_n(\boldsymbol{R})\,. \tag{2.32}$$

The time dependence of such a state is *periodic*, with angular frequency $\omega_n = E_n/\hbar$. As regards the *probability density*,

$$\begin{aligned} |\Psi(\boldsymbol{R},t)|^2 &= \Psi^\star(\boldsymbol{R},t)\Psi(\boldsymbol{R},t) \\ &= e^{iE_nt/\hbar}\Psi_n^\star(\boldsymbol{R})e^{-iE_nt/\hbar}\Psi_n(\boldsymbol{R}) \\ &= \Psi_n^\star(\boldsymbol{R})\Psi_n(\boldsymbol{R}) = |\Psi_n(\boldsymbol{R})|^2 \,. \end{aligned} \tag{2.33}$$

Thus, the probability density does not depend on time. In other words, the system does not evolve and is said to be in *a stationary state*. To each stationary state corresponds a definite value of the energy of the system, E_n.

(ii) At $t = 0$, *the wavefunction is a linear combination of the eigenfunctions* given by Eq. (2.29). Such a linear combination is often called a *coherent superposition* or also a *wavepacket*.

In other words,

$$\Psi(\boldsymbol{R}, t = 0) = \sum_n c_n \Psi_n(\boldsymbol{R}) \,, \tag{2.34}$$

in the *discrete case* and

$$\Psi(\boldsymbol{R}, t = 0) = \int c(E)\Psi_E(\boldsymbol{R})dE \,, \tag{2.35}$$

in the *continuous case*. For systems with both discrete and continuous portions in their spectra the expression is a sum of Eqs. (2.34) and (2.35).

In Eq. (2.34),

$$c_n = \int \Psi_n^\star(\boldsymbol{R})\Psi(\boldsymbol{R}, t = 0)d\boldsymbol{R} \,, \tag{2.36}$$

is the *overlap* between the wavefunction and the eigenfunction $\Psi_n(\boldsymbol{R})$ (see Eq. (2.21)).

Equation (2.34) can be seen as a particular case of the aforementioned superposition principle. The physical situation described by the wavefunction of Eq. (2.34) has no classical counterpart. In classical mechanics, the system has a single well-defined energy. On the other hand, if the energy of the quantum system described by Eq. (2.34) is measured, several different outcomes, E_n, are possible. Quantum mechanics can only predict *the probability of measuring the value E_n, which is equal to $|c_n|^2$* (see Eq. (2.24) the Born probability rule). As already explained, if the energy is measured and if the value $E_{n'}$ is obtained, it means that the wavefunction $\sum_n c_n \Psi_n(\boldsymbol{R})$ has been reduced to $\Psi_{n'}(\boldsymbol{R})$. Thus, the quantum system described by Eq. (2.34) has no well-defined energy. However, the *expectation value* of the Hamiltonian operator, $\langle\Psi|\hat{H}|\Psi\rangle$, gives the mean energy that is constant over time for an isolated system (*law of energy conservation*).

Now, inserting Eq. (2.34) into Eq. (2.30), we obtain

$$\begin{aligned}\Psi(\boldsymbol{R},t) &= e^{-i\hat{H}t/\hbar}\sum_n c_n\Psi_n(\boldsymbol{R})\,,\\ &= \sum_n c_n e^{-i\hat{H}t/\hbar}\Psi_n(\boldsymbol{R})\,,\\ &= \sum_n c_n e^{-iE_nt/\hbar}\Psi_n(\boldsymbol{R})\,.\end{aligned} \tag{2.37}$$

This wavefunction is no longer stationary and the probability density $|\Psi(\boldsymbol{R},t)|^2$ exhibits an *interference pattern*. Indeed

$$\begin{aligned}|\Psi(\boldsymbol{R},t)|^2 &= \Psi^\star(\boldsymbol{R},t)\Psi(\boldsymbol{R},t)\\ &= \Big(\sum_l c_l^\star e^{iE_lt/\hbar}\Psi_l^\star(\boldsymbol{R})\Big)\Big(\sum_n c_n e^{-iE_nt/\hbar}\Psi_n(\boldsymbol{R})\Big)\\ &= \sum_l\sum_n c_l^\star c_n e^{-i(E_n-E_l)t/\hbar}\Psi_l^\star(\boldsymbol{R})\Psi_n(\boldsymbol{R})\\ &= \sum_n |c_n|^2|\Psi_n(\boldsymbol{R})|^2 + \sum_n\sum_{l,l\neq n} c_l^\star c_n e^{-i(E_n-E_l)t/\hbar}\Psi_l^\star(\boldsymbol{R})\Psi_n(\boldsymbol{R})\,.\end{aligned} \tag{2.38}$$

Equation (2.38) has thus two terms:

– The first term, which does not depend on time.

– The second one called the "*interference term*" is the result of the interferences between the different eigenstates. The *probability density* now *depends on time* and does so through the second term only. Again, this originates from having *a coherent superposition* of eigenstates with different energies. Such a wavefunction is often referred to as *a wavepacket*. In practice, a wavepacket often corresponds to a wavefunction that is localized in space.

Creating experimentally such a coherent superposition and preserving it is a difficult task. Since a system is never isolated, it interacts with its environment that dissipates the quantum coherence: thus, a *loss of coherence* between the different quantum states occurs and is called *quantum decoherence*. The quantum interference terms vanish locally, i.e. in the system, and the latter is formally equivalent to a classical statistical mixture or an "*incoherent*" mixture of states rather than a single coherent quantum superposition of them. The effect of the environment is not included in the Hamiltonian operator introduced above and for which the eigenstates have been defined.

Let us return to our assembly of one thousand Schrödinger's cats. After interaction with the environment the assembly of one thousand cats simply corresponds to a statistical assembly of roughly five hundred cats that are dead and five hundred cats that are alive. The different states do not interact or "interfere" together to create non-classical behaviors. Since a cat is a large-scale system, the interaction with the environment will occur extremely fast. Therefore, even if it were possible to generate such a quantum superposition of cats, the quantum decoherence would be so fast that it would be impossible to measure the corresponding quantum properties. The difficulty and the importance of creating a quantum superposition has been emphasized by the 2012 Nobel prize in Physics awarded to Serge Haroche and David Jeffrey Wineland. For instance, Serge Haroche was able to create a coherent state of light in a cavity cooled to 0.8 K and perform a step-by-step measurement of the wavefunction collapse by interaction with circular Rydberg atoms of Rubidium, i.e. excited atoms with electrons that have very high principal quantum numbers. He produced actual movies of the decoherence process in progress until the system became a Fock state of well-defined energy with a definite number of photons [17, 18].

The *systematic control of quantum coherence* is seen as one of the main challenges in basic energy sciences since such a control may lead to highly desirable materials and devices such as *quantum computers* [19]. In the same manner, the exploitation of this quantum coherence in molecular processes may lead to a *more efficient chemistry*. Indeed, in traditional chemistry, the quantum states involved in the chemical process are, in general, populated in a incoherent way described as a "*mixed state*" in quantum statistical mechanics. Since the invention of lasers, which emit light *coherently* thanks to the process of stimulated emission, it is possible to create coherent superpositions of molecular quantum states. Such coherent superpositions are called a "*pure state*" in quantum statistical mechanics and their systematic use might strongly increase the efficiency and the control of the corresponding chemical processes and at a much lower energy cost. In practice, quantum coherence can be preserved during a time that is sufficient for a complete rearrangement of the molecular system. In this context and as aforementioned in the introduction, it is worth noting that recent experiments provided observation of *long-lived electronic quantum coherence*, after excitation by laser pulses, for energy transfer processes in light-harvesting complexes of biological systems such as photosynthetic systems at ambient temperature and in the condensed phase [20].

These coherent superpositions correspond to what we called a *wavepacket* above. *The central object of study in this book is the use of wavepackets in the context of chemistry.* As explained, molecular wavepackets can be created by lasers and exploited to guide chemical processes. In addition, from their intrinsic mathematical nature, wavepackets are very efficient tools for *theoreticians to compute several physical observables, independently of the possibility to create the former experimentally*. In simulations, artificial molecular wavepackets can be generated with the MCTDH package for instance. Their propagations can lead to quantities such as *absorption spectra* or *cross sections* including all the most subtle quantum effects that may occur in molecular processes.

2.2 Observables and Correspondence Principle

We have seen in the previous section that, in order to treat quantum mechanically a physical quantity $\mathcal{A}$ pertaining to a given system, we have to rely on a corresponding observable $\hat{A}$, which is a Hermitian linear operator acting on the wavefunctions that define the states of the system. *What are the rules to be used to carry out the correspondence* $\mathcal{A} \rightarrow \hat{A}$?

2.2.1 Observables Corresponding to Common Physical Quantities

Let us begin by considering a particle of mass m, of (Cartesian) *position coordinates* $\boldsymbol{r} = (x, y, z)$ and *classical momentum* $\boldsymbol{P} = (P_x, P_y, P_z)$. Concerning the corresponding *position observable* $\hat{\boldsymbol{r}} = (\hat{x}, \hat{y}, \hat{z})$, it is simply a multiplication of the wavefunction by $\boldsymbol{r} = (x, y, z)$. Concerning the momentum properties, the corresponding *momentum observable* is[7]

$$\begin{aligned}\hat{\boldsymbol{P}} = (\hat{P}_x, \hat{P}_y, \hat{P}_z) &= (-i\hbar\frac{\partial}{\partial x}, -i\hbar\frac{\partial}{\partial y}, -i\hbar\frac{\partial}{\partial z}) \\ &= -i\hbar\nabla\,. \end{aligned} \tag{2.39}$$

Concerning other physical properties that in classical mechanics are functions $A(\boldsymbol{r}, \boldsymbol{P})$ of the position and momentum variables, *a correspondence principle* is introduced, which consists in choosing for quantum mechanics the same functions of the position and momentum observables. So, to the classical quantity $A(\boldsymbol{r}, \boldsymbol{P})$ corresponds the observable $\hat{A}(\hat{\boldsymbol{r}}, \hat{\boldsymbol{P}})$. See Table 2.1 for the expressions of the observables corresponding to common physical quantities. For instance, the observable corresponding to the classical *angular momentum*[8] $\boldsymbol{L} = \boldsymbol{r} \times \boldsymbol{P}$ is given by $\hat{\boldsymbol{L}} = -i\hbar\boldsymbol{r} \times \nabla$ that is

$$\hat{L}_x = -i\hbar(y\frac{\partial}{\partial z} - z\frac{\partial}{\partial y})\,, \tag{2.40}$$

and the cyclic permutations

$$\begin{aligned}\hat{L}_y &= -i\hbar(z\frac{\partial}{\partial x} - x\frac{\partial}{\partial z}) \\ \hat{L}_z &= -i\hbar(x\frac{\partial}{\partial y} - y\frac{\partial}{\partial x})\,. \end{aligned} \tag{2.41}$$

[7] ∇ denotes the gradient operator, which here is defined as a row vector. See also Sect. 3.2.3 for a more comprehensive definition of gradient operators.

[8] In the present book, $\times$ denotes the vector product.

Table 2.1 Observables corresponding to common physical quantities [12, 13]

Physical quantity $\mathcal{A}$	Observable $\hat{A}$
Position $x, y, z, \boldsymbol{r}$	Multiplication by $x, y, z, \boldsymbol{r}$
Momentum $P_x, P_y, P_z, \boldsymbol{P}$	$\hat{P}_x = -i\hbar\frac{\partial}{\partial x}$, $\hat{P}_y = -i\hbar\frac{\partial}{\partial y}$, $\hat{P}_z = -i\hbar\frac{\partial}{\partial z}$, $\hat{\boldsymbol{P}} = -i\hbar\nabla$ (gradient)
Kinetic energy $T = \frac{1}{2m}(P_x^2 + P_y^2 + P_z^2)$	$\hat{T} = -\frac{\hbar^2}{2m}(\frac{\partial^2}{\partial x^2} + \frac{\partial^2}{\partial y^2} + \frac{\partial^2}{\partial z^2})$
$T = \frac{1}{2m}\boldsymbol{P}^2$	$\hat{T} = -\frac{\hbar^2}{2m}\nabla^2 = -\frac{\hbar^2}{2m}\Delta$ (Laplacian)
Potential energy $V(\boldsymbol{r})$	Multiplication by $V(\boldsymbol{r})$
Total energy $E = T + V(\boldsymbol{r})$	Hamiltonian $\hat{H} = -\frac{\hbar^2}{2m}\Delta + V(\boldsymbol{r})$
Angular momentum $\boldsymbol{L} = \boldsymbol{r} \times \boldsymbol{P}$	$\hat{\boldsymbol{L}} = \hat{\boldsymbol{r}} \times \hat{\boldsymbol{P}} = -i\hbar\boldsymbol{r} \times \nabla$

Let us notice and recall that the position and momentum observables, $\hat{\boldsymbol{r}}$ and $\hat{\boldsymbol{P}}$, *do not commute*. Indeed,

$$\hat{x}\hat{P}_x\Psi = \hat{x}(\hat{P}_x\Psi) = -ix\hbar\frac{\partial\Psi}{\partial x}, \tag{2.42}$$

and

$$\hat{P}_x\hat{x}\Psi = \hat{P}_x(\hat{x}\Psi) = -i\hbar\frac{\partial}{\partial x}(x\Psi) = -ix\hbar\frac{\partial\Psi}{\partial x} - i\hbar\Psi. \tag{2.43}$$

Introducing the *commutator* of two operators $\hat{A}$ and $\hat{B}$ defined above

$$\left[\hat{A}, \hat{B}\right] = \hat{A}\hat{B} - \hat{B}\hat{A}, \tag{2.44}$$

we see that

$$\left[\hat{x}, \hat{P}_x\right] = \hat{x}\hat{P}_x - \hat{P}_x\hat{x} = i\hbar\hat{I}, \tag{2.45}$$

where $\hat{I}$ is the *identity operator*. Moreover, it is clear that $\hat{x}$ and $\hat{P}_y$ commute. Denoting the components of $\hat{\boldsymbol{r}}$ and $\hat{\boldsymbol{P}}$ by $\hat{x}_i$ and $\hat{P}_i$, $i = 1, 2, 3$, we obtain the well-known fundamental *commutation relations*:

$$\left[\hat{x}_i, \hat{x}_j\right] = \left[\hat{P}_i, \hat{P}_j\right] = 0, \tag{2.46}$$

and

$$\left[\hat{x}_i, \hat{P}_k\right] = i\hbar\delta_{ik}. \tag{2.47}$$

Owing to these fundamental commutation relations, it is worth noting that the correspondence between physical quantities and operators is simple when the quantity is a function of either the position or the momentum alone. In contrast, when the quantity is a function of both $\hat{\boldsymbol{r}}$ and $\hat{\boldsymbol{P}}$, some care has to be taken since products of (non commuting) operators depend in general on the order in which they are written.

The correspondence principle is rather easily extended to the case of a collection of particles such as molecular systems made up of nuclei and electrons. We will describe these systems with the help of a set $\boldsymbol{R}$ of *well-adapted curvilinear coordinates* and their canonical conjugate momenta $\boldsymbol{P}$. Particular attention will be paid to the way of applying the correspondence principle to $T(\boldsymbol{R}, \boldsymbol{P})$, the classical Hamiltonian kinetic energy, where the ordering problem for $\hat{\boldsymbol{R}}$ and $\hat{\boldsymbol{P}}$ mentioned above will crop up.

The properties and explicit expressions of the momentum operators $\hat{\boldsymbol{P}}$, corresponding to the classical momenta conjugate to the generalized coordinates $\boldsymbol{R}$ will be detailed in Chap. 6. Concerning the energy observable, the Hamiltonian operator $\hat{H}(\boldsymbol{R}) = \hat{T}(\boldsymbol{R}) + \hat{V}(\boldsymbol{R})$ describing a confined system, i.e. a *bound system*, has a *discrete spectrum*. In contrast, for *unbound systems*, the energy spectrum is *continuous*.

2.2.2 Angular Momentum Operators

We now recapitulate some properties of the angular momentum operators. For a more detailed account see for instance Refs. [7, 9, 10, 13].

The *angular momentum observable* of a particle according to the correspondence principle (see Table 2.1) is

$$\hat{\boldsymbol{L}} = -i\hbar \boldsymbol{r} \times \nabla . \tag{2.48}$$

The three components $\hat{L}_x$, $\hat{L}_y$, $\hat{L}_z$ of this (vector) observable, given by Eqs. (2.40) and (2.41), do not commute. Indeed, after a simple calculation, one obtains

$$\begin{aligned} \left[\hat{L}_x, \hat{L}_y\right] &= i\hbar\hat{L}_z \\ \left[\hat{L}_y, \hat{L}_z\right] &= i\hbar\hat{L}_x \\ \left[\hat{L}_z, \hat{L}_x\right] &= i\hbar\hat{L}_y , \end{aligned} \tag{2.49}$$

which can be summarized as

$$\hat{\boldsymbol{L}} \times \hat{\boldsymbol{L}} = i\hbar\hat{\boldsymbol{L}} . \tag{2.50}$$

For systems of N particles with *position and momentum operators* $\hat{\boldsymbol{r}}_i$, $\hat{\boldsymbol{P}}_i$ with $i = 1, \ldots, N$, the total angular momentum is

$$\hat{\boldsymbol{J}} = \sum_{i=1}^{N} \hat{\boldsymbol{L}}_i = \sum_{i=1}^{N} \hat{\boldsymbol{r}}_i \times \hat{\boldsymbol{P}}_i . \tag{2.51}$$

$\hat{\boldsymbol{J}}$ also satisfies the three commutation relations (Eq. (2.50)), which leads to take them as a definition of an *angular (vector) observable* $\hat{\boldsymbol{J}}$ with the following fundamental commutation relations between its components

$$\hat{\boldsymbol{J}} \times \hat{\boldsymbol{J}} = i\hbar\hat{\boldsymbol{J}}\,. \tag{2.52}$$

The observable $\hat{\boldsymbol{J}}^2 = \hat{J}_x^2 + \hat{J}_y^2 + \hat{J}_z^2$, which is associated to the square of the angular momentum, commutes with each component of $\hat{\boldsymbol{J}}$. Thus, the two operators $\hat{\boldsymbol{J}}^2$ and $\hat{J}_z$ must have common eigenfunctions, denoted Ψ_{jm}, that obey the eigenvalue equations

$$\hat{\boldsymbol{J}}^2\Psi_{jm} = j(j+1)\hbar^2\Psi_{jm}\,, \tag{2.53}$$

and

$$\hat{J}_z\Psi_{jm} = m\hbar\Psi_{jm}\,, \tag{2.54}$$

with eigenvalues, $j(j+1)\hbar^2$ and $m\hbar$, respectively. j is called the *angular momentum quantum number* and m is called (for historical reasons) the *magnetic quantum number*. Let us also introduce the two operators

$$\hat{J}_+ = \hat{J}_x + i\hat{J}_y\,, \tag{2.55}$$

and

$$\hat{J}_- = \hat{J}_x - i\hat{J}_y\,. \tag{2.56}$$

It can be shown that their action on Ψ_{jm} is

$$\hat{J}_+\Psi_{jm} = \hbar\sqrt{j(j+1) - m(m+1)}\Psi_{j,m+1}\,, \tag{2.57}$$

and

$$\hat{J}_-\Psi_{jm} = \hbar\sqrt{j(j+1) - m(m-1)}\Psi_{j,m-1}\,. \tag{2.58}$$

For obvious reasons, $\hat{J}_+$ is called *raising operator* and $\hat{J}_-$ *lowering operator*. Through purely *algebraic* manipulations, Dirac was able to show the following quantization of j and m:

- j is a positive (or zero) integer or half integer:

$$j = 0, \frac{1}{2}, 1, \frac{3}{2}, 2, \frac{5}{2}, \ldots \tag{2.59}$$

- m is an integer or half integer, and for a given j, the only possible values of m are the $2j + 1$ numbers:

$$m = -j, -j + 1, \ldots, j - 1, j\,. \tag{2.60}$$

References

1. Haroche S, Raimond JM (2006) Exploring the quantum: atoms, cavities, and photons. Oxford University Press
2. Baggott J (2004) Beyond measure. Oxford University Press, Oxford
3. Laloë F (2012) Do we really understand quantum mechanics?. Cambridge University Press, Cambridge
4. d'Espagnat B (2006) On physics and philosophy. Princeton University Press, Princeton
5. Brickmont J (2016) Making sense of quantum mechanics. Springer
6. Dirac PAM (1958) The principles of quantum mechanics, 4th edn. Oxford University Press, Oxford
7. Messiah A (1962) Quantum mechanics, vol 1. Wiley, New York
8. Landau LD, Lifchitz EM (1979) Quantum mechanics. Pergamon Press
9. Bransden B, Joachain C (1989) Introduction to quantum mechanics. Longman Scientific & Technical, Harlow
10. Cohen-Tannoudji C, Diu B, Laloë F (1992) Quantum mechanics. Wiley-VCH
11. Merzbacher E (1998) Quantum mechanics. Wiley, New York
12. Bransden BH, Joachain CJ (2000) Quantum mechanics. Prentice Hall, Upper Saddle River, New Jersey
13. Basdevant JL, Dalibard J (2005) Quantum mechanics. Springer, Heidelberg
14. Le Bellac M (2006) Quantum physics. Cambridge University Press, Cambridge
15. Sakurai JJ, Napolitano JJ (2010) Modern quantum mechanics. Pearson, London
16. Basdevant JL (2007) Lectures on quantum mechanics. Springer, Heidelberg
17. Gleyzes SB, Kuhr S, Guerlin C, Bernu J, Déglise S, Hoff UB, Brune M, Raimond JM, Haroche S (2007) Quantum jumps of light recording the birth and death of a photon in a cavity. Nature 446:297–300
18. Guerlin C, Bernu J, Deléglise S, Sayrin CM, Gleyzes SB, Kuhr S, Brune M, Raimond JM, Haroche S (2007) Progressive field-state collapse and quantum non-demolition photon counting. Nature 448:889893
19. Fleming GR, Ratner MA (2008) Grand challenges in basic energy sciences. Phys Today 61:28
20. Engel GS, Calhoun TR, Read EL, Ahn T-K, Mancal T, Cheng Y-C, Blankenship RE, Fleming GR (2007) Evidence for wavelike energy transfer through quantum coherence in photosynthetic systems. Nature 446:782–786

Chapter 3
Molecular Hamiltonian Operators

In the present book, we will consider molecular systems either isolated or in interaction with external electromagnetic fields. In a bottom-up approach, which we will try to follow here, a molecule, or more generally a *molecular system*, is regarded as *a collection of electrons and nuclei in interaction with each other and possibly with external fields*.

Ideally, the motion of charged particles in the presence of electromagnetic fields is described by *quantum electrodynamics* [1]. In practice, the problem has to be simplified and several approximations are in order. First, we assume that the motions of the particles under study do not affect the external fields, which thus appear as driving fields. This corresponds to the first-order term in the perturbative treatment of electrodynamics (with the *Coulomb gauge condition*) [1, 2]. At higher orders, the dynamics of particles and fields are inextricably mixed [1–3] and these higher orders are usually neglected in non-relativistic quantum chemistry and molecular physics by introducing the *Coulomb Hamiltonian*, which is an *effective* Hamiltonian derived from electrodynamics by integrating out some effects of the radiation vacuum field.

In the following, except in Chap. 6, an operator $\hat{A}$ will be denoted simply A to ease notations.

3.1 The Non-relativistic Coulomb Hamiltonian Operator

For a collection of N nuclei and n electrons, let $\boldsymbol{R} = (\boldsymbol{R}_1, \ldots, \boldsymbol{R}_\alpha, \ldots, \boldsymbol{R}_N)$ and $\boldsymbol{r} = (\boldsymbol{r}_1, \ldots, \boldsymbol{r}_i, \ldots, \boldsymbol{r}_n)$ denote the *position vectors* of the coordinates of the nuclei and the electrons, respectively. The so-called *non-relativistic Coulomb molecular Hamiltonian operator* may be written as (see the correspondence principle in Sect. 2.2):

$$H^{mol}(\boldsymbol{r}, \boldsymbol{R}) = V^{el-el}(\boldsymbol{r}) + V^{nu-nu}(\boldsymbol{R}) + V^{nu-el}(\boldsymbol{r}, \boldsymbol{R}) + T^{el}(\boldsymbol{r}) + T^{nu}(\boldsymbol{R}) . \quad (3.1)$$

F. Gatti et al., *Applications of Quantum Dynamics in Chemistry*,
Lecture Notes in Chemistry 98, DOI 10.1007/978-3-319-53923-2_3

The various terms occurring are

- the *electrostatic repulsion between the electrons,*

$$V^{el-el}(\boldsymbol{r}) = \sum_{j>i}^{n} \frac{e^2}{4\pi\varepsilon_0 ||\boldsymbol{r}_i - \boldsymbol{r}_j||}, \quad (3.2)$$

  with:

  - e, the elementary charge, i.e. the opposite of the electric charge carried by a single electron,
  - ε_0, the vacuum permittivity,
  - $||\boldsymbol{r}_i - \boldsymbol{r}_j||$, the distance between electrons i and j;

- the *electrostatic repulsion between the nuclei,*

$$V^{nu-nu}(\boldsymbol{R}) = \sum_{\beta>\alpha}^{N} \frac{Z_\alpha Z_\beta e^2}{4\pi\varepsilon_0 ||\boldsymbol{R}_\alpha - \boldsymbol{R}_\beta||}, \quad (3.3)$$

  with Z_α and Z_β the atomic numbers of nuclei α and β, and $||\boldsymbol{R}_\alpha - \boldsymbol{R}_\beta||$ the distance between them;

- the *electrostatic attraction between the electrons and the nuclei,* which is responsible for the stability of molecules,

$$V^{nu-el}(\boldsymbol{r}, \boldsymbol{R}) = -\sum_{i}^{n} \sum_{\alpha}^{N} \frac{Z_\alpha e^2}{4\pi\varepsilon_0 ||\boldsymbol{R}_\alpha - \boldsymbol{r}_i||}; \quad (3.4)$$

- the *kinetic energy operator for the electrons,*

$$T^{el}(\boldsymbol{r}) = -\frac{\hbar^2}{2m_e} \sum_{i}^{n} \left(\frac{\partial^2}{\partial x_i^2} + \frac{\partial^2}{\partial y_i^2} + \frac{\partial^2}{\partial z_i^2}\right), \quad (3.5)$$

  with:

  - m_e, the electron mass,
  - x_i, y_i, z_i, the Cartesian coordinates of $\boldsymbol{r}_i$ in the (Galilean) axis system fixed in the laboratory, also called the *Laboratory-Fixed (LF) frame*;

- the *kinetic energy operator for the nuclei,*

$$T^{nu}(\boldsymbol{R}) = -\sum_{\alpha}^{N} \frac{\hbar^2}{2M_\alpha} \left(\frac{\partial^2}{\partial X_\alpha^2} + \frac{\partial^2}{\partial Y_\alpha^2} + \frac{\partial^2}{\partial Z_\alpha^2}\right), \quad (3.6)$$

  with M_α, the mass of nucleus α and X_α, Y_α, Z_α the Cartesian coordinates of $\boldsymbol{R}_\alpha$ in the LF frame.

The *molecular Hamiltonian operator* in Eq. (3.1) is the *quantum-mechanical observable* corresponding to the Coulomb part of the classical Hamiltonian of a set of N nuclei and n electrons in the Coulomb gauge.[1] In particular, it is assumed that the electrostatic interaction between the charges is instantaneous and that the kinetic energy operator corresponds to the quantum-mechanical counterpart of the non-relativistic kinetic energy. In addition, only the electrostatic interactions between the particles are taken into account (since the electrons and nuclei move, some magnetic forces are also present but will not be considered here).

Obviously, the present approach is not well-suited to treat strong relativistic effects but has the great advantage that the Coulomb potential energy clearly separates out [1]. For the slow-moving particles in bound states, which are studied in molecular physics, this *Coulomb Hamiltonian operator* is an excellent approximation, and relativistic and magnetic effects can be treated as corrections. For instance, it may be necessary to add relativistic corrections when considering electrons in atoms with high atomic numbers since, in that case, electrons, especially the core electrons close to the nucleus, may reach relativistic velocities. These relativistic corrections can be derived from the Dirac equation, which describes the motion of the electrons in a coherent way with both the principles of quantum mechanics and the theory of special relativity [3]. For instance, it may be necessary to add *spin-orbit terms* describing the interactions between the electron spin and the magnetic field generated by the electron orbit around the nucleus [4]. This may be important when considering electronic states with a total spin that is different from zero. For free radicals, even in the electronic ground state, the spin-orbit coupling may not be negligible.

The *time-independent Schrödinger equation* corresponding to the molecular Hamiltonian reads (a discrete spectrum being assumed, see Eq. (2.29) of Sect. 2.1.4),

$$H^{mol}(\boldsymbol{r}, \boldsymbol{R})\, \Psi_l^{mol}(\boldsymbol{r}, \boldsymbol{R}) = E_l \Psi_l^{mol}(\boldsymbol{r}, \boldsymbol{R})\,. \tag{3.7}$$

It provides a prescription for getting the molecular *eigenenergies*, E_l, and *eigenfunctions*, $\Psi_l^{mol}(\boldsymbol{r}, \boldsymbol{R})$.

The *corresponding time-dependent Schrödinger equation* reads (see Eq. (2.28) of Sect. 2.1.4)

$$H^{mol}(\boldsymbol{r}, \boldsymbol{R})\, \Psi^{mol}(\boldsymbol{r}, \boldsymbol{R}, t) = i\hbar \frac{\partial \Psi^{mol}(\boldsymbol{r}, \boldsymbol{R}, t)}{\partial t}\,. \tag{3.8}$$

It provides a prescription for describing the evolution in time of a molecular wavefunction.

These equations are extremely difficult to solve and several approximations and separations of variables are required to make them easier to handle and to understand. Moreover, the molecular Hamiltonian operators as given in Eq. (3.1) *correspond to mere shapeless heaps of nuclei and electrons*, with nuclear and electronic coordinates referred to a LF frame. They do not involve *molecular parameters* such as *bond*

[1] See the correspondence principle in Sect. 2.2.

lengths and *angles*, and the only quantities occurring are the masses and the charges of the nuclei and the electrons that make up the molecular system. Such seemingly harmless concepts as *electronic state*, *molecular orbital*, *molecular shape* and *potential energy surface*, *rotational* and *internal motion*, *Coriolis coupling* come about because of *approximations and separation of variables* (Ref. [5]) that will be studied in more detail in the following sections. However, let us first sketch the most important one i.e. *the separation of the electronic* and *nuclear motions*.

3.2 Separation of the Electronic and Nuclear Motions

3.2.1 General Overview

Let us split the molecular Hamiltonian operator of Eq. (3.1) as follows:

$$H^{mol}(\boldsymbol{r}, \boldsymbol{R}) = T^{nu}(\boldsymbol{R}) + H^{el}(\boldsymbol{r}; \boldsymbol{R}), \tag{3.9}$$

where

$$H^{el}(\boldsymbol{r}; \boldsymbol{R}) = V^{el-el}(\boldsymbol{r}) + V^{nu-nu}(\boldsymbol{R}) + V^{nu-el}(\boldsymbol{r}, \boldsymbol{R}) + T^{el}(\boldsymbol{r}) \tag{3.10}$$

denotes what is traditionally called, in spite of the presence of the repulsion potential, $V^{nu-nu}(\boldsymbol{R})$, the *electronic Hamiltonian* for a given nuclear configuration $\boldsymbol{R}$. For this Hamiltonian, the nuclear positions are considered simply as parameters since the nuclear kinetic energy term $T^{nu}(\boldsymbol{R})$ has been left out. For this reason, this electronic Hamiltonian is often called the *clamped-nucleus* Hamiltonian. This can be seen as an attempted separation of the electronic and nuclear motions based on the (classical) physical picture that *the rapidly moving light electrons instantaneously follow the slowly moving heavy nuclei*. "Clamping the nuclei" can be regarded as a way to introduce "by hand" the shape of the molecular systems but, although intuitively plausible, it may lead to some subtle conceptual problems concerning the structure of molecules [6, 7]. Some attempts to extract elements of molecular structure from all-particle (nuclei and electrons) wave functions have been realized [8].

Let us now turn to the solution of the *time-dependent molecular Schrödinger equation* Eq. (3.8). Using an orthonormal basis set of nuclear functions, $\{\varphi_\lambda^{nu}(\boldsymbol{R})\}$, and an orthonormal basis set of electronic functions, $\{\Phi_m^{el}(\boldsymbol{r}; \boldsymbol{R})\}$, we express the solution as a standard expansion

$$\Psi^{mol}(\boldsymbol{r}, \boldsymbol{R}, t) = \sum_m \sum_\lambda d_{m\lambda}(t)\, \varphi_\lambda^{nu}(\boldsymbol{R})\, \Phi_m^{el}(\boldsymbol{r}; \boldsymbol{R}), \tag{3.11}$$

with

$$\int \varphi_{\lambda}^{nu\star}(\boldsymbol{R})\, \varphi_{\mu}^{nu}(\boldsymbol{R}) d\boldsymbol{R} = \delta_{\lambda\mu}\,, \tag{3.12}$$

and[2]

$$\int \Phi_{m}^{el\star}(\boldsymbol{r};\boldsymbol{R})\, \Phi_{n}^{el}(\boldsymbol{r};\boldsymbol{R}) d\boldsymbol{r} = \langle \Phi_{m}^{el};\boldsymbol{R}|\Phi_{n}^{el};\boldsymbol{R}\rangle_{\boldsymbol{r}} = \delta_{mn}\,, \tag{3.13}$$

for each nuclear configuration $\boldsymbol{R}$. Note that the electronic basis functions, $\Phi_{m}^{el}(\boldsymbol{r};\boldsymbol{R})$, as for now quite general, are however parametrized by the nuclear coordinates, $\boldsymbol{R}$. This reflects the asymmetric treatment of the electrons and the nuclei alluded to above and will enable a drastic reduction of the number of electronic functions to be used as shown below.

We rewrite Eq. (3.11) as follows[3]:

$$\Psi^{mol}(\boldsymbol{r},\boldsymbol{R},t) = \sum_{m} \Psi_{m}(\boldsymbol{R},t)\, \Phi_{m}^{el}(\boldsymbol{r};\boldsymbol{R})\,, \tag{3.14}$$

with

$$\Psi_{m}(\boldsymbol{R},t) = \sum_{\lambda} d_{m\lambda}(t)\, \varphi_{\lambda}^{nu}(\boldsymbol{R})\,. \tag{3.15}$$

In this approach, the time dependence of the molecular wavefunction, $\Psi^{mol}(\boldsymbol{r},\boldsymbol{R},t)$, is carried by *the nuclear wavefunctions*, $\Psi_{m}(\boldsymbol{R},t)$, associated with the electronic basis function $\Phi_{m}^{el}(\boldsymbol{r};\boldsymbol{R})$.

Next, introducing Eq. (3.14) into the time-dependent Schrödinger equation (3.8), multiplying, on the left, both sides of this equation by $\Phi_{n}^{el\star}(\boldsymbol{r};\boldsymbol{R})$ and integrating over the coordinates of the electrons, yields[4]

$$\sum_{m} \langle \Phi_{n}^{el};\boldsymbol{R}|H^{mol}(\boldsymbol{R})\Psi_{m}(\boldsymbol{R},t)|\Phi_{m}^{el};\boldsymbol{R}\rangle_{\boldsymbol{r}} = \\ i\hbar \sum_{m} \langle \Phi_{n}^{el};\boldsymbol{R}|\Phi_{m}^{el};\boldsymbol{R}\rangle_{\boldsymbol{r}}\, \frac{\partial \Psi_{m}(\boldsymbol{R},t)}{\partial t}\,. \tag{3.16}$$

In view of the orthonormality of the electronic basis functions (Eq. (3.13)) we obtain the following system of *coupled differential equations*

---

[2]A bracket notation $\langle|\rangle_{\boldsymbol{r}}$ is used to indicate an integration over the electronic coordinates, $\boldsymbol{r}$, alone.

[3]Also known as the Born expansion (see Eq. (3.56)).

[4]As in Eq. (3.13) a Dirac bracket notation $\langle|\rangle_{\boldsymbol{r}}$ has been used for the integration over the electronic coordinates, $\boldsymbol{r}$, which have thus been left out in H^{mol}, Φ_{m}^{el} and H^{el}.

$$\sum_m \langle \Phi_n^{el}; \boldsymbol{R} | H^{mol}(\boldsymbol{R}) | \Phi_m^{el}; \boldsymbol{R} \rangle_{\boldsymbol{r}} \; \Psi_m(\boldsymbol{R}, t) = i\hbar \frac{\partial \Psi_n(\boldsymbol{R}, t)}{\partial t} . \tag{3.17}$$

Now, substituting into the previous equation the expression of $H^{mol}(\boldsymbol{r}, \boldsymbol{R})$ as given in Eq. (3.9), we obtain the time-dependent Schrödinger equation in the form of the following *system of coupled differential equation* (in bracket notation for the electronic functions),

$$\sum_m (T_{nm}(\boldsymbol{R}) + V_{nm}(\boldsymbol{R}))\Psi_m(\boldsymbol{R}, t) = i\hbar \frac{\partial \Psi_n(\boldsymbol{R}, t)}{\partial t} , \tag{3.18}$$

with[5]

$$\begin{aligned} T_{nm}(\boldsymbol{R})\Psi_m(\boldsymbol{R}, t) &= \langle \Phi_n^{el}; \boldsymbol{R} | T^{nu}(\boldsymbol{R}) | \; \Phi_m^{el}; \boldsymbol{R} \rangle_{\boldsymbol{r}} \Psi_m(\boldsymbol{R}, t) \\ &= \int \Phi_n^{el\star}(\boldsymbol{r}; \boldsymbol{R}) T^{nu}(\boldsymbol{R}) \Phi_m^{el}(\boldsymbol{r}; \boldsymbol{R}) d\boldsymbol{r} \Psi_m(\boldsymbol{R}, t) . \end{aligned} \tag{3.19}$$

and

$$V_{nm}(\boldsymbol{R}) = \langle \Phi_n^{el}; \boldsymbol{R} | H^{el}(\boldsymbol{R}) | \Phi_m^{el}; \boldsymbol{R} \rangle_{\boldsymbol{r}} . \tag{3.20}$$

In order to highlight the general structure of these coupled equations (see Eq. (3.17)), we will, occasionally, write them out in matrix form. So, for instance, if we restrict ourselves to only two electronic states, i.e.

$$\Psi^{mol}(\boldsymbol{r}, \boldsymbol{R}, t) = \Psi_1(\boldsymbol{R}, t)\Phi_1^{el}(\boldsymbol{r}; \boldsymbol{R}) + \Psi_2(\boldsymbol{R}, t)\Phi_2^{el}(\boldsymbol{r}; \boldsymbol{R}) , \tag{3.21}$$

we obtain for the time-dependent Schrödinger equation:

$$\begin{bmatrix} T_{11}(\boldsymbol{R}) + V_{11}(\boldsymbol{R}) & T_{12}(\boldsymbol{R}) + V_{12}(\boldsymbol{R}) \\ T_{21}(\boldsymbol{R}) + V_{21}(\boldsymbol{R}) & T_{22}(\boldsymbol{R}) + V_{22}(\boldsymbol{R}) \end{bmatrix} \begin{bmatrix} \Psi_1(\boldsymbol{R}, t) \\ \Psi_2(\boldsymbol{R}, t) \end{bmatrix} = i\hbar \frac{\partial}{\partial t} \begin{bmatrix} \Psi_1(\boldsymbol{R}, t) \\ \Psi_2(\boldsymbol{R}, t) \end{bmatrix} . \tag{3.22}$$

As regards the time-independent Schrödinger equation, the eigenfunction $\Psi_l^{mol}(\boldsymbol{r}, \boldsymbol{R})$ associated with the molecular energy E_l, can also be expanded in the same nuclear and electronic basis sets, $\{\varphi_\lambda^{nu}(\boldsymbol{R})\}$ and $\{\Phi_m^{el}(\boldsymbol{r}; \boldsymbol{R})\}$:

---

[5]In order to avoid all ambiguities and misunderstanding, it is worth noting that, in spite of its appearance, the matrix element $T_{nm}(\boldsymbol{R})$ is not a purely multiplicative operator (i.e. a pure number) but still contains differential operators with respect to $\boldsymbol{R}$, acting on the nuclear functions $\Psi_m(\boldsymbol{R}, t)$. On the contrary, since $H^{el}(\boldsymbol{r}; \boldsymbol{R})$ does not contain differential operators with respect to $\boldsymbol{R}$, the matrix element $V_{nm}(\boldsymbol{R})$ is a pure multiplicative operator.

$$\Psi_l^{mol}(\boldsymbol{r}, \boldsymbol{R}) = \sum_m \sum_\lambda d_{\lambda,l}^m \, \varphi_\lambda^{nu}(\boldsymbol{R}) \, \Phi_m^{el}(\boldsymbol{r}; \boldsymbol{R}) \,, \tag{3.23}$$

which parallels the expansion given in Eq. (3.11), except that the time-dependent coefficients, $d_{m\lambda}(t)$, are to be replaced by the *time-independent coefficients*, $d_{\lambda,l}^m$, since for each molecular eigenfunction, $\Psi_l^{mol}(\boldsymbol{r}, \boldsymbol{R})$, indexed by l, a different expression is needed. This implies that Eqs. (3.15) and (3.14) have to be rewritten, respectively, as

$$\Psi_m^l(\boldsymbol{R}) = \sum_\lambda d_{\lambda,l}^m \varphi_\lambda^{nu}(\boldsymbol{R}) \,, \tag{3.24}$$

and

$$\Psi_l^{mol}(\boldsymbol{r}, \boldsymbol{R}) = \sum_m \Psi_m^l(\boldsymbol{R}) \Phi_m^{el}(\boldsymbol{r}; \boldsymbol{R}) \,. \tag{3.25}$$

Now, for two electronic states only, say states 1 and 2, we obtain

$$\begin{bmatrix} T_{11}(\boldsymbol{R}) + V_{11}(\boldsymbol{R}) & T_{12}(\boldsymbol{R}) + V_{12}(\boldsymbol{R}) \\ T_{21}(\boldsymbol{R}) + V_{21}(\boldsymbol{R}) & T_{22}(\boldsymbol{R}) + V_{22}(\boldsymbol{R}) \end{bmatrix} \begin{bmatrix} \Psi_1^l(\boldsymbol{R}) \\ \Psi_2^l(\boldsymbol{R}) \end{bmatrix} = E_l \begin{bmatrix} \Psi_1^l(\boldsymbol{R}) \\ \Psi_2^l(\boldsymbol{R}) \end{bmatrix} . \tag{3.26}$$

3.2.2 *The Adiabatic Electronic Basis Set*

Let us now define a specific electronic basis set frequently used in molecular physics and quantum chemistry, where the electronic basis functions are the *eigenfunctions* of Eq. (3.10). They are called *adiabatic electronic states* and form the so-called *adiabatic basis set*. They are denoted $\Phi_m^{el/ad}(\boldsymbol{r}; \boldsymbol{R})$ and thus, by definition,

$$H^{el}(\boldsymbol{r}; \boldsymbol{R}) \, \Phi_m^{el/ad}(\boldsymbol{r}; \boldsymbol{R}) = E_m^{el}(\boldsymbol{R}) \, \Phi_m^{el/ad}(\boldsymbol{r}; \boldsymbol{R}) \,, \tag{3.27}$$

which entails that

$$\begin{aligned} V_{nm}^{ad}(\boldsymbol{R}) &= \langle \Phi_n^{el/ad}; \boldsymbol{R} | H^{el}(\boldsymbol{R}) | \Phi_m^{el/ad}; \boldsymbol{R} \rangle_{\boldsymbol{r}} \\ &= \delta_{nm} E_m^{el}(\boldsymbol{R}) \,, \end{aligned} \tag{3.28}$$

in view of Eq. (3.27) and the orthonormality relation Eq. (3.13). It is worth noting that the electronic energy, $E_m^{el}(\boldsymbol{R})$, is a function of the nuclear coordinates, $\boldsymbol{R}$, and as such, is commonly regarded as a *potential energy surface* (PES) for the nuclear motion. For a detailed discussion of the notion of PES see Refs. [6] and [7].

The expansion of the time-dependent molecular wavefunction (Eq. 3.14) particularizes into[6]

$$\Psi^{mol}(\boldsymbol{r}, \boldsymbol{R}, t) = \sum_m \Psi_m^{ad}(\boldsymbol{R}, t)\Phi_m^{el/ad}(\boldsymbol{r}; \boldsymbol{R}) \,, \tag{3.29}$$

with

$$\Psi_m^{ad}(\boldsymbol{R}, t) = \sum_\lambda d_{m\lambda}^{ad}(t)\, \varphi_\lambda^{nu}(\boldsymbol{R}) \,. \tag{3.30}$$

So that the system of coupled differential equations (3.18) becomes

$$\sum_m (T_{nm}^{ad}(\boldsymbol{R}) + \delta_{nm} E_m^{el}(\boldsymbol{R}))\Psi_m^{ad}(\boldsymbol{R}, t) = i\hbar \frac{\partial \Psi_n^{ad}(\boldsymbol{R}, t)}{\partial t} \,, \tag{3.31}$$

with

$$T_{nm}^{ad}(\boldsymbol{R}) = \langle \Phi_n^{el/ad}; \boldsymbol{R} | T^{nu}(\boldsymbol{R}) | \Phi_m^{el/ad}; \boldsymbol{R} \rangle_{\boldsymbol{r}} \,. \tag{3.32}$$

In particular, Eqs. (3.22) and (3.26) become, respectively,

$$\begin{bmatrix} T_{11}^{ad}(\boldsymbol{R}) + E_1^{el}(\boldsymbol{R}) & T_{12}^{ad}(\boldsymbol{R}) \\ T_{21}^{ad}(\boldsymbol{R}) & T_{22}^{ad}(\boldsymbol{R}) + E_2^{el}(\boldsymbol{R}) \end{bmatrix} \begin{bmatrix} \Psi_1^{ad}(\boldsymbol{R}, t) \\ \Psi_2^{ad}(\boldsymbol{R}, t) \end{bmatrix} = i\hbar \frac{\partial}{\partial t} \begin{bmatrix} \Psi_1^{ad}(\boldsymbol{R}, t) \\ \Psi_2^{ad}(\boldsymbol{R}, t) \end{bmatrix} , \tag{3.33}$$

and

$$\begin{bmatrix} T_{11}^{ad}(\boldsymbol{R}) + E_1^{el}(\boldsymbol{R}) & T_{12}^{ad}(\boldsymbol{R}) \\ T_{21}^{ad}(\boldsymbol{R}) & T_{22}^{ad}(\boldsymbol{R}) + E_2^{el}(\boldsymbol{R}) \end{bmatrix} \begin{bmatrix} \Psi_1^{l,ad}(\boldsymbol{R}) \\ \Psi_2^{l,ad}(\boldsymbol{R}) \end{bmatrix} = E_l \begin{bmatrix} \Psi_1^{l,ad}(\boldsymbol{R}) \\ \Psi_2^{l,ad}(\boldsymbol{R}) \end{bmatrix} . \tag{3.34}$$

The use of the adiabatic basis set allows one to introduce the notion of a PES and simplifies the coupled differential equations by suppressing the potential couplings: see Eq. (3.28). Nevertheless, a main drawback remains, i.e. the equations are still coupled through the adiabatic coupling terms, $T_{nm}^{ad}(\boldsymbol{R})$ (for $n \neq m$). The next step in our attempt to solve the molecular Schrödinger equations and to develop a more physically-intuitive picture of molecular systems is to see *under what circumstances the adiabatic kinetic coupling terms can be approximately neglected.*

[6]Equations (3.29) and (3.30) clearly show that the nuclear wavefunction $\Psi_m^{ad}(\boldsymbol{R}, t)$ depends on the choice of the electronic basis functions, in particular here the adiabatic electronic basis set.

3.2.3 The Born-Oppenheimer and Adiabatic Approximations

The Born-Oppenheimer (BO) approximation is a milestone in the theory of molecules and, even in cases where it fails, it remains a reference to which we compare and in terms of which we discuss this failure [9]. This approximation takes advantage of the great difference in masses of electrons and nuclei. Because of this difference, it is assumed that the nuclei move relatively slowly and may be treated as stationary while the electrons can respond almost instantaneously to the motion of the nuclei.

To understand more clearly the mathematical conditions corresponding to the BO approximation, let us first remark that the nuclear kinetic energy operator is essentially a second derivative with respect to the nuclear variables $\boldsymbol{R}$,

$$T^{nu}(\boldsymbol{R}) = -\frac{\hbar^2}{2M}\frac{\partial^2}{\partial \boldsymbol{R}^2}, \tag{3.35}$$

where $\frac{\partial}{\partial \boldsymbol{R}} = \nabla_{\boldsymbol{R}}$ is the gradient with respect to nuclear coordinates and M corresponds to a nuclear mass (for the discussion, we do not need to specify anything about its actual value). For a given (Cartesian) coordinate vector, $\boldsymbol{R} = (X, Y, Z)$, the gradient operator is defined as

$$\frac{\partial}{\partial \boldsymbol{R}} = \nabla_{\boldsymbol{R}} = (\frac{\partial}{\partial X}, \frac{\partial}{\partial Y}, \frac{\partial}{\partial Z}) \tag{3.36}$$

and the scalar products as

$$\frac{\partial}{\partial \boldsymbol{R}_1} \cdot \frac{\partial}{\partial \boldsymbol{R}_2} = \nabla_{\boldsymbol{R}_1} \cdot \nabla_{\boldsymbol{R}_2} = \frac{\partial^2}{\partial X_1 \partial X_2} + \frac{\partial^2}{\partial Y_1 \partial Y_2} + \frac{\partial^2}{\partial Z_1 \partial Z_2} \tag{3.37}$$

and

$$\frac{\partial^2}{\partial \boldsymbol{R}^2} = \nabla_{\boldsymbol{R}} \cdot \nabla_{\boldsymbol{R}} = \nabla^2_{\boldsymbol{R}} = \frac{\partial^2}{\partial X^2} + \frac{\partial^2}{\partial Y^2} + \frac{\partial^2}{\partial Z^2}. \tag{3.38}$$

The action of $T^{nu}(\boldsymbol{R})$ on a product $\Psi^{ad}_m(\boldsymbol{R}, t)\Phi^{el/ad}_m(\boldsymbol{r}; \boldsymbol{R})$ results in three terms,

$$\begin{aligned} T^{nu}(\boldsymbol{R})(\Psi^{ad}_m(\boldsymbol{R}, t)\Phi^{el/ad}_m(\boldsymbol{r}; \boldsymbol{R})) = &- \frac{\hbar^2}{2M}(\Phi^{el/ad}_m(\boldsymbol{r}; \boldsymbol{R}))(\frac{\partial^2}{\partial \boldsymbol{R}^2}\Psi^{ad}_m(\boldsymbol{R}, t)) \\ &- \frac{\hbar^2}{M}(\frac{\partial}{\partial \boldsymbol{R}}\Phi^{el/ad}_m(\boldsymbol{r}; \boldsymbol{R})).(\frac{\partial}{\partial \boldsymbol{R}}\Psi^{ad}_m(\boldsymbol{R}, t)) \\ &- \frac{\hbar^2}{2M}(\frac{\partial^2}{\partial \boldsymbol{R}^2}\Phi^{el/ad}_m(\boldsymbol{r}; \boldsymbol{R}))(\Psi^{ad}_m(\boldsymbol{R}, t)). \end{aligned} \tag{3.39}$$

Now, in mathematical terms, the fact that the nuclei move relatively slowly while the electrons respond almost instantaneously to the motion of the nuclei implies that, very often, we can assume

$$\frac{\partial \Phi_m^{el/ad}(\boldsymbol{r}; \boldsymbol{R})}{\partial \boldsymbol{R}} \approx \boldsymbol{0}\,, \tag{3.40}$$

and

$$\frac{\partial^2 \Phi_m^{el/ad}(\boldsymbol{r}; \boldsymbol{R})}{\partial^2 \boldsymbol{R}} \approx \boldsymbol{0}\,, \tag{3.41}$$

In view of Eq. (3.39), the above conditions also imply that

$$T^{nu}(\boldsymbol{R})\Phi^{el/ad}(\boldsymbol{r}; \boldsymbol{R}) \approx \Phi^{el/ad}(\boldsymbol{r}; \boldsymbol{R})T^{nu}(\boldsymbol{R})\,. \tag{3.42}$$

Obviously, the fact that the nuclei are much heavier than the electrons does not guarantee that Eqs. (3.40) and (3.41) are satisfied for all nuclear configurations, $\boldsymbol{R}$. More explicitly, as we will see in Sect. 3.2.5, when the *adiabatic electronic states are close in energy*, i.e.

$$E_n^{el}(\boldsymbol{R}) \approx E_m^{el}(\boldsymbol{R})\,, \tag{3.43}$$

the coupling elements, $T_{nm}^{ad}(\boldsymbol{R})$, *diverge*.

Now, assuming that the BO approximation is strictly valid, i.e.

$$T^{nu}(\boldsymbol{R})\Phi^{el/ad}(\boldsymbol{r}; \boldsymbol{R}) = \Phi^{el/ad}(\boldsymbol{r}; \boldsymbol{R})T^{nu}(\boldsymbol{R})\,, \tag{3.44}$$

implies that

$$T^{nu}(\boldsymbol{R})\Psi_m^{ad}(\boldsymbol{R}, t)\Phi_m^{el/ad}(\boldsymbol{r}; \boldsymbol{R}) = \Phi_m^{el/ad}(\boldsymbol{r}; \boldsymbol{R})T^{nu}(\boldsymbol{R})\Psi_m^{ad}(\boldsymbol{R}, t)\,, \tag{3.45}$$

so that, inserting Eq. (3.45) into Eq. (3.32) yields

$$T_{nm}^{ad/BO}(\boldsymbol{R})\Psi_m^{ad}(\boldsymbol{R}, t) = \langle \Phi_n^{el/ad}; \boldsymbol{R} | \Phi_m^{el/ad}; \boldsymbol{R} \rangle_{\boldsymbol{r}}\, T^{nu}(\boldsymbol{R})\Psi_m^{ad}(\boldsymbol{R}, t)\,, \tag{3.46}$$

or, in view of the orthonormality of the (adiabatic) electronic basis functions,

$$T_{nm}^{ad/BO}(\boldsymbol{R}) = \delta_{nm}\, T^{nu}(\boldsymbol{R})\,. \tag{3.47}$$

In addition, as already pointed out (see Eq. (3.28))

$$V_{nm}^{ad}(\boldsymbol{R}) = \delta_{nm}\, E_m^{el}(\boldsymbol{R})\,. \tag{3.48}$$

Thus, by means of Eq. (3.47), the BO approximation decouples the systems of the (still) coupled differential equations (3.18), (3.33) and (3.34) as follows:

- Equation (3.18) becomes

$$(T^{nu}(\boldsymbol{R}) + E_m^{el}(\boldsymbol{R}))\Psi_m^{ad}(\boldsymbol{R}, t) = i\hbar\frac{\partial\Psi_m^{ad}(\boldsymbol{R}, t)}{\partial t} \quad (m = 0, 1, \ldots)\,, \tag{3.49}$$

where m labels the electronic states.

- Equation (3.33) becomes

$$\begin{aligned}(T^{nu}(\boldsymbol{R}) + E_1^{el}(\boldsymbol{R}))\Psi_1^{ad}(\boldsymbol{R}, t) &= i\hbar\frac{\partial\Psi_1^{ad}(\boldsymbol{R}, t)}{\partial t}\\ (T^{nu}(\boldsymbol{R}) + E_2^{el}(\boldsymbol{R}))\Psi_2^{ad}(\boldsymbol{R}, t) &= i\hbar\frac{\partial\Psi_2^{ad}(\boldsymbol{R}, t)}{\partial t}\,.\end{aligned} \tag{3.50}$$

Equation (3.50) clearly highlights the fact that, within the BO approximation, the electronic states are decoupled and the adiabatic nuclear wavefunctions, $\Psi_1^{ad}(\boldsymbol{R}, t)$ and $\Psi_2^{ad}(\boldsymbol{R}, t)$ evolve separately on either $E_1^{el}(\boldsymbol{R})$ or $E_2^{el}(\boldsymbol{R})$, which are *potential energy surfaces (PES) provided by the electrons through the electronic Schrödinger equation* (3.27). It follows that the solution of the time-dependent Schrödinger equation can be written in a product form:

$$\Phi_m^{el/ad}(\boldsymbol{r}; \boldsymbol{R})\Psi_m^{ad}(\boldsymbol{R}, t)\,. \tag{3.51}$$

Figure 3.1 shows an illustration of the evolution of a nuclear wavefunction $\Psi_m^{ad}(\boldsymbol{R}, t)$ within the Born-Oppenheimer approximation (here for $m = 0$, i.e. the electronic ground state).

- Similarly, Eq. (3.34) becomes

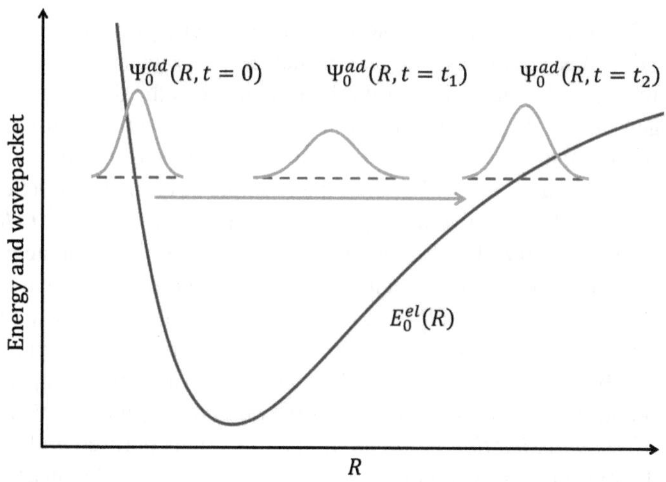

Fig. 3.1 Evolution of a wavepacket in an electronic state within the Born-Oppenheimer approximation

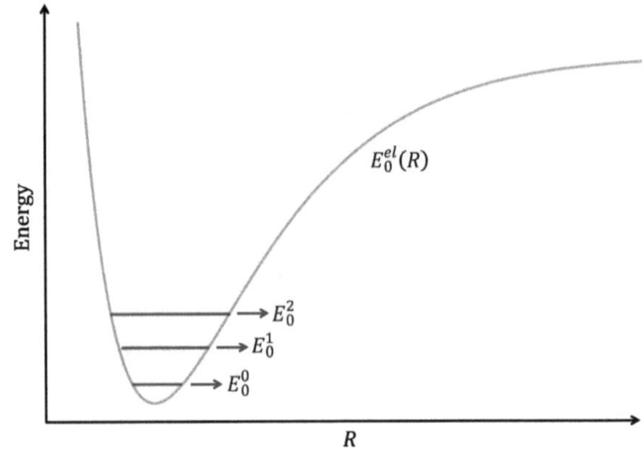

Fig. 3.2 Example of potential energy surface of the electronic ground state, $E_0^{el}(R)$, and of the first three vibrational levels $E_0^{l_0}$ in the electronic ground state for a system such as H_2. R is the distance between the two nuclei

$$(T^{nu}(\boldsymbol{R}) + E_1^{el}(\boldsymbol{R}))\Psi_1^{l_1,\,ad}(\boldsymbol{R}) = E_1^{l_1}\Psi_1^{l_1,\,ad}(\boldsymbol{R})$$
$$(T^{nu}(\boldsymbol{R}) + E_2^{el}(\boldsymbol{R}))\Psi_2^{l_2,\,ad}(\boldsymbol{R}) = E_2^{l_2}\Psi_2^{l_2,\,ad}(\boldsymbol{R})\,. \tag{3.52}$$

In view of the decoupling of Eq. (3.34), the set of total eigenvalues $\{E_l, (l = 0, 1, \ldots)\}$ is worth splitting into $\{E_1^{l_1}, (l_1 = 0, 1, \ldots)\}$ and $\{E_2^{l_2}, (l_2 = 0, 1, \ldots)\}$. 1 and 2 are, of course, electronic indices whereas l_1 and l_2 are nuclear indices, for instance, indices labeling rovibrational levels of a molecule within a given electronic state. The solutions of the time-independent Schrödinger equation can also be written in product forms:

$$\Phi_m^{el/ad}(\boldsymbol{r}; \boldsymbol{R})\Psi_m^{l_m,\,ad}(\boldsymbol{R})\,. \tag{3.53}$$

In Fig. 3.2, we present an example of a potential energy surface: $E_0^{el}(R)$ is a Morse function that could describe, for instance, the potential energy surface of the electronic ground state of H_2. In the latter case, R is the distance between the two hydrogen nuclei. E_0^0 corresponds to energy of the vibrational ground state in the electronic ground state and E_0^1 and E_0^2 to the energies of the first and second excited vibrational states. We will discuss the example of H_2 in detail in Sect. 3.5. In particular, Eq. (3.169) will give an approximate expression of $\Phi_0^{el/ad}(\boldsymbol{r}; \boldsymbol{R})$ around the equilibrium geometry of the molecules, Eq. (3.172) an approximate expression of $E_0^{el}(R)$ and Eq. (3.177) approximate values of the purely vibrational levels $E_0^{l_0}$. In the low-energy domain, $E_0^{el}(R)$ can be considered as harmonic. The vibrational eigenfunctions of an harmonic potential are the Hermite functions. Figure 3.3 gives the first three vibrational eigenfunctions, $\Psi_0^{l,\,ad}(\boldsymbol{R})$, within a harmonic approximation.

At this point, it is worth mentioning that the *adiabatic approximation*, which is close to the *BO approximation* to the point that it is often confused with it, consists in writing the molecular wavefunction as a single product from the onset

$$\Psi_{ml}^{mol}(\boldsymbol{r}, \boldsymbol{R}) = \Psi_m^l(\boldsymbol{R})\Phi_m^{el/ad}(\boldsymbol{r}; \boldsymbol{R})\,, \tag{3.54}$$

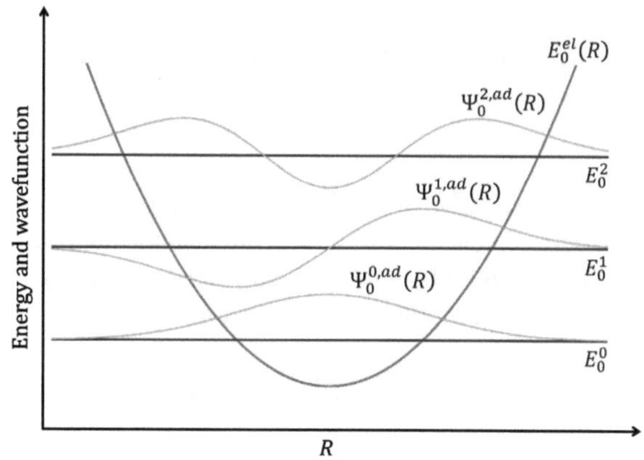

Fig. 3.3 Examples of vibrational eigenfunctions $\Psi_0^{l_0,ad}(\boldsymbol{R})$ in the case of a harmonic potential

and

$$\Psi_m^{mol}(\boldsymbol{r}, \boldsymbol{R}, t) = \Psi_m(\boldsymbol{R}, t)\Phi_m^{el/ad}(\boldsymbol{r}; \boldsymbol{R}) , \tag{3.55}$$

where the first equation applies to the time-independent picture with l denoting a (ro-)vibrational state, and the second equation is for a time-dependent wavepacket. Introducing these single-products of nuclear and electronic functions into the corresponding Schrödinger equations automatically yields the decoupled equations of Eqs. (3.50) and (3.52), except that $T^{nu}(\boldsymbol{R})$ is replaced by $T_{mm}^{ad}(\boldsymbol{R})$, which, as will be seen in the next section, is a minor difference with respect to the BO approximation.

3.2.4 Breakdown of the Born-Oppenheimer Approximation *

We will show that the Born-Oppenheimer approximation breaks down when the electronic states are close in energy, i.e. if $E_n^{el}(\boldsymbol{R}) \approx E_m^{el}(\boldsymbol{R})$, because $T_{nm}^{ad}(\boldsymbol{R})$ (in the adiabatic basis) diverges, which we now discuss in more detail. This singular behavior reflects the presence of strong vibronic couplings between both electronic states n and m through the nuclear motion (this may, of course, concern more than one pair of states).

If the Born-Oppenheimer approximation is not valid, the molecular states can no longer be written as single products of nuclear and electronic contributions. This applies to both eigenstates and time-dependent wavefunctions. In contrast, we must write them as *so-called Born expansions* (see Eqs. (3.14) and (3.29)),

$$\Psi(\boldsymbol{r}, \boldsymbol{R}, t) = \sum_m \Psi_m^{el/ad}(\boldsymbol{R}, t)\Phi_m^{el/ad}(\boldsymbol{r}; \boldsymbol{R}) . \tag{3.56}$$

The vibrational parts are thus solutions of a system of coupled differential equations as shown in Eq. (3.18).

Let us now detail the expression of $T^{ad}_{nm}(\boldsymbol{R})$. Let us start from Eq. (3.39) and sum over m, overlap with the electronic state n, and remove the nuclear wavepacket. This yields

$$T^{ad}_{nm}(\boldsymbol{R}) = \delta_{nm} T^{nu}(\boldsymbol{R}) + \hbar^2 \Lambda_{nm}(\boldsymbol{R}) \,, \tag{3.57}$$

where the so-called non-adiabatic couplings $\Lambda_{nm}(\boldsymbol{R})$ describe the dynamical interaction between the electronic structure and the nuclear motion (note that some authors define $\Lambda_{nm}(\boldsymbol{R})$ with a minus sign and/or containing $\hbar^2$). They split into a derivative operator and a multiplicative operator,

$$\Lambda_{nm}(\boldsymbol{R}) = -\boldsymbol{F}_{nm}(\boldsymbol{R}).\frac{1}{M}\frac{\partial}{\partial \boldsymbol{R}} - \frac{1}{2M} G_{nm}(\boldsymbol{R}) \,. \tag{3.58}$$

where the first-order non-adiabatic couplings, which are in the nuclear coordinate space, are defined as

$$\boldsymbol{F}_{nm}(\boldsymbol{R}) = \big\langle \Phi^{el/ad}_n ; \boldsymbol{R} \big| \frac{\partial}{\partial \boldsymbol{R}} \Phi^{el/ad}_m ; \boldsymbol{R} \big\rangle_{\boldsymbol{r}} \,, \tag{3.59}$$

and the second-order non-adiabatic scalar couplings as

$$G_{nm}(\boldsymbol{R}) = \big\langle \Phi^{el/ad}_n ; \boldsymbol{R} \big| \frac{\partial^2}{\partial \boldsymbol{R}^2} \Phi^{el/ad}_m ; \boldsymbol{R} \big\rangle_{\boldsymbol{r}} \,. \tag{3.60}$$

As a consequence, the coupled equations can be recast as

$$(T^{nu}(\boldsymbol{R}) + E^{el}_n(\boldsymbol{R}))\Psi_n(\boldsymbol{R},t) + \sum_m \hbar^2 \Lambda_{nm}(\boldsymbol{R})\Psi_m(\boldsymbol{R},t) = i\hbar \frac{\partial \Psi_n(\boldsymbol{R},t)}{\partial t} \,. \tag{3.61}$$

This shows that neglecting $\Lambda_{nm}(\boldsymbol{R})$ is equivalent to the *Born-Oppenheimer approximation* (see Eq. (3.49)). Neglecting $\Lambda_{nm}(\boldsymbol{R})$ for $n \neq m$ while keeping $\Lambda_{nn}(\boldsymbol{R})$ is equivalent to the *adiabatic approximation* mentioned in the previous section.

*3.2.5 The Off-Diagonal Hellmann-Feynman Theorem **

The diagonal Hellmann-Feynman theorem is widely used in practice to calculate forces (slopes in the PES) analytically in quantum chemistry. The off-diagonal Hellmann-Feynman theorem is a natural extension that provides analytic relationships for the first-order non-adiabatic couplings, for $n \neq m$,

$$\boldsymbol{F}_{nm}(\boldsymbol{R}) = \frac{\langle \Phi^{el/ad}_n ; \boldsymbol{R} | \frac{\partial}{\partial \boldsymbol{R}} H^{el}(\boldsymbol{R}) | \Phi^{el/ad}_m ; \boldsymbol{R} \rangle_{\boldsymbol{r}}}{E^{el}_m(\boldsymbol{R}) - E^{el}_n(\boldsymbol{R})} \,, \tag{3.62}$$

where $E_m(\boldsymbol{R}) - E_n(\boldsymbol{R})$ is the energy gap between the two states. This shows that $T_{nm}^{ad}(\boldsymbol{R})$ diverges when the electronic states become degenerate.

The derivation of this formula is not complicated, it relies on the assumption that the electronic states are the exact eigenstates of $H^{el}(\boldsymbol{R})$. Let us first write, for the first two electronic adiabatic states,

$$H^{el}(\boldsymbol{r};\boldsymbol{R})\Phi_1^{el/ad}(\boldsymbol{r};\boldsymbol{R}) = E_1^{el}(\boldsymbol{R})\Phi_1^{el/ad}(\boldsymbol{r};\boldsymbol{R})\,, \tag{3.63}$$

and

$$H^{el}(\boldsymbol{r};\boldsymbol{R})\Phi_2^{el/ad}(\boldsymbol{r};\boldsymbol{R}) = E_2^{el}(\boldsymbol{R})\Phi_2^{el/ad}(\boldsymbol{r};\boldsymbol{R})\,. \tag{3.64}$$

In other words,

$$E_1^{el}(\boldsymbol{R}) = \langle\Phi_1^{el/ad};\boldsymbol{R}|H^{el}(\boldsymbol{R})|\Phi_1^{el/ad};\boldsymbol{R}\rangle_{\boldsymbol{r}}\,, \tag{3.65}$$

$$E_2^{el}(\boldsymbol{R}) = \langle\Phi_2^{el/ad};\boldsymbol{R}|H^{el}(\boldsymbol{R})|\Phi_2^{el/ad};\boldsymbol{R}\rangle_{\boldsymbol{r}}\,, \tag{3.66}$$

$$0 = \langle\Phi_1^{el/ad};\boldsymbol{R}|H^{el}(\boldsymbol{R})|\Phi_2^{el/ad};\boldsymbol{R}\rangle_{\boldsymbol{r}}\,. \tag{3.67}$$

Applying the gradient, $\frac{\partial}{\partial\boldsymbol{R}}$, to the off-diagonal element gives 0, since $\langle\Phi_1^{el/ad};\boldsymbol{R}|H^{el}(\boldsymbol{R})|\Phi_2^{el/ad};\boldsymbol{R}\rangle_{\boldsymbol{r}}$ is identically zero for all $\boldsymbol{R}$. In other words,

$$\boldsymbol{0} = \frac{\partial}{\partial\boldsymbol{R}}\langle\Phi_1^{el/ad};\boldsymbol{R}|H^{el}(\boldsymbol{R})|\Phi_2^{el/ad};\boldsymbol{R}\rangle_{\boldsymbol{r}}\,. \tag{3.68}$$

We thus obtain

$$\begin{aligned}\boldsymbol{0} = {} & \langle\frac{\partial}{\partial\boldsymbol{R}}\Phi_1^{el/ad};\boldsymbol{R}|H^{el}(\boldsymbol{R})|\Phi_2^{el/ad};\boldsymbol{R}\rangle_r + \langle\Phi_1^{el/ad};\boldsymbol{R}|\frac{\partial}{\partial\boldsymbol{R}}H^{el}(\boldsymbol{R})|\Phi_2^{el/ad};\boldsymbol{R}\rangle_r \\ & + \langle\Phi_1^{el/ad};\boldsymbol{R}|H^{el}(\boldsymbol{R})|\frac{\partial}{\partial\boldsymbol{R}}\Phi_2^{el/ad};\boldsymbol{R}\rangle_{\boldsymbol{r}}\,. \end{aligned} \tag{3.69}$$

Using the hermiticity of the Hamiltonian operator, we get

$$\begin{aligned}\boldsymbol{0} = {} & E_2^{el}(\boldsymbol{R})\langle\frac{\partial}{\partial\boldsymbol{R}}\Phi_1^{el/ad};\boldsymbol{R}|\Phi_2^{el/ad};\boldsymbol{R}\rangle_r \\ & + E_1^{el}(\boldsymbol{R})\langle\Phi_1^{el/ad};\boldsymbol{R}|\frac{\partial}{\partial\boldsymbol{R}}\Phi_2^{el/ad};\boldsymbol{R}\rangle_r + \langle\Phi_1^{el/ad};\boldsymbol{R}|\frac{\partial}{\partial\boldsymbol{R}}H^{el}(\boldsymbol{R})|\Phi_2^{el/ad};\boldsymbol{R}\rangle_r\,. \end{aligned} \tag{3.70}$$

On the other hand we have

$$\frac{\partial}{\partial\boldsymbol{R}}\langle\Phi_1^{el/ad};\boldsymbol{R}|\Phi_2^{el/ad};\boldsymbol{R}\rangle_r = \langle\frac{\partial}{\partial\boldsymbol{R}}\Phi_1^{el/ad};\boldsymbol{R}|\Phi_2^{el/ad};\boldsymbol{R}\rangle_r + \langle\Phi_1^{el/ad};\boldsymbol{R}|\frac{\partial}{\partial\boldsymbol{R}}\Phi_2^{el/ad};\boldsymbol{R}\rangle_r\,, \tag{3.71}$$

and

$$\frac{\partial}{\partial \boldsymbol{R}}\langle\Phi_1^{el/ad};\boldsymbol{R}|\Phi_2^{el/ad};\boldsymbol{R}\rangle_{\boldsymbol{r}}=\boldsymbol{0}\,, \tag{3.72}$$

since the eigenfunctions are orthogonal for all $\boldsymbol{R}$. Thus,

$$\langle\frac{\partial}{\partial \boldsymbol{R}}\Phi_1^{el/ad};\boldsymbol{R}|\Phi_2^{el/ad};\boldsymbol{R}\rangle_{\boldsymbol{r}}=-\langle\Phi_1^{el/ad};\boldsymbol{R}|\frac{\partial}{\partial \boldsymbol{R}}\Phi_2^{el/ad};\boldsymbol{R}\rangle_{\boldsymbol{r}}\,. \tag{3.73}$$

This leads to

$$\boldsymbol{0}=(E_1^{el}(\boldsymbol{R})-E_2^{el}(\mathbf{R}))\langle\Phi_1^{el/ad};\boldsymbol{R}|\frac{\partial}{\partial \boldsymbol{R}}\Phi_2^{el/ad};\boldsymbol{R}\rangle_{\boldsymbol{r}}+\langle\Phi_1^{el/ad};\boldsymbol{R}|\frac{\partial}{\partial \boldsymbol{R}}H^{el}(\boldsymbol{R})|\Phi_2^{el/ad};\boldsymbol{R}\rangle_{\boldsymbol{r}}\,. \tag{3.74}$$

Making use of Eq. (3.59) for the special case $n=1$ and $m=2$, the above equations can be written as

$$\boldsymbol{F}_{12}(\boldsymbol{R})=\frac{\langle\Phi_1^{el/ad};\boldsymbol{R}|\frac{\partial}{\partial \boldsymbol{R}}H^{el}(\boldsymbol{R})|\Phi_2^{el/ad};\boldsymbol{R}\rangle_{\boldsymbol{r}}}{E_2^{el}(\boldsymbol{R})-E_1^{el}(\boldsymbol{R})}\,. \tag{3.75}$$

which proves Eq. (3.62), having set $n=1$ and $m=2$.

3.2.6 The Electronic Basis Sets: An Overview

The molecular Hamiltonian matrix in the electronic basis set, $\Phi_n^{el}(\boldsymbol{r};\boldsymbol{R})$, can be split into two terms: one corresponding to the kinetic energy operator for the nuclei (*kinetic energy operator matrix*):

$$T_{nm}(\boldsymbol{R})=\langle\Phi_n^{el};\boldsymbol{R}|T^{nu}(\boldsymbol{R})|\Phi_m^{el};\boldsymbol{R}\rangle_{\boldsymbol{r}}\,, \tag{3.76}$$

and one for the electronic Hamiltonian (*potential energy operator matrix*):

$$V_{nm}(\boldsymbol{R})=\langle\Phi_n^{el};\boldsymbol{R}|H^{el}(\boldsymbol{R})|\Phi_m^{el};\boldsymbol{R}\rangle_{\boldsymbol{r}}\,. \tag{3.77}$$

For the sake of simplicity, let us consider the case of two electronic states only. The coupled equations read

$$\begin{bmatrix} T_{11}(\boldsymbol{R})+V_{11}(\boldsymbol{R}) & T_{12}(\boldsymbol{R})+V_{12}(\boldsymbol{R}) \\ T_{21}(\boldsymbol{R})+V_{21}(\boldsymbol{R}) & T_{22}(\boldsymbol{R})+V_{22}(\boldsymbol{R}) \end{bmatrix}\begin{bmatrix}\Psi_1(\boldsymbol{R},t)\\ \Psi_2(\boldsymbol{R},t)\end{bmatrix}=i\hbar\frac{\partial}{\partial t}\begin{bmatrix}\Psi_1(\boldsymbol{R},t)\\ \Psi_2(\boldsymbol{R},t)\end{bmatrix}\,. \tag{3.78}$$

Now, three cases must be clearly distinguished (the structure of the Hamiltonian matrix is summarized in Table 3.1):

Table 3.1 Hamiltonian matrix for two electronic states in different electronic basis sets and different approximations (the matrix for the adiabatic approximation is not given)

General case:

$$\begin{bmatrix} T_{11}(\boldsymbol{R}) + V_{11}(\boldsymbol{R}) & T_{12}(\boldsymbol{R}) + V_{12}(\boldsymbol{R}) \\ T_{12}(\boldsymbol{R}) + V_{12}(\boldsymbol{R}) & T_{22}(\boldsymbol{R}) + V_{22}(\boldsymbol{R}) \end{bmatrix}$$

with

$$T_{nm}(\boldsymbol{R}) = \langle \Phi_n^{el}; \boldsymbol{R} | T^{nu}(\boldsymbol{R}) | \Phi_m^{el}; \boldsymbol{R} \rangle_r$$

and

$$V_{nm}(\boldsymbol{R}) = \langle \Phi_n^{el}; \boldsymbol{R} | H^{el}(\boldsymbol{R}) | \Phi_m^{el}; \boldsymbol{R} \rangle_r$$

Adiabatic basis set:

$$\begin{bmatrix} T_{11}^{ad}(\boldsymbol{R}) + E_1^{el}(\boldsymbol{R}) & T_{12}^{ad}(\boldsymbol{R}) \\ T_{21}^{ad}(\boldsymbol{R}) & T_{22}^{ad}(\boldsymbol{R}) + E_2^{el}(\boldsymbol{R}) \end{bmatrix}$$

Adiabatic basis set with BO approximation:

$$\begin{bmatrix} T^{nu}(\boldsymbol{R}) + E_1^{el}(\boldsymbol{R}) & 0 \\ 0 & T^{nu}(\boldsymbol{R}) + E_2^{el}(\boldsymbol{R}) \end{bmatrix}$$

Diabatic basis set:

$$\begin{bmatrix} T^{nu}(\boldsymbol{R}) + V_1^{dia}(\boldsymbol{R}) & V_{12}^{dia}(\boldsymbol{R}) \\ V_{21}^{dia}(\boldsymbol{R}) & T^{nu}(\boldsymbol{R}) + V_2^{dia}(\boldsymbol{R}) \end{bmatrix}$$

- *In the adiabatic basis set and without the Born-Oppenheimer approximation*, the coupled equations read

$$\begin{bmatrix} T_{11}^{ad}(\boldsymbol{R}) + E_1^{el}(\boldsymbol{R}) & T_{12}^{ad}(\boldsymbol{R}) \\ T_{21}^{ad}(\boldsymbol{R}) & T_{22}^{ad}(\boldsymbol{R}) + E_2^{el}(\boldsymbol{R}) \end{bmatrix} \begin{bmatrix} \Psi_1^{ad}(\boldsymbol{R},t) \\ \Psi_2^{ad}(\boldsymbol{R},t) \end{bmatrix} = i\hbar\frac{\partial}{\partial t}\begin{bmatrix} \Psi_1^{ad}(\boldsymbol{R},t) \\ \Psi_2^{ad}(\boldsymbol{R},t) \end{bmatrix}, \tag{3.79}$$

with $E_1^{el}(\boldsymbol{R})$ and $E_2^{el}(\boldsymbol{R})$ the potential energy surfaces of states 1 and 2 that can be calculated with *quantum chemistry* methods. $T_{12}^{ad}(\boldsymbol{R})$ diverges when the potential energy surfaces are degenerate.

- *In the adiabatic basis set and within the adiabatic approximation*, $T_{12}^{ad}(\boldsymbol{R})$ is assumed to be equal to 0 and the equations are no longer coupled. They read

$$\begin{bmatrix} T_{11}^{ad}(\boldsymbol{R}) + E_1^{el}(\boldsymbol{R}) & 0 \\ 0 & T_{22}^{ad}(\boldsymbol{R}) + E_2^{el}(\boldsymbol{R}) \end{bmatrix} \begin{bmatrix} \Psi_1^{ad}(\boldsymbol{R},t) \\ \Psi_2^{ad}(\boldsymbol{R},t) \end{bmatrix} = i\hbar\frac{\partial}{\partial t}\begin{bmatrix} \Psi_1^{ad}(\boldsymbol{R},t) \\ \Psi_2^{ad}(\boldsymbol{R},t) \end{bmatrix}. \tag{3.80}$$

- *In the adiabatic basis set and within the Born-Oppenheimer approximation*, $T_{nn}^{ad/BO}(\boldsymbol{R})$ is assumed to be equal to $T^{nu}(\boldsymbol{R})$. The uncoupled equations read

$$\begin{bmatrix} T^{nu}(\boldsymbol{R}) + E_1^{el}(\boldsymbol{R}) & 0 \\ 0 & T^{nu}(\boldsymbol{R}) + E_2^{el}(\boldsymbol{R}) \end{bmatrix} \begin{bmatrix} \Psi_1^{ad/BO}(\boldsymbol{R},t) \\ \Psi_2^{ad/BO}(\boldsymbol{R},t) \end{bmatrix} = i\hbar\frac{\partial}{\partial t}\begin{bmatrix} \Psi_1^{ad/BO}(\boldsymbol{R},t) \\ \Psi_2^{ad/BO}(\boldsymbol{R},t) \end{bmatrix}. \tag{3.81}$$

- We will see in Chap. 4, especially in Sect. 4.3, when the coupling between the electronic functions can no longer be neglected, one cannot work in the adiabatic electronic basis set due to the divergence of the kinetic couplings. One then switches to a basis set of *diabatic* electronic functions, $\Phi_n^{el/dia}(\boldsymbol{r};\boldsymbol{R})$. There is a unitary transformation between the adiabatic and diabatic basis functions and in the diabatic basis set we assume that

$$T_{12}^{dia}(\boldsymbol{R}) = T_{21}^{dia}(\boldsymbol{R}) = \langle \Phi_1^{el/dia};\boldsymbol{R}|T^{nu}(\boldsymbol{R})|\Phi_2^{el/dia};\boldsymbol{R}\rangle_{\boldsymbol{r}} \approx 0\,, \tag{3.82}$$

and $T_{11}^{dia}(\boldsymbol{R}) = T_{22}^{dia}(\boldsymbol{R}) \approx T^{nu}(\boldsymbol{R})$ so that the set of coupled equations becomes

$$\begin{bmatrix} T^{nu}(\boldsymbol{R}) + V_1^{dia}(\boldsymbol{R}) & V_{12}^{dia}(\boldsymbol{R}) \\ V_{21}^{dia}(\boldsymbol{R}) & T^{nu}(\boldsymbol{R}) + V_2^{dia}(\boldsymbol{R}) \end{bmatrix} \begin{bmatrix} \Psi_1^{dia}(\boldsymbol{R},t) \\ \Psi_2^{dia}(\boldsymbol{R},t) \end{bmatrix} = i\hbar\frac{\partial}{\partial t}\begin{bmatrix} \Psi_1^{dia}(\boldsymbol{R},t) \\ \Psi_2^{dia}(\boldsymbol{R},t) \end{bmatrix}. \tag{3.83}$$

The functions $V_1^{dia}(\boldsymbol{R})$, $V_2^{dia}(\boldsymbol{R})$ and $V_{12}^{dia}(\boldsymbol{R})$ are obtained by a *diabatization* procedure: an example is given in Sect. 4.4. Let us just notice here that diabatic states vary smoothly as functions of the nuclear coordinates and often have a clear interpretation from a chemical point of view.

3.3 Molecular Coordinates

The aim of the present section is to describe in some detail coordinate changes that facilitate the separation of variables, in particular, the separation of the total center of mass (Sects. 3.3.1 and 3.3.2), the description of the shape and the rotation of a molecular system by means of internal and orientation coordinates, respectively, (Sect. 3.3.3) and the notion of potential energy surface (PES) (Sect. 3.3.4). We will also revisit the separation of the electronic and nuclear motions and, at the same time, discuss the *approximations which support some common and closer-to-the-intuition concepts and parameters used in chemistry and molecular physics* and alluded to at the end of Sect. 3.1. In this respect, an important result of the present section is Eq. (3.129), where a clear separation of the nuclear kinetic energy, expressed in terms of shape and rotation coordinates, and the electronic kinetic energies is obtained subject to two (small) approximations. It is worth noting here that two methods exist for changing coordinates in a Hamiltonian operator:

(1) Starting from the *quantum* Hamiltonian operator in Cartesian coordinates and changing to generalized curvilinear coordinates using the *chain rule*.

(2) Starting from the *classical* Hamiltonian expressed in generalized curvilinear coordinates and using the *quantum mechanical postulates* to change to the corresponding *quantum* Hamiltonian operator.

The first procedure will be mainly used in the present chapter, while the second one, as well as a mix of both procedures, will be developed in Chap. 6.

*3.3.1 Separation of the Total Center of Mass and Internal Motions of the Molecular System **

Let us consider again the non-relativistic Coulomb Hamiltonian operator for a molecular system regarded as a collection of N nuclei and n electrons (see Eq. (3.1)). We follow rather closely Ref. [4]. For a careful approach see Ref. [10].

Given an arbitrary laboratory-fixed (LF) Cartesian axis system (X_{LF}, Y_{LF}, Z_{LF}) assumed to be Galilean, let the coordinates of *all* particles (N nuclei plus n electrons) be denoted collectively as

$$\tilde{\boldsymbol{R}}_{LF} = (\tilde{\boldsymbol{R}}^1_{LF}, \ldots, \tilde{\boldsymbol{R}}^r_{LF}, \ldots, \tilde{\boldsymbol{R}}^{N_T=N+n}_{LF})$$

$$= (\tilde{X}^1_{LF}, \tilde{Y}^1_{LF}, \tilde{Z}^1_{LF}, \ldots, \tilde{X}^r_{LF}, \tilde{Y}^r_{LF}, \tilde{Z}^r_{LF}, \ldots, \tilde{X}^{N_T}_{LF}, \tilde{Y}^{N_T}_{LF}, \tilde{Z}^{N_T}_{LF}) \,. \tag{3.84}$$

If we want to distinguish between electrons and nuclei, the position variables, $\tilde{\boldsymbol{R}}^{r}_{LF}$ ($r = 1, \ldots, N, N+1, \ldots, N_T = N + n$), may be split up into two sets, one set consisting of N coordinate vectors, $\boldsymbol{R}^{\alpha}_{LF}$ ($\alpha = 1, \ldots, N$) describing the nuclei and the other set of n coordinate vectors, $\boldsymbol{r}^{i}_{LF}$ ($i = 1, \ldots, n$) describing the electrons as in the previous sections. Thus, the total non-relativistic kinetic operator given by $T^{el}(\boldsymbol{r}) + T^{nu}(\boldsymbol{R})$ of Eq. (3.1) can now be written as

$$T(\tilde{\boldsymbol{R}}_{LF}) = -\frac{\hbar^2}{2}\sum_{r=1}^{N_T}\frac{1}{m_r}\left(\frac{\partial^2}{\partial \tilde{X}^{r\,2}_{LF}} + \frac{\partial^2}{\partial \tilde{Y}^{r\,2}_{LF}} + \frac{\partial^2}{\partial \tilde{Z}^{r\,2}_{LF}}\right), \tag{3.85}$$

with m_r the mass of particle r.

Using the gradient operator notation (see Eq. (3.36)), we have

$$T(\tilde{\boldsymbol{R}}_{LF}) = -\frac{\hbar^2}{2}\sum_{r=1}^{N_T}\frac{1}{m_r}\nabla^2_{\tilde{\boldsymbol{R}}^{r}_{LF}}\,. \tag{3.86}$$

An explicit separation of the total center of mass and the internal motions is possible because the Coulomb interactions are translationally invariant and various choices of equivalent internal coordinates are possible. A usual way to proceed is to introduce a space-fixed (SF) axis system (X_{SF}, Y_{SF}, Z_{SF}) *with origin at the total molecular center of mass* (CM) and axes parallel to (X_{LF}, Y_{LF}, Z_{LF}), so that

$$\tilde{\boldsymbol{R}}^{r}_{SF} = \tilde{\boldsymbol{R}}^{r}_{LF} - \boldsymbol{R}^{CM}_{LF} \quad (r = 1, \ldots, N_T) \tag{3.87}$$

where

$$\boldsymbol{R}^{CM}_{LF} = \frac{1}{M_T}\sum_{r=1}^{N_T} m_r \tilde{\boldsymbol{R}}^{r}_{LF} \tag{3.88}$$

is the coordinate vector of the (total) center of mass in the LF axis system and

$$M_T = \sum_{r=1}^{N_T} m_r \tag{3.89}$$

is the total mass of the molecular system. To separate the translational kinetic energy, we have to write the total LF kinetic energy operator in terms of the $3N_T$ coordinates

$$\tilde{\boldsymbol{R}}_{SF} = (\tilde{\boldsymbol{R}}^{CM}_{LF}, \tilde{\boldsymbol{R}}^{2}_{SF}, \ldots, \tilde{\boldsymbol{R}}^{N_T}_{SF}), \tag{3.90}$$

where the coordinates of particle 1, for instance, have been eliminated upon using the center of mass relation

$$\tilde{\boldsymbol{R}}^1_{SF} = -\frac{1}{m_1}\sum_{r=2}^{N_T} m_r \tilde{\boldsymbol{R}}^r_{SF} \,. \tag{3.91}$$

Therefore, applying to the total LF kinetic energy operator (Eq. (3.86)) the chain rule (we refer to complement Sect. 3.6.3 for a quick reminder concerning the chain rule) for the LF → SF coordinate transformation

$$(\tilde{\boldsymbol{R}}^1_{LF}, \tilde{\boldsymbol{R}}^2_{LF}, \ldots, \tilde{\boldsymbol{R}}^{N_T}_{LF}) \rightarrow (\tilde{\boldsymbol{R}}^{CM}_{LF}, \tilde{\boldsymbol{R}}^2_{SF}, \ldots, \tilde{\boldsymbol{R}}^{N_T}_{SF}) \,, \tag{3.92}$$

yields, after a slightly tedious but straightforward calculation, the following expression for the total kinetic energy operator

$$T^{Tot} = T^{CM}(\boldsymbol{R}^{CM}_{LF}) + T^0(\tilde{\boldsymbol{R}}^2_{SF}, \ldots, \tilde{\boldsymbol{R}}^{N_T}_{SF}) + T'(\tilde{\boldsymbol{R}}^2_{SF}, \ldots, \tilde{\boldsymbol{R}}^{N_T}_{SF}) \,, \tag{3.93}$$

where:

$$T_{CM}(\boldsymbol{R}^{CM}_{LF}) = -\frac{\hbar^2}{2M_T}\nabla^2_{\boldsymbol{R}^{CM}_{LF}} \,, \tag{3.94}$$

$$T^0(\tilde{\boldsymbol{R}}^2_{SF}, \ldots, \tilde{\boldsymbol{R}}^{N_T}_{SF}) = -\frac{\hbar^2}{2}\sum_{r=2}^{N_T}\frac{1}{m_r}\nabla^2_{\tilde{\boldsymbol{R}}^r_{SF}} \,, \tag{3.95}$$

and

$$T'(\tilde{\boldsymbol{R}}^2_{SF}, \ldots, \tilde{\boldsymbol{R}}^{N_T}_{SF}) = +\frac{\hbar^2}{2M_T}\sum_{r,s=2}^{N_T}\nabla_{\tilde{\boldsymbol{R}}^r_{SF}} \cdot \nabla_{\tilde{\boldsymbol{R}}^s_{SF}} \,. \tag{3.96}$$

Moreover, the electrostatic potential energy given by

$$V(\boldsymbol{R}^{SF}) = \sum_{r=1, r<s}^{N_T}\frac{C_r C_s e^2}{4\pi\epsilon_0 ||\tilde{\boldsymbol{R}}^r_{SF} - \tilde{\boldsymbol{R}}^s_{SF}||} \,, \tag{3.97}$$

with $C_{r=\alpha} = Z_\alpha$ for nucleus α and $C_{r=i} = -1$ for the electrons, is translationally invariant so that the Hamiltonian operator[7]

$$H^{SF}_{rve} = T^0 + T' + V \tag{3.98}$$

also called *the spin-free rovibronic Hamiltonian* [4] is translationally invariant and gives rise to the *rovibronic Schrödinger equation*

$$H_{rve}\Psi_{rve} = \mathcal{E}_{rve}\Psi_{rve} \,. \tag{3.99}$$

[7] rve stands for rotational, vibrational, electronic.

The total center of mass of the molecular system has now been separated but as a result of this coordinate change, the kinetic energy operator (3.93) is no longer diagonal in the particle indices owing to the presence of T' (see Eq. (3.96)) where the indices are mixed, which means that nuclear and electronic motions are coupled. *In order to highlight the structure of the kinetic energy operators, we refer to Sect.* 3.6.1 *where a simple example of a molecular system made of two electrons and three nuclei is given, together with matrix expressions that clearly bring out the coupling terms.*

3.3.2 Nuclear Center of Mass Coordinates *

Let all particles, electrons and nuclei, be referred to the *nuclear center of mass* and space-fixed (NSF) axis system $(X_{NSF}, Y_{NSF}, Z_{NSF})$ parallel to the (X_{SF}, Y_{SF}, Z_{SF}) system but with the origin at the *nuclear center of mass*, so that

$$\tilde{\boldsymbol{R}}^r_{NSF} = \tilde{\boldsymbol{R}}^r_{SF} - \boldsymbol{R}^{NCM}_{SF} \quad (r = 2, \ldots, N_T)\,, \tag{3.100}$$

where

$$\boldsymbol{R}^{NCM}_{SF} = \frac{1}{M_N} \sum_{\alpha=1}^{N} m_\alpha \boldsymbol{R}^\alpha_{SF} \tag{3.101}$$

is the coordinate vector of the *nuclear center of mass* in the SF frame and

$$M_N = \sum_{\alpha=1}^{N} m_\alpha \tag{3.102}$$

is the *total nuclear mass* of the molecular system. By the very definition of the *total* center of mass, we have in the SF frame

$$-\sum_{\alpha=1}^{N} m_\alpha \boldsymbol{R}^\alpha_{SF} = m_e \sum_{i=1}^{n} \boldsymbol{r}^i_{SF}\,, \tag{3.103}$$

or in view of Eq. (3.101)

$$-\boldsymbol{R}^{NCM}_{SF} = \frac{m_e}{M_N} \sum_{i=1}^{n} \boldsymbol{r}^i_{SF}\,. \tag{3.104}$$

So that Eq. (3.100) may be written as

$$\tilde{\boldsymbol{R}}^r_{NSF} = \tilde{\boldsymbol{R}}^r_{SF} + \frac{m_e}{M_N} \sum_{i=1}^{n} \boldsymbol{r}^i_{SF} \quad (r = 2, \ldots, N_T). \tag{3.105}$$

Distinguishing between nuclei and electrons, we obtain

$$\boldsymbol{R}^\alpha_{NSF} = \boldsymbol{R}^\alpha_{SF} + \frac{m_e}{M_N} \sum_{i=1}^{n} \boldsymbol{r}^i_{SF} \quad (\alpha = 2, \ldots, N), \tag{3.106}$$

and

$$\boldsymbol{r}^i_{NSF} = \boldsymbol{r}^i_{SF} + \frac{m_e}{M_N} \sum_{i=1}^{n} \boldsymbol{r}^i_{SF} \quad (i = 1, \ldots, n). \tag{3.107}$$

Applying to the Space-Fixed and spin-free rovibronic Hamiltonian, $H^{SF}_{rve} = T^0 + T' + V$ (See Eqs. (3.95)–(3.98)), the chain rule (again we refer to complement Sect. 3.6.3 for the chain rule) for the SF → NSF coordinate transformation

$$(\tilde{\boldsymbol{R}}^2_{SF}, \ldots, \tilde{\boldsymbol{R}}^{N_T}_{SF}) \rightarrow (\boldsymbol{R}^2_{NSF}, \ldots, \boldsymbol{R}^N_{NSF}, \boldsymbol{r}^1_{NSF}, \ldots, \boldsymbol{r}^n_{NSF}) = (\boldsymbol{R}_{NSF}, \boldsymbol{r}_{NSF}) \tag{3.108}$$

yields

$$H^{NSF}_{rve}(\boldsymbol{R}_{NSF}, \boldsymbol{r}_{NSF}) = T^{nu}(\boldsymbol{R}_{NSF}) + T^{el}(\boldsymbol{r}_{NSF}) + V(\boldsymbol{r}_{NSF}, \boldsymbol{R}_{NSF}) \tag{3.109}$$

with

$$T^{nu}(\boldsymbol{R}_{NSF}) = -\frac{\hbar^2}{2} \sum_{\alpha=2}^{N} \frac{1}{m_\alpha} \nabla^2_{\boldsymbol{R}^\alpha_{NSF}} + \frac{\hbar^2}{2M_N} \sum_{\alpha,\beta=2}^{N} \nabla_{\boldsymbol{R}^\alpha_{NSF}} \cdot \nabla_{\boldsymbol{R}^\beta_{NSF}}, \tag{3.110}$$

$$T^{el}(\boldsymbol{r}_{NSF}) = -\frac{\hbar^2}{2m_e} \sum_{i=1}^{n} \nabla^2_{\boldsymbol{r}^i_{NSF}} - \frac{\hbar^2}{2M_N} \sum_{i,j=1}^{n} \nabla_{\boldsymbol{r}^i_{NSF}} \cdot \nabla_{\boldsymbol{r}^j_{NSF}}, \tag{3.111}$$

and

$$V(\boldsymbol{r}_{NSF}, \boldsymbol{R}_{NSF}) = \sum_{j>i}^{n} \frac{e^2}{4\pi\varepsilon_0 ||\boldsymbol{r}^i_{NSF} - \boldsymbol{r}^j_{NSF}||} - \sum_{i}^{n} \sum_{\alpha}^{N} \frac{Z_\alpha e^2}{4\pi\varepsilon_0 ||\boldsymbol{R}^\alpha_{NSF} - \boldsymbol{r}^i_{NSF}||}$$
$$+ \sum_{\beta>\alpha}^{N} \frac{Z_\alpha Z_\beta e^2}{4\pi\varepsilon_0 ||\boldsymbol{R}^\alpha_{NSF} - \boldsymbol{R}^\beta_{NSF}||}. \tag{3.112}$$

From these Eqs. (3.109–3.111), it is obvious that the kinetic energies of the nuclei and the electrons are separated. The second term on the right-hand side of Eq. (3.111) often called the *mass polarization* it term, gives rise to coupling of the electronic motions but this kinetic energy correlation term is a nuclear mass dependent contribution and can be neglected in all but the most precise works in spectroscopy.

The second term on the right-hand side of Eq. (3.110) gives rise to coupling of the nuclear motions but, unlike the above electronic mass polarization terms, cannot be systematically neglected but can nevertheless be reduced and even eliminated by a clever choice of nuclear curvilinear coordinates (see next Sect. 3.3.3).

Henceforth, the electronic mass polarization term is left out, so that the common expression of the NSF spin-free rovibronic Hamiltonian operator is given by Eqs. (3.109), (3.110), and (3.112), the electronic kinetic energy operator being the diagonal expression

$$T^{el}(\boldsymbol{r}_{NSF}) = -\frac{\hbar^2}{2m_e}\sum_{i=1}^{n}\nabla^2_{r^i_{NSF}} . \tag{3.113}$$

Finally, we refer to Sect. 3.6.2 *where the SF → NSF coordinate change is explicitly given in the case of a molecular system made of two electrons and three nuclei.*

3.3.3 *Internal (Shape/Deformation) and Rotational Coordinates* *

In order to describe the rotation and deformation of the molecular system, motions supposed to be carried by the nuclei only, it is necessary to introduce a *Body-Fixed* (BF) frame, i.e. a BF axis system (X_{BF}, Y_{BF}, Z_{BF}), with origin at *the nuclear center of mass*, and with an orientation relative to the nuclear axis system (NSF), this orientationbeing defined by the positions of the nuclei only. These axes should be called nuclear axes but they have come to be called Body-Fixed (BF) axes. Again, their orientation is determined by the coordinates of the nuclei only, the coordinates of the electrons being not involved. This is called defining or embedding a frame fixed in the body and a usual way to proceed is to consider the BF nuclear coordinates $(X^{\alpha}_{BF}, Y^{\alpha}_{BF}, Z^{\alpha}_{BF})$ related to the NSF nuclear coordinates $(X^{\alpha}_{NSF}, Y^{\alpha}_{NSF}, Z^{\alpha}_{NSF})$ by

$$\begin{bmatrix} X^{\alpha}_{BF} \\ Y^{\alpha}_{BF} \\ Z^{\alpha}_{BF} \end{bmatrix} = \begin{bmatrix} C_{XX} & C_{XY} & C_{XZ} \\ C_{YX} & C_{YY} & C_{YZ} \\ C_{ZX} & C_{ZY} & C_{ZZ} \end{bmatrix} \begin{bmatrix} X^{\alpha}_{NSF} \\ Y^{\alpha}_{NSF} \\ Z^{\alpha}_{NSF} \end{bmatrix} , \tag{3.114}$$

where $\alpha = 2, \ldots, N$ and $\boldsymbol{C}$ is an orthogonal matrix, the elements of which are *direction cosines* connecting the BF and NSF axes. These matrix elements are functions of three Euler angles $\Theta = (\alpha, \beta, \gamma)$, which are the *orientation variables* defining the position of the BF frame with respect to the NSF frame.

There are twelve ways of specifying Euler angles and we use here the one of Zare [11] where α and β are the ordinary spherical angles of the $\boldsymbol{e}_z^{BF}$ vector in the frame and γ is the angle measuring a clockwise rotation about $\boldsymbol{e}_z^{BF}$. In other words, for a given vector $\boldsymbol{R}$, the relation between the SF- and BF-components of this vector is given by

$$\begin{bmatrix} X_{BF} \\ Y_{BF} \\ Z_{BF} \end{bmatrix} = \begin{bmatrix} \cos\gamma & \sin\gamma & 0 \\ -\sin\gamma & \cos\gamma & 0 \\ 0 & 0 & 1 \end{bmatrix} \times \begin{bmatrix} \cos\beta & 0 & -\sin\beta \\ 0 & 1 & 0 \\ \sin\beta & 0 & \cos\beta \end{bmatrix}$$

$$\times \begin{bmatrix} \cos\alpha & \sin\alpha & 0 \\ -\sin\alpha & \cos\alpha & 0 \\ 0 & 0 & 1 \end{bmatrix} \begin{bmatrix} X_{NSF} \\ Y_{NSF} \\ Z_{NSF} \end{bmatrix}. \tag{3.115}$$

The corresponding transformation from the NSF frame to the BF frame is shown in Fig. 3.4. The first two Euler angles, α and β, are the two spherical angles of any vector parallel to the z^{BF} axis. α corresponds to a rotation about the z^{NSF} axis, $\alpha \in [0, 2\pi[$ and the frame obtained from this rotation only and starting from the NSF frame can be called the $E1$ frame (for one Euler angle). β corresponds to a rotation about the y^{E1} axis and $\beta \in [0, \pi]$. This rotation starting from the $E1$ frame leads to the $E2$ frame (for two Euler angles). We note that $y^{E1} = y^{E2}$ by construction. Finally, the angle γ corresponds to a rotation about the z^{E2} axis. $\gamma \in [0, 2\pi[$. By definition, we thus have $z^{BF} = z^{E2}$.

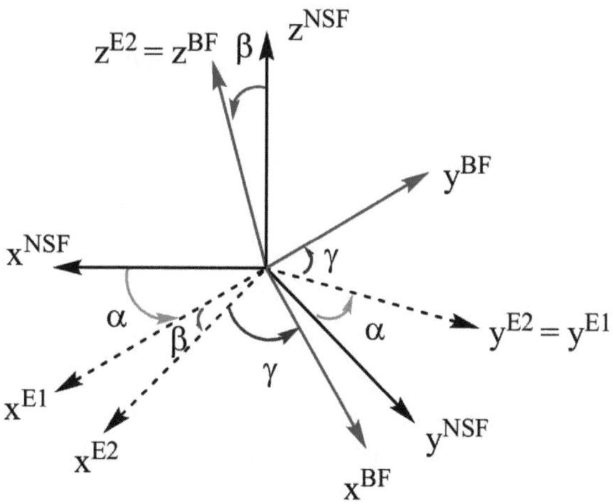

Fig. 3.4 Definition of the three Euler angles that determine the BF frame with respect to the NSF frame

As mentioned above, we require that the matrix $\boldsymbol{C}$ *be specified entirely in terms of the nuclear coordinates* $\boldsymbol{R}_{NSF}$, i.e.

$$\boldsymbol{C} = \tilde{\boldsymbol{C}}(\Theta(\boldsymbol{R}_{NSF})) = \boldsymbol{C}(\boldsymbol{R}_{NSF})\,. \tag{3.116}$$

Moreover, since the $\boldsymbol{R}_{BF} = (\boldsymbol{R}^2_{BF}, \boldsymbol{R}^3_{BF}, \ldots, \boldsymbol{R}^N_{BF})$ are fixed in the body, not all their $3N-3$ components are independent for there must be three relations among them. Hence, the Cartesian coordinates X^α_{BF}, Y^α_{BF}, and Z^α_{BF} ($\alpha = 2, \ldots, N$) must be expressible in terms of $3N-6$ independent internal shape coordinates $\boldsymbol{q} = \{q^1, \ldots, q^{3N-6}\}$.

Similarly, the BF electronic Cartesian coordinates $x^i_{BF}, y^i_{BF}, z^i_{BF}$ ($i = 1, \ldots, n$) are related to the NSF electronic Cartesian coordinates by

$$\begin{bmatrix} x^i_{BF} \\ y^i_{BF} \\ z^i_{BF} \end{bmatrix} = \boldsymbol{C}(\boldsymbol{R}_{NSF}) \begin{bmatrix} x^i_{NSF} \\ y^i_{NSF} \\ z^i_{NSF} \end{bmatrix}, \tag{3.117}$$

with $i = 1, \ldots, n$ and where $\boldsymbol{C}(\boldsymbol{R}_{NSF})$ is, of course, the same orthogonal matrix as in Eqs. (3.115) and (3.116).

To change the coordinates in the molecular Hamiltonian operator from $(\boldsymbol{R}^2_{NSF}, \ldots, \boldsymbol{R}^N_{NSF}, \boldsymbol{r}^1_{NSF}, \ldots, \boldsymbol{r}^n_{NSF})$ to the internal (shape), rotational and BF electronic coordinates $(\boldsymbol{q}, \Theta, \boldsymbol{r}_{BF}) = (q^1, \ldots, q^{3N-6}, \alpha, \beta, \gamma, x^1_{BF}, \ldots, z^n_{BF})$, we can apply the chain rule to express the derivatives of the $\boldsymbol{R}^\alpha_{NSF}$ and $\boldsymbol{r}^i_{NSF}$ in terms of these new coordinates. The transformation from $(\boldsymbol{R}_{NSF}, \boldsymbol{r}_{NSF})$ to $(\boldsymbol{q}, \Theta, \boldsymbol{r}_{BF})$ is assumed to be (locally) invertible so that, together with the specific role of the nuclei as carrying the deformation and rotation motions of the molecular system, we may write

$$\boldsymbol{q} = \boldsymbol{q}(\boldsymbol{R}_{NSF})\,, \tag{3.118}$$

$$\Theta = \Theta(\boldsymbol{R}_{NSF})\,, \tag{3.119}$$

$$\boldsymbol{r}_{BF} = \boldsymbol{r}_{BF}(\boldsymbol{R}_{NSF}, \boldsymbol{r}_{NSF})\,, \tag{3.120}$$

this last expression resulting from Eq. (3.117).

Using the chain rule and a rather obvious matrix notation, we can write the expressions of

$$\nabla_{\boldsymbol{R}_{NSF}} = (\nabla_{\boldsymbol{R}^2_{NSF}}, \ldots, \nabla_{\boldsymbol{R}^N_{NSF}})\,, \tag{3.121}$$

and

$$\nabla_{\boldsymbol{r}_{NSF}} = (\nabla_{\boldsymbol{r}^1_{NSF}}, \ldots, \nabla_{\boldsymbol{r}^n_{NSF}}) \,, \tag{3.122}$$

in terms of $\nabla_{\boldsymbol{q}}$, ∇_{Θ}, $\nabla_{\boldsymbol{r}_{BF}}$ as follows

$$\begin{bmatrix} \nabla_{\boldsymbol{R}_{NSF}} \\ \hline \nabla_{\boldsymbol{r}_{NSF}} \end{bmatrix} = \left[\begin{array}{c|c|c} \frac{\partial \boldsymbol{q}}{\partial \boldsymbol{R}_{NSF}} & \frac{\partial \Theta}{\partial \boldsymbol{R}_{NSF}} & \frac{\partial \boldsymbol{r}_{BF}}{\partial \boldsymbol{R}_{NSF}} \\ \hline 0 & 0 & \frac{\partial \boldsymbol{r}_{BF}}{\partial \boldsymbol{r}_{NSF}} \end{array}\right] \cdot \begin{bmatrix} \nabla_{\boldsymbol{q}} \\ \hline \nabla_{\Theta} \\ \hline \nabla_{\boldsymbol{r}_{BF}} \end{bmatrix} , \tag{3.123}$$

since $\frac{\partial \boldsymbol{q}}{\partial \boldsymbol{r}_{NSF}} = 0$ and $\frac{\partial \Theta}{\partial \boldsymbol{r}_{NSF}} = 0$ in view of Eqs. (3.118) and (3.119). In addition, owing to Eq. (3.117)

$$\frac{\partial \boldsymbol{r}^i_{BF}}{\partial \boldsymbol{r}^i_{NSF}} = \boldsymbol{C} \,, \tag{3.124}$$

so that, substituting

$$\nabla_{\boldsymbol{r}^i_{NSF}} = \frac{\partial \boldsymbol{r}^i_{BF}}{\partial \boldsymbol{r}^i_{NSF}} \cdot \nabla_{\boldsymbol{r}^i_{BF}} = \boldsymbol{C} \cdot \nabla_{\boldsymbol{r}^i_{BF}} \tag{3.125}$$

into the electronic kinetic energy (without the mass polarization *terms*),

$$T^{el}(\boldsymbol{r}_{NSF}) = -\frac{\hbar^2}{2m_e} \sum_{i=1}^{n} \nabla^2_{\boldsymbol{r}^i_{NSF}} \,, \tag{3.126}$$

yields

$$T^{el}(\boldsymbol{r}_{BF}) = -\frac{\hbar^2}{2m_e} \sum_{i=1}^{n} \nabla^2_{\boldsymbol{r}^i_{BF}} \,. \tag{3.127}$$

Concerning the nuclei, since $\frac{\partial \boldsymbol{r}_{BF}}{\partial \boldsymbol{R}_{NSF}} \neq 0$, we see from Eq. (3.123) that derivatives with respect to the electronic coordinates are introduced into the expression of $\nabla_{\boldsymbol{R}_{NSF}}$ and hence in the expression for T^{nu}. In other words [4], although by using $(X_{NSF}, Y_{NSF}, Z_{NSF})$ coordinates, we achieve a complete separation of the electronic and nuclear coordinates in the kinetic energy operator $T^{el} + T^{nu}$, when we change to (X_{BF}, Y_{BF}, Z_{BF}) coordinates (in order to introduce the rotation and deformation coordinates), we introduce the electronic coordinates back into T^{nu}. The effect of the nuclear-electronic coupling terms introduced into T^{nu} is, however, generally small and these terms are generally neglected in the rotation-deformation Hamiltonian that is obtained by using a BF frame and $(\boldsymbol{q}, \Theta, \boldsymbol{r}_{BF})$ coordinates.

From a mathematical point of view, we see that the derivatives with respect to electronic coordinates occur in T^{nu} because the $\boldsymbol{r}_{BF}$ coordinates of the electrons depend on the $\boldsymbol{R}_{NSF}$ coordinates of the nuclei through the dependence of the matrix $\boldsymbol{C}$ on the nuclear coordinates (see Eqs. (3.117) and (3.120)).

From a physical point of view, we are now referring the electrons to the BF frame that rotates with the nuclei so that the electrons are subject to centrifugal and Coriolis forces in this axis system. A fully detailed and careful derivation of the Hamiltonian has been given by Sutcliffe [10].

As mentioned above for the simplification of the rotation-deformation Hamiltonian, it is well-worth neglecting the coupling between electrons and nuclei, i.e. neglecting in Eq. (3.123) the matrix $\frac{\partial \boldsymbol{r}_{BF}}{\partial \boldsymbol{R}_{NSF}}$, so that

$$\left[\nabla_{\boldsymbol{R}_{NSF}}\right] = \left[\frac{\partial \boldsymbol{q}}{\partial \boldsymbol{R}_{NSF}} \;\; \frac{\partial \Theta}{\partial \boldsymbol{R}_{NSF}}\right]\begin{bmatrix}\nabla_{\boldsymbol{q}}\\ \nabla_{\Theta}\end{bmatrix}. \tag{3.128}$$

If the coupling between electrons and nuclei as well as the polarization terms are neglected, the coordinate change from $(\boldsymbol{R}_{NSF}, \boldsymbol{r}_{NSF})$ to $(\boldsymbol{q}, \Theta, \boldsymbol{r}_{BF})$ yields the following expression for the molecular Hamiltonian operator

$$\begin{aligned} H^{mol}(\boldsymbol{r}_{BF}, \boldsymbol{q}, \Theta) = {} & T^{nu}(\boldsymbol{q}, \Theta) + T^{el}(\boldsymbol{r}_{BF}) \\ & + V^{el-el}(\boldsymbol{r}_{BF}) + V^{nu-nu}(\boldsymbol{q}) + V^{nu-el}(\boldsymbol{r}_{BF}, \boldsymbol{q}), \end{aligned} \tag{3.129}$$

with $T^{el}(\boldsymbol{r}_{BF})$ given by Eq. (3.127).

$T^{nu}(\boldsymbol{q}, \Theta)$ can be derived by substituting Eq. (3.128) into Eq. (3.110), i.e. by using the *chain rule* straightforwardly (first procedure for changing coordinates in a Hamiltonian operator). $T^{nu}(\boldsymbol{q}, \Theta)$ can also be obtained by applying the *second procedure for changing coordinates in a Hamiltonian operator* to the coordinate change $(\boldsymbol{R}_{NSF}) \rightarrow (\boldsymbol{q}, \Theta)$, starting from the classical nuclear kinetic energy[8]

$$T^{nu} = \frac{1}{2}\sum_{\alpha=1}^{N} m_\alpha \dot{\boldsymbol{R}}^\alpha_{NSF} \cdot \dot{\boldsymbol{R}}^\alpha_{NSF}, \tag{3.130}$$

with the constraints $\sum_\alpha m_\alpha \boldsymbol{R}^\alpha_{NSF} = 0$ since the NSF axis system has origin at the nuclear center of mass.

---

[8]The second procedure may equivalently be applied to the coordinate change $(\boldsymbol{R}^1_{LF}, \ldots, \boldsymbol{R}^\alpha_{LF}, \ldots, \boldsymbol{R}^N_{LF}) \rightarrow (\boldsymbol{R}^{NCM}_{LF}, \boldsymbol{q}, \Theta)$ starting from the classical nuclear kinetic energy in the LF frame: $T^{nu} = \frac{1}{2}\sum_{\alpha=1}^{N} m_\alpha \dot{\boldsymbol{R}}^\alpha_{LF} \cdot \dot{\boldsymbol{R}}^\alpha_{LF}$.

The separation of the electronic and nuclear motions can be taken up again and Eqs. (3.9) and (3.10) be rewritten in terms of the coordinates $(\boldsymbol{q}, \Theta, \boldsymbol{r}_{BF})$ by splitting the molecular Hamiltonian operator as follows

$$H^{mol}(\boldsymbol{r}_{BF}, \boldsymbol{q}, \Theta) = H^{el}(\boldsymbol{r}_{BF}; \boldsymbol{q}) + T^{nu}(\boldsymbol{q}, \Theta), \tag{3.131}$$

with

$$H^{el}(\boldsymbol{r}_{BF}; \boldsymbol{q}) = V^{el-el}(\boldsymbol{r}_{BF}) + V^{nu-nu}(\boldsymbol{q}) + V^{nu-el}(\boldsymbol{r}_{BF}, \boldsymbol{q}) + T^{el}(\boldsymbol{r}_{BF}) \tag{3.132}$$

the *electronic Hamiltonian* for a given set of *relative* nuclear positions given by $\boldsymbol{q}$. Now Eqs. (3.14), (3.18), (3.19) and (3.20) can be recast in a more explicit way

$$\begin{aligned} \Psi(\boldsymbol{r}_{BF}, \boldsymbol{q}, \Theta, t) &= \sum_m \sum_\lambda d_{m\lambda}(t) \varphi_\lambda^{nu}(\boldsymbol{q}, \Theta) \Phi_m^{el}(\boldsymbol{r}_{BF}; \boldsymbol{q}) \\ &= \sum_m \Psi_m(\boldsymbol{q}, \Theta, t) \Phi_m^{el}(\boldsymbol{r}_{BF}; \boldsymbol{q}), \end{aligned} \tag{3.133}$$

and

$$\sum_m (T_{nm}(\boldsymbol{q}, \Theta) + V_{nm}(\boldsymbol{q})) \Psi_m(\boldsymbol{q}, \Theta, t) = i\hbar \frac{\partial \Psi_n(\boldsymbol{q}, \Theta, t)}{\partial t}, \tag{3.134}$$

with

$$T_{nm}(\boldsymbol{q}, \Theta) = \langle \Phi_n^{el}; \boldsymbol{q} | T^{nu}(\boldsymbol{q}, \Theta) | \Phi_m^{el}; \boldsymbol{q} \rangle_{\boldsymbol{r}_{BF}}, \tag{3.135}$$

and

$$V_{nm}(\boldsymbol{q}) = \langle \Phi_n^{el}; \boldsymbol{q} | H^{el}(\boldsymbol{q}) | \Phi_m^{el}; \boldsymbol{q} \rangle_{\boldsymbol{r}_{BF}}. \tag{3.136}$$

3.3.4 The Potential Energy Surface and the Nuclear Hamiltonian Operator

Let us now examine more closely the concept of potential energy for the nuclear motion. As already pointed out, non-adiabatic couplings can be neglected when the energy gap between electronic states is large (adiabatic and BO approximation). In this situation, it is legitimate to use the adiabatic representation introduced in

Sect. 3.2.2, which corresponds to the eigenfunctions of the electronic Hamiltonian. Due to the invariance of the Coulomb potentials of Eq. (3.1) with respect to any global rotation and translation of the molecule, it appears that the adiabatic electronic energies (the eigenvalues of the electronic Hamiltonian) *depend only on 3N-6 internal coordinates*[9] *that describe the shape of the molecule (the molecular geometry),* $\boldsymbol{q}$, i.e.

$$H^{el}(\boldsymbol{r}_{BF};\boldsymbol{q})\Phi_m^{el/ad}(\boldsymbol{r}_{BF};\boldsymbol{q}) = E_m^{el}(\boldsymbol{q})\Phi_m^{el/ad}(\boldsymbol{r}_{BF};\boldsymbol{q})\,. \tag{3.137}$$

for a given electronic state labeled[10] m. In practice, one gives a value to $\boldsymbol{q}$ and computes the corresponding electronic energy, $E_m^{el}(\boldsymbol{q})$, at a certain level of theory (see Sect. 3.5.5) with a *quantum chemistry* package such as, for example, GAUSSIAN [12] or MOLPRO [13]. In other words, the electronic Schrödinger equation must be solved repeatedly, as many times as there are values of $\boldsymbol{q}$ (molecular geometries) to be explored.

Up to this point, $\boldsymbol{q}$ has played the role of a parameter with respect to the electronic Hamiltonian and its eigensolutions. Connecting the electronic energies (eigenvalues) smoothly with respect to continuous variations of $\boldsymbol{q}$ turns $E_m^{el}(\boldsymbol{q})$ effectively into a function of $\boldsymbol{q}$ now considered as a variable. This function, usually referred to as the *potential energy surface* plays the role of a potential energy for the nuclear motion, as the operator $H^{el}(\boldsymbol{r}_{BF};\boldsymbol{q})$ has been replaced by a function of $\boldsymbol{q}$ to be added to the nuclear kinetic energy operator *within the BO approximation* (see Eq. 3.49):

$$(T^{nu}(\boldsymbol{q},\Theta) + E_m^{el}(\boldsymbol{q}))\Psi_m^{ad}(\boldsymbol{q},\Theta,t) = i\hbar\frac{\partial\Psi_m^{ad}(\boldsymbol{q},\Theta,t)}{\partial t}\,. \tag{3.138}$$

To emphasize this identification, we will write $E_m^{el}(\boldsymbol{q})$ as $V(\boldsymbol{q})$ in the following (or $V_m(\boldsymbol{q})$ when the label of the adiabatic electronic state is to be specified). The potential energy surface, $V(\boldsymbol{q})$, is thus a function of the $3N-6$ internal nuclear coordinates only.

3.4 Interaction with External Fields

In this book, a semi-classical approach is adopted whereby the particles making up the molecule (nuclei and electrons) are subjected to the rules of quantum mechanics while the external radiation fields are taken to be classical. In other words, the Hamiltonian of a molecule in interaction with a radiation can be split into two terms [2]:

$$H = H^{mol}(\boldsymbol{r},\boldsymbol{R}) + H^{ext}\,. \tag{3.139}$$

where $H^{mol}(\boldsymbol{r},\boldsymbol{R})$ is the molecular Hamiltonian operator introduced in the previous sections, which depends on the coordinates of the electrons, $\boldsymbol{r}$, and of the nuclei, $\boldsymbol{R}$,

---

[9]For diatomic molecules ($N=2$), the number of internal nuclear coordinates is equal to $3N-5=1$.

[10]Traditionally, $m = 0, 1, 2, \ldots$ and $m = 0$ corresponds to the ground state.

but not on the external electromagnetic fields, and H^{ext} is an operator describing the interaction of classical electromagnetic fields with charged particles. This semi-classical approach compared to fully quantum electrodynamics theories virtually gives identical results provided that the external fields are strong enough for any reciprocal influence on them by the molecular system to be negligible [2]. However, this semi-classical approach cannot describe processes such as spontaneous emission.

The usual way to tackle the semi-classical approach is to start from the *classical Hamiltonian expression* of the interaction of an electromagnetic field with a charged particle and to apply the correspondence rule to obtain a quantum interaction operator (see e.g. Chaps. 11 and 12 in Bransden and Joachain [14], Chap. 13 in Cohen-Tannoudji et al. [15] and Chap. 5 in Schatz and Ratner [16]). Next, particularizing to the case of a monochromatic plane wave and taking into account the fact that the wavelength of the radiation is much larger than the size of the molecule, the interaction operator can be written, after some manipulations and further approximations, as a dipole interaction with the electric field, the so-called *dipole approximation*. In this approach, the electric and magnetic fields must be expressed in terms of scalar and vector potentials and well-chosen gauges are to be used.

However, the dipole approximation can also be obtained more straightforwardly as follows (see, e.g. Chap. 3 in Levine [17]). For a *monochromatic light wave* of *frequency* ν and *wavelength* λ, traveling in the z direction and linearly polarized in the xy plane, the space and time variations of the *electric field* $\boldsymbol{E}$ and the *magnetic field* $\boldsymbol{B}$ are

$$\boldsymbol{E}(z,t) = \boldsymbol{i}\xi_x(z,t) = \boldsymbol{i}\xi_x^0 \cos(\omega t - kz)\,, \tag{3.140}$$

and

$$\boldsymbol{B}(z,t) = \boldsymbol{j}\mathcal{B}_y(z,t) = \boldsymbol{j}\mathcal{B}_y^0 \cos(\omega t - kz)\,, \tag{3.141}$$

where $\boldsymbol{i}$ and $\boldsymbol{j}$ are unit vectors along x and y, ξ_x^0 and $\mathcal{B}_y^0$ are the maximum magnitudes of $\boldsymbol{E}$ and $\boldsymbol{B}$, $\omega = 2\pi\nu$ is the *angular frequency* and $k = \frac{2\pi}{\lambda}$ is the *wavenumber.*

The wavelength λ is the distance between successive crests of $\boldsymbol{E}$ and the frequency ν is the number of crests that pass a fixed point in space each second.

Thus, we have

$$\lambda\nu = c\,, \tag{3.142}$$

where c is the *speed of light.*

In addition, from Maxwell's equations of electromagnetism it can be shown that

$$\frac{\mathcal{B}_y^0}{\xi_x^0} = \frac{1}{c}\,, \tag{3.143}$$

in International System of Units (SI) and the radiation intensity is given by $S = \frac{c\epsilon_0}{2}(\xi_x^0)^2$ in SI units (Watt/m^2).

The electric and magnetic fields of the radiation will interact with the molecular electrons and nuclei giving a time-dependent perturbation. In classical electromagnetism a charged particle, for instance an electron of velocity $\boldsymbol{v}$ moving in electric and magnetic fields experiences the *Lorentz force*

$$\boldsymbol{F} = -e(\boldsymbol{E} + \boldsymbol{v} \times \boldsymbol{B}), \tag{3.144}$$

in SI units. From this equation and in view of Eq. (3.143), the ratio of the magnetic to the electric force is $\frac{v}{c}$.

For the ground-state hydrogen atom the electron velocity is $v = \alpha c = \frac{c}{137}$ where α is the *fine structure constant*. This is also the order of magnitude for valence electrons of other atoms and also for electrons in molecules. The nuclei being heavier their velocities are even smaller and thus to a good approximation, we can ignore the interaction between $\boldsymbol{B}$ and the electrons and nuclei of a molecule, and consider only the effect of $\boldsymbol{E}$.

In order for the radiation to be correctly described by means of a time-dependent perturbation approach, the effect of the radiation electric field must be a small addition to the molecular Hamiltonian. To estimate the internal electric field acting on an electron in an atom, let us calculate the field at a distance of 1 Bohr:

$$\xi = \frac{1}{4\pi\epsilon_0}\frac{e}{a_0^2} \sim 10^{11}\frac{V}{m}, \tag{3.145}$$

in SI units. If we convert this field into a radiation intensity, we obtain 400×10^{18} W/m^2 or 400×10^{14} W/cm^2, which is an enormous laser flux. Thus, provided we do not use high-power lasers, we are justified in considering the radiation as a small perturbation. The potential energy of interaction between a system of charged particles and the electric field is

$$V = -\sum_i q_i x_i \xi_x, \tag{3.146}$$

where x_i and q_i are the x coordinates and charges of the particles (nuclei and electrons). We thus have for the perturbation part of the Hamiltonian

$$H^{ext}(t) = -\xi_x^0 \sum_i q_i x_i \cos\left(\omega t - \frac{2\pi z_i}{\lambda}\right). \tag{3.147}$$

One last step before reaching the dipole approximation is to introduce the *long-wave approximation*. Indeed, for transitions e.g. between different molecular electronic states, the wavelengths usually lie in the ultraviolet, typically $\lambda \approx 10^3$ Å. For vibrational and rotational transitions the wavelengths are even larger. The size of molecules is about 10–100 Å, so that the wave length is usually much larger than the size of the molecules and the space variation of the electric field is negligible: $\frac{z_i}{\lambda} \approx 0$. With this further approximation, Eq. (3.147) becomes

$$H^{ext}(t) = -\xi_x^0 \sum_i q_i x_i \cos \omega t \,. \tag{3.148}$$

Now, noticing that $\mu_x = \sum_i q_i x_i$ is the x component of the dipole moment operator and $\xi_x(t) = \xi_x^0 \cos \omega t$ is the x component of the external electric field, using the vector notation, assuming that the molecular system is neutral and reverting to a separate notation for electrons and nuclei as used previously, we may write $H^{ext}(t)$ in the dipole approximation quite generally as

$$H^{ext} = -\boldsymbol{\mu}(\boldsymbol{r}, \boldsymbol{R}) \cdot \boldsymbol{E} \,, \tag{3.149}$$

with $\boldsymbol{\mu}(\boldsymbol{r}, \boldsymbol{R})$ the *dipole vector of the molecule*, and $\boldsymbol{E}$ the *external field*.

In a molecule, the dipole moment can be split into two parts, one corresponding to the electrons, $\boldsymbol{\mu}^{el}(\boldsymbol{r})$, and one corresponding to the nuclei, $\boldsymbol{\mu}^{nu}(\boldsymbol{R})$. The electric dipole moment vector of the molecule thus reads

$$\begin{aligned} \boldsymbol{\mu}(\boldsymbol{r}, \boldsymbol{R}) &= -e \sum_i \boldsymbol{r}_i + e \sum_\alpha Z_\alpha \boldsymbol{R}_\alpha \\ &= \boldsymbol{\mu}^{el}(\boldsymbol{r}) + \boldsymbol{\mu}^{nu}(\boldsymbol{R}) \,. \end{aligned} \tag{3.150}$$

Let us now integrate over the *adiabatic states* introduced in Sect. 3.2.2:

$$\langle \Phi_n^{el/ad}(\boldsymbol{R}) | \boldsymbol{\mu}(\boldsymbol{R}) | \Phi_l^{el/ad}(\boldsymbol{R}) \rangle_r = \boldsymbol{\mu}_{nl}(\boldsymbol{R}) = \boldsymbol{\mu}_{nl}^{el}(\boldsymbol{R}) + \delta_{nl}\, \boldsymbol{\mu}_n^{nu}(\boldsymbol{R}) \,. \tag{3.151}$$

In the previous equation, $\boldsymbol{\mu}_{nl}^{el}(\boldsymbol{R})$ and $\boldsymbol{\mu}_n^{nu}(\boldsymbol{R})$ are vectors. The orientation of these vectors depend on the Euler angles that describe the overall rotation of the system. However, the components of the vectors themselves depend on the $3N - 6$ internal nuclear coordinates $\boldsymbol{q}$ only, i.e.

$$\begin{aligned} \boldsymbol{\mu}_{nl}^{el}(\boldsymbol{R}) &= \boldsymbol{\mu}_{nl}^{el}(\boldsymbol{q}) \,, \\ \boldsymbol{\mu}_n^{nu}(\boldsymbol{R}) &= \boldsymbol{\mu}_n^{nu}(\boldsymbol{q}) \,. \end{aligned} \tag{3.152}$$

Several remarks can be made on Eq. (3.151):

- If $n \neq l$, the matrix elements will induce *transitions between different electronic states*.[11] In view of (3.151), it is clear that only the contribution coming from the electrons, i.e. from $\boldsymbol{\mu}_{el}(\boldsymbol{r})$, will play a role in these transitions. In addition, the fact that the $\boldsymbol{\mu}_{nl}^{el}(\boldsymbol{q})$ matrix elements are functions of $\boldsymbol{q}$ comes only from the adiabatic electronic states depending on the nuclear coordinates. Now, the derivatives of $\Phi_n^{el/ad}(\boldsymbol{r}; \boldsymbol{R})$ with respect to $\boldsymbol{R}$ are often small. In particular, in the Born-

[11] These transitions generally correspond to wavelengths in the ultraviolet-visible domain.

Oppenheimer approximation, these derivatives are assumed to be equal to zero (see Eqs. (3.40) and (3.41)). This allows us to make approximations, e.g. to assume that $\boldsymbol{\mu}_{nl}^{el}(\boldsymbol{q})$ is independent on $\boldsymbol{q}$. Starting from the vibrational ground state in the electronic ground state,[12] when considering the electronic transition to state l, we can assume that $\boldsymbol{\mu}_{0l}^{el}(\boldsymbol{q}) \approx \boldsymbol{\mu}_{0l}^{el}(\boldsymbol{q}_{eq})$, $\boldsymbol{q}_{eq}$ corresponding to the equilibrium geometry of the molecule in the electronic ground state. This approximation is called the *Condon approximation*.

- If $n = l$, the matrix elements will induce transitions between different vibrational states.[13] For those transitions, both nuclei and electrons contribute. Here, $\boldsymbol{\mu}_n^{nu}(\boldsymbol{q})$ and $\boldsymbol{\mu}_{nn}(\boldsymbol{q})$ cannot be assumed to be constant.

Let us be more specific and let us assume that $\Psi_i^{mol}(\boldsymbol{r}, \boldsymbol{R})$ and $\Psi_f^{mol}(\boldsymbol{r}, \boldsymbol{R})$ are two solutions of the molecular time-independent Schrödinger equation (see Eq. (3.7)). When the molecule is in interaction with an external field $\boldsymbol{E}$, the coupling between $\Psi_i^{mol}(\boldsymbol{r}, \boldsymbol{R})$ and $\Psi_f^{mol}(\boldsymbol{r}, \boldsymbol{R})$ is given by the matrix element:

$$\int \Psi_f^{mol\star}(\boldsymbol{r}, \boldsymbol{R}) \left(-\boldsymbol{\mu}(\boldsymbol{r}, \boldsymbol{R}) \cdot \boldsymbol{E}\right) \Psi_i^{mol}(\boldsymbol{r}, \boldsymbol{R}) d\boldsymbol{r} d\boldsymbol{R}$$
$$= -\left(\int \Psi_f^{mol\star}(\boldsymbol{r}, \boldsymbol{R}) \boldsymbol{\mu}(\boldsymbol{r}, \boldsymbol{R}) \Psi_i^{mol}(\boldsymbol{r}, \boldsymbol{R}) d\boldsymbol{r} d\boldsymbol{R}\right) \cdot \boldsymbol{E} \,. \tag{3.153}$$

Thus, for a given external field, the coupling between the two eigenfunctions is proportional to

$$\boldsymbol{d}_{fi} = \int \Psi_f^{mol\star}(\boldsymbol{r}, \boldsymbol{R}) \boldsymbol{\mu}(\boldsymbol{r}, \boldsymbol{R}) \Psi_i^{mol}(\boldsymbol{r}, \boldsymbol{R}) d\boldsymbol{r} d\boldsymbol{R} \,, \tag{3.154}$$

which is the *transition dipole moment* between initial state i and final state f. Using first order perturbation theory, one can show that the square of the magnitude of this transition dipole moment,

$$\left| \int \Psi_f^{mol\star}(\boldsymbol{r}, \boldsymbol{R}) \boldsymbol{\mu}(\boldsymbol{r}, \boldsymbol{R}) \Psi_i^{mol}(\boldsymbol{r}, \boldsymbol{R}) d\boldsymbol{r} d\boldsymbol{R} \right|^2 \,, \tag{3.155}$$

is proportional to the intensity of the transition between initial state i and final state f[14]: see Eq. (9.35) in Sect. 9.5.

Thus, *there is resonant absorption of light by the molecule if the energy of the photon corresponds to the energy difference between two eigenvalues of the molecule. The efficiency of this absorption is linked to the transition dipole moment between the*

---

[12] We assume that the ground state potential energy surface has at least one local minimum.

[13] These transitions generally correspond to wavelengths in the infrared domain. The matrix elements can also induce transitions between rotational states corresponding to wavelengths in the microwave domain.

[14] See, for instance, Chap. 13 in Ref. [15].

two corresponding eigenstates. In particular, *if for symmetry reasons the transition dipole moment is equal to zero, the transition is forbidden.*

Within the Born-Oppenheimer approximation, the eigenfunctions of the molecular Hamiltonian operator can be written as a product: $\Phi_m^{el/ad}(\boldsymbol{r};\boldsymbol{R})\Psi_m^l(\boldsymbol{R})$. Let us assume that the molecule is first in the vibrational ground state of its electronic ground state: $\Phi_0^{el/ad}(\boldsymbol{r};\boldsymbol{R})\Psi_0^0(\boldsymbol{R})$. Two case must be distinguished:

- First, the final state is a vibrational state, let us say state f, of the *electronic ground state*. We thus consider purely *(ro)vibrational transitions*. The corresponding transition dipole moment reads

$$\begin{aligned}
\boldsymbol{d}_{(0f)(00)} &= \int \Phi_0^{el/ad\star}(\boldsymbol{r};\boldsymbol{R})\Psi_0^{f\star}(\boldsymbol{R})\boldsymbol{\mu}(\boldsymbol{r},\boldsymbol{R})\Phi_0^{el/ad}(\boldsymbol{r};\boldsymbol{R})\Psi_0^0(\boldsymbol{R})d\boldsymbol{r}d\boldsymbol{R}\,,\\
&= \int \Psi_0^{f\star}(\boldsymbol{R})\left(\int \Phi_0^{el/ad\star}(\boldsymbol{r};\boldsymbol{R})\boldsymbol{\mu}(\boldsymbol{r},\boldsymbol{R})\Phi_0^{el/ad}(\boldsymbol{r};\boldsymbol{R})d\boldsymbol{r}\right)\Psi_0^0(\boldsymbol{R})d\boldsymbol{R}\,,\\
&= \int \Psi_0^{f\star}(\boldsymbol{R})\langle\Phi_0^{el/ad\star}(\boldsymbol{R})|\boldsymbol{\mu}(\boldsymbol{R})|\Phi_0^{el/ad}(\boldsymbol{R})\rangle_{\boldsymbol{r}}\Psi_0^0(\boldsymbol{R})d\boldsymbol{R}\,,\\
&= \int \Psi_0^{f\star}(\boldsymbol{R})\left(\boldsymbol{\mu}_{00}^{el}(\boldsymbol{R})+\boldsymbol{\mu}_0^{nu}(\boldsymbol{R})\right)\Psi_0^0(\boldsymbol{R})d\boldsymbol{R}\,,\\
&= \int \Psi_0^{f\star}(\boldsymbol{R})\boldsymbol{\mu}_{00}(\boldsymbol{R})\Psi_0^0(\boldsymbol{R})d\boldsymbol{R}\,.
\end{aligned} \tag{3.156}$$

  We will see in Chap. 9 (more precisely in Sect. 9.5), that propagating the wavepackets $\mu_{00}^{\lambda}(\boldsymbol{R})\Psi_0^0(\boldsymbol{R})$, with μ_{00}^{λ} ($\lambda = x, y, z$) the Body-Fixed components of the dipole vector, will allow us to obtain the infrared spectrum of the molecule.
- Second, the final state is a vibrational state, let us say state f, in *another electronic state*, let us say electronic state 1. The corresponding transition dipole moment reads

$$\begin{aligned}
\boldsymbol{d}_{(1f)(00)} &= \int \Phi_1^{el/ad\star}(\boldsymbol{r};\boldsymbol{R})\Psi_1^{f\star}(\boldsymbol{R})\boldsymbol{\mu}(\boldsymbol{r},\boldsymbol{R})\Phi_0^{el/ad}(\boldsymbol{r};\boldsymbol{R})\Psi_0^0(\boldsymbol{R})d\boldsymbol{r}d\boldsymbol{R}\\
&= \int \Psi_1^{f\star}(\boldsymbol{R})\left(\int \Phi_1^{el/ad\star}(\boldsymbol{r};\boldsymbol{R})\boldsymbol{\mu}(\boldsymbol{r},\boldsymbol{R})\Phi_0^{el/ad}(\boldsymbol{r};\boldsymbol{R})d\boldsymbol{r}\right)\Psi_0^0(\boldsymbol{R})d\boldsymbol{R}\\
&= \int \Psi_1^{f\star}(\boldsymbol{R})\langle\Phi_1^{el/ad\star}(\boldsymbol{R})|\boldsymbol{\mu}(\boldsymbol{R})|\Phi_0^{el/ad}(\boldsymbol{R})\rangle_{\boldsymbol{r}}\Psi_0^0(\boldsymbol{R})d\boldsymbol{R}\\
&= \int \Psi_1^{f\star}(\boldsymbol{R})\boldsymbol{\mu}_{10}^{el}(\boldsymbol{R})\Psi_0^0(\boldsymbol{R})d\boldsymbol{R}\\
&= \int \Psi_1^{f\star}(\boldsymbol{R})\boldsymbol{\mu}_{10}(\boldsymbol{R})\Psi_0^0(\boldsymbol{R})d\boldsymbol{R}\,.
\end{aligned} \tag{3.157}$$

  Since $\boldsymbol{\mu}_{10}(\boldsymbol{R})$ is a vector, the transition dipole moment has three components:

$$d^{\lambda}_{(1f)(00)} = \int \Psi_1^{f\,\star}(\boldsymbol{R})\mu^{\lambda}_{10}(\boldsymbol{q})\Psi_0^0(\boldsymbol{R})d\boldsymbol{R}\,, \tag{3.158}$$

with $\lambda = x, y, z$. If we assume $\mu^{\lambda}_{01}(\boldsymbol{q}) \approx \mu^{\lambda}_{01}(\boldsymbol{q}_{eq})$, $\boldsymbol{q}_{eq}$ corresponding to the equilibrium geometry of the molecule in the electronic ground state (Condon approximation), we obtain

$$d^{\lambda}_{(1f)(00)} = \mu^{\lambda}_{01}(\boldsymbol{q}_{eq}) \int \Psi_1^{f\,\star}(\boldsymbol{R})\Psi_0^0(\boldsymbol{R})d\boldsymbol{R}\,. \tag{3.159}$$

We will see in Chap. 10, that propagating the wavepacket[15] $\Psi_0^0(\boldsymbol{R})$ on the electronic state 1 will lead to the photoabsorption spectrum from state 0 to state 1. $|\int \Psi_1^{f\,\star}(\boldsymbol{R})\Psi_0^0(\boldsymbol{R})d\boldsymbol{R}|^2$ are called *Franck-Condon factors* .

3.5 Illustration with the H_2 Molecule

Let us now give some examples of the electronic basis functions $\Phi_m^{el/ad}(\boldsymbol{r}; \boldsymbol{R})$ of Sect. 3.2.2, the potential energy surfaces $V_m(\boldsymbol{q})$ of Sect. 3.3.4 and the rovibrational levels of a molecule within a given electronic state $E^m_{l_1}$ of Sect. 3.2.3. For this, let us consider the smallest neutral molecule. The dihydrogen molecule, H_2, is a four-body system made of two nuclei and two electrons. There is no exact analytical solution to the corresponding Schrödinger equation. However, the Born-Oppenheimer (BO) approximation can be used to first separate the electronic problem from the nuclear one. Second, as regards the resolution of the electronic Schrödinger equation, Eq. (3.137), the simplest approximation consists in ignoring explicit electron correlation upon treating electron repulsion as a mean field for each electron. Two-electron wavefunctions (at any fixed geometry) can thus be written as so-called Hartree products of one-electron wavefunctions called molecular orbitals (MOs). This is known as the orbital (one-electron) approximation and yields the concept of electronic configuration. Further, owing to the fermionic character of electrons, Hartree products must be antisymmetrized with respect to permutation of the space and spin variables of the electrons. This is readily formulated in terms of Slater determinants built on spin-orbitals. In this context, applying the variational principle to such an ansatz corresponds to mathematically finding the best solution with the corresponding form. In other words, the variational principle guarantees that the final result is as close as possible to the exact solution. Technically, the optimal MOs can be obtained as solutions of Fock equations using a self-consistent-field (SCF) procedure. This description is known as the Hartree-Fock (HF) approximation. In most practical applications, the MOs are expanded in terms of atomic orbitals (AOs), using the approach known as linear combination of atomic orbitals (LCAO). Solving Fock

[15] $\Psi_0^0(\boldsymbol{R})$ is an eigenstate for the Hamiltonian operator of the electronic ground state but a wavepacket for the Hamiltonian operator of the electronic state 1.

equations in this representation, produces what are known as the Roothaan equations for the expansion coefficients. Finally, a single Slater determinant may prove to be a poor description. More sophisticated (post-Hartree-Fock) treatments should thus be used, whereby electron correlation is recovered upon mixing several Slater determinants in the electronic wavefunction (configuration-interaction methods reflecting the multiconfigurational or multideterminental character of the electronic wavefunction) or by accounting for their influence perturbatively. We refer to textbooks on quantum chemistry for further details (e.g., Ref. [19]). In what follows, we will illustrate such concepts qualitatively by applying this methodology to the valence states of H_2, i.e. the first solutions, $\Phi_m^{el/ad}(\boldsymbol{r}_{BF}; \boldsymbol{q})$, of Eq. (3.137).

3.5.1 Atomic Orbitals

Qualitative descriptions often limit the set of atomic orbitals (AOs) to valence orbitals only: in our case a $1s$ ($n = 1, l = 0, m = 0$) on each H atom. For a single H atom, the $1s$ orbital is represented as a function of the position vector of the electron with respect to the nucleus, $\boldsymbol{r}$. Its expression in terms of spherical coordinates centered at the nucleus reads

$$\Phi_{1s}(r, \theta, \varphi) = Y_0^0(\theta, \varphi) R_{10}(r) , \tag{3.160}$$

where

$$\begin{aligned} Y_0^0(\theta, \varphi) &= \frac{1}{\sqrt{4\pi}} , \\ R_{10}(r) &= \frac{2}{a_0^{3/2}} \exp\left(-\frac{r}{a_0}\right) , \end{aligned} \tag{3.161}$$

are the normalized angular and radial parts, and a_0 is the Bohr constant that plays the role of the atomic unit (a.u.) of length ($a_0 = 0.529177$ Å). A plot of the corresponding wavefunction is shown on Fig. 3.5.

The optimal AOs used to expand the MOs of H_2 within the LCAO method are not identical to the AOs of a single H atom. They now must account for the repulsion between both electrons in an average manner. Various kinds of AO basis sets have been developed over the years (for further details, we refer to textbooks on quantum chemistry, e.g., Ref. [19]). Here, let us consider Slater-type orbitals, where the nuclear charge, Z, is replaced by an effective reduced parameter, $\zeta = \frac{Z-\sigma}{a_0}$, that accounts for the average effect of the other electron through the Slater shielding constant, σ. Let a and b label the two nuclei, such that $\boldsymbol{r}_a$ or $\boldsymbol{r}_b$ denote the position vectors of a single electron originating from either H_a or H_b, respectively. The corresponding AO functions read, in their own spherical coordinates,

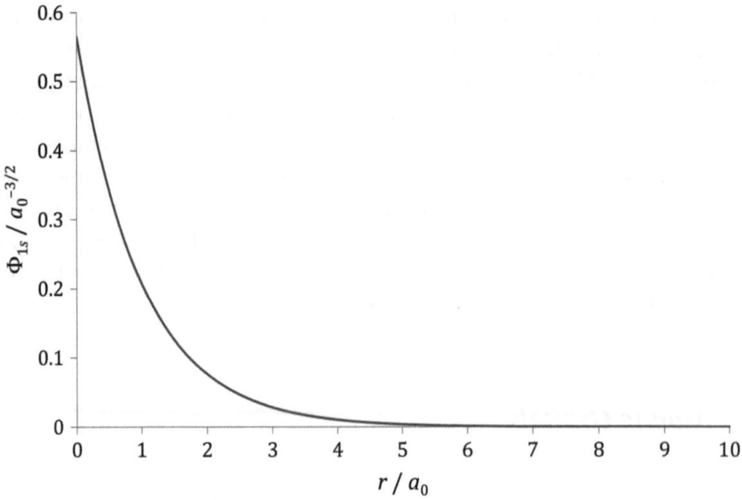

Fig. 3.5 1*s* AO of a single H atom as a function of the electron-nucleus distance

$$\Phi_{1s_a}(r_a, \theta_a, \varphi_a) = \frac{1}{\sqrt{\pi}} \zeta^{3/2} \exp^{-\zeta r_a} ,$$
$$\Phi_{1s_b}(r_b, \theta_b, \varphi_b) = \frac{1}{\sqrt{\pi}} \zeta^{3/2} \exp^{-\zeta r_b} . \tag{3.162}$$

Here, $Z = 1$ and we use a typical value $\sigma = 0.30$ for the shielding of a 1s electron on another 1s electron (hence, $\zeta = 0.70\, a_0^{-1}$).

3.5.2 Molecular Orbitals

Let us now expand the valence MOs in terms of the 1s_a and 1s_b AOs using the LCAO approach. Only two independent combinations can be generated and each must belong to an irreducible representation of the symmetry point group of the molecule: here, $\mathcal{D}_{\infty\mathrm{h}}$. In this case, there are only two choices for the MOs: the normalized sum (totally symmetric), Φ_{σ_g}, and the normalized difference (antisymmetric with respect to inversion), Φ_{σ_u},

$$\Phi_{\sigma_g} = \frac{\Phi_{1sa} + \Phi_{1sb}}{\sqrt{2(1+S)}} ,$$
$$\Phi_{\sigma_u} = \frac{\Phi_{1sa} - \Phi_{1sb}}{\sqrt{2(1-S)}} . \tag{3.163}$$

The MO σ_g is bonding whereas σ_u is anti-bonding between the two nuclei. One easily checks that the MOs form a pair of orthonormal wavefunctions[16]:

$$\begin{aligned} \langle \sigma_g | \sigma_g \rangle &= \frac{\langle 1s_a | 1s_a \rangle + \langle 1s_b | 1s_b \rangle + \langle 1s_a | 1s_b \rangle + \langle 1s_b | 1s_a \rangle}{2(1+S)} = 1\,, \\ \langle \sigma_u | \sigma_u \rangle &= \frac{\langle 1s_a | 1s_a \rangle + \langle 1s_b | 1s_b \rangle - \langle 1s_a | 1s_b \rangle - \langle 1s_b | 1s_a \rangle}{2(1-S)} = 1\,, \\ \langle \sigma_g | \sigma_u \rangle = \langle \sigma_u | \sigma_g \rangle &= \frac{\langle 1s_a | 1s_a \rangle - \langle 1s_b | 1s_b \rangle - \langle 1s_a | 1s_b \rangle + \langle 1s_b | 1s_a \rangle}{2\sqrt{1-S^2}} = 0\,, \end{aligned} \tag{3.164}$$

using $\langle 1s_a | 1s_a \rangle = \langle 1s_b | 1s_b \rangle = 1$ (note that only real-valued orbitals are considered here). S is the overlap between the two 1s AOs,

$$S = \langle 1s_a | 1s_b \rangle = \langle 1s_b | 1s_a \rangle\,. \tag{3.165}$$

It depends parametrically on R, the bond length (distance between the two nuclei). At the equilibrium geometry, $R = 0.741$ Å $= 1.40\,a_0$ [20]. Such values yield $S = 0.86$ for the Slater-type orbitals considered here.

In the previous expressions of the AOs, r_a was originated from H_a and r_b from H_b. However, cylindrical coordinates are more balanced to describe the relative positions of the particles in the molecule. Let us define z as the inter-nuclear axis, such that $z_a = -R/2$ and $z_b = +R/2$. The azimuthal angle, φ, is unchanged. The other coordinates satisfy

$$\begin{aligned} z &= r\cos\theta = r_a \cos\theta_a - R/2 = r_b \cos\theta_b + R/2\,, \\ \rho &= r\sin\theta = r_a \sin\theta_a = r_b \sin\theta_b\,. \end{aligned} \tag{3.166}$$

The MOs read thus

$$\begin{aligned} \Phi_{\sigma_g}(\rho, z, \varphi) &= \frac{1}{\sqrt{\pi}} \zeta^{3/2} \frac{\exp\left(-\zeta\sqrt{\rho^2 + (z+R/2)^2}\right) + \exp\left(-\zeta\sqrt{\rho^2 + (z-R/2)^2}\right)}{\sqrt{2(1+S)}}\,, \\ \Phi_{\sigma_u}(\rho, z, \varphi) &= \frac{1}{\sqrt{\pi}} \zeta^{3/2} \frac{\exp\left(-\zeta\sqrt{\rho^2 + (z+R/2)^2}\right) - \exp\left(-\zeta\sqrt{\rho^2 + (z-R/2)^2}\right)}{\sqrt{2(1-S)}}\,. \end{aligned} \tag{3.167}$$

The AOs and MOs are plotted on Figs. 3.6 and 3.7 along z for $\rho = 0$, respectively. The bonding and anti-bonding MOs are represented on Figs. 3.8 and 3.9, respectively, over the (ρ, z)-plane for any arbitrary value of φ. Their energetic representation is shown on Fig. 3.10.

[16] In this section, we have dropped the subscript $\boldsymbol{r}$ on brackets, as there no ambiguity: integration is performed over the electronic coordinates, $\boldsymbol{r}$. In other words $\langle \cdots | \cdots \rangle$ implicitly means $\langle \cdots | \cdots \rangle_{\boldsymbol{r}}$.

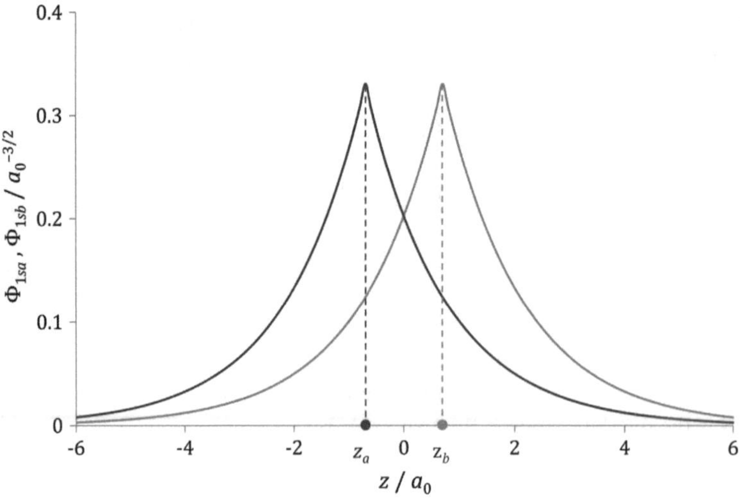

Fig. 3.6 Right and left 1*s* AOs along the internuclear axis of the H_2 molecule

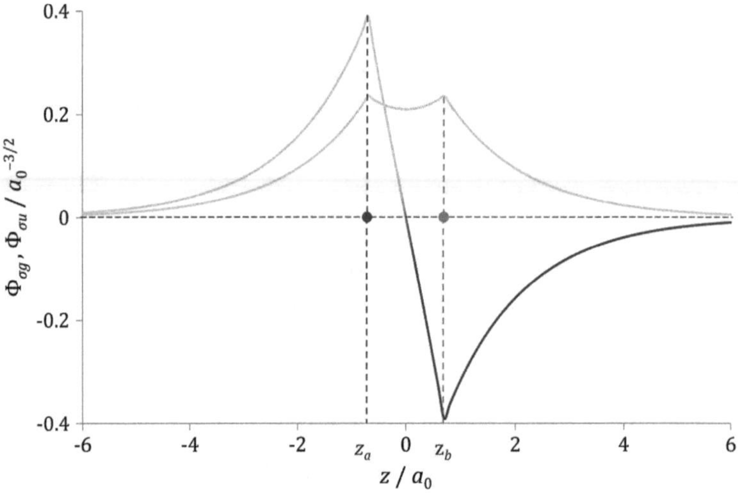

Fig. 3.7 Bonding (*yellow*) and antibonding (*yellow* and *purple*) MOs along the internuclear axis of the H_2 molecule

3.5.3 *Electronic Configurations*

The concept of electronic configuration is an approximate description of the many-electron wavefunction, whereby each MO can be occupied by zero, one, or two electrons, according to the Pauli exclusion principle. The distribution of electrons in orbitals is expected to minimize the energy of the ground state upon filling the

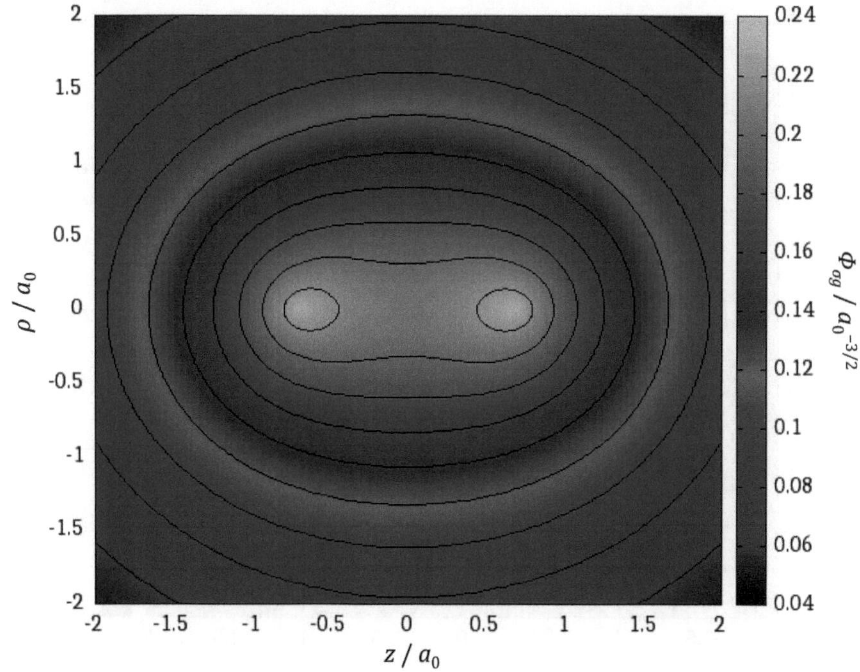

Fig. 3.8 Bonding MO over any plane containing the internuclear axis. Isocontours are spaced at $0.02\, a_0^{-3/2}$. A negative value of ρ is defined as the absolute value of ρ with φ changed for $\varphi + \pi$

lowest-energy MOs first, or to reflect the nature of the excited states where one or a few electrons are promoted to higher-energy MOs that are vacant in the ground-state configuration (single, double excitations, etc.).

As a first approximation, the square modulus of an MO occupied by an electron gives the density of probability of finding this electron in space. In Fig. 3.8, we observe that the corresponding electron density is high between the two nuclei. This is the reason why it is said to be bonding: a configuration with a pair of electrons in this MO will lead to some significant shielding of the repulsion between the nuclei, which is at the origin of the σ chemical bond in H_2. A classical picture would describe this situation as the alternation of a positive charge (the first nucleus), a pair of negative charges (the two electrons), and another positive charge (the second nucleus). In contrast, Fig. 3.9 shows that occupying the anti-bonding MO with an electron would yield a small negative-charge density in the region between the two positive charges, no longer large enough to prevent their repulsion to dominate over the electron shielding.

Around the equilibrium geometry, the ground state of H_2 is well-represented with σ_g^2, i.e., two electrons in the bonding MO. This is the prototype of a covalent σ-bond. Electronic configurations are often represented on the so-called MO diagram (see Fig. 3.10): the valence electrons are shown in the outer region of the diagram as

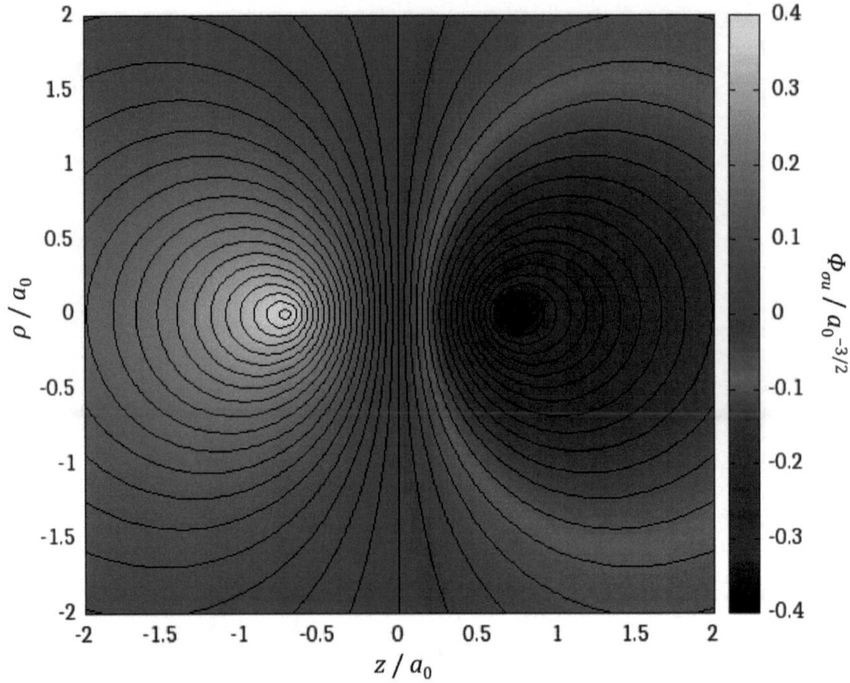

Fig. 3.9 Anti-bonding MO over any plane containing the internuclear axis. Isocontours are spaced at $0.02\, a_0^{-3/2}$. A negative value of ρ is defined as the absolute value of ρ with φ changed for $\varphi + \pi$

occupying the valence AOs of the separate atoms, while in the inner region they fill the MOs according to the configuration under consideration.

At this stage, the ground-state energy at the equilibrium geometry, $V_0(R_e) = E_0^{el}(R_e)$, hence the energetic stabilization due to the creation of a σ chemical bond, can be estimated semi-empirically using the extended Hückel method, for instance [18]. In this, approximate MOs and their energies are obtained as eigensolutions of a simplified Fock operator involving semi-empirical integrals as parameters, and the total energy is expressed ultimately as a sum of such terms arising from occupied orbitals. This qualitative approach is often used as a starting point for generating a guess for further ab initio calculations in quantum chemistry.

3.5.4 Electronic Wavefunctions

A more rigorous description of the electronic states must account for the spin of the electrons. As electrons are fermions, total (spin and space) electronic wavefunctions must be antisymmetric with respect to the permutation of any two electrons. A con-

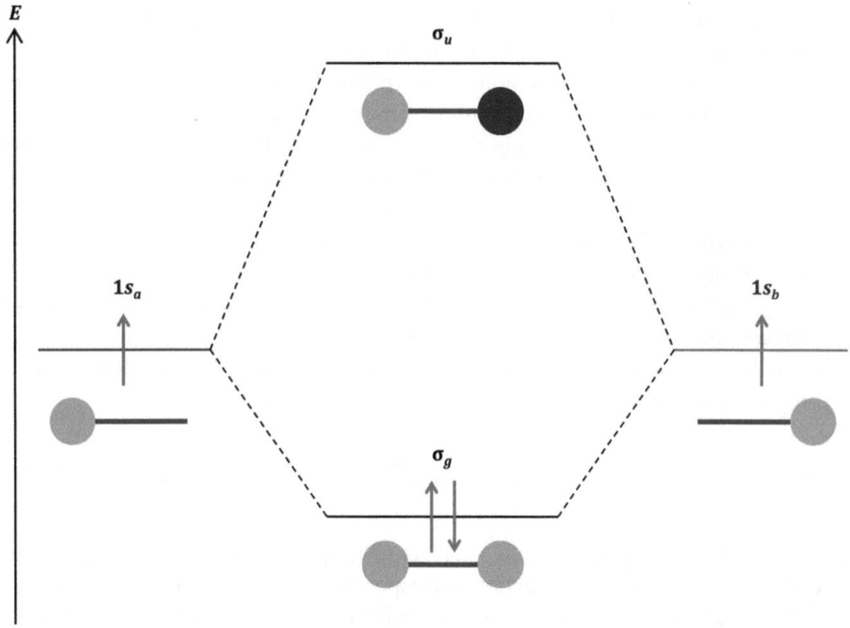

Fig. 3.10 MO diagram of H_2 with electron filling corresponding to a σ_g^2 configuration

venient way of achieving this condition is obtained when using Slater determinants built on spin-orbitals.

First, each MO is associated to a pair of spin-orbitals,

$$\Phi^{\alpha}_{\sigma_{g/u}}(\boldsymbol{r}, \sigma) = \Phi_{\sigma_{g/u}}(\boldsymbol{r})\alpha(\sigma)\,,$$
$$\Phi^{\beta}_{\sigma_{g/u}}(\boldsymbol{r}, \sigma) = \Phi_{\sigma_{g/u}}(\boldsymbol{r})\beta(\sigma)\,, \tag{3.168}$$

where $\alpha(\sigma)$ and $\beta(\sigma)$ are the eigenfunctions of the one-electron spin operators, s^2 and s_z with eigenvalues $s(s+1)\hbar^2$ and $m_s\hbar$ ($s = \frac{1}{2}$; $m_s = +\frac{1}{2}$ and $m_s = -\frac{1}{2}$, respectively). Then, Slater determinants are built on spin-orbitals according to electronic configurations. When the Hartree-Fock method is adequate for an electronic wavefunction, it means that it is well-approximated with a single Slater determinant. For example, around the equilibrium geometry, a good approximation of the ground electronic state of H_2 is obtained as[17]:

$$\Phi_0^{el/ad}(\boldsymbol{r}_1, \sigma_1, \boldsymbol{r}_2, \sigma_2) \approx \Phi_{\sigma_g}(\boldsymbol{r}_1)\Phi_{\sigma_g}(\boldsymbol{r}_2)\frac{\alpha(\sigma_1)\beta(\sigma_2) - \beta(\sigma_1)\alpha(\sigma_2)}{\sqrt{2}}\,, \tag{3.169}$$

[17]Note that this spin function is an eigenfunction of the total spin operator for eigenvalue zero (singulet). The separation of an electronic wavefunction into a symmetric (antisymmetric) spatial part and a antisymmetric (symmetric) spin part is in general only possible for two-electron wavefunctions.

with Φ_{σ_g} coming from Eq. (3.163). Note that the parametric dependence of $\Phi_0^{el/ad}$ on R is not explicitly indicated here to simplify the notations.

In practice, applying the variational principle to the Schrödinger equation yields the Fock equations that are solved using a self-consistent-field procedure. This corresponds mathematically to finding the best solution with the form of Eq. (3.169). In other words, the variational principle guarantees that the final result is as close as possible to the exact solution. As a consequence, this level of description can provide a correct solution only if the exact wavefunction is well-approximated with a single Slater determinant. However, this is never really true and, in general, the Hartree-Fock approach gives quantitative results for the electronic ground state around the equilibrium geometry only.

In much the same way as for the ground state, the second excited valence state is well-represented with a σ_u^2 configuration, which corresponds to

$$\Phi_2^{el/ad}(\boldsymbol{r}_1, \sigma_1, \boldsymbol{r}_2, \sigma_2) \approx \Phi_{\sigma_u}(\boldsymbol{r}_1)\Phi_{\sigma_u}(\boldsymbol{r}_2)\frac{\alpha(\sigma_1)\beta(\sigma_2) - \beta(\sigma_1)\alpha(\sigma_2)}{\sqrt{2}} . \qquad (3.170)$$

In contrast, the configuration associated with the first excited valence state is an open-shell, $\sigma_g^1\sigma_u^1$. Now, this situation corresponds to four possible Slater determinants that must be combined in order to build electronic wavefunctions of well-defined total spin. Such spin-adapted combinations of Slater determinants are called configuration-state functions. The singlet state is well-approximated with the following combination of two Slater determinants, which after expansion reads

$$\Phi_{1(S)}^{el/ad}(\boldsymbol{r}_1, \sigma_1, \boldsymbol{r}_2, \sigma_2) \approx \frac{\Phi_{\sigma_g}(\boldsymbol{r}_1)\Phi_{\sigma_u}(\boldsymbol{r}_2) + \Phi_{\sigma_u}(\boldsymbol{r}_1)\Phi_{\sigma_g}(\boldsymbol{r}_2)}{\sqrt{2}}\frac{\alpha(\sigma_1)\beta(\sigma_2) - \beta(\sigma_1)\alpha(\sigma_2)}{\sqrt{2}} .$$

The related triplet state corresponds to the following three components,

$$\begin{aligned}
&\Phi_{1(T_-)}^{el/ad}(\boldsymbol{r}_1, \sigma_1, \boldsymbol{r}_2, \sigma_2) \approx \frac{\Phi_{\sigma_g}(\boldsymbol{r}_1)\Phi_{\sigma_u}(\boldsymbol{r}_2) - \Phi_{\sigma_u}(\boldsymbol{r}_1)\Phi_{\sigma_g}(\boldsymbol{r}_2)}{\sqrt{2}}\beta(\sigma_1)\beta(\sigma_2) , \\
&\Phi_{1(T_0)}^{el/ad}(\boldsymbol{r}_1, \sigma_1, \boldsymbol{r}_2, \sigma_2) \approx \\
&\frac{\Phi_{\sigma_g}(\boldsymbol{r}_1)\Phi_{\sigma_u}(\boldsymbol{r}_2) - \Phi_{\sigma_u}(\boldsymbol{r}_1)\Phi_{\sigma_g}(\boldsymbol{r}_2)}{\sqrt{2}}\frac{\alpha(\sigma_1)\beta(\sigma_2) + \beta(\sigma_1)\alpha(\sigma_2)}{\sqrt{2}} , \\
&\Phi_{1(T_+)}^{el/ad}(\boldsymbol{r}_1, \sigma_1, \boldsymbol{r}_2, \sigma_2) \approx \frac{\Phi_{\sigma_g}(\boldsymbol{r}_1)\Phi_{\sigma_u}(\boldsymbol{r}_2) - \Phi_{\sigma_u}(\boldsymbol{r}_1)\Phi_{\sigma_g}(\boldsymbol{r}_2)}{\sqrt{2}}\alpha(\sigma_1)\alpha(\sigma_2) .
\end{aligned} \qquad (3.171)$$

As already mentioned, the expressions given in Eqs. (3.169), (3.5.4), (3.170), and (3.171) are correct around the equilibrium geometry. However, Slater determinants are over-constrained when the bond distance is increased. At infinity (dissociation), they correlate to mixtures of the ionic and neutral eigenstates of the separate atoms. This problem is known as a lack of left-right static correlation in the Hartree-Fock ansatz. Configuration-interaction methods can be used to write electronic wavefunctions as mixtures of several configurations with coefficients that depend on the nuclear coordinates. In addition, we only considered valence states here, whereas Rydberg states may correspond to lower energies, depending on the bond distance. Their description requires additional orbitals. In summary, quantitative results, especially for excited electronic states, often require basis sets that are larger than simply the set of valence orbitals and post-Hartree-Fock methods where the multideterminental character of the wavefunction is accounted for.

3.5.5 Potential Energy Surfaces

Once a level of theory is determined as adequate for the problem under consideration, electronic energies can be calculated at various molecular geometries. These are directly identified with the potential energy surfaces for the nuclear motion $V_m(\boldsymbol{q})$ of Sect. 3.3.4. As an example, the potential energy curves of the three singlet valence states and the triplet valence state of H_2, $V_0(R)$, $V_{1(S)}(R)$, $V_2(R)$, and $V_{1(T)}(R)$, are shown on Fig. 3.11. These were obtained using a configuration-interaction (post-Hartree-Fock) method such that both the equilibrium geometry and the dissociation limit are described adequately. The curve giving the Hartree-Fock energy of the ground state is also shown for comparison: the description is correct around the equilibrium geometry but lies between the ground state and the excited states at infinity.

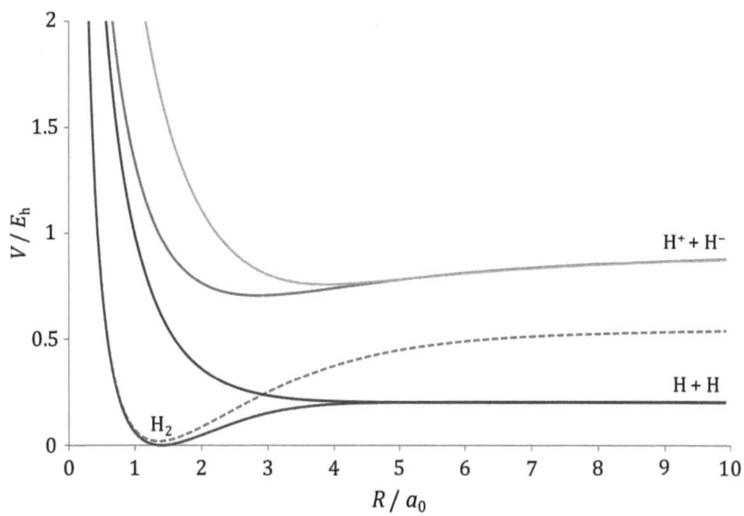

Fig. 3.11 Potential energy curves of H_2 against the internuclear distance calculated with a configuration-interaction (post-Hartree-Fock) method (*blue* ground state; *red* triplet valence state; *green* first excited singlet valence state; *yellow* second excited singlet valence state; *dashed cyan* ground state calculated with the Hartree-Fock method shown for comparison). E_h denotes the Hartree constant that plays role of the atomic unit (a.u.) of energy ($E_h = 27.2114$ eV)

3.5.6 Vibrational Eigenstates in the Electronic Ground State

A closer look at the ground-state potential energy curve (see Fig. 3.11) shows three characteristic regions: a short-range repulsive wall where the internuclear repulsion dominates; a long-range dissociative asymptote that becomes an attractive slope when the two nuclei get closer due to the formation of the chemical bond by the paired electrons; a stable well characterized by an optimal distance when both effects equilibrate. This kind of spoon-shaped curve is often represented analytically with a Morse function. The Morse curve shown on Fig. 3.12 was obtained from a three-parameter fitting procedure based on the equilibrium distance, R_e, the curvature (force constant) at this point, k_e, and the energy difference between the asymptote and the minimum, D_e, such that

$$V^{Morse}(R) = D_e(1 - e^{-\sqrt{\frac{k_e}{2D_e}}(R-R_e)})^2 . \tag{3.172}$$

When only small vibrations around the equilibrium position are to be described, a cruder approximation can be considered upon expanding the potential energy function to second order,

$$V^{harm}(R) = \frac{k_e}{2}(R - R_e)^2 . \tag{3.173}$$

This is known as a harmonic approximation. The corresponding curve is the approximating parabola of the Morse curve at $R = R_e$ (see Fig. 3.12).

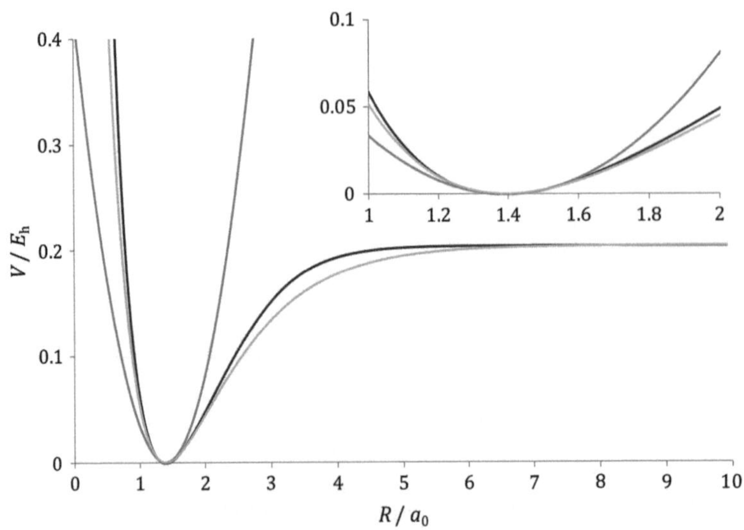

Fig. 3.12 Ground-state potential energy curves of H_2 against the internuclear distance (*blue* reference calculated with a configuration-interaction (post-Hartree-Fock) method—see Fig. 3.11; *green* Morse approximation; *red* harmonic approximation)

Both Morse and harmonic oscillators are problems for which exact quantum solutions are known analytically. In both cases, the fundamental angular frequency is obtained as

$$\omega_e = \sqrt{\frac{k_e}{\mu}}, \tag{3.174}$$

where $\mu = M_H/2$ is the reduced mass of the system. Experimentally, $\bar{\nu}_e = \frac{\omega_e}{2\pi c} = 4401.213\ \text{cm}^{-1}$ [21]. The energies of the harmonic levels ($v = 0, 1, 2, \ldots$) are given as

$$E_v^{harm} = \hbar\omega_e(v + \frac{1}{2}), \tag{3.175}$$

with no upper bound, whereas those of the Morse oscillator read

$$E_v^{Morse} = \hbar\omega_e(v + \frac{1}{2}) - \frac{(\hbar\omega_e(v + \frac{1}{2}))^2}{4D_e}. \tag{3.176}$$

Here, there is a maximum to the vibrational quantum number[18]: $v_{\max} = \lfloor \frac{2D_e}{\hbar\omega_e} \rfloor$, where $\lfloor \ldots \rfloor$ is the floor function (integer part). For $v > v_{\max}$, Eq. 3.176 becomes invalid because the condition $E_v^{Morse} > E_{v-1}^{Morse}$ is no longer satisfied. The energy spectrum is continuous for $E \geq D_e$, thus reflecting the property that levels are quantized (bound states) below the dissociation limit only ($E_0^{Morse} < \cdots < E_{v_{\max}}^{Morse} < D_e$).

The energy spacing between levels is now uneven, which is an effect known as anharmonicity. Usually, a vibrational spectrum obtained from infrared and Raman experiments is fitted to the following anharmonic form [20],

$$E_v^{anharm} = \hbar\omega_e(v + \frac{1}{2}) - \hbar\omega_e x_e(v + \frac{1}{2})^2 + \hbar\omega_e y_e(v + \frac{1}{2})^3 + \cdots, \tag{3.177}$$

where $\omega_e x_e$, $\omega_e y_e$, etc. are referred to as anharmonicity constants. For H_2, $\bar{\nu}_e x_e = \frac{\omega_e x_e}{2\pi c} = 121.336\ \text{cm}^{-1}$ and $\bar{\nu}_e y_e = \frac{\omega_e y_e}{2\pi c} = 0.8129\ \text{cm}^{-1}$ [21]. Within a Morse approximation, one obtains $y_e = 0$ and

$$x_e = \frac{\hbar\omega_e}{4D_e}. \tag{3.178}$$

Using this relationship yields an estimate of the dissociation energy, $D_e = 4.95$ eV $= 0.182\ E_h$ (in good agreement with more accurate experimental determinations). With these values, we can determine $v_{\max} = \lfloor 18.1 \rfloor = 18$ (see Fig. 3.13).

The values given by Eq. (3.177) correspond to approximate values of the purely vibrational levels $E_0^{l_0}$ of Eq. (3.52) of the H_2 molecule within its electronic ground state.

[18] If $\frac{2D_e}{\hbar\omega_e}$ occurs to be an integer, $v_{\max} = \frac{2D_e}{\hbar\omega_e} - 1$, because $E_{v_{\max}+1}^{Morse} = E_{v_{\max}}^{Morse}$, and only the first of the two seemingly-degenerate levels is physical. This is unlikely in actual cases but could happen when considering a numerical model.

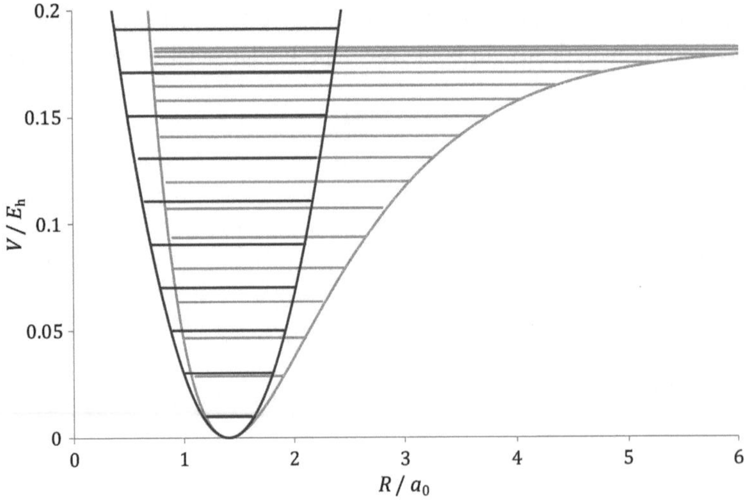

Fig. 3.13 Ground-state potential energy curves of H_2 against the internuclear distance and vibrational energy levels (*green* Morse approximation, $v = 0$–18; *red* harmonic approximation, $v = 0$–9). The potential energy functions correspond to the experimental data: $R_e = 1.40\ a_0$, $D_e = 0.182\ E_h$, $\hbar\omega_e = 0.0201\ E_h$

Fig. 3.14 Space fixed frame for a system with three nuclei, A, B, and C, and two electrons, for instance H_3^+. CM_T denotes the total center of mass and O the origin of the LF frame

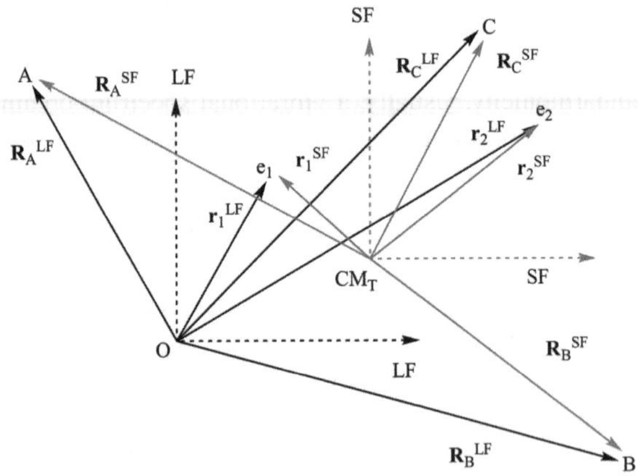

3.6 Complements to Chapter 3

3.6.1 *Example of Space-Fixed (SF) Coordinates* *

We illustrate the general structure of the kinetic energy operators introduced in Sect. 3.3.1 by explicit calculations in the case of a molecular system consisting of three atoms A, B and C and two electrons: see Fig. 3.14. Using a matrix notation, we express the coordinate change $\tilde{\boldsymbol{R}}_{LF} \rightarrow \tilde{\boldsymbol{R}}_{SF}$, i.e. from Eq. (3.84) to (3.90) as

$$\begin{bmatrix} \tilde{\boldsymbol{R}}^1_{SF} = \boldsymbol{R}^{CM}_{LF} \\ \tilde{\boldsymbol{R}}^2_{SF} = \boldsymbol{R}^{B}_{SF} \\ \tilde{\boldsymbol{R}}^3_{SF} = \boldsymbol{R}^{C}_{SF} \\ \tilde{\boldsymbol{R}}^4_{SF} = \boldsymbol{r}^{1}_{SF} \\ \tilde{\boldsymbol{R}}^5_{SF} = \boldsymbol{r}^{2}_{SF} \end{bmatrix} = \begin{bmatrix} \frac{m_A}{M_T} & \frac{m_B}{M_T} & \frac{m_C}{M_T} & \frac{m_e}{M_T} & \frac{m_e}{M_T} \\ -\frac{m_A}{M_T} & 1-\frac{m_B}{M_T} & -\frac{m_C}{M_T} & -\frac{m_e}{M_T} & -\frac{m_e}{M_T} \\ -\frac{m_A}{M_T} & -\frac{m_B}{M_T} & 1-\frac{m_C}{M_T} & -\frac{m_e}{M_T} & -\frac{m_e}{M_T} \\ -\frac{m_A}{M_T} & -\frac{m_B}{M_T} & -\frac{m_C}{M_T} & 1-\frac{m_e}{M_T} & -\frac{m_e}{M_T} \\ -\frac{m_A}{M_T} & -\frac{m_B}{M_T} & -\frac{m_C}{M_T} & -\frac{m_e}{M_T} & 1-\frac{m_e}{M_T} \end{bmatrix} \cdot \begin{bmatrix} \boldsymbol{R}^{A}_{LF} = \tilde{\boldsymbol{R}}^1_{LF} \\ \boldsymbol{R}^{B}_{LF} = \tilde{\boldsymbol{R}}^2_{LF} \\ \boldsymbol{R}^{C}_{LF} = \tilde{\boldsymbol{R}}^3_{LF} \\ \boldsymbol{r}^{1}_{LF} = \tilde{\boldsymbol{R}}^4_{LF} \\ \boldsymbol{r}^{2}_{LF} = \tilde{\boldsymbol{R}}^5_{LF} \end{bmatrix}, \tag{3.179}$$

or

$$\tilde{\boldsymbol{R}}^T_{SF} = \boldsymbol{A}^{SF\leftarrow LF} \cdot \tilde{\boldsymbol{R}}^T_{LF} \,. \tag{3.180}$$

Applying the chain rule for this change of coordinates yields[19]

$$\begin{bmatrix} \nabla_{\boldsymbol{R}^A_{LF}} \\ \nabla_{\boldsymbol{R}^B_{LF}} \\ \nabla_{\boldsymbol{R}^C_{LF}} \\ \nabla_{\boldsymbol{r}^1_{LF}} \\ \nabla_{\boldsymbol{r}^2_{LF}} \end{bmatrix} = \boldsymbol{A}^{SF\leftarrow LF^T} \cdot \begin{bmatrix} \nabla_{\boldsymbol{R}^{CM}_{LF}} \\ \nabla_{\boldsymbol{R}^B_{SF}} \\ \nabla_{\boldsymbol{R}^C_{SF}} \\ \nabla_{\boldsymbol{r}^1_{SF}} \\ \nabla_{\boldsymbol{r}^2_{SF}} \end{bmatrix}, \tag{3.181}$$

or

$$\nabla^T_{\tilde{\boldsymbol{R}}_{LF}} = \boldsymbol{A}^{SF\leftarrow LF^T} \cdot \nabla^T_{\tilde{\boldsymbol{R}}_{SF}} \,. \tag{3.182}$$

Rewriting the LF-frame kinetic energy operator of Eq. (3.86) as

$$T(\tilde{\boldsymbol{R}}_{LF}) = -\frac{\hbar^2}{2} \nabla_{\tilde{\boldsymbol{R}}_{LF}} \cdot \boldsymbol{m}^{-1} \cdot \nabla^T_{\tilde{\boldsymbol{R}}_{LF}} \,, \tag{3.183}$$

---

[19]See Eqs. (3.203) and (3.204) in Sect. 3.6.3.

where

$$\boldsymbol{m}^{-1} = \begin{bmatrix} \frac{1}{m_A} & 0 & 0 & 0 & 0 \\ 0 & \frac{1}{m_B} & 0 & 0 & 0 \\ 0 & 0 & \frac{1}{m_C} & 0 & 0 \\ 0 & 0 & 0 & \frac{1}{m_e} & 0 \\ 0 & 0 & 0 & 0 & \frac{1}{m_e} \end{bmatrix}, \tag{3.184}$$

and inserting Eq. (3.182) into (3.183) yields the following expression for the SF-frame kinetic energy operator

$$T(\tilde{\boldsymbol{R}}_{SF}) = -\frac{\hbar^2}{2} \nabla_{\tilde{\boldsymbol{R}}_{SF}} \cdot \boldsymbol{A}^{SF\leftarrow LF} \cdot \boldsymbol{m}^{-1} \cdot \boldsymbol{A}^{SF\leftarrow LF^T} \cdot \nabla^T_{\tilde{\boldsymbol{R}}_{SF}}, \tag{3.185}$$

with

$$\boldsymbol{A}^{SF\leftarrow LF} \cdot \boldsymbol{m}^{-1} \cdot \boldsymbol{A}^{SF\leftarrow LF^T} = \begin{bmatrix} \frac{1}{M_T} & 0 & 0 & 0 & 0 \\ 0 & \frac{1}{M_B} - \frac{1}{M_T} & -\frac{1}{M_T} & -\frac{1}{M_T} & -\frac{1}{M_T} \\ 0 & -\frac{1}{M_T} & \frac{1}{M_C} - \frac{1}{M_T} & -\frac{1}{M_T} & -\frac{1}{M_T} \\ 0 & -\frac{1}{M_T} & -\frac{1}{M_T} & \frac{1}{m_e} - \frac{1}{M_T} & -\frac{1}{M_T} \\ 0 & -\frac{1}{M_T} & -\frac{1}{M_T} & -\frac{1}{M_T} & \frac{1}{m_e} - \frac{1}{M_T} \end{bmatrix}. \tag{3.186}$$

Equation (3.185) is the matrix expression of Eqs. (3.93)–(3.96) in the particular case of a molecular system with three nuclei and two electrons. The structure of matrix (3.186) clearly indicates that *the motion of the center of mass is separated,* but also that *the nuclear and electronic motions are coupled.*

3.6.2 Example of Nuclear Space-Fixed (NSF) Coordinates *

Let us illustrate the expressions of the kinetic energy operators in a nuclear-center-of-mass frame, once again in the case of a molecular system consisting of three atoms A, B and C and two electrons. As in the previous example (see Sect. 3.6.1) M_T is the total mass of the system. Here, in addition, we introduce $M_N = M_A + M_B + M_C$, the total mass of the nuclei. Let us start by writing in matrix form the coordinate change SF → NSF given by Eq. (3.100) together with $\boldsymbol{R}_{LF}^{CM}$, the LF coordinate vector of the total center of mass regarded as the first NSF vector, $\tilde{\boldsymbol{R}}_1^{NSF}$:

$$
\begin{bmatrix} \tilde{\boldsymbol{R}}_{NSF}^{1} = \boldsymbol{R}_{LF}^{CM} \\ \tilde{\boldsymbol{R}}_{NSF}^{2} = \boldsymbol{R}_{NSF}^{B} \\ \tilde{\boldsymbol{R}}_{NSF}^{3} = \boldsymbol{R}_{NSF}^{C} \\ \tilde{\boldsymbol{R}}_{NSF}^{4} = \boldsymbol{r}_{NSF}^{1} \\ \tilde{\boldsymbol{R}}_{NSF}^{5} = \boldsymbol{r}_{NSF}^{2} \end{bmatrix} = \begin{bmatrix} 1\,0\,0 & 0 & 0 \\ 0\,1\,0 & \frac{m_e}{M_N} & \frac{m_e}{M_N} \\ 0\,0\,1 & \frac{m_e}{M_N} & \frac{m_e}{M_N} \\ 0\,0\,0 & 1+\frac{m_e}{M_N} & \frac{m_e}{M_N} \\ 0\,0\,0 & \frac{m_e}{M_N} & 1+\frac{m_e}{M_N} \end{bmatrix} \cdot \begin{bmatrix} \boldsymbol{R}_{LF}^{CM} = \tilde{\boldsymbol{R}}_{SF}^{1} \\ \boldsymbol{R}_{SF}^{B} = \tilde{\boldsymbol{R}}_{SF}^{2} \\ \boldsymbol{R}_{SF}^{C} = \tilde{\boldsymbol{R}}_{SF}^{3} \\ \boldsymbol{r}_{SF}^{1} = \tilde{\boldsymbol{R}}_{SF}^{4} \\ \boldsymbol{r}_{SF}^{2} = \tilde{\boldsymbol{R}}_{SF}^{5} \end{bmatrix}, \tag{3.187}
$$

or

$$
\tilde{\boldsymbol{R}}_{NSF}^{T} = \boldsymbol{A}^{NSF\leftarrow SF} \cdot \tilde{\boldsymbol{R}}_{SF}^{T}. \tag{3.188}
$$

Substituting Eq. (3.180) into Eq. (3.188) yields

$$
\tilde{\boldsymbol{R}}_{NSF}^{T} = \boldsymbol{A}^{NSF\leftarrow SF} \cdot \boldsymbol{A}^{SF\leftarrow LF} \cdot \tilde{\boldsymbol{R}}_{SF}^{T}, \tag{3.189}
$$

or

$$
\tilde{\boldsymbol{R}}_{NSF}^{T} = \boldsymbol{A}^{NSF\leftarrow LF} \cdot \tilde{\boldsymbol{R}}_{SF}^{T}, \tag{3.190}
$$

with

$$
\boldsymbol{A}^{NSF\leftarrow LF} = \boldsymbol{A}^{NSF\leftarrow SF} \cdot \boldsymbol{A}^{SF\leftarrow LF} = \begin{bmatrix} \frac{m_A}{M_T} & \frac{m_B}{M_T} & \frac{m_C}{M_T} & \frac{m_e}{M_T} & \frac{m_e}{M_T} \\ -\frac{m_A}{M_N} & 1-\frac{m_B}{M_N} & -\frac{m_C}{M_N} & 0 & 0 \\ -\frac{m_A}{M_N} & -\frac{m_B}{M_N} & 1-\frac{m_C}{M_N} & 0 & 0 \\ -\frac{m_A}{M_N} & -\frac{m_B}{M_N} & -\frac{m_C}{M_N} & 1 & 0 \\ -\frac{m_A}{M_N} & -\frac{m_B}{M_N} & -\frac{m_C}{M_N} & 0 & 1 \end{bmatrix}. \tag{3.191}
$$

Applying the chain rule gives

$$\nabla^T_{\tilde{\boldsymbol{R}}_{LF}} = \boldsymbol{A}^{NSF \leftarrow LF^T} \cdot \nabla^T_{\tilde{\boldsymbol{R}}_{NSF}} . \tag{3.192}$$

Abbreviating $\boldsymbol{A}^{NSF \leftarrow LF}$ in $\boldsymbol{A}$ and substituting this equation into Eq. (3.183) yields the following expression for the kinetic energy operator in Cartesian NSF coordinates

$$T(\tilde{\boldsymbol{R}}_{NSF}) = -\frac{\hbar^2}{2} \nabla_{\tilde{\boldsymbol{R}}_{NSF}} \cdot \boldsymbol{A} \cdot \boldsymbol{m}^{-1} \cdot \boldsymbol{A}^T \cdot \nabla^T_{\tilde{\boldsymbol{R}}_{NSF}} , \tag{3.193}$$

with

$$\boldsymbol{A} \cdot \boldsymbol{m}^{-1} \cdot \boldsymbol{A}^T = \begin{bmatrix} \frac{1}{M_T} & 0 & 0 & 0 & 0 \\ 0 & \frac{1}{m_B} - \frac{1}{M_N} & -\frac{1}{M_N} & 0 & 0 \\ 0 & -\frac{1}{M_N} & \frac{1}{m_C} - \frac{1}{M_N} & 0 & 0 \\ 0 & 0 & 0 & \frac{1}{m_e} + \frac{1}{M_N} & \frac{1}{M_N} \\ 0 & 0 & 0 & \frac{1}{M_N} & \frac{1}{m_e} + \frac{1}{M_N} \end{bmatrix} . \tag{3.194}$$

Equation (3.193) is the matrix form of Eqs. (3.110) and (3.111) in the particular case of a molecular system with three nuclei and two electrons. The structure of the matrix (3.194) clearly indicates that *the motion of the total center of mass* separates from the internal motions, that *the motions of the nuclei are separated from those of the electrons* but unfortunately that *mass polarization terms containing the inverse of* M_N, the total mass of the nuclei, crop up in the kinetic energy of the electrons. If the mass polarization terms are neglected, Eq. (3.194) reduces to

$$\boldsymbol{A} \cdot \boldsymbol{m}^{-1} \cdot \boldsymbol{A}^T = \begin{bmatrix} \frac{1}{M_T} & 0 & 0 & 0 & 0 \\ 0 & \frac{1}{m_B} - \frac{1}{M_N} & -\frac{1}{M_N} & 0 & 0 \\ 0 & -\frac{1}{M_N} & \frac{1}{m_C} - \frac{1}{M_N} & 0 & 0 \\ 0 & 0 & 0 & \frac{1}{m_e} & 0 \\ 0 & 0 & 0 & 0 & \frac{1}{m_e} \end{bmatrix} . \tag{3.195}$$

3.6.3 Chain Rule

To explain the chain rule let us consider the coordinate change from the coordinates $(Q_1, \dots, Q_n)$ to the coordinates $(q_1, \dots, q_n)$ where $q_i = f_i(Q_1, \dots, Q_n)$. By this coordinate change, a function $\Phi(Q_1, \dots, Q_n)$ becomes

$$\tilde{\Phi}(q_1, \dots, q_n) = \Phi(f_1(Q_1, \dots, Q_n), \dots, f_n(Q_1, \dots, Q_n)), \tag{3.196}$$

so that

$$\frac{\partial \Phi}{\partial Q_j} = \sum_i \frac{\partial \tilde{\Phi}}{\partial q_i} \frac{\partial f_i}{\partial Q_j}, \tag{3.197}$$

hence the chain rule

$$\frac{\partial}{\partial Q_j} = \sum_i \frac{\partial q_i}{\partial Q_j} \frac{\partial}{\partial q_i} \tag{3.198}$$

In the interesting particular case of a linear coordinate change such as

$$q_i = \sum_j A_{ij} Q_j ,$$
$$\frac{\partial q_i}{\partial Q_j} = A_{ij} , \tag{3.199}$$

the chain rule particularize into

$$\frac{\partial}{\partial Q_j} = \sum_i A_{ij} \frac{\partial}{\partial q_i} . \tag{3.200}$$

It is sometimes convenient to write these relations in a matrix form. Thus, for the coordinate change

$$\begin{bmatrix} q_1 \\ q_2 \\ \vdots \\ q_n \end{bmatrix} = \begin{bmatrix} A_{11} & A_{12} & \cdots & A_{1n} \\ A_{21} & A_{22} & \cdots & A_{2n} \\ \vdots & \vdots & \ddots & \vdots \\ A_{n1} & A_{n2} & \cdots & A_{nn} \end{bmatrix} \cdot \begin{bmatrix} Q_1 \\ Q_2 \\ \vdots \\ Q_n \end{bmatrix}, \tag{3.201}$$

or

$$\boldsymbol{q}^T = \boldsymbol{A} \cdot \boldsymbol{Q}^T . \tag{3.202}$$

The matrix form of the chain rule reads

$$\begin{bmatrix} \frac{\partial}{\partial Q_1} \\ \frac{\partial}{\partial Q_2} \\ \vdots \\ \frac{\partial}{\partial Q_n} \end{bmatrix} = \begin{bmatrix} A_{11} & A_{21} & \cdots & A_{n1} \\ A_{12} & A_{22} & \cdots & A_{n2} \\ \vdots & \vdots & \ddots & \vdots \\ A_{1n} & A_{2n} & \cdots & A_{nn} \end{bmatrix} \cdot \begin{bmatrix} \frac{\partial}{\partial q_1} \\ \frac{\partial}{\partial q_2} \\ \vdots \\ \frac{\partial}{\partial q_n} \end{bmatrix}, \tag{3.203}$$

or

$$\frac{\partial}{\partial \boldsymbol{Q}}^T = \boldsymbol{A}^T \cdot \frac{\partial}{\partial \boldsymbol{q}}^T . \tag{3.204}$$

References

1. Cohen-Tannoudji C, Dupont-Roc J, Grynberg G (1997) Photons and atoms: introduction to quantum electrodynamics. Wiley Professional
2. Craig DP, Thirunamachandran T (1998) Molecular quantum electrodynamics. Dover Publications
3. Saue T (2011) Relativistic Hamiltonians for chemistry: a primer. Chem Phys Chem 12:3077
4. Bunker PR, Jensen P (1998) Molecular symmetry and spectroscopy, 2nd edn. NRC Research Press, Ottawa
5. Bunker PR and Jensen P (2005) Fundamentals of molecular symmetry, iop, Bristol, Philadelphia
6. Sutcliffe BT (1999) The idea of the potential energy surface. In: Sax AF (ed.) Potential energy surfaces. Springer, Berlin, Heidelberg, p 61
7. Sutcliffe BT, Wooley RG (2005) Molecular strucutre calculations without clamping the nuclei. Phys Chem Chem Phys 7:3664
8. Mátyus E, Hutter J, Müller-Herold U, Reicher M (2011) Extracting elements of molecular structure from all-particle wave function. J Chem Phys 135:04302
9. Cederbaum LS (2004) Born-oppenheimer approximation and beyond. In: Domcke W, Yarkony DR, Köppel H (eds) Conical intersections. World Scientific Co., Singapore, pp 3
10. Sutcliffe BT (2000) The decoupling of electronic and nuclear motions in the isolated molecule Schrödinger hamiltonian. Adv Chem Phys 114:1
11. Zare RN (1988) Angular momentum. Wiley, New York
12. Frisch MJ, Trucks GW, Schlegel HB, Scuseria GE, Robb MA, Cheeseman JR, Montgomery Jr JA, Vreven T, Kudin KN, Burant JC, Millam JM, Iyengar SS, Tomasi J, Barone V, Mennucci B, Cossi M, Scalmani G, Rega N, Petersson GA, Nakatsuji H, Hada M, Ehara M, Toyota K, Fukuda R, Hasegawa J, Ishida M, Nakajima T, Honda Y, Kitao O, Nakai H, Klene M, Li X, Knox JE, Hratchian HP, Cross JB, Bakken V, Adamo C, Jaramillo J, Gomperts R, Stratmann RE, Yazyev O, Austin AJ, Cammi R, Pomelli C, Ochterski JW, Ayala PY, Morokuma K, Voth GA, Salvador P, Dannenberg JJ, Zakrzewski VG, Dapprich S, Daniels AD, Strain MC, Farkas O, Malick DK, Rabuck AD, Raghavachari K, Foresman JB, Ortiz JV, Cui Q, Baboul AG, Clifford S, Cioslowski J, Stefanov BB, Liu G, Liashenko A, Piskorz P, Komaromi I, Martin RL, Fox DJ, Keith T, Al-Laham MA, Peng CY, Nanayakkara A, Challacombe M, Gill PMW, Johnson B, Chen W, Wong MW, Gonzalez C, Pople JA (2003) Gaussian 03, Revision C.02. Gaussian, Inc., Pittsburgh PA
13. Werner H-J, Knowles PJ, Knizia G, Manby FR, Schütz M, Celani P, Korona T, Lindh R, Mitrushenkov A, Rauhut G, Shamasundar KR, Adler TB, Amos RD, Bernhardsson A, Berning A, Cooper DL, Deegan MJO, Dobbyn AJ, Eckert F, Goll E, Hampel C, Hesselmann A, Hetzer G, Hrenar T, Jansen G, Köppl C, Liu Y, Lloyd AW, Mata RA, May AJ, McNicholas SJ, Meyer W, Mura ME, Nicklass A, O'Neill DP, Palmieri P, Pflüger K, Pitzer R, Reiher M, Shiozaki T, Stoll H, Stone AJ, Tarroni R, Thorsteinsson T, Wang M, Wolf A (2010) Molpro, version 2010.1, a package of ab initio programs. http://www.molpro.net

14. Bransden BH, Joachain CJ (2000) Quantum mechanics. Prentice Hall, Upper Saddle River, New Jersey
15. Cohen-Tannoudji C, Diu B, Laloe F (1992) Quantum mechanics. Wiley-VCH
16. Schatz GC, Ratner MA (2002) Quantum mechanics in chemistry. Dover Publications, New York
17. Levine IN (1975) Molecular spectroscopy. Wiley, New-York, London
18. Szabo A, Ostlund NS (1996) Modern quantum chemistry. Dover, Mineola, New York
19. Herzberg GH (1966) Electronic spectra of polyatomic molecules. Van Nostrand Reihnold, Toronto
20. Irikura KK (2007) Experimental vibrational zero-point energies: diatomic molecules. J Phys Chem 36:389
21. Herzberg G (1992) Molecular spectra and molecular structure. Krieger Pub Co

Chapter 4
Vibronic Couplings

Photoinduced processes (photochemical and photophysical) often involve vibronic couplings that are responsible of ultrafast radiationless decay processes from an excited electronic state to a lower-energy one (typically, internal conversion, between same-spin electronic states, or intersystem crossing for different spins; note that corresponding light-emitting processes are called fluorescence and phosphorescence, respectively). In such a situation, the excess energy first given to the molecule through light absorption is converted into electronic excitation and then transformed into vibrational excitation. Chemiluminescence occurs in the reverse situation, when vibrational excitation (heat) is transformed into electronic excitation through internal conversion to a higher-energy electronic state that further relaxes upon light emission. *Such processes are governed by so-called non-adiabatic couplings* (interactions between the electronic structure and the nuclear motion) that are, by definition, *beyond the Born-Oppenheimer (adiabatic) approximation* (see Sect. 3.2.4). As proved from the off-diagonal Hellmann-Feynman theorem (see Eq. (3.75)), their effect becomes significant when the energy difference between two electronic states is small, that is to say of the same order of magnitude as vibrational energies. The non-adiabatic coupling elements even diverge when the energy difference vanishes, i.e., when two electronic states are degenerate at what is called *a conical intersection* (the generalization of a Jahn-Teller crossing to a lesser-symmetrical situation). For a comprehensive treatment and for references to the extensive literature on such concepts, one may refer the reader to the following selected list of books and reviews [1–21].

4.1 Conical Intersections

Conical intersections (see Fig. 4.1) are special topographic features that occur in potential energy surfaces at geometries where two electronic states are degenerate. As their name indicates, the local shape of the two surfaces in the vicinity of the intersection point is *a double cone*. This will be justified in what follows.

F. Gatti et al., *Applications of Quantum Dynamics in Chemistry*,
Lecture Notes in Chemistry 98, DOI 10.1007/978-3-319-53923-2_4

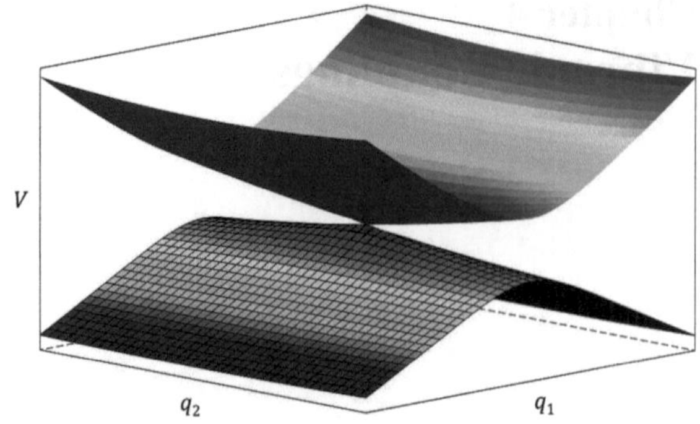

Fig. 4.1 Generic conical intersection between two potential energy surfaces V_1 (*lower one*) and V_2 (*upper one*). The (q_1, q_2)-plane is spanned by two specific nuclear displacements called branching-space vectors, further discussed below

Conical intersections can occur for symmetry reasons (for instance in the case of a Jahn-Teller crossing between states belonging to a degenerate irreducible representation of E-type; see Chap. 7 and some details in the next section) or simply be accidental [18]. For the sake of simplicity, we will restrict our discussion to the prototypical case of a two-state problem (this is the most usual situation in non-adiabatic photochemistry).

Before going further, let us consider the Hellmann-Feynman expression of the first-order non-adiabatic coupling (see Sect. 3.2.5 and Refs. [15, 22]), here in terms of internal coordinates, $\boldsymbol{q} = (..., q_j, ...)$, and corresponding vector components,

$$F_j(\boldsymbol{q}) = \langle \Phi_1^{el/ad}; \boldsymbol{q} | \partial_j \Phi_2^{el/ad}; \boldsymbol{q} \rangle = \frac{\langle \Phi_1^{el/ad}; \boldsymbol{q} | \partial_j H^{el}(\boldsymbol{q}) | \Phi_2^{el/ad}; \boldsymbol{q} \rangle}{V_2(\boldsymbol{q}) - V_1(\boldsymbol{q})}, \tag{4.1}$$

where $\partial_j(...)(\boldsymbol{q})$ stands for the value of the partial derivative at $\boldsymbol{q}$, $\left[\frac{\partial}{\partial q_j}(...)\right]_{\boldsymbol{q}}$. $\boldsymbol{F}(\boldsymbol{q})$ diverges when $V_2(\boldsymbol{q}) = V_1(\boldsymbol{q})$, that is if $\boldsymbol{q}$ is the locus of a conical intersection (note, however, that the numerator is finite). To simplify notations, integration over the electron coordinates, $\boldsymbol{r}$, is implicitly assumed in the present chapter when using Dirac's bracket notations. Also, $\boldsymbol{F}$ simply denotes $\boldsymbol{F}_{12}$, as there is no ambiguity in the two-state case, and electronic states and energies, $\Phi^{el/ad}$ and V, are considered in the adiabatic (Born-Oppenheimer) representation. The above equation shows that the non-adiabatic coupling term becomes large (infinite) when the energy difference becomes small (zero). This is the reason why *the Born-Oppenheimer approximation breaks down when approaching geometries where electronic states get close in energy*. Conical intersections are thus points that are representative of regions where significant probability of transfer of electronic population can occur through ultrafast radiationless decay. As such, these points are keys for describing non-adiabatic photochemical mechanisms.

Let us now examine in more detail the local shape of the potential energy surfaces in the vicinity of a conical intersection. For a two-state problem there are two adiabatic eigenstates of interest to be considered, $|\Phi_1^{el/ad}; \boldsymbol{q}\rangle$ and $|\Phi_2^{el/ad}; \boldsymbol{q}\rangle$. The matrix representation of $H^{el}(\boldsymbol{q})$ in the adiabatic basis set is, by definition, diagonal,

$$V(q) = \begin{bmatrix} V_1(q) & 0 \\ 0 & V_2(q) \end{bmatrix} = \Sigma(q)\mathbf{1} + \begin{bmatrix} -\Delta(q) & 0 \\ 0 & \Delta(q) \end{bmatrix}, \tag{4.2}$$

where

$$\Sigma(q) = \frac{V_1(q) + V_2(q)}{2}, \tag{4.3}$$

$$\Delta(q) = \frac{V_2(q) - V_1(q)}{2} \geq 0. \tag{4.4}$$

Note that we are considering a labelling convention such that $V_2(q) \geq V_1(q)$, whereas the reverse one is sometimes found in the literature. Separating the half-trace part, $\Sigma(q)\begin{bmatrix} 1 & 0 \\ 0 & 1 \end{bmatrix}$, and the traceless part, $\Delta(q)\begin{bmatrix} -1 & 0 \\ 0 & 1 \end{bmatrix}$, is a common trick when diagonalising a (2×2)-matrix, the use of which will become more apparent below.

Let us now consider as a working basis set a pair of electronic states, $|\Phi_1; q\rangle$ and $|\Phi_2; q\rangle$, which are assumed to span the same Hilbert space at any q as the two adiabatic eigenstates of interest, $|\Phi_1^{el/ad}; q\rangle$ and $|\Phi_2^{el/ad}; q\rangle$. The matrix representation of $H^{el}(q)$ in this basis set, $H(q)$, is not necessarily diagonal.[1] If the states are chosen real-valued, the Hamiltonian matrix is real symmetric. Again, separating the half-trace and the traceless parts yields

$$H(q) = \begin{bmatrix} H_{11}(q) & H_{12}(q) \\ H_{21}(q) & H_{22}(q) \end{bmatrix} = S(q)\mathbf{1} + \begin{bmatrix} -D(q) & W(q) \\ W(q) & D(q) \end{bmatrix}, \tag{4.5}$$

where

$$S(q) = \frac{H_{11}(q) + H_{22}(q)}{2}, \tag{4.6}$$

$$D(q) = \frac{H_{22}(q) - H_{11}(q)}{2}, \tag{4.7}$$

$$W(q) = H_{12}(q) = H_{21}(q). \tag{4.8}$$

Diagonalising $H(q)$ can be achieved as a similarity transformation involving a rotation of the states through an angle $\varphi(q)$, such that

$$|\Phi_1^{el/ad}; q\rangle = \cos\varphi(q)|\Phi_1; q\rangle + \sin\varphi(q)|\Phi_2; q\rangle, \tag{4.9}$$

$$|\Phi_2^{el/ad}; q\rangle = -\sin\varphi(q)|\Phi_1; q\rangle + \cos\varphi(q)|\Phi_2; q\rangle. \tag{4.10}$$

In other words, both matrices are related through

[1]Throughout this chapter, the matrix $H(q)$ will refer exclusively to the representation of the electronic Hamiltonian $H^{el}(q)$. It should not be confused with the matrix representation of the molecular Hamiltonian. The two quantities differ by the nuclear KEO matrix (including the non-adiabatic couplings and corrections, i.e., $T^{nu}\mathbf{1} + \mathbf{\Lambda}$ (see Sect. 3.2.4 for the definition of $\mathbf{\Lambda}$).

$$V(q) = U^\dagger(q)H(q)U(q)\,, \tag{4.11}$$

where the unitary (here, real orthogonal) matrix

$$U(q) = \begin{bmatrix} \cos\varphi(q) & -\sin\varphi(q) \\ \sin\varphi(q) & \cos\varphi(q) \end{bmatrix}, \tag{4.12}$$

is such that its columns give the adiabatic states into the working basis set and $U^\dagger(q)$ is the adjoint matrix (here, its transpose) characterizing the inverse transformation. As the identity matrix is unaffected ($U^\dagger(q)\mathbf{1}U(q) = \mathbf{1}$), the half-trace is preserved trough this transformation, such that

$$S(q) = \Sigma(q)\,. \tag{4.13}$$

Only the traceless part is to be transformed in practice. From

$$\Delta(q)\begin{bmatrix} -1 & 0 \\ 0 & 1 \end{bmatrix} = U^\dagger(q)\begin{bmatrix} -D(q) & W(q) \\ W(q) & D(q) \end{bmatrix}U(q)\,, \tag{4.14}$$

one gets two independent equations (adiabaticity conditions) in the following form,

$$\Delta(q) = \cos 2\varphi(q)D(q) - \sin 2\varphi(q)W(q)\,, \tag{4.15}$$

$$0 = \sin 2\varphi(q)D(q) + \cos 2\varphi(q)W(q)\,, \tag{4.16}$$

where we used $\cos^2\varphi(q) - \sin^2\varphi(q) = 1$ and $2\sin\varphi(q)\cos\varphi(q) = \sin 2\varphi(q)$. Explicit solutions are then readily obtained,

$$\cos 2\varphi(q) = \frac{D(q)}{\Delta(q)}\,, \quad \sin 2\varphi(q) = -\frac{W(q)}{\Delta(q)}\,, \tag{4.17}$$

with

$$\Delta(q) = \sqrt{D^2(q) + W^2(q)}\,. \tag{4.18}$$

The eigenvalues thus read

$$V_{1,2}(q) = \Sigma(q) \pm \Delta(q) = S(q) \pm \sqrt{D^2(q) + W^2(q)}\,. \tag{4.19}$$

Note that the minus sign in the sine expression comes from assuming $V_2(q) \geq V_1(q)$ (the reverse convention thus yields a formula with a plus sign, as sometimes found in the literature).

We will now show more precisely why geometries where two adiabatic eigenstates are degenerate are called conical intersections. Let us consider $q = q_X$ as a point where $|\Phi_1^{el/ad}; q_X\rangle$ and $|\Phi_2^{el/ad}; q_X\rangle$ are degenerate. In other words, $V_2(q_X) = V_1(q_X)$, i.e., $\Delta(q_X) = 0$. *Achieving this condition implies the fulfillment of two a priori*

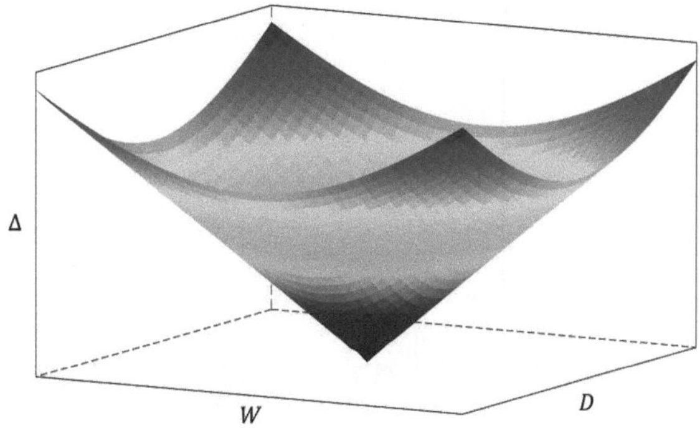

Fig. 4.2 Conical shape of $\Delta = \sqrt{D^2 + W^2}$ represented as a function of D and W

independent mathematical conditions: $D(\boldsymbol{q}_X) = 0$ and $W(\boldsymbol{q}_X) = 0$. Lifting degeneracy means increasing the value of $\Delta(\boldsymbol{q})$, which is thus achieved upon increasing either the values of $D(\boldsymbol{q})$ or $W(\boldsymbol{q})$, or both together, as illustrated on Fig. 4.2 when plotting Δ in the (D, W)-frame. We can already notice that Δ and -2φ behave formally as the polar coordinates (radius and angle[2]) to be associated to D and W viewed as Cartesian coordinates. This will be made more explicit in the next section on the prototypical case of a Jahn-Teller crossing.

From an intuitive perspective, we can already understand that *lifting degeneracy from a conical intersection requires constraints involving at least two independent degrees of freedom.* Doing so to first order in $\boldsymbol{q}$ from $\boldsymbol{q}_X$ will thus correspond to releasing the constraints $D(\boldsymbol{q}_X) = 0$ and $W(\boldsymbol{q}_X) = 0$ upon following the local gradients of $D(\boldsymbol{q})$ and $W(\boldsymbol{q})$ at $\boldsymbol{q} = \boldsymbol{q}_X$. When they describe *two linearly-independent directions* (which is the most common situation), the energy difference increases linearly within the two-dimensional plane spanned by these two gradients. This plane is called *the branching plane or branching space of the conical intersection* (more precisely, the branching plane is tangent to the branching space at the conical intersection). As a consequence, *degeneracy can be preserved locally along the remaining* $3N - 8$ *degrees of freedom* (where $N > 2$ is the number of atoms) that define the so-called intersection space or *seam* [11, 12, 14]. From a global

---

[2]Often, the equation defining the angle is simply given as

$$\varphi(\boldsymbol{q}) = -\frac{1}{2}\arctan\frac{W(\boldsymbol{q})}{D(\boldsymbol{q})}.$$

However, some information is missing regarding the relative signs of $D(\boldsymbol{q})$ and $W(\boldsymbol{q})$ and the corresponding domains of variation of the angle. A constant value should be added according to the corresponding quadrants if a continuity condition is required. A more accurate formulation is

$$\varphi(\boldsymbol{q}) = -\frac{1}{2}\mathrm{Arg}(D(\boldsymbol{q}) + iW(\boldsymbol{q})).$$

where $-\pi < \mathrm{Arg}(x + iy) \leq \pi$ is the principal value of the argument of the complex number $x + iy$. The corresponding function is sometimes denoted atan2(y, x) and this is how it is known in computer programming.

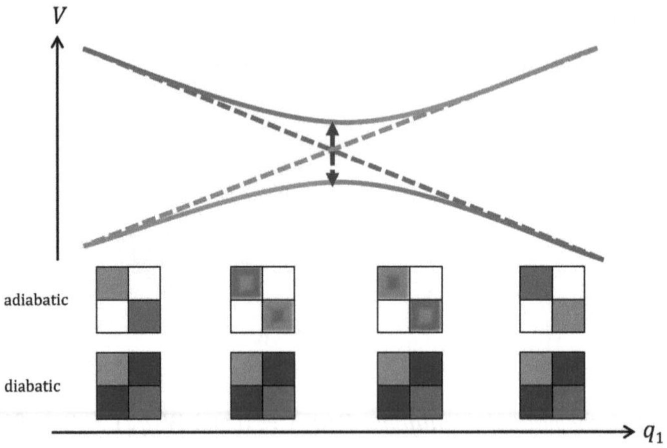

Fig. 4.3 Typical shape of an avoided crossing between two adiabatic potential energy curves (*plain lines*), V_2 (*upper one*) and V_1 (*lower one*), along a nuclear coordinate q_1. The *dashed lines* represent the diabatic energies curves while the *blue arrow* corresponds to twice the value of the off-diagonal diabatic coupling term at the crossing point (the *orange* and *green* color code refers to the diabatic states; see below for further details on the diabatic representation). The (2×2)-matrices are a pictorial representation of the electronic Hamiltonian in both basis sets that illustrates how the adiabatic representation is related to the diabatic representation along q_1

perspective, this subspace[3] is in general curved and can be made of one or several open or closed sets depending on the case. Each point within the seam corresponds to a specific conical intersection with its own branching plane defined locally as the plane orthogonal to the seam at this point. In theoretical photochemistry, one usually optimises the lowest-energy point within the seam [17], as this minimum-energy conical intersection is the most likely to be accessed, hence the most representative of the process under study. For diatoms ($N = 2$), there is a single degree of freedom (the bond length) and potential energy curves cannot cross unless the coupling is zero for symmetry reasons (this is referred to as the Wigner non-crossing rule). In that case, the shape of the curves is known as *an avoided crossing* (see Fig. 4.3). This pattern can be found also in polyatomic molecules when considering a pathway along which the states become near degenerate (in general, such an avoided crossing is the signature of a conical intersection nearby).

Note that the Wigner non-crossing rule (which, again, should concern diatoms exclusively) has led to many misinterpretations of curve crossings and corresponding photochemical mechanisms in polyatomic systems. This made conical intersections viewed as unusual singularities for quite a while. In fact, *conical intersections are now perceived as the rule rather than the exception*. This is even more often the

[3]Note that two hypersurfaces that depend on $3N - 6$ degrees of freedom can, in principle, intersect along a $(3N - 7)$-dimensional hyperline. In contrast, two adiabatic potential energy surfaces can only intersect along the $(3N - 8)$-dimensional seam, which is a direct consequence of the possible coupling between the degenerate states along certain directions.

case when excited electronic states are considered, where conical intersections are more likely to occur due to state congestion in terms of energy (the density of states increases with the energy). *Ground-state thinking tends to give prevalence to avoided crossings, as they correspond to the lower-energy pathways* (transition barriers that are typical of electron redistribution between the reactant and the product). However, *most photochemical mechanisms directly involve conical intersections that act as funnels through which the reaction path can go.* In addition to this, we should mention that Jahn-Teller crossings, mostly encountered in small and/or very symmetrical systems, have induced some prejudice about the understanding of conical intersections. Although it is true that symmetry may induce degeneracy (in E-type electronic states for example), this does not mean that accidental conical intersections are unlikely. In contrast, they are very common in photochemical reactivity, even when there is no symmetry at all. This said, interpreting them in the context of a more symmetrical prototype (often related to some typical Jahn-Teller crossing) is useful, as this provides some rationale for two states to be degenerate (see, e.g., Ref. [18]).

A first-order description of the local behavior of the functions $V_1(\boldsymbol{q})$ and $V_2(\boldsymbol{q})$ around $\boldsymbol{q}_X$ raises specific mathematical issues. The common half-trace function, $S(\boldsymbol{q}) = \Sigma(\boldsymbol{q})$, can be assumed to be regular in most situations and will not be further discussed. However, the function $\Delta(\boldsymbol{q})$ is singular: it cannot be differentiated at $\boldsymbol{q}_X$ because of the square-root.[4] In other words, the local derivative, $\partial_j \Delta(\boldsymbol{q}_X)$, is ill-defined. The shapes of the potential energy surfaces in the vicinity of $\boldsymbol{q}_X$ show a two-dimensional cusp at $\boldsymbol{q}_X$, which cannot be described mathematically in terms of ordinary local derivatives. An important remark must be made at this stage: the singular behaviors of both $\partial_j \Delta(\boldsymbol{q})$ and $F_j(\boldsymbol{q})$ (see Eq. 4.1) at a conical intersection are due to the fact that the lower-energy state (number 1) and the higher-energy state (number 2) abruptly swap their respective chemical natures when $\boldsymbol{q}$ varies smoothly across a conical intersection.

A rigorous formulation of this apparent issue can be derived from a generalisation of the Hellmann-Feynman theorem to degenerate situations [15] or, similarly, within the framework of degenerate perturbation theory [22]. Here, we will adopt a more intuitive picture that conveys the same basic ideas. As already pointed out, achieving $\Delta(\boldsymbol{q}_X) = 0$ implies two simultaneous conditions: $D(\boldsymbol{q}_X) = 0$ and $W(\boldsymbol{q}_X) = 0$. If so, the Hamiltonian matrix is diagonal when using the working basis set at this point,

$$\boldsymbol{H}(\boldsymbol{q}_X) = S(\boldsymbol{q}_X)\mathbf{1} = \Sigma(\boldsymbol{q}_X)\mathbf{1} = \boldsymbol{V}(\boldsymbol{q}_X) . \tag{4.20}$$

[4] Here, ill-defined local derivatives must be replaced by well-defined directional derivatives. Let us consider a first-order, two-dimensional case: $\Delta(x, y) = \sqrt{(ax)^2 + (bx)^2}$. This function satisfies $\Delta(0, 0) = 0$, $\Delta(x, 0) = |ax|$, and $\Delta(0, y) = |by|$. It behaves as a two-dimensional absolute-value function and is represented by an elliptic cone in the (x, y)-frame, the apex of which is located at $(0, 0)$. The left- and right-derivatives are well-defined and satisfy $\lim_{\varepsilon \to 0^+} \frac{\Delta(\varepsilon,0)-\Delta(0,0)}{\varepsilon} = -\lim_{\varepsilon \to 0^-} \frac{\Delta(\varepsilon,0)-\Delta(0,0)}{\varepsilon} = |a|$ and $\lim_{\varepsilon \to 0^+} \frac{\Delta(0,\varepsilon)-\Delta(0,0)}{\varepsilon} = -\lim_{\varepsilon \to 0^-} \frac{\Delta(0,\varepsilon)-\Delta(0,0)}{\varepsilon} = |b|$. They are opposite on both sides. However, $\lim_{\varepsilon \to 0} \frac{\Delta(\varepsilon,0)-\Delta(-\varepsilon,0)}{2\varepsilon} = \lim_{\varepsilon \to 0} \frac{\Delta(0,\varepsilon)-\Delta(0,-\varepsilon)}{2\varepsilon} = 0$, which shows that a two-point formula is inadequate here: the local derivatives are ill-defined because of the cusp at the origin.

In other words, both $|\Phi_1; \boldsymbol{q}_X\rangle$ and $|\Phi_2; \boldsymbol{q}_X\rangle$ form a possible pair of degenerate orthogonal eigenstates. The transformation angle $\varphi(\boldsymbol{q}_X)$ is then arbitrary and can take any value. This simply reflects the property that any linear combination of two degenerate eigenstates also is an eigenstate. Now, although the local derivatives of $\Delta(\boldsymbol{q})$ are ill-defined at $\boldsymbol{q}_X$, we will assume in what follows that both $D(\boldsymbol{q})$ and $W(\boldsymbol{q})$ are regular at this point. Since the operator $H^{el}(\boldsymbol{q})$ has no physical reason not to be regular with respect to variations of $\boldsymbol{q}$, this means that $|\Phi_1; \boldsymbol{q}\rangle$ and $|\Phi_2; \boldsymbol{q}\rangle$ are specifically chosen so as to vary smoothly with $\boldsymbol{q}$ around $\boldsymbol{q}_X$, as opposed to $|\Phi_1^{el/ad}; \boldsymbol{q}\rangle$ and $|\Phi_2^{el/ad}; \boldsymbol{q}\rangle$ that change brutally by construction.

Let us clarify this point upon considering that a specific pair of degenerate orthogonal eigenstates is known at $\boldsymbol{q}_X$ with no ambiguity (for example as the result of an actual quantum-chemistry calculation) and denote them $|\overline{\Phi}_1^{el/ad}; \boldsymbol{q}_X\rangle$ and $|\overline{\Phi}_2^{el/ad}; \boldsymbol{q}_X\rangle$. The bar is here to remind us that they correspond to a well-determined choice of states, while any other pair of rotated eigenstates (for an arbitrary angle φ_{ref}, now fixed and not to be confused with $\varphi(\boldsymbol{q}_X)$) would work just as well,

$$|\Phi_1^{el/ad}; \boldsymbol{q}_X\rangle = \cos\varphi_{\text{ref}}|\overline{\Phi}_1^{el/ad}; \boldsymbol{q}_X\rangle + \sin\varphi_{\text{ref}}|\overline{\Phi}_2^{el/ad}; \boldsymbol{q}_X\rangle\,, \tag{4.21}$$

$$|\Phi_2^{el/ad}; \boldsymbol{q}_X\rangle = -\sin\varphi_{\text{ref}}|\overline{\Phi}_1^{el/ad}; \boldsymbol{q}_X\rangle + \cos\varphi_{\text{ref}}|\overline{\Phi}_2^{el/ad}; \boldsymbol{q}_X\rangle\,. \tag{4.22}$$

The adiabatic states that are known (with the bar) thus correspond here to $\varphi_{\text{ref}} = 0$ by convention (note that $\varphi_{\text{ref}} = \pi$ yields equivalent states up to an unphysical sign[5]). We have already proved that the working states are eigenstates at $\boldsymbol{q}_X$. We can thus consider a well-determined pair of working states upon making the following identification: $|\overline{\Phi}_1; \boldsymbol{q}_X\rangle = |\overline{\Phi}_1^{el/ad}; \boldsymbol{q}_X\rangle$ and $|\overline{\Phi}_2; \boldsymbol{q}_X\rangle = |\overline{\Phi}_2^{el/ad}; \boldsymbol{q}_X\rangle$ (again, the bar reminds us that the arbitrary mixing angle between degenerate states, φ_{ref}, is conventionally fixed to zero for the reference eigenstates). It must be stressed that the working states, $|\overline{\Phi}_1; \boldsymbol{q}\rangle$ and $|\overline{\Phi}_2; \boldsymbol{q}\rangle$, are eigenstates at $\boldsymbol{q} = \boldsymbol{q}_X$ but not necessarily elsewhere. The only requirement is that they vary smoothly with $\boldsymbol{q}$ around $\boldsymbol{q}_X$ such that the functions entering the corresponding Hamiltonian matrix, $\overline{\boldsymbol{H}}(\boldsymbol{q})$, are regular.

We now make the strong hypothesis that $|\overline{\Phi}_1; \boldsymbol{q}\rangle$ and $|\overline{\Phi}_2; \boldsymbol{q}\rangle$ do not vary with $\boldsymbol{q}$ to first order around $\boldsymbol{q}_X$. In other words, for any $\boldsymbol{q} = \boldsymbol{q}_X + \delta\boldsymbol{q}$ (where $\delta\boldsymbol{q}$ is a first-order infinitesimal displacement), we get $|\overline{\Phi}_1; \boldsymbol{q}\rangle \equiv |\overline{\Phi}_1; \boldsymbol{q}_X\rangle$ and $|\overline{\Phi}_2; \boldsymbol{q}\rangle \equiv |\overline{\Phi}_2; \boldsymbol{q}_X\rangle$. Such states are often referred to as *crude adiabatic*, as they are adiabatic states (eigenstates) for $\boldsymbol{q} = \boldsymbol{q}_X$ only and fixed elsewhere (here, to first order only but this concept can be extended to any value of $\boldsymbol{q}$). In the spirit of first-order perturbation theory, they play the role of zero-order states such that the local gradients of $\overline{D}(\boldsymbol{q})$ and $\overline{W}(\boldsymbol{q})$ at $\boldsymbol{q}_X$ can be expressed with no difficulty as

$$\partial_j\overline{D}(\boldsymbol{q}_X) = \frac{\langle\overline{\Phi}_2; \boldsymbol{q}_X|\partial_j H^{el}(\boldsymbol{q}_X)|\overline{\Phi}_2; \boldsymbol{q}_X\rangle - \langle\overline{\Phi}_1; \boldsymbol{q}_X|\partial_j H^{el}(\boldsymbol{q}_X)|\overline{\Phi}_1; \boldsymbol{q}_X\rangle}{2}\,,$$

$$\partial_j\overline{W}(\boldsymbol{q}_X) = \langle\overline{\Phi}_1; \boldsymbol{q}_X|\partial_j H^{el}(\boldsymbol{q}_X)|\overline{\Phi}_2; \boldsymbol{q}_X\rangle\,. \tag{4.23}$$

[5]This apparent double-valuedness issue is related to what is known as the geometrical Berry phase (see Refs. [7, 16, 23, 24] for further details).

It must be noted that the above formulae reflect the fact that the crude adiabatic states do not vary with $\boldsymbol{q}$ (at least not to first order from $\boldsymbol{q}_X$).[6] Indeed,

$$\begin{aligned} \partial_j \overline{H}^{(\alpha\beta)}(\boldsymbol{q}_X) &= \partial_j \langle \overline{\Phi}_\alpha; \boldsymbol{q}_X | H^{el}(\boldsymbol{q}_X) | \overline{\Phi}_\beta; \boldsymbol{q}_X \rangle \\ &= \underbrace{\langle \partial_j \overline{\Phi}_\alpha; \boldsymbol{q}_X | H^{el}(\boldsymbol{q}_X) | \overline{\Phi}_\beta; \boldsymbol{q}_X \rangle}_{=0} \\ &+ \langle \overline{\Phi}_\alpha; \boldsymbol{q}_X | \partial_j H^{el}(\boldsymbol{q}_X) | \overline{\Phi}_\beta; \boldsymbol{q}_X \rangle \\ &+ \underbrace{\langle \overline{\Phi}_\alpha; \boldsymbol{q}_X | H^{el}(\boldsymbol{q}_X) | \partial_j \overline{\Phi}_\beta; \boldsymbol{q}_X \rangle}_{=0} \\ &= \langle \overline{\Phi}_\alpha; \boldsymbol{q}_X | \partial_j H^{el}(\boldsymbol{q}_{X)} | \overline{\Phi}_\beta; \boldsymbol{q}_X \rangle \, . \end{aligned} \tag{4.24}$$

Such expressions are regular by construction and similar in form to Hellmann-Feynman derivatives (and calculated as such in practice). At this stage, we have what we need to express a well-defined linear expansion of $\overline{\boldsymbol{H}}(\boldsymbol{q})$ around $\boldsymbol{q}_X$. Let us denote

$$\overline{x}_j^{(1)} = \partial_j \overline{D}(\boldsymbol{q}_X) \, , \tag{4.25}$$

and

$$\overline{x}_j^{(2)} = \partial_j \overline{W}(\boldsymbol{q}_X) \, . \tag{4.26}$$

We get, to first order,

$$\overline{\boldsymbol{H}}(\boldsymbol{q}_X + \delta \boldsymbol{q}) \approx (S(\boldsymbol{q}_X) + \sum_j \partial_j S(\boldsymbol{q}_X) \delta q_j) \mathbf{1} + \sum_j \begin{bmatrix} -\overline{x}_j^{(1)} & \overline{x}_j^{(2)} \\ \overline{x}_j^{(2)} & \overline{x}_j^{(1)} \end{bmatrix} \delta q_j \, . \tag{4.27}$$

The corresponding adiabatic matrix to first order reads thus

---

[6]In fact, the same result can be obtained with a less restrictive hypothesis. It is enough to consider that both $|\overline{\Phi}_1; \boldsymbol{q}\rangle$ and $|\overline{\Phi}_2; \boldsymbol{q}\rangle$ vary smoothly with $\boldsymbol{q}$ around $\boldsymbol{q}_X$ such that $\langle \overline{\Phi}_\alpha; \boldsymbol{q}_X | \partial_j \Phi_\beta; \boldsymbol{q}_X \rangle$ take finite values (as opposed to $\langle \Phi_1^{el/ad}; \boldsymbol{q} | \partial_j \Phi_2^{el/ad}; \boldsymbol{q} \rangle$ that diverges when $\boldsymbol{q}$ tends to $\boldsymbol{q}_X$). Using a Hellmann-Feynman-like formula, we get

$$\partial_j \overline{H}^{(\alpha\beta)}(\boldsymbol{q}_X) = \underbrace{(V_\alpha(\boldsymbol{q}_X) - V_\beta(\boldsymbol{q}_X))}_{=0} \langle \overline{\Phi}_\alpha; \boldsymbol{q}_X | \partial_j \overline{\Phi}_\beta; \boldsymbol{q}_X \rangle + \langle \overline{\Phi}_\alpha; \boldsymbol{q}_X | \partial_j H^{el}(\boldsymbol{q}_{X)} | \overline{\Phi}_\beta; \boldsymbol{q}_X \rangle \, ,$$

where we used, first, the fact that $\overline{\Phi}$-type states coincide with the eigenstates at $\boldsymbol{q}_X$ and, second, that they are assumed to be orthonormal,

$$\langle \partial_j \overline{\Phi}_\alpha; \boldsymbol{q}_X | \overline{\Phi}_\beta; \boldsymbol{q}_X \rangle + \langle \overline{\Phi}_\alpha; \boldsymbol{q}_X | \partial_j \overline{\Phi}_\beta; \boldsymbol{q}_X \rangle = \partial_j \langle \overline{\Phi}_\alpha; \boldsymbol{q}_X | \overline{\Phi}_\beta; \boldsymbol{q}_X \rangle = 0 \, .$$

In addition, it is worth noticing that, for $\alpha, \beta = 1, 2$ (restriction to two states), there is no explicit involvement of the other possible eigenstates (there are no couplings $\langle \overline{\Phi}_1; \boldsymbol{q}_X | \partial_j \overline{\Phi}_\gamma; \boldsymbol{q}_X \rangle$ or $\langle \overline{\Phi}_2; \boldsymbol{q}_X | \partial_j \overline{\Phi}_\gamma; \boldsymbol{q}_X \rangle$ for $\gamma \geq 3$ in the above expressions). This holds to first order but a second-order expansion would require considering such terms.

Fig. 4.4 Linearised double cone represented in the branching space

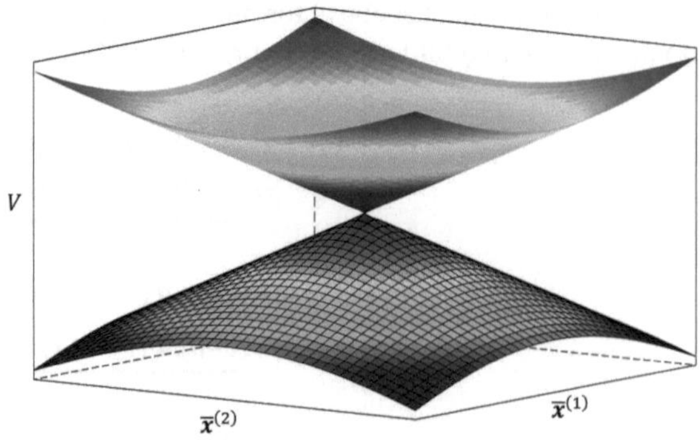

$$V(q_X + \delta q) \approx (\Sigma(q_X) + \sum_j \partial_j \Sigma(q_X)\delta q_j)\mathbf{1} + \begin{bmatrix} -\Delta^{(1)}(\delta q) & 0 \\ 0 & \Delta^{(1)}(\delta q) \end{bmatrix}, \quad (4.28)$$

where

$$\Delta^{(1)}(\delta q) = \sqrt{(\sum_j \overline{x}_j^{(1)}\delta q_j)^2 + (\sum_j \overline{x}_j^{(2)}\delta q_j)^2}, \quad (4.29)$$

and $\Sigma(q_X) = S(q_X)$ and $\partial_j \Sigma(q_X) = \partial_j S(q_X)$ (preservation of the trace to all orders). *The expression of the linearised energy half-difference,* $\Delta^{(1)}(\delta q)$, *involves both gradient-type vectors,* $\overline{x}^{(1)}$ *and* $\overline{x}^{(2)}$, *usually called gradient difference and derivative coupling.* They span the so-called branching space (or branching plane) over which degeneracy is lifted to first order in q from q_X. $\Delta^{(1)}(\delta q)$ increases only if the displacement δq contains some contribution over one or both of them, whereas orthogonal directions preserve degeneracy and are thus tangent to the seam. We have now proved more formally that the local shape of both potential energy surfaces within the branching space is a double cone the apex of which is at q_X, hence the name conical intersection (see Fig. 4.4). Note that some authors [16] denote them $g = \overline{x}^{(1)}$ and $h = \overline{x}^{(2)}$ and thus call the branching space the gh-plane.

A legitimate question could be raised at this stage: the above expression of $\Delta^{(1)}(\delta q)$ has been obtained from the vectors $\overline{x}^{(1)}$ and $\overline{x}^{(2)}$ corresponding to some specific choice of mixing angle (namely, $\varphi_{\text{ref}} = 0$), while the value of this angle should be unphysical. In fact, we will show that the angle is irrelevant indeed. The original crude adiabatic states can be mixed arbitrarily to form another, but equivalent, working basis set with no bar (rotated through φ_{ref} with respect to the reference ones) at $q = q_X$: $|\Phi_1; q_X\rangle = |\Phi_1^{el/ad}; q_X\rangle$ and $|\Phi_2; q_X\rangle = |\Phi_2^{el/ad}; q_X\rangle$. Keeping this angle constant at other points, we then have

$$|\Phi_1; q\rangle = \cos\varphi_{\text{ref}}|\overline{\Phi}_1; q\rangle + \sin\varphi_{\text{ref}}|\overline{\Phi}_2; q\rangle, \quad (4.30)$$

$$|\Phi_2; q\rangle = -\sin\varphi_{\text{ref}}|\overline{\Phi}_1; q\rangle + \cos\varphi_{\text{ref}}|\overline{\Phi}_2; q\rangle, \quad (4.31)$$

and both Hamiltonian matrices are related through

$$\boldsymbol{H}(\boldsymbol{q}) = \boldsymbol{U}_{\text{ref}}^{\dagger}\overline{\boldsymbol{H}}(\boldsymbol{q})\boldsymbol{U}_{\text{ref}}\,, \tag{4.32}$$

where

$$\boldsymbol{U}_{\text{ref}} = \begin{bmatrix} \cos\varphi_{\text{ref}} & -\sin\varphi_{\text{ref}} \\ \sin\varphi_{\text{ref}} & \cos\varphi_{\text{ref}} \end{bmatrix}. \tag{4.33}$$

As the angle is constant ($\boldsymbol{q}$-independent), the functions that define $\boldsymbol{H}(\boldsymbol{q})$ are as regular as those involved in $\overline{\boldsymbol{H}}(\boldsymbol{q})$. Further,

$$|\Phi_1^{el/ad};\boldsymbol{q}\rangle = \cos\overline{\varphi}(\boldsymbol{q})|\overline{\Phi}_1;\boldsymbol{q}\rangle + \sin\overline{\varphi}(\boldsymbol{q})|\overline{\Phi}_2;\boldsymbol{q}\rangle\,, \tag{4.34}$$

$$|\Phi_2^{el/ad};\boldsymbol{q}\rangle = -\sin\overline{\varphi}(\boldsymbol{q})|\overline{\Phi}_1;\boldsymbol{q}\rangle + \cos\overline{\varphi}(\boldsymbol{q})|\overline{\Phi}_2;\boldsymbol{q}\rangle\,, \tag{4.35}$$

where

$$\overline{\varphi}(\boldsymbol{q}) = \varphi(\boldsymbol{q}) + \varphi_{\text{ref}}\,. \tag{4.36}$$

Similarly, we can define

$$\overline{\boldsymbol{U}}(\boldsymbol{q}) = \boldsymbol{U}_{\text{ref}}\boldsymbol{U}(\boldsymbol{q}) = \begin{bmatrix} \cos\overline{\varphi}(\boldsymbol{q}) & -\sin\overline{\varphi}(\boldsymbol{q}) \\ \sin\overline{\varphi}(\boldsymbol{q}) & \cos\overline{\varphi}(\boldsymbol{q}) \end{bmatrix}, \tag{4.37}$$

such that

$$\boldsymbol{V}(\boldsymbol{q}) = \overbrace{\boldsymbol{U}^{\dagger}(\boldsymbol{q})\,\boldsymbol{U}_{\text{ref}}^{\dagger}}^{\overline{\boldsymbol{U}}^{\dagger}(\boldsymbol{q})}\!\!\underbrace{\phantom{\boldsymbol{U}}\overline{\boldsymbol{H}}(\boldsymbol{q})\boldsymbol{U}_{\text{ref}}}_{\boldsymbol{H}(\boldsymbol{q})}\!\!\overbrace{\phantom{\boldsymbol{U}}\boldsymbol{U}(\boldsymbol{q})}^{\overline{\boldsymbol{U}}(\boldsymbol{q})} \tag{4.38}$$

The resulting branching-space vectors (with no bar) now read

$$\begin{aligned} x_j^{(1)} &= \partial_j D(\boldsymbol{q}_\text{X}) \\ &= \frac{\langle\Phi_2;\boldsymbol{q}_\text{X}|\partial_j H^{el}(\boldsymbol{q}_\text{X})|\Phi_2;\boldsymbol{q}_\text{X}\rangle - \langle\Phi_1;\boldsymbol{q}_\text{X}|\partial_j H^{el}(\boldsymbol{q}_\text{X})|\Phi_1;\boldsymbol{q}_\text{X}\rangle}{2}\,, \end{aligned} \tag{4.39}$$

$$\begin{aligned} x_j^{(2)} &= \partial_j W(\boldsymbol{q}_\text{X}) \\ &= \langle\Phi_1;\boldsymbol{q}_\text{X}|\partial_j H^{el}(\boldsymbol{q}_\text{X})|\Phi_2;\boldsymbol{q}_\text{X}\rangle\,. \end{aligned} \tag{4.40}$$

Using

$$\begin{bmatrix} -x_j^{(1)} & x_j^{(2)} \\ x_j^{(2)} & x_j^{(1)} \end{bmatrix} = \boldsymbol{U}_{\text{ref}}^{\dagger}\begin{bmatrix} -\overline{x}_j^{(1)} & \overline{x}_j^{(2)} \\ \overline{x}_j^{(2)} & \overline{x}_j^{(1)} \end{bmatrix}\boldsymbol{U}_{\text{ref}}\,, \tag{4.41}$$

they are obtained from the original ones according to

$$x_j^{(1)} = \cos 2\varphi_{\text{ref}}\overline{x}_j^{(1)} - \sin 2\varphi_{\text{ref}}\overline{x}_j^{(2)}\,, \tag{4.42}$$

$$x_j^{(2)} = \sin 2\varphi_{\text{ref}} \overline{x}_j^{(1)} + \cos 2\varphi_{\text{ref}} \overline{x}_j^{(2)} . \tag{4.43}$$

When $\varphi_{\text{ref}} = 0$ (or π), we get the original branching-space vectors, $x_j^{(1)} = \overline{x}_j^{(1)}$ and $x_j^{(2)} = \overline{x}_j^{(2)}$. It is straightforward to show that the trigonometric functions of φ_{ref} actually disappear when considering the expression of the linearised energy half-difference, which can equally be expressed with the rotated vectors (with no bar) as

$$\Delta^{(1)}(\delta \boldsymbol{q}) = \sqrt{(\sum_j x_j^{(1)} \delta q_j)^2 + (\sum_j x_j^{(2)} \delta q_j)^2} . \tag{4.44}$$

This, again, shows that degeneracy is lifted to first order from $\boldsymbol{q}_{\text{X}}$ along any direction contained within the two-dimensional plane spanned by the branching-space vectors for any value of φ_{ref}.[7] We have defined the branching-space vectors from a derivation based on crude adiabatic $\overline{\Phi}$-states as formal intermediates playing the role of zero-order states in the context of degenerate perturbation theory. From an operational point of view, $\overline{x}_j^{(1)}$ and $\overline{x}_j^{(2)}$ are simply determined from Hellmann-Feynman-type formulae based on a specific pair of degenerate adiabatic $\overline{\Phi}^{el/ad}$-states (the states actually calculated and for which we have set $\varphi_{\text{ref}} = 0$ by convention). In other words, they are obtained in practice from

$$\overline{x}_j^{(1)} = \frac{\langle \overline{\Phi}_2^{el/ad}; \boldsymbol{q}_{\text{X}} | \partial_j H^{el}(\boldsymbol{q}_{\text{X}}) | \overline{\Phi}_2^{el/ad}; \boldsymbol{q}_{\text{X}} \rangle - \langle \overline{\Phi}_1^{el/ad}; \boldsymbol{q}_{\text{X}} | \partial_j H^{el}(\boldsymbol{q}_{\text{X}}) | \overline{\Phi}_1^{el/ad}; \boldsymbol{q}_{\text{X}} \rangle}{2} ,$$
$$\overline{x}_j^{(2)} = \langle \overline{\Phi}_1^{el/ad}; \boldsymbol{q}_{\text{X}} | \partial_j H^{el}(\boldsymbol{q}_{\text{X}}) | \overline{\Phi}_2^{el/ad}; \boldsymbol{q}_{\text{X}} \rangle . \tag{4.45}$$

[7] The working basis set (with the bar) was chosen as a pair of crude adiabatic states for which $\varphi_{\text{ref}} = 0$ corresponded by convention to a specific pair of degenerate adiabatic states obtained from an actual calculation. This arbitrary angle occurs to get fixed implicitly from symmetry considerations when both degenerate states potentially belong to different irreducible representations (see Chap. 7). However, in a general situation, the value of φ_{ref} can be fixed for convenience through an extra constraint according to context. For example, it can be used to make the two branching-space vectors orthogonal. Alternatively, it may be convenient to fix the value of φ_{ref} to change from the original $\overline{\boldsymbol{H}}(\boldsymbol{q})$ to an equivalent $\boldsymbol{H}(\boldsymbol{q})$ according to a condition such that the off-diagonal term is zero at some reference geometry, $\boldsymbol{q} = \boldsymbol{q}_{\text{ref}}$, where the states are not degenerate (often, the Franck-Condon point). This ensures coincidence with the adiabatic representation at this point in addition to coincidence at the conical intersection $\boldsymbol{q}_{\text{X}}$. The entries of both equivalent Hamiltonian matrices are related through

$$D(\boldsymbol{q}) = \cos 2\varphi_{\text{ref}} \overline{D}(\boldsymbol{q}) - \sin 2\varphi_{\text{ref}} \overline{W}(\boldsymbol{q}) ,$$
$$W(\boldsymbol{q}) = \sin 2\varphi_{\text{ref}} \overline{D}(\boldsymbol{q}) + \cos 2\varphi_{\text{ref}} \overline{W}(\boldsymbol{q}) .$$

Setting $W(\boldsymbol{q}_{\text{ref}}) = 0$ with $D(\boldsymbol{q}_{\text{ref}}) = \Delta(\boldsymbol{q}_{\text{ref}}) > 0$ thus yields

$$\cos 2\varphi_{\text{ref}} = \frac{\overline{D}(\boldsymbol{q}_{\text{ref}})}{\Delta(\boldsymbol{q}_{\text{ref}})} , \quad \sin 2\varphi_{\text{ref}} = -\frac{\overline{W}(\boldsymbol{q}_{\text{ref}})}{\Delta(\boldsymbol{q}_{\text{ref}})} .$$

This is also useful in the context of degenerate perturbation theory in order to identify $\boldsymbol{V}(\boldsymbol{q}_{\text{X}}+\delta\boldsymbol{q})$ to $\boldsymbol{H}(\boldsymbol{q}_{\text{X}}+\delta\boldsymbol{q})$ defined as the the result of the diagonalisation of $\overline{\boldsymbol{H}}(\boldsymbol{q}_{\text{X}}+\delta\boldsymbol{q})$ upon setting $\boldsymbol{q}_{\text{ref}} = \boldsymbol{q}_{\text{X}}+\delta\boldsymbol{q}$.

The gradient difference is often expressed as $\overline{x}_j^{(1)} = \frac{\partial_j V_2(\boldsymbol{q}_X) - \partial_j V_1(\boldsymbol{q}_X)}{2} = \partial_j \Delta(\boldsymbol{q}_X)$, as if local derivatives were well-behaved with no further prescription. However, as we just stressed, there are hidden subtleties behind the actual meaning of the adiabatic gradient difference and first-order non-adiabatic coupling when both states are degenerate. Let us consider an infinitesimal displacement $\boldsymbol{q} = \boldsymbol{q}_X + \delta\boldsymbol{q} \rightarrow \boldsymbol{q}_X$ (i.e., $\delta\boldsymbol{q} \rightarrow \boldsymbol{0}$), such that $\Delta(\boldsymbol{q}) \rightarrow \Delta(\boldsymbol{q}_X) = 0$ and the transformation angle giving the adiabatic $\Phi^{el/ad}$-states in terms of the reference $\overline{\Phi}$-states is fixed, $\overline{\varphi}(\boldsymbol{q}) = \varphi(\boldsymbol{q}) + \varphi_{\text{ref}} \equiv \varphi_{\text{ref}}$ (i.e., $\varphi(\boldsymbol{q}) \equiv 0$ with respect to the rotated Φ-states). Then, the adiabatic derivatives along this direction satisfy

$$\partial_j \Delta(\boldsymbol{q}) \rightarrow \cos 2\varphi_{\text{ref}} \overline{x}_j^{(1)} - \sin 2\varphi_{\text{ref}} \overline{x}_j^{(2)} = x_j^{(1)}\,, \tag{4.46}$$

$$2\Delta(\boldsymbol{q}) F_j(\boldsymbol{q}) \rightarrow \sin 2\varphi_{\text{ref}} \overline{x}_j^{(1)} + \cos 2\varphi_{\text{ref}} \overline{x}_j^{(2)} = x_j^{(2)}\,, \tag{4.47}$$

where $x_j^{(1)}$ and $x_j^{(2)}$ are obtained from $\overline{x}_j^{(1)}$ and $\overline{x}_j^{(2)}$ via a rotation through $-2\varphi_{\text{ref}}$. In other words, in the vicinity of $\boldsymbol{q}_X$, the actual adiabatic gradient difference and the numerator of the first-order non-adiabatic coupling depend on the value of the polar angle within the branching plane and rotate with it, as illustrated in Fig. 4.5. Again, this can be formulated rigorously within the framework of degenerate perturbation theory (considering $\sum_j \partial_j H^{el}(\boldsymbol{q}_X)\delta q_j$ as the perturbation to be diagonalised first). In any case, the important result of this derivation is that none of the two vectors makes

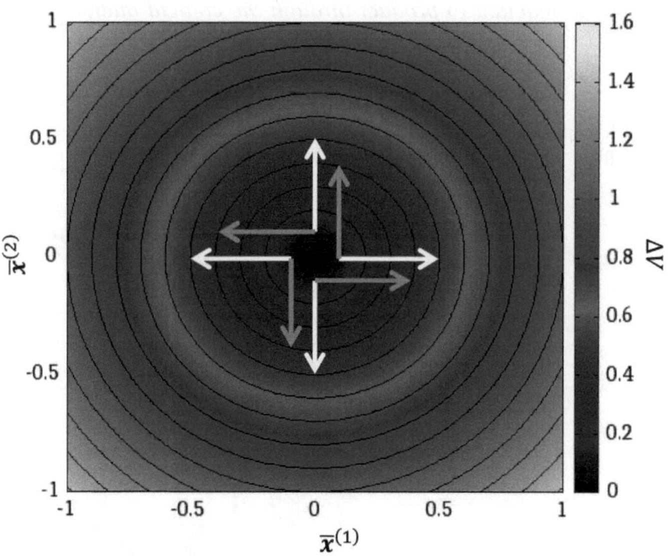

Fig. 4.5 Gradient difference (*white arrow*) and non-adiabatic coupling (*green arrow*) rotating within the branching space. Four different origins are considered in the vicinity of the conical intersection at the center. Contour plots correspond to increasing values of $\Delta(\boldsymbol{q})$

sense on its own. The only meaningful quantity is the branching plane spanned by the pair of them, which is well-defined as long as both vectors are not parallel.

Now, let us examine in more detail the local shape of the double cone and its implications on photochemical reactivity. Conical intersections are often classified into two main types: *sloped or peaked* (and intermediate cases in between). Intuitively, this concept is easy to grasp if one considers two parabolas that cross along a single coordinate. A sloped pattern is usually observed when both minima occur on the same side of the crossing, whereas a peaked pattern corresponds typically to the situation where there is a minimum on one side and a minimum on the other side. Each minimum belongs to a different adiabatic surface in the sloped case, while in the peaked case they both belong to the lower-energy adiabatic surface. This is illustrated on Fig. 4.6. When considering a full-dimensional picture, the sloped/peaked character is determined upon comparing the relative direction of the energy gradients at the crossing point. Roughly speaking, almost-parallel gradients yield a sloped topography, and almost-antiparallel gradients, a peaked topography. For a more rigorous definition and a detailed analysis, we refer to Ref. [14]. The consequence of the topography on the photochemical reactivity is, again, quite intuitive: *a sloped pattern tends to drive the system back to its original position because both gradients are similar. This ultimately favours photostability* (regeneration of the reactant in the ground state, once the excited state is depopulated). In contrast, *a peaked pattern allows some branching between the reactant (original minimum) and the photoproduct in the ground state (additional minimum).* In the latter case, *the photochemical reaction path from reactant to product through the conical intersection may be an*

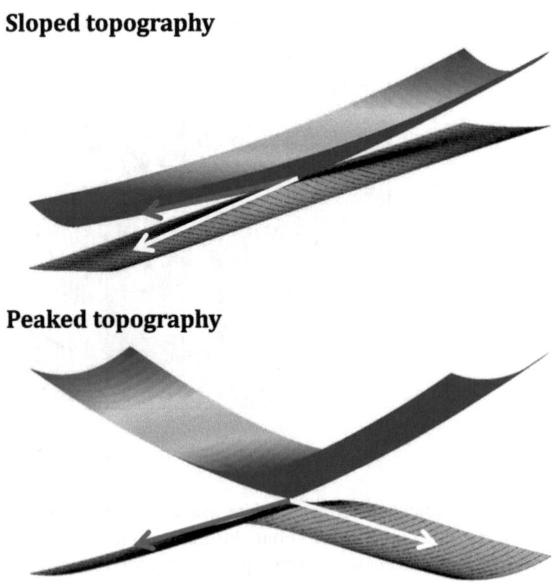

Fig. 4.6 Conical intersections with sloped or peaked topographies

alternative route for a chemical reaction that would be impossible or difficult to achieve thermally.

4.2 The Jahn-Teller Prototype *

The Jahn-Teller effect is well-known in inorganic and organometallic chemistry but occurs also in organic chemistry when dealing with open-shell systems (radicals, ions, or excited states). It is encountered when degenerate orbitals are not fully occupied. In such a situation, if the electronic state is degenerate, breaking the symmetry to lift the degeneracy of the orbitals lowers the energy of the system, which means—somewhat counterintuitively—that the structure with less symmetry is more stable. There are many famous examples of such situations. The simplest one is H + H_2 (with three electrons), but many examples have been reported in the literature [5, 6]. A prototypical example is the benzene cation for which the Jahn-Teller effect takes place in $\mathcal{D}_{6h}$ symmetry between the degenerate electronic ground state, ${}^2E_{1g}$, and the four ${}^2E_{2g}$ twofold-degenerate normal modes of vibration (see, e.g., Ref. [25]). In this situation, the lower surface shows three equivalent minima connected through three equivalent transition structures. This typical threefold shape is often referred to as the *warped Mexican hat*[8] as shown on Fig. 4.7. In fact, this reflects the occurrence of a Jahn-Teller crossing at the high-symmetry point, which is the archetypal case of conical intersection induced by symmetry for a twofold-degenerate irreducible representation in a non-Abelian point group (a point group with degenerate irreducible representations; see Chap. 7).

In what follows, we will consider a two-dimensional description that will help to clarify some general aspects involved in the description of the branching space in terms of polar coordinates. Let us consider a specific pair of branching-space vectors, $\overline{\boldsymbol{x}}^{(1)}$ and $\overline{\boldsymbol{x}}^{(2)}$, q_1 and q_2 displacements along them from the conical intersection where $q_1 = q_2 = 0$ (they are to be considered as two distinct components of an irreducible representation of E-type). We assume that they can be discriminated in an Abelian subgroup (with only non-degenerate irreducible representations; see Chap. 7) where q_1^2 and q_2^2 are totally symmetric while q_1q_2 is not. The prototypical two-dimensional quadratic expansion of $\overline{\boldsymbol{H}}(q_1, q_2)$ reads

$$\overline{\boldsymbol{H}}(q_1, q_2) = d\mathbf{1} + a(q_1^2 + q_2^2)\mathbf{1} + b\begin{bmatrix} -q_1 & q_2 \\ q_2 & q_1 \end{bmatrix} + c\begin{bmatrix} -(q_1^2 - q_2^2) & -2q_1q_2 \\ -2q_1q_2 & q_1^2 - q_2^2 \end{bmatrix}, \quad (4.48)$$

where a, b, c, and d are four real numbers. The corresponding eigenvalues are obtained as

[8]The *Mexican hat* model is similar but simpler: the lower surface is isotropic around the apex of the cone, as further discussed below.

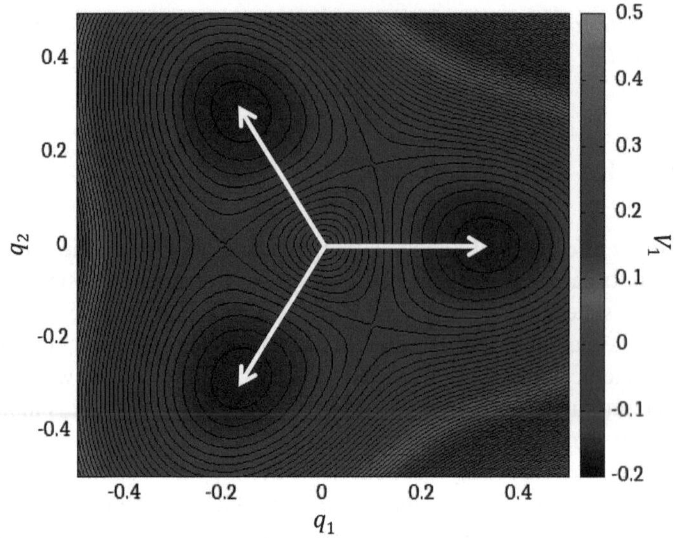

Fig. 4.7 Typical ground-state potential energy surface (contour plot) for a *warped Mexican hat*-type Jahn-Teller case. The *white arrows* denote directions leading from the apex of the cone to the three equivalent minima

$$V_{1,2}(q_1, q_2) = d + a(q_1^2 + q_2^2) \pm \sqrt{b^2(q_1^2 + q_2^2) + 2bcq_1(q_1^2 - 3q_2^2) + c^2(q_1^2 + q_2^2)^2}\,. \tag{4.49}$$

These have a threefold symmetry in the (q_1, q_2)-plane, which is more obvious when using polar coordinates,

$$q_1 = \rho \cos\theta\,, \tag{4.50}$$

$$q_2 = \rho \sin\theta\,, \tag{4.51}$$

such that

$$\overline{\boldsymbol{H}}(\rho, \theta) = d\mathbf{1} + a\rho^2\mathbf{1} + b\rho \begin{bmatrix} -\cos\theta & \sin\theta \\ \sin\theta & \cos\theta \end{bmatrix} + c\rho^2 \begin{bmatrix} -\cos 2\theta & -\sin 2\theta \\ -\sin 2\theta & \cos 2\theta \end{bmatrix}, \tag{4.52}$$

and

$$V_{1,2}(\rho, \theta) = d + a\rho^2 \pm \rho\sqrt{b^2 + 2bc\rho\cos 3\theta + c^2\rho^2}\,. \tag{4.53}$$

Note that a partially-linear expansion (when $c = 0$) yields isotropic eigenvalues, which is called the Mexican hat model, as opposed to the threefold warped Mexican hat model.

Let us now focus on the fully-linear prototype: $a = c = 0$. We can also set $d = 0$ for simplicity and assume $b > 0$. The corresponding Hamiltonian matrix now reads

$$\overline{\boldsymbol{H}}(q_1, q_2) = b \begin{bmatrix} -q_1 & q_2 \\ q_2 & q_1 \end{bmatrix} . \tag{4.54}$$

We define the underlying crude adiabatic states $|\overline{\Phi}_1\rangle = |\overline{\Phi}_1^{el/ad}; 0, 0\rangle$ and $|\overline{\Phi}_2\rangle = |\overline{\Phi}_2^{el/ad}; 0, 0\rangle$ (as already pointed out, a first-order description does not require explicit knowledge of their behaviours with respect to variations of the nuclear coordinates, as long as they are regular, so that they can be considered as fixed). In the adiabatic representation, in the vicinity of $(q_1, q_2) = (0, 0)$, the adiabatic states $|\Phi_1^{el/ad}; q_1, q_2\rangle$ and $|\Phi_2^{el/ad}; q_1, q_2\rangle$ yield

$$\boldsymbol{V}(q_1, q_2) = b \begin{bmatrix} -\sqrt{q_1^2 + q_2^2} & 0 \\ 0 & \sqrt{q_1^2 + q_2^2} \end{bmatrix} . \tag{4.55}$$

Both basis sets are related by a rotation through an angle $\overline{\varphi}(q_1, q_2)$ (modulo 2π, *a priori*),

$$\overline{\boldsymbol{U}}(q_1, q_2) = \begin{bmatrix} \cos\overline{\varphi}(q_1, q_2) & -\sin\overline{\varphi}(q_1, q_2) \\ \sin\overline{\varphi}(q_1, q_2) & \cos\overline{\varphi}(q_1, q_2) \end{bmatrix} , \tag{4.56}$$

such that

$$|\Phi_1^{el/ad}; q_1, q_2\rangle = \cos\overline{\varphi}(q_1, q_2)|\overline{\Phi}_1\rangle + \sin\overline{\varphi}(q_1, q_2)|\overline{\Phi}_2\rangle , \tag{4.57}$$

$$|\Phi_2^{el/ad}; q_1, q_2\rangle = -\sin\overline{\varphi}(q_1, q_2)|\overline{\Phi}_1\rangle + \cos\overline{\varphi}(q_1, q_2)|\overline{\Phi}_2\rangle , \tag{4.58}$$

which yields

$$\cos 2\overline{\varphi}(q_1, q_2) = \frac{q_1}{\sqrt{q_1^2 + q_2^2}} , \quad \sin 2\overline{\varphi}(q_1, q_2) = -\frac{q_2}{\sqrt{q_1^2 + q_2^2}} . \tag{4.59}$$

We can thus make a correspondence with the previously-defined polar coordinates in the (q_1, q_2)-plane: $\rho = \sqrt{q_1^2 + q_2^2}$ and $\theta = -2\overline{\varphi}(q_1, q_2)$ (modulo 2π). Note that $\overline{\varphi}(q_1, q_2)$ and $\overline{\varphi}(q_1, q_2) + \pi$ actually correspond to the same point in space (θ is transformed into $\theta - 2\pi$). Using polar coordinates, the electronic Hamiltonian matrices can be recast as

$$\overline{\boldsymbol{H}}(\rho, \theta) = b \begin{bmatrix} -\rho\cos\theta & \rho\sin\theta \\ \rho\sin\theta & \rho\cos\theta \end{bmatrix} , \tag{4.60}$$

and

$$\boldsymbol{V}(\rho, \theta) = b \begin{bmatrix} -\rho & 0 \\ 0 & \rho \end{bmatrix} . \tag{4.61}$$

The gradient of $\overline{\boldsymbol{H}}(q_1, q_2)$ satisfies at any point

$$\nabla\overline{\boldsymbol{H}}(q_1, q_2) = b\begin{bmatrix} -\boldsymbol{u}_1 & \boldsymbol{u}_2 \\ \boldsymbol{u}_2 & \boldsymbol{u}_1 \end{bmatrix}, \tag{4.62}$$

where we introduced the unit vectors in the (q_1, q_2)-plane[9]: $\boldsymbol{u}_1 = \nabla q_1$ and $\boldsymbol{u}_2 = \nabla q_2$. Differentiating the adiabaticity conditions,

$$q_1 \cos 2\overline{\varphi}(q_1, q_2) - q_2 \sin 2\overline{\varphi}(q_1, q_2) = \sqrt{q_1^2 + q_2^2}\,, \tag{4.63}$$

$$q_1 \sin 2\overline{\varphi}(q_1, q_2) + q_2 \cos 2\overline{\varphi}(q_1, q_2) = 0\,, \tag{4.64}$$

yields (at any point but the origin),

$$\underbrace{2\nabla\overline{\varphi}(q_1, q_2)}_{=-\nabla\theta=-\frac{1}{\rho}\boldsymbol{u}_\theta}\underbrace{(-q_1 \sin 2\overline{\varphi}(q_1, q_2) - q_2 \cos 2\overline{\varphi}(q_1, q_2))}_{=0}$$

$$+\underbrace{\boldsymbol{u}_1 \cos 2\overline{\varphi}(q_1, q_2) - \boldsymbol{u}_2 \sin 2\overline{\varphi}(q_1, q_2)}_{=\nabla\rho=\boldsymbol{u}_\rho} = \underbrace{\nabla\sqrt{q_1^2 + q_2^2}}_{=\nabla\rho=\boldsymbol{u}_\rho}\,, \tag{4.65}$$

$$\underbrace{2\nabla\overline{\varphi}(q_1, q_2)}_{=-\nabla\theta=-\frac{1}{\rho}\boldsymbol{u}_\theta}\underbrace{(q_1 \cos 2\overline{\varphi}(q_1, q_2) - q_2 \sin 2\overline{\varphi}(q_1, q_2))}_{=\sqrt{q_1^2+q_2^2}=\rho}$$

$$+\underbrace{\boldsymbol{u}_1 \sin 2\overline{\varphi}(q_1, q_2) + \boldsymbol{u}_2 \cos 2\overline{\varphi}(q_1, q_2)}_{=\rho\nabla\theta=\boldsymbol{u}_\theta} = \boldsymbol{0}\,, \tag{4.66}$$

where the polar unit vectors satisfy

$$\boldsymbol{u}_\rho = \nabla\rho = \frac{q_1}{\sqrt{q_1^2 + q_2^2}}\boldsymbol{u}_1 + \frac{q_2}{\sqrt{q_1^2 + q_2^2}}\boldsymbol{u}_2 = \cos\theta\,\boldsymbol{u}_1 + \sin\theta\,\boldsymbol{u}_2\,, \tag{4.67}$$

$$\boldsymbol{u}_\theta = \rho\nabla\theta = -\frac{q_2}{\sqrt{q_1^2 + q_2^2}}\boldsymbol{u}_1 + \frac{q_1}{\sqrt{q_1^2 + q_2^2}}\boldsymbol{u}_2 = -\sin\theta\,\boldsymbol{u}_1 + \cos\theta\,\boldsymbol{u}_2\,. \tag{4.68}$$

For such vectors to be defined when approaching the pole implies to make ρ tend to zero for a given value of θ (directional derivative instead of local derivative, as aforementioned). The adiabatic gradient difference (with no bar) thus satisfies

$$\boldsymbol{x}^{(1)} = b\nabla\rho = b\boldsymbol{u}_\rho\,, \tag{4.69}$$

[9]This notation reflects that $\partial_{q_1} q_1 = 1$ and $\partial_{q_2} q_1 = 0$ while $\partial_{q_1} q_2 = 0$ and $\partial_{q_2} q_2 = 1$.

which thus rotate in the (q_1, q_2)-plane depending on the value of θ. Note that $\Delta(\rho, \theta) = b\rho$ and

$$\frac{\partial \Delta(\rho,\theta)}{\partial \rho} = b\,, \quad \frac{\partial \Delta(\rho,\theta)}{\partial \theta} = 0\,, \tag{4.70}$$

which shows that the energy difference behaves as an isotropic two-dimensional cone of radial slope b. The first-order non-adiabatic coupling vector is obtained from

$$\begin{aligned} |\nabla \Phi_2^{el/ad}; q_1, q_2\rangle &= -\nabla\overline{\varphi}(q_1,q_2)(\cos\overline{\varphi}(q_1,q_2)|\overline{\Phi}_1\rangle + \sin\overline{\varphi}(q_1,q_2)|\overline{\Phi}_2\rangle) \\ &= -\nabla\overline{\varphi}(q_1,q_2)|\Phi_1^{el/ad}; q_1,q_2\rangle\,, \end{aligned} \tag{4.71}$$

such that

$$\langle \Phi_1^{el/ad}; q_1, q_2|\nabla \Phi_2^{el/ad}; q_1, q_2\rangle = -\nabla\overline{\varphi}(q_1,q_2) = \frac{1}{2}\nabla\theta = \frac{1}{2\rho}\boldsymbol{u}_\theta\,. \tag{4.72}$$

Note that

$$\langle \Phi_1^{el/ad}; q_1, q_2|\frac{\partial}{\partial\rho}\Phi_2^{el/ad}; q_1, q_2\rangle = 0\,, \quad \langle \Phi_1^{el/ad}; q_1, q_2|\frac{\partial}{\partial\theta}\Phi_2^{el/ad}; q_1, q_2\rangle = \frac{1}{2}\,, \tag{4.73}$$

which shows that the mixing between both states does not depend on the radius and depends uniformly on the angle (and the half value shows that the states turn half more slowly than the polar angle). The partial derivatives expressed in polar coordinates are regular, but there is a still singularity of the total gradient at the pole. Indeed,

$$\nabla = \frac{\partial}{\partial q_1}\boldsymbol{u}_1 + \frac{\partial}{\partial q_2}\boldsymbol{u}_2 = \frac{\partial}{\partial\rho}\boldsymbol{u}_\rho + \frac{1}{\rho}\frac{\partial}{\partial\theta}\boldsymbol{u}_\theta\,. \tag{4.74}$$

In summary, the branching-space vectors in the crude adiabatic representation (chosen as a conventional reference) are fixed and read

$$\overline{\boldsymbol{x}}^{(1)} = b\boldsymbol{u}_1\,, \tag{4.75}$$
$$\overline{\boldsymbol{x}}^{(2)} = b\boldsymbol{u}_2\,, \tag{4.76}$$

while in the adiabatic representation (states allowed to mix around the apex of the double cone), they depend on θ,

$$\boldsymbol{x}^{(1)} = b\boldsymbol{u}_\rho = \cos\theta\overline{\boldsymbol{x}}^{(1)} + \sin\theta\overline{\boldsymbol{x}}^{(2)}\,, \tag{4.77}$$
$$\boldsymbol{x}^{(2)} = b\boldsymbol{u}_\theta = -\sin\theta\overline{\boldsymbol{x}}^{(1)} + \cos\theta\overline{\boldsymbol{x}}^{(2)}\,, \tag{4.78}$$

A similar treatment in terms of polar coordinates can be applied to any type of conical intersection, provided the branching-plane coordinates q_1 and q_2 are duly scaled and expressed along vectors $\boldsymbol{x}^{(1)}$ and $\boldsymbol{x}^{(2)}$ rotated so as to be orthogonal.

Jahn-Teller crossings are very specific types of conical intersections induced by symmetry. However, many molecules that present conical intersections involved in their photochemical reactivity can be understood upon making a comparison with a Jahn-Teller analogue involving the same number of electrons (see, e.g., Ref. [18]).

4.3 Diabatic Representations

Up to now, we have considered as a working basis set a pair of electronic states, $|\Phi_1; \boldsymbol{q}\rangle$ and $|\Phi_2; \boldsymbol{q}\rangle$, with only three requirements: they are orthogonal, span the same Hilbert space as the two adiabatic eigenstates of interest, $|\Phi_1^{el/ad}; \boldsymbol{q}\rangle$ and $|\Phi_2^{el/ad}; \boldsymbol{q}\rangle$, and vary smoothly with $\boldsymbol{q}$ (in particular around a conical intersection where they also are eigenstates by construction). We shall now be more specific upon introducing the concept of diabatic states.

As already mentioned above, $\Delta(\boldsymbol{q})$ exhibits a two-dimensional cusp when $\boldsymbol{q}$ is the locus of a conical intersection. In addition, the components of the first-order non-adiabatic coupling vector, $\boldsymbol{F}(\boldsymbol{q})$, diverge. Solving the Schrödinger equation for the nuclear wavepackets requires the various quantities entering the Hamiltonian operator to be represented as explicit functions of $\boldsymbol{q}$ (and corresponding first and second derivatives) that will be involved in integrals over $\boldsymbol{q}$ when evaluating matrix elements. As a consequence, although the adiabatic representation is uniquely-defined and is a good starting point, it is impractical as such for quantum dynamics simulations when the effect of a conical intersection and non-adiabatic couplings are to be considered. Smoother functions, easier to handle numerically, can be obtained upon considering an alternative electronic basis set called diabatic. Transformations from adiabatic states to diabatic states are called *diabatisations.*

Formally, *diabatic states are defined such that the corresponding kinetic-coupling operators,*

$$\begin{aligned}\Lambda_{nm}^{dia}(\boldsymbol{R}) = &-\frac{1}{M}\langle\Phi_n^{el/dia}; \boldsymbol{R}|\frac{\partial}{\partial \boldsymbol{R}}\Phi_m^{el/dia}; \boldsymbol{R}\rangle_{\boldsymbol{r}}\cdot\frac{\partial}{\partial \boldsymbol{R}} \\ &-\frac{1}{2M}\langle\Phi_n^{el/dia}; \boldsymbol{R}|\frac{\partial^2}{\partial \boldsymbol{R}^2}\Phi_m^{el/dia}; \boldsymbol{R}\rangle_{\boldsymbol{r}} \approx 0\end{aligned} \tag{4.79}$$

(see Sect. 3.2.4), *vanish.*

Instead, *the electronic Hamiltonian matrix is no longer diagonal, as the diabatic states are not eigenstates*. The couplings between the electronic states are now represented by the off-diagonal entries

$$H_{\alpha\beta}^{dia}(\boldsymbol{q}) = \langle\Phi_\alpha^{el/dia}; \boldsymbol{q}|\ H^{el}(\boldsymbol{q})|\Phi_\beta^{el/dia}; \boldsymbol{q}\rangle\ . \tag{4.80}$$

For this reason, they are *called potential couplings*, as opposed to the kinetic couplings that arise in the adiabatic representation. This concept was introduced in 1935

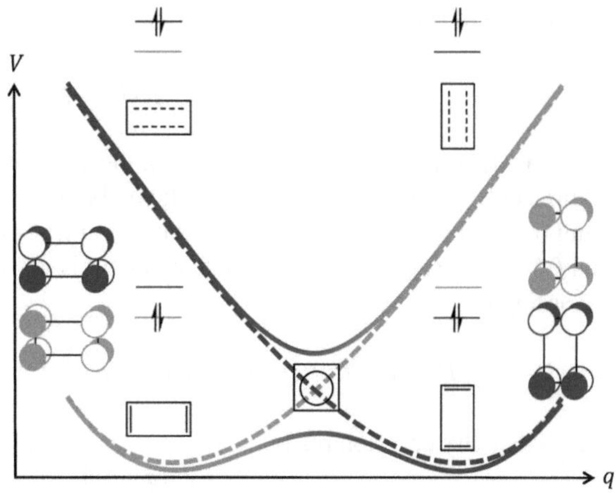

Fig. 4.8 Avoided crossing in the cyclobutadiene molecule. The diabatic potentials are in *green* and *orange*: they follow the evolution of the molecular orbitals shown on the figure. On the other hand the adiabatic potential energy surfaces imply a complete change of the dominant electronic configurations around the transition state

by Polanyi [26] and Hellmann and Syrkin [27] and further generalised by Smith and Baer [28–30]. Originally used essentially in the context of inelastic scattering in molecular physics, diabatic states have gradually become essential tools in non-adiabatic photochemistry.

As opposed to adiabatic states, diabatic states are not eigenstates of any operator in particular. Their definition is not unique and, as shown by Mead and Truhlar [31], the diabaticity criterion (see below), which is a local condition, cannot be achieved globally (except for a diatom or for the ideal case of a finite Hilbert space). In the general case, a complete (thus infinite) basis set of electronic states is required for integrating the condition of diabaticity over the whole space spanned by the nuclear coordinates. However, it is possible to find states that make the non-adiabatic couplings negligible and with no significant effect on the dynamics of the molecule; such states are called quasidiabatic and often referred to as diabatic for simplicity. On the other hand, *diabatic states often have a clear interpretation from a chemical point of view*: this is illustrated by the avoided crossing in the cyclobutadiene molecule shown on Fig. 4.8. The diabatic potentials correspond to the curves in green and orange. The lower adiabatic potential has two minima corresponding to a distortion of the molecule that stabilizes the π molecular orbitals depicted in Fig. 4.8: we see that *the diabatic functions follow the evolution of the molecular orbitals and correspond to one dominant electronic configuration. On the other hand, the adiabatic functions imply a deep change in the dominant electronic configuration around the geometry corresponding to the transition state.*

Formally, the diabatic and adiabatic basis sets can be considered both as orthonormal and spanning the same Hilbert space at all $\boldsymbol{q}$ (at least to a given order around some reference geometry). They are thus transformed into each other through a unitary transformation, $\boldsymbol{U}^{dia}(\boldsymbol{q})$,

$$\boldsymbol{U}^{dia\dagger}(\boldsymbol{q})\boldsymbol{U}^{dia}(\boldsymbol{q}) = \boldsymbol{U}^{dia}(\boldsymbol{q})\boldsymbol{U}^{dia\dagger}(\boldsymbol{q}) = \mathbf{1}\,, \tag{4.81}$$

such that both Hamiltonian matrices are related through a similarity transformation,

$$\boldsymbol{V}(\boldsymbol{q}) = \boldsymbol{U}^{dia\dagger}(\boldsymbol{q})\boldsymbol{H}^{dia}(\boldsymbol{q})\boldsymbol{U}^{dia}(\boldsymbol{q}), \tag{4.82}$$

where $\boldsymbol{V}(\boldsymbol{q})$ is diagonal. In this definition, the columns of $\boldsymbol{U}^{dia}(\boldsymbol{q})$ correspond to the components of the adiabatic states in the diabatic basis set. $\boldsymbol{F}_j(\boldsymbol{q})$ is the j-th vector component of the first-order non-adiabatic coupling matrix in the adiabatic basis set and $\boldsymbol{f}_j^{dia}(\boldsymbol{q})$ the same quantity in the diabatic basis set. They transform into each other according to

$$\boldsymbol{F}_j(\boldsymbol{q}) = \boldsymbol{U}^{dia\dagger}(\boldsymbol{q})\boldsymbol{f}_j^{dia}(\boldsymbol{q})\boldsymbol{U}^{dia}(\boldsymbol{q}) + \boldsymbol{U}^{dia\dagger}(\boldsymbol{q})\partial_j\boldsymbol{U}^{dia}(\boldsymbol{q}). \tag{4.83}$$

The diabaticity criterion reads

$$\boldsymbol{f}_j^{dia}(\boldsymbol{q}) \approx \boldsymbol{0}, \tag{4.84}$$

so that the unitary transformation must fulfill (assuming that the first-order non-adiabatic coupling is known)

$$\boldsymbol{F}_j(\boldsymbol{q}) \approx \boldsymbol{U}^{dia\dagger}(\boldsymbol{q})\partial_j\boldsymbol{U}^{dia}(\boldsymbol{q}). \tag{4.85}$$

As already mentioned, in a two-state problem, the unitary transformation can be chosen as a real rotation,

$$\boldsymbol{U}^{dia}(\boldsymbol{q}) = \begin{bmatrix} \cos\varphi^{dia}(\boldsymbol{q}) & -\sin\varphi^{dia}(\boldsymbol{q}) \\ \sin\varphi^{dia}(\boldsymbol{q}) & \cos\varphi^{dia}(\boldsymbol{q}) \end{bmatrix}, \tag{4.86}$$

where

$$|\Phi_1^{el/ad}; \boldsymbol{q}\rangle = \cos\varphi^{dia}(\boldsymbol{q})|\Phi_1^{el/dia}; \boldsymbol{q}\rangle + \sin\varphi^{dia}(\boldsymbol{q})|\Phi_2^{el/dia}; \boldsymbol{q}\rangle, \tag{4.87}$$

$$|\Phi_2^{el/ad}; \boldsymbol{q}\rangle = -\sin\varphi^{dia}(\boldsymbol{q})|\Phi_1^{el/dia}; \boldsymbol{q}\rangle + \cos\varphi^{dia}(\boldsymbol{q})|\Phi_2^{el/dia}; \boldsymbol{q}\rangle. \tag{4.88}$$

The first derivative is simply obtained as

$$\partial_j\boldsymbol{U}^{dia}(\boldsymbol{q}) = -\partial_j\varphi^{dia}(\boldsymbol{q}) \begin{bmatrix} \sin\varphi^{dia}(\boldsymbol{q}) & \cos\varphi^{dia}(\boldsymbol{q}) \\ -\cos\varphi^{dia}(\boldsymbol{q}) & \sin\varphi^{dia}(\boldsymbol{q}) \end{bmatrix}, \tag{4.89}$$

such that

$$\boldsymbol{U}^{dia\dagger}(\boldsymbol{q})\partial_j\boldsymbol{U}^{dia}(\boldsymbol{q}) = -\partial_j\varphi^{dia}(\boldsymbol{q}) \begin{bmatrix} 0 & 1 \\ -1 & 0 \end{bmatrix}. \tag{4.90}$$

The condition of diabaticity is thus achieved if the mixing angle, $\varphi^{dia}(\boldsymbol{q})$, satisfies at all $\boldsymbol{q}$,

$$F_j(\boldsymbol{q}) \approx -\partial_j\varphi^{dia}(\boldsymbol{q}). \tag{4.91}$$

where we recall that $\boldsymbol{F}$ now simply stands for $\boldsymbol{F}_{12}$ (two-state problem). This differential relationship has led to methods for determining $\varphi^{dia}(\boldsymbol{q})$ based on the integration of the coupling vector $\boldsymbol{F}(\boldsymbol{q})$ [32]. This kind of explicit procedure is tractable for small systems but becomes rapidly tedious as the size of the system increases. However, for semiclassical dynamics, it is still interesting, as the integration can be achieved along the one-dimensional path followed by a trajectory $\boldsymbol{q}(t)$ [33].

The development of various diabatisation formalisms was an active field of research in the 1980 s and has recently become central again with the advent of quantum dynamics methods able to treat large molecular systems. Many approaches, based on different criteria, have been proposed to build quasidiabatic states and/or Hamiltonians. Some were formulated in terms of effective Hamiltonians in the framework of quasidegenerate perturbation theory, within a model subspace of states having the same eigenvalues as the targeted adiabatic ones [34–40]. Other methods were based on the construction of states yielding mathematical or physical properties that are smooth functions of $\boldsymbol{q}$ [41, 42]. A detailed review of diabatisation methods is beyond the scope of this chapter. We shall concentrate here on the approach called diabatisation by ansatz, which is an energy-based diabatisation. Vibronic-coupling Hamiltonian models [13, 19, 20] are typical examples that have been used with success in concert with MCTDH calculations to produce absorption and photoelectron spectra in molecules made of about ten atoms.

4.4 Diabatisation by Ansatz

The energy-based approach named diabatisation by ansatz relies on the construction of a Hamiltonian matrix, $\boldsymbol{H}^{dia}(\boldsymbol{q})$, made of smooth and simple analytic functions of $\boldsymbol{q}$. The values of the various parameters involved in these functions are adjusted through a fitting procedure so that the eigenvalues of $\boldsymbol{H}^{dia}(\boldsymbol{q})$ are as close as possible to the ab initio adiabatic energies over a sample of relevant molecular geometries. The non-adiabatic couplings are not used from the onset in a diabatisation by ansatz. However, they will be correct by construction around a conical intersection if the double cone is described correctly to first order in the model. In other words, the physical information carried by the couplings is contained in the local topography of the conical intersection. Indeed, as already shown, when $\boldsymbol{q} \to \boldsymbol{q}_X$ the adiabatic gradient difference and first-order non-adiabatic coupling span the branching plane: $\partial_j \Delta(\boldsymbol{q}) \to x_j^{(1)}$ and $2\Delta(\boldsymbol{q})F_j(\boldsymbol{q}) \to x_j^{(2)}$ to first order in $\boldsymbol{q}$ (see above). This ensures that the effect of the non-adiabatic couplings will be treated adequately in regions where they are significant (around conical intersections). In addition, even if there is no direct check of how much diabatic the states are, the smoothness and simplicity of the entries of $\boldsymbol{H}^{dia}(\boldsymbol{q})$ as functions of $\boldsymbol{q}$ guarantee indirectly that the underlying states vary as little as possible with $\boldsymbol{q}$.

Let us now illustrate this approach with a two-dimensional model,

$$\boldsymbol{H}^{dia}(x, y) = \left[H_{\alpha\beta}^{dia}(x, y)\right], \tag{4.92}$$

where we assume a real symmetric matrix, i.e., $H^{dia}_{\alpha\beta}(x, y) = H^{dia}_{\beta\alpha}(x, y)$. The functions entering the diabatic Hamiltonian matrix are regular by construction. This implies that they can be expanded as Taylor series around some reference geometry, (x_0, y_0),

$$\begin{aligned} H^{dia}_{\alpha\beta}(x, y) = H^{dia}_{\alpha\beta}(x_0, y_0) &+ \left(\frac{\partial H^{dia}_{\alpha\beta}}{\partial x}\right)_0 (x - x_0) + \left(\frac{\partial H^{dia}_{\alpha\beta}}{\partial y}\right)_0 (y - y_0) \\ &+ \frac{1}{2}\left(\frac{\partial^2 H^{dia}_{\alpha\beta}}{\partial x^2}\right)_0 (x - x_0)^2 + \frac{1}{2}\left(\frac{\partial^2 H^{dia}_{\alpha\beta}}{\partial y^2}\right)_0 (y - y_0)^2 \\ &+ \left(\frac{\partial^2 H^{dia}_{\alpha\beta}}{\partial x \partial y}\right)_0 (x - x_0)(y - y_0) + \ldots, \end{aligned} \tag{4.93}$$

The values of $H^{dia}_{\alpha\beta}(x_0, y_0)$ and of the local derivatives at the reference point, $\left(\frac{\partial H^{dia}_{\alpha\beta}}{\partial x}\right)_0$, $\left(\frac{\partial H^{dia}_{\alpha\beta}}{\partial y}\right)_0$, etc. are then adjusted self-consistently until the eigenvalues[10] of $\boldsymbol{H}^{dia}(x, y)$ fit the adiabatic energies (in general calculated ab initio with a quantum chemistry method) within some tolerance threshold over a representative sample of values of (x, y). The quality of the fitted model is measured with the root-mean-square deviation (least-square procedure) between the model and the data. Sometimes, the parameters are further readjusted to reproduce experimental data obtained from spectroscopy. Alternatively to a Taylor expansion truncated to some given order, it is possible to use predetermined analytic expressions such as, e.g., Morse functions, repulsive exponentials, or periodic functions, according to the nature of the coordinates and whether the state is bound or dissociative. Again, such functions depend parametrically on a set of characteristic numbers (dissociation asymptote, harmonic curvature, anharmonic constant, period, etc.) that are to be adjusted through a fitting procedure. The typical shapes of diabatic and adiabatic surfaces around a conical intersection is illustrated schematically on Fig. 4.9. Note that a cut along x for a non-zero value of y implies that the off-diagonal coupling term, H^{dia}_{12}, is not zero, which corresponds to an avoided crossing such as shown on Fig. 4.3.

When the entries of $\boldsymbol{H}^{dia}(\boldsymbol{q})$ are expressed as truncated Taylor expansions around the Franck-Condon point (typically, as linear or quadratic functions of normal coordinates originated from the ground-state equilibrium geometry), this matrix is referred to as a vibronic-coupling Hamiltonian model [13, 19, 20]. Let us first consider a simple example in the form of a two-state, two-mode model (q_1 and q_2 will be considered as a pair of dimensionless normal coordinates below) with quadratic diagonal entries and linear off-diagonal entries. Typically, the various terms are separated as follows,

---

[10]Note that, as already shown, a two-state problem gets explicit solutions for the eigenvalues but this is no longer the case for problems involving more states. If so, the eigenvalues are to be calculated numerically using some diagonalisation procedure (for example the Jacobi algorithm).

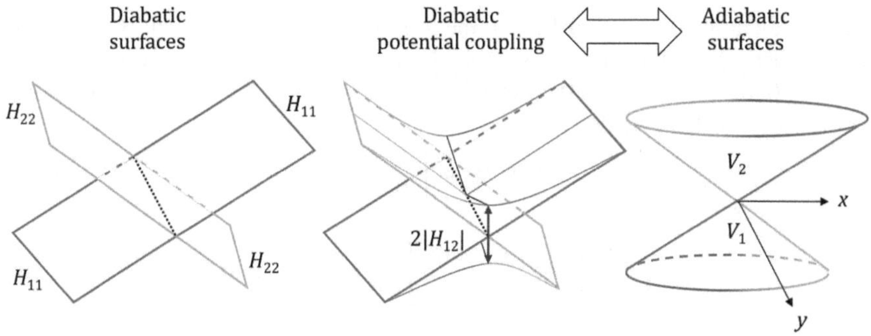

Fig. 4.9 Shapes of diabatic (H; the superscript dia has been dropped for notation simplicity), and adiabatic (V) surfaces along two coordinates x and y spanning the two-dimensional branching space of a conical intersection (linear model)

$$\boldsymbol{H}^{dia}(q_1, q_2) = \left(\frac{\hbar\omega_1}{2}q_1^2 + \frac{\hbar\omega_2}{2}q_2^2\right)\mathbf{1} + \begin{bmatrix} \varepsilon_1 & 0 \\ 0 & \varepsilon_2 \end{bmatrix} + \begin{bmatrix} \kappa^{(1)}q_1 & \lambda q_2 \\ \lambda q_2 & \kappa^{(2)}q_1 \end{bmatrix}. \tag{4.94}$$

The parameters are related to the derivatives according to

$$\boldsymbol{H}^{dia}(0, 0) = \begin{bmatrix} \varepsilon_1 & 0 \\ 0 & \varepsilon_2 \end{bmatrix}, \tag{4.95}$$

$$\left(\frac{\partial \boldsymbol{H}^{dia}}{\partial q_1}\right)_0 = \begin{bmatrix} \kappa^{(1)} & 0 \\ 0 & \kappa^{(2)} \end{bmatrix}, \quad \left(\frac{\partial \boldsymbol{H}^{dia}}{\partial q_2}\right)_0 = \begin{bmatrix} 0 & \lambda \\ \lambda & 0 \end{bmatrix}, \tag{4.96}$$

$$\left(\frac{\partial^2 \boldsymbol{H}^{dia}}{\partial q_1^2}\right)_0 = \hbar\omega_1\mathbf{1}, \quad \left(\frac{\partial^2 \boldsymbol{H}^{dia}}{\partial q_2^2}\right)_0 = \hbar\omega_2\mathbf{1}, \quad \left(\frac{\partial^2 \boldsymbol{H}^{dia}}{\partial q_1 \partial q_2}\right)_0 = \mathbf{0}. \tag{4.97}$$

If $\kappa^{(1)} = \kappa^{(2)} = 0$, we have two equivalent paraboloids centred at the origin and with harmonic angular frequencies ω_1 and ω_2. Non-zero parameters $\kappa^{(1)}$ and $\kappa^{(2)}$ characterize the slopes at the origin when q_1 is varied (hence, the shift of both minima with respect to the origin). The energy difference $\varepsilon_2 - \varepsilon_1$ is the vertical transition energy at the origin. The parameter λ characterises the strength of the coupling when q_2 is varied. The first coordinate is called the tuning mode and the second coordinate the coupling mode. The eigenvalues read

$$V_{1,2}(q_1, q_2) = \frac{\hbar\omega_1}{2}q_1^2 + \frac{\hbar\omega_2}{2}q_2^2 + \frac{\varepsilon_1 + \varepsilon_2}{2} + \frac{\kappa^{(1)} + \kappa^{(2)}}{2}q_1 \pm\sqrt{\left(\frac{\varepsilon_2 - \varepsilon_1}{2} + \frac{\kappa^{(2)} - \kappa^{(1)}}{2}q_1\right)^2 + (\lambda q_2)^2}. \tag{4.98}$$

The off-diagonal term vanishes at the origin, such that we can consider that the adiabatic and diabatic states coincide. Assuming $\varepsilon_2 > \varepsilon_1$, we can identify $|\Phi_1^{el/dia}; 0, 0\rangle = |\Phi_1^{el/ad}; 0, 0\rangle$ and $|\Phi_2^{el/dia}; 0, 0\rangle = |\Phi_2^{el/ad}; 0, 0\rangle$. In other words, $\varphi^{dia}(0, 0) = 0$. This yields

$$\boldsymbol{V}(0, 0) = \begin{bmatrix} \varepsilon_1 & 0 \\ 0 & \varepsilon_2 \end{bmatrix}, \tag{4.99}$$

$$\left(\frac{\partial \boldsymbol{V}}{\partial q_1}\right)_0 = \begin{bmatrix} \kappa^{(1)} & 0 \\ 0 & \kappa^{(2)} \end{bmatrix}, \quad \left(\frac{\partial \boldsymbol{V}}{\partial q_2}\right)_0 = \boldsymbol{0}, \tag{4.100}$$

$$\left(\frac{\partial^2 \boldsymbol{V}}{\partial q_1^2}\right)_0 = \hbar\omega_1 \boldsymbol{1}, \quad \left(\frac{\partial^2 \boldsymbol{V}}{\partial q_2^2}\right)_0 = \hbar\omega_2 \boldsymbol{1} + 2\frac{\lambda^2}{\varepsilon_2 - \varepsilon_1} \begin{bmatrix} -1 & 0 \\ 0 & 1 \end{bmatrix}, \quad \left(\frac{\partial^2 \boldsymbol{V}}{\partial q_1 \partial q_2}\right)_0 = \boldsymbol{0}. \tag{4.101}$$

The zero- and first-order diagonal parameters can thus be directly identified to the ab initio energies and gradients. The second derivatives are equal along the tuning mode but are altered along the coupling mode with a term involving the square of the coupling constant added to the diabatic curvature of the higher state and removed from the diabatic curvature of the lower state. In cases where $\hbar\omega_2 \boldsymbol{1} - 2\frac{\lambda^2}{\varepsilon_2 - \varepsilon_1}$ is negative (i.e., $2\frac{\lambda^2}{\varepsilon_2 - \varepsilon_1} > \hbar\omega_2$, which occurs when the coupling is strong and/or the energy difference is small), the origin exhibits a negative curvature along q_2, such that a double-well pattern can occur in the lower-energy surface,[11] as shown on Fig. 4.10. This is known as a second-order Jahn-Teller effect [5, 6]. This simple example also allows us to locate the conical intersection easily. It must correspond to $H_{12}^{dia}(q_1, q_{2X}) = 0$, hence $q_{2X} = 0$. Further, $H_{22}^{dia}(q_{1X}, 0) = H_{11}^{dia}(q_{1X}, 0)$ yields $q_{1X} = -\frac{\varepsilon_2 - \varepsilon_1}{\kappa^{(2)} - \kappa^{(1)}}$ (note that $\kappa^{(2)} = \kappa^{(1)}$ implies parallel curves that never cross). The branching-space vectors can be taken as $\boldsymbol{x}^{(1)} = \left[\frac{\kappa^{(2)} - \kappa^{(1)}}{2}, 0\right]$ and $\boldsymbol{x}^{(2)} = [0, \lambda]$ upon assuming that φ_{ref} is determined so as to correspond to the direction from the conical intersection to the origin (as already mentioned, one can calculate the value of φ_{ref} from $\boldsymbol{q}_{\text{ref}} = \boldsymbol{0}$, the Franck-Condon point, to rotate the branching-space vectors obtained from an actual calculation, $\overline{\boldsymbol{x}}^{(1)}$ and $\overline{\boldsymbol{x}}^{(2)}$, with the bar, in order to produce the effectively-used $\boldsymbol{x}^{(1)}$ and $\boldsymbol{x}^{(2)}$, with no bar). Again, let us stress that the values of the parameters can also be fitted numerically such that the eigenvalues of $\boldsymbol{H}^{dia}(q_1, q_2)$ are as close as possible to $V_1(q_1, q_2)$ and $V_2(q_1, q_2)$ over a selected sample of points.

In most application cases treated with a vibronic-coupling Hamiltonian model, point-group symmetry (see Chap. 7) was used to simplify the procedure determining which parameters vanish for symmetry reasons. A prototypical example is the butatriene cation, $C_4H_4^+$, for which such a model was used in MCTDH calculations to calculate a photoelectron spectrum in good agreement with experiments [43].

---

[11] This situation looks similar to an avoided crossing with respect to the adiabatic curves, see Fig. 4.3, except that now the strongly-coupled diabatic states produce curves that do not cross but rather coincide with the adiabatic ones at the origin; a $\frac{\pi}{4}$-rotation, sometimes referred to as a Nikitin transformation, would generate alternative diabatic states that cross and are weakly coupled. This would swap the roles of q_1 and q_2 as tuning and coupling modes, respectively.

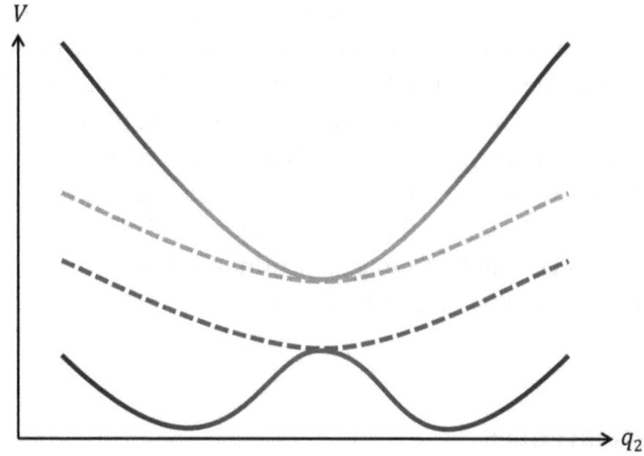

Fig. 4.10 Typical shape of a second-order Jahn-Teller effect in the strong-coupling regime for two adiabatic potential energy curves (*plain lines*), V_2 (*upper one*) and V_1 (*lower one*), along a nuclear coordinate q_2. The *dashed lines* represent the diabatic energies curves (the *orange* and *green* color code refers to the diabatic states)

The corresponding 18-dimensional vibronic-coupling Hamiltonian model is given in Chap. 7 where details about how point-group symmetry is to be applied to this case are provided.

Finally, let us again notice that *diabatic states often have a clear interpretation from a chemical point of view*. In the case of the butatriene cation, the dominant configurations of $X^2B_{2g}(D_0)$ and $A^2B_{2u}(D_1)$ at the Franck-Condon point are [closed shells]$b_{2u}^2b_{2g}^1$ and [closed shells]$b_{2u}^1b_{2g}^2$, respectively. The corresponding molecular orbitals are depicted on Fig. 7.9 in Chap. 7. The two diabatic states are chosen such that $|\Phi_1^{el/dia}; \boldsymbol{q}\rangle$ and $|\Phi_2^{el/dia}; \boldsymbol{q}\rangle$ coincide with the adiabatic ones, $X^2B_{2g}(D_0)$ and $A^2B_{2u}(D_1)$ at the Franck-Condon point ($\boldsymbol{q} = \boldsymbol{0}$) and keep belonging to the B_{2g}

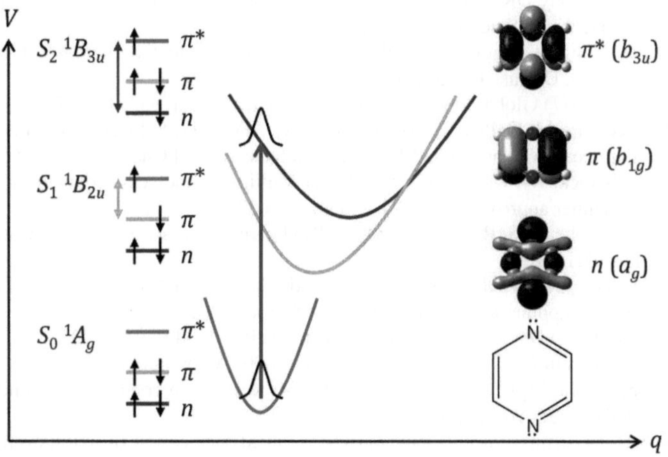

Fig. 4.11 Conical intersection, molecular orbitals, and dominant configurations involved in the *diabatic* $^1B_{3u}$ and $^1B_{2u}$ excited states of pyrazine (the convention for the body-fixed axes of $\mathcal{D}_{2h}$ are such that at the ground-state equilibrium geometry the molecule lies in the (y, z)-plane, with both nitrogen atoms on the z-axis)

and B_{2u} irreducible representations, respectively, for other geometries $\boldsymbol{q}$ as long as $\mathcal{D}_{2h}$ symmetry is preserved. This means that they preserve their dominant electronic configurations [closed shells]$b_{2u}^2 b_{2g}^1$ and [closed shells]$b_{2u}^1 b_{2g}^2$ at all geometries.

Another well-known example is pyrazine [44] for which the photoabsorption spectrum is affected by a conical intersection between the $^1B_{3u}$ and $^1B_{2u}$ excited states. Their dominant configurations are [closed shells]$n^1\pi^{*1}$ and [closed shells]$\pi^1\pi^{*1}$, respectively, where the n molecular orbital corresponds to lone pairs on the N-nuclei while the π and π^* molecular orbitals involve the π-system of the conjugated ring. This is summarised on Fig. 4.11.

References

1. Michl J, Bonacic-Koutecky V (1990) Electronic aspects of organic photochemistry. Wiley, New York
2. Klessinger M, Michl J (1994) Excited states and photochemistry of organic molecules. VCH, New York
3. Turro NJ, Ramamurthy V, Scaiano JC (2009) Principles of molecular photochemistry: an introduction. University Science Books, Sausalito, CA
4. Olivucci M (2005) Computational photochemistry. Elsevier, Amsterdam
5. Bersuker IB (2006) The Jahn-Teller effect. Cambridge University Press, Cambridge
6. Köppel H, Yarkony DR, Barentzen H (eds) (2009) Springer series in chemical physics: the Jahn-Teller effect: fundamentals and implications for physics and chemistry, vol 97. Springer, Heidelberg
7. Baer M (2006) Beyond Born-Oppenheimer: electronic nonadiabatic coupling terms and conical intersections. Wiley, Hoboken, NJ
8. Domcke W, Yarkony DR, Köppel H (eds) (2004) Conical intersections: electronic strucutre, dynamics and spectroscopy. World Scientific, New Jersey
9. Domcke W, Yarkony DR, Köppel H (eds) (2004) Conical intersections: theory, computation and experiment. World Scientific, New Jersey
10. Herzberg G, Longuet-Higgins HC (1963) Intersection of potential energy surfaces in polyatomic molecules. Discuss Faraday Soc 35:77
11. Davidson ER (1977) Global topology of triatomic potential surfaces. J Am Chem Soc 99:397
12. Desouter-Lecomte M, Galloy C, Lorquet JC, Pires MV (1979) Nonadiabatic interactions in unimolecular decay. V. Conical and Jahn-Teller intersections. J Chem Phys 71:3661
13. Köppel H, Domcke W, Cederbaum LS (1984) Multimode molecular dynamics beyond the Born-Oppenheimer approximation. Adv Chem Phys 57:59
14. Atchity GJ, Xantheas SS, Ruendenberg K (1991) Potential energy surfaces near intersections. J Chem Phys 95:1862
15. Lengsfield BH, Yarkony DR (1992) Non-adiabatic interactions between potential energy surfaces: theory and applications. Adv Chem Phys 82 (part 2), 1
16. Yarkony DR (1996) Diabolical conical intersections. Rev Mod Phys 68:985
17. Robb MA, Garavelli M, Olivucci M, Bernardi F (2000) A computational strategy for organic photochemistry. In: Lipkowitz KB, Boyd DB (eds) Reviews in computational chemistry, vol 15. Wiley-VCH, New York, pp 87
18. Robb MA (2014) In this molecule there must be a conical intersection. In: Williams IH, Williams NH (eds) Advances in physical organic chemistry, vol 48. Academic Press, pp 189
19. Worth GA, Cederbaum LS (2004) Beyond Born-Oppenheimer: conical intersections and their impact on molecular dynamics. Ann Rev Phys Chem 55:127
20. Worth GA, Meyer H-D, Köppel H, Cederbaum LS (2008) Using the MCTDH wavepacket propagation method to describe multimode non-adiabatic dynamics. Int Rev Phys Chem 27:569

21. Lasorne B, Worth GA, Robb MA (2011) Excited-state dynamics. WIREs: Comput Mol Sci 1:460
22. Alon OE, Cederbaum LS (2003) Hellmann-Feynman theorem at degeneracies. Phys Rev B 68:033105
23. Berry MV (1984) Quantal phase factors accompanying adiabatic changes. Proc R Soc A 392:45
24. Mead CA (1992) The geometric phase in molecular systems. Rev Mod Phys 64:51
25. Döscher M, Köppel H, Szalay PG (2002) Multistate vibronic interactions in the benzene radical cation. I. Electronic structure calculations. J Chem Phys 117:2645
26. Ogg RA, Polanyi M (1935) Diabatic reactions and primary chemiluminescence. Trans Faraday Soc 31:1375
27. Hellmann H, Syrkin JK (1935) Problem of anomalous small steric factors in chemical kinetics. Acta Physicochim URSS 2:433
28. Smith FT (1969) Diabatic and adiabatic representations for atomic collision problems. Phys Rev 179:111
29. Baer M (1975) Adiabatic and diabatic representations for atom-molecule collisions: treatment of the collinear arrangement. Chem Phys Lett 35:112
30. Baer M (1976) Adiabatic and diabatic representations for atom-diatom collisions: treatment of the three-dimensional case. Chem Phys 15:49
31. Mead CA, Truhlar DG (1982) Conditions for the definition of a strictly diabatic electronic basis for molecular systems. J Chem Phys 77:6090
32. Desouter-Lecomte M, Dehareng D, Leyhnihant B, Praet MT, Lorquet AJ, Lorquet JC (1985) Nonadiabatic unimolecular reactions of polyatomic molecules. J Phys Chem 89:214
33. Tully JC (1990) Molecular dynamics with electronic transitions. J Chem Phys 93:1061
34. Spiegelmann F, Malrieu JP (1984) The use of effective-Hamiltonians for the treatment of avoided crossings. 1. Adiabatic potential curves. J Phys B 17:1235
35. Spiegelmann F, Malrieu JP (1984) The use of effective-Hamiltonians for the treatment of avoided crossings. 2. Nearly diabatic potential curves. J Phys B 17:1259
36. Desouter-Lecomte M, Dehareng D, Lorquet JC (1987) Constructing approximately diabatic states from LCAO-SCF-CL calculations. J Chem Phys 86:1429
37. Cimiraglia R, Malrieu JP, Persico M, Spiegelmann F (1985) Quasi-diabatic states and dynamical couplings from ab initio CI calculations—a new proposal. J Phys B 18:3073
38. Gadea FX, Pelissier M (1990) Approximately diabatic states—a relation between effective Hamiltonian techniques and explicit cancellation of the derivative coupling. J Chem Phys 93:545
39. Pacher T, Cederbaum LS, Köppel H (1988) Approximately diabatic states from block diagonalization of the electronic Hamiltonian. J Chem Phys 89:7367
40. Pacher T, Cederbaum LS, Köppel H (1993) Adiabatic and quasidiabatic states in a gauge theoretical framework. Adv Chem Phys 84:293
41. Werner H-J, Meyer W (1981) MCSCF study of the avoided curve crossing of the two lowest Σ_1^+ states of LiF. J Chem Phys 74:5802
42. Hirsch G, Buenker RJ, Petrongolo C (1990) Ab initio study of NO_2. 2. Nonadiabatic coupling between the two lowest $^2A'$ states and the construction of a diabatic representation. Mol Phys 70:835
43. Cattarius C, Worth GA, Meyer H-D, Cederbaum LS (2001) All mode dynamics at the conical intersection of an octa-atomic molecule: multi-configuration time-dependent Hartree (MCTDH) investigation on the butatriene cation. J Chem Phys 115:2088
44. Raab A, Worth G, Meyer H-D, Cederbaum LS (1999) Molecular dynamics of pyrazine after excitation to the S_2 electronic state using a realistic 24-mode model Hamiltonian. J Chem Phys 110:936

Chapter 5
Choosing the Set of Coordinates for the Nuclei

Once the potential energy surfaces have been calculated, the Schrödinger equation for the nuclei has to be solved. Within the Born-Oppenheimer approximation, the nuclear Hamiltonian operators simply reads:

$$H = (T^{nu}(\boldsymbol{q}, \boldsymbol{\Theta}) + V_m(\boldsymbol{q})) , \tag{5.1}$$

with $T^{nu}(\boldsymbol{q}, \boldsymbol{\Theta})$ the nuclear kinetic energy operator and $V_m(\boldsymbol{q})$ the PES of the working electronic state. $T^{nu}(\boldsymbol{q}, \boldsymbol{\Theta})$ and $V_m(\boldsymbol{q})$ were defined in Chap. 3. If the Born-Oppenheimer approximation is no longer valid, several coupled electronic states have to be taken into. As explained in Sect. 4.3, an electronic diabatic representation must be introduced and the potential coupling, V_{12}^{dia}, must be included. In the case of two electronic states only, the nuclear Hamiltonian operator thus reads

$$H = T^{nu}(\boldsymbol{q}, \boldsymbol{\Theta})\,\mathbf{1} + \begin{bmatrix} V_1^{dia} & V_{12}^{dia} \\ V_{12}^{dia} & V_2^{dia} \end{bmatrix} . \tag{5.2}$$

Before solving the nuclear Schrödinger equations, one key issue in molecular quantum dynamics is the choice of the $3N$-6 internal nuclear coordinates, $\boldsymbol{q}$, and the three Euler angles, $\boldsymbol{\Theta}$, introduced in Chap. 3. $\boldsymbol{q}$ describe the shape of the molecule (the molecular geometry) and $\boldsymbol{\Theta}$ parametrize the BF frame and thus the overall rotation of the molecule.

A well-adapted set of coordinates leads to a nuclear Hamiltonian operator that is as separable as possible, i.e. that reduces as much as possible the coupling terms between the coordinates [1]. If one set of coordinates tends to increase these coupling terms, the number of basis functions used to solve the nuclear Schrödinger equation will increase dramatically. In practice, a set of coordinates that makes the Hamiltonian operator completely separable is rarely found. We present in this chapter several families of nuclear coordinates often used in quantum dynamics.

F. Gatti et al., *Applications of Quantum Dynamics in Chemistry*,
Lecture Notes in Chemistry 98, DOI 10.1007/978-3-319-53923-2_5

5.1 Normal Coordinates

We first show that *when only small-amplitude motions occur around a given equilibrium geometry, molecules can be considered as vibrating in a quasi-harmonic way*. The well-known rectilinear normal coordinates [2] then make the Hamiltonian operator almost separable (it is exactly separable for infinitesimal displacements). Thus, simple harmonic motions can serve as a mathematical model to describe the vibrations of the molecule and *the use of normal coordinates will be optimal in this low-energy domain.*

For an N-atom molecular system let $\tilde{m}_\alpha$, and $\tilde{X}^\alpha$, $\tilde{Y}^\alpha$, and $\tilde{Z}^\alpha$ ($\alpha = 1, \ldots, N$) be, respectively, the mass and the Laboratory-Fixed (LF) Cartesian coordinates of the α-th atom. Let us define μ and X^μ as follows: ($\mu = 1, \ldots, 3N$) and for ($\alpha = 1, \ldots, N$)

$$\begin{aligned} X^\mu &= \tilde{X}^\alpha \,;\, \text{if } \mu = 3\alpha - 2\,, \\ X^\mu &= \tilde{Y}^\alpha \,;\, \text{if } \mu = 3\alpha - 1\,, \\ X^\mu &= \tilde{Z}^\alpha \,;\, \text{if } \mu = 3\alpha\,, \end{aligned} \tag{5.3}$$

and $m_\mu = m_{\mu+1} = m_{\mu+2} = \tilde{m}_\alpha$.

Let us work within the Born-Oppenheimer approximation and in the electronic ground state and assume that there is an equilibrium geometry. For small values of the displacements around this equilibrium geometry, the potential energy surface of the electronic ground state, V, may be expressed as a Taylor series in the displacements X_i:

$$V = V_0 + \frac{1}{2} \sum_{i,j=1}^{3N} \left(\frac{\partial^2 V}{\partial X_i \partial X_j}\Big|_0 \right) X_i X_j + \text{higher order terms.} \tag{5.4}$$

X_i denote the ordinary $3N$ Cartesian coordinates of the N nuclei referred to a laboratory-fixed frame minus their values at the equilibrium geometry. If we choose the zero of energy so that the energy of the equilibrium configuration is zero, V_0 can be eliminated. Let us now introduce the mass weighted coordinates, x_i, such as

$$x_i = \sqrt{m_i} X_i\,, \tag{5.5}$$

with m_i the mass of the nucleus being displaced along X_i because the weighted coordinates simplify the expression for the kinetic energy. For sufficiently small amplitudes of vibration, the higher terms in the potential energy surface can be neglected, and the Hamiltonian operator can be written as:

$$H = \frac{1}{2} \sum_{i,j=1}^{3N} \left(\frac{\partial^2 V}{\partial x_i \partial x_j}\Big|_0 \right) x_i x_j - \frac{\hbar^2}{2} \sum_{i=1}^{3N} \frac{\partial^2}{\partial x_i^2}\,. \tag{5.6}$$

The difficult terms when solving the time-independent or time-dependent Schrödinger equation are the cross terms in the potential, i.e. the terms with $i \neq j$. However, within the harmonic approximation it is possible to find a linear combination of Cartesian coordinates that leads to a Hamiltonian operator in which there are no cross terms. Indeed, diagonalising the matrix $\boldsymbol{F}$, where

$$F_{ij} = \left(\frac{\partial^2 V}{\partial x_i \partial x_j}\bigg|_0\right) = \left(\frac{1}{m_i m_j}\right)^{1/2} \left(\frac{\partial^2 V}{\partial X_i \partial X_j}\bigg|_0\right) \tag{5.7}$$

gives $3N$ eigenvalues λ_l and $3N$ eigenvectors $\{(c_{1l}, \ldots, c_{3Nl}); l = 1, \ldots, 3N\}$ defining $3N$ new coordinates: $Q_l = \sum_{i=1}^{3N} c_{il} x_i$. For non-linear molecules, six eigenvalues are zero and correspond to the overall translation and the overall rotation of the molecule (for linear molecules, only five eigenvalues are equal to zero). If we remove the terms corresponding to overall translation and overall rotation, the Hamiltonian operator has the following form in terms of the $3N$-6 ($3N$-5 for linear molecules) remaining eigenvectors [3]:

$$H = \frac{1}{2} \sum_{l=1}^{3N-6} \lambda_l \, Q_l^2 - \frac{\hbar^2}{2} \sum_{l=1}^{3N-6} \frac{\partial^2}{\partial Q_l^2} \,. \tag{5.8}$$

λ_l is also equal to $\left(\frac{\partial^2 V}{\partial Q_l^2}\Big|_0\right)$. This new expression of the operator has the important property to be separable in the set of coordinates Q_l with $l = 1, \ldots, 3N-6$, i.e. it is a sum of one-dimensional operators. Q_l are said to be the normal coordinates of the system. The corresponding normal modes of vibration, thus vibrate in an independent way. As for any separable Hamiltonian operator, the eigenfunctions, Ψ_m^{vib} are simply a product of the eigenfunctions of each mode:

$$\Psi_m^{vib} = \prod_{l=1}^{3N-6} \Psi_{v_l}(Q_l) \,, \tag{5.9}$$

with[1]

$$\left(\frac{1}{2}\lambda_l \, Q_l^2 - \frac{\hbar^2}{2} \frac{\partial^2}{\partial Q_l^2}\right) \Psi_{v_l}(Q_l) = E_{v_l} \Psi_{v_l}(Q_l) \,. \tag{5.10}$$

Equation (5.10) corresponds to the well-known harmonic oscillator problem with the eigenvalues $E_{v_l} = \hbar\omega_l(v_l + \frac{1}{2})$ and $\omega_l = \lambda_l^{1/2}$. $v_l = 0, 1, 2, \ldots$ is the vibrational quantum number. The eigenfunctions are the Hermite functions. The total vibrational energy is the sum over the eigenvalues of the one-dimensional problems, i.e.

$$E_m^{vib} = E(v_1, v_2, v_3, \ldots) = \hbar\omega_1(v_1 + \frac{1}{2}) + \hbar\omega_2(v_2 + \frac{1}{2}) + \hbar\omega_3(v_3 + \frac{1}{2}) + \ldots \,. \tag{5.11}$$

[1] This equation appears as a very particular case of Eq. (3.52) of Sect. 3.2.3.

Normal coordinates are optimal coordinates for small vibrations since they lead to a Hamiltonian operator without any cross term between the coordinates such as $K_{ij} Q_i Q_j$ with $i \neq j$, i.e. without any *correlation* between the coordinates. The ($3N$-6) quantum mechanical problem simplifies into $3N$-6 quantum mechanical problems of one dimension. The same would be true for the time-dependent Schrödinger equation: the solution of the latter with the Hamiltonian operator of Eq. (5.8) is also a product of one-dimensional time-dependent functions. Solving the vibrational Schrödinger equation (time-dependent or time-independent) using Cartesian coordinates as in Eq. (5.6) would be much more complicated. We arrive at the important conclusion that *a set of coordinates that makes the Hamiltonian operator as separable as possible greatly simplifies the resolution of the Schrödinger equation. Such a set of coordinates is said to be "adapted to the physics of the problem".*

For most physical situations, it is not possible to find a set of coordinates that makes the Hamiltonian operator separable. However, *the "correlation", i.e. the numerical amplitude of the cross terms, can be much larger in one set of coordinates compared to another. The former is said to be an unsuitable set of coordinates since it leads to artificial correlation and will drastically complicate the numerical resolution of the Schrödinger equation.* At this level, the reader could think that the solution would be to find for each process the "optimal" set of coordinates that minimizes correlation. Unfortunately, the situation is more complicated since the "optimal" set of coordinates can lead to a kinetic energy operator that is so complicated or to a number of terms in the potential that is so large that the advantage of using this set of coordinates is lost. Thus, in practice, one simply tries to find a set of coordinates that leads to moderate correlation.

Some rules of thumb can be brought up. As already said, the normal coordinates are optimal in the low-energy domain of systems with a well-defined single minimum and this is the reason why normal coordinates play a central role in traditional infrared spectroscopy [2, 4, 5]. But at higher excitation energies, vibrations become more marked by anharmonicity. Thus, under these conditions it is no longer possible to neglect the "higher terms" in the Taylor expansion of Eq. (5.4). The anharmonicity can be treated first as a perturbation of the harmonic model and a description in normal coordinates remains pertinent as a "zero-order" description: in particular, some additional terms can be added to Eq. (5.11) [5]. In a normal mode of vibration, all the atoms of the molecules often move in a concerted way. In other words, the normal mode generally does not correspond to the vibration of, let us say, only two atoms, but of all the atoms at the same time.

5.2 Examples of Valence and Jacobi Coordinates for Tri-atomic Systems

Higher in energy, the situation can change and some chemical bonds may vibrate in a more independent way: some vibrations are more localized and this situation is

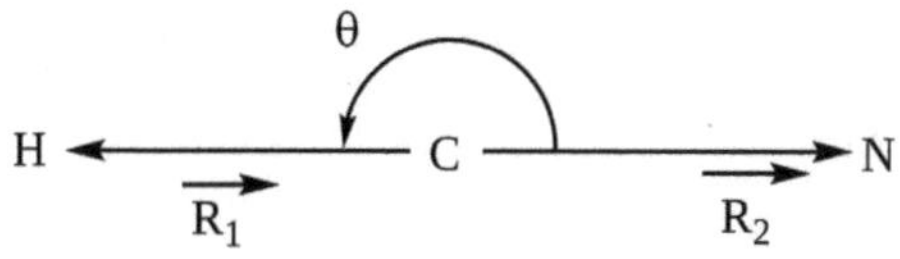

Fig. 5.1 Valence coordinates for HCN. R_1 is the length of the CH bond, R_2 is the length of the CN bond and θ is the angle between the two chemical bonds

sometimes referred to as *the local-mode regime*[2] [6]. In the latter case, the distances between the atoms and the angles between the chemical bonds may describe more naturally the physical situation than normal coordinates [7]. It is thus necessary to switch from a description in terms of *rectilinear coordinates*, i.e. coordinates that are linear combinations of Cartesian coordinates such as normal coordinates, to a description in terms of *curvilinear* coordinates, i.e. coordinates involving angles and that cannot be expressed as linear combinations of Cartesian coordinates (with coefficients that are constant). More generally speaking, it is natural to think that curvilinear coordinates are better-adapted to describe *motions of large amplitudes*. For instance, it is natural to think that a rotation within a molecule is better-described with angles than with Cartesian coordinates. *The coordinates involving the distances between the atoms and the angles between the chemical bonds are often called internal or valence coordinates.* The curvilinear coordinates have the disadvantage that they lead to complicated expression of the nuclear kinetic energy operator (See Chap. 6).

Now even higher in energy the molecular system can explore two or several minima. For those cases, valence coordinates may no longer be adapted and *Jacobi coordinates* that describe the motion of one atom or the center of mass of a group of atoms with respect to the center of mass of another group of atoms can be more adapted. Let us consider an example to illustrate our point: the molecule of hydrogen cyanide as shown on Fig. 5.1. The depicted geometry corresponds to the most stable isomer. Low in energy, normal coordinates are optimal but, higher in energy and below the transition state between the two isomers, the description with the curvilinear valence coordinates of Fig. 5.1 is more adapted: R_1 is the length of the CH bond, R_2 is the length of the CN bond and θ is the angle between the two chemical bonds. The introduction of the angle θ allows us to describe in a more natural way the bending vibration of the system when the harmonic approximation becomes too crude. Now, even higher in energy, the HCN/CNH isomerization can occur and the descriptions in normal and curvilinear valence coordinates become inadequate to the dynamics on the CNH side. Generally speaking, describing isomerizing species must allow the localization of certain atoms or centers of mass of atomic groups with respect to one another. For HCN/CNH, such coordinates are the (curvilinear) Jacobi ones as shown in Fig. 5.2: R_2 is the length of the CN vector but R_1 is now the length of the vector joining the center of mass of CN to the H atom and θ is now the angle

[2] We will describe in detail the transition from the normal-mode regime to the local-mode regime in H_2O in Chap. 9.

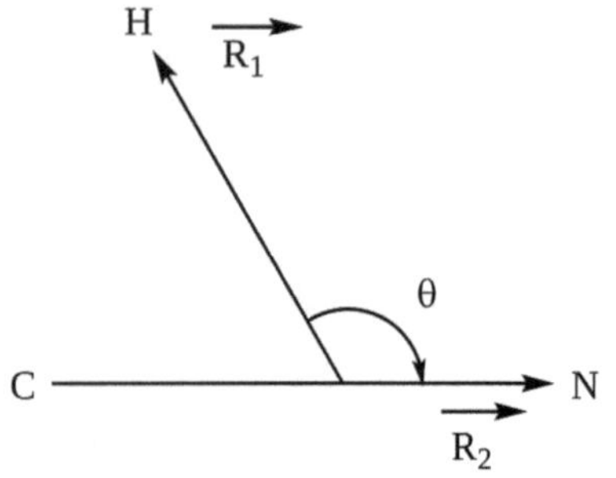

Fig. 5.2 Jacobi coordinates for HCN/CNH. R_2 is the length of the CN vector, R_1 is the length of the vector joining the center of mass of CN to the H atom. θ is the angle between these two vectors

between these two vectors. Such a set of coordinates is less adapted to describe the vibrations in the HCN or the CNH potential wells but is globally more adapted to describe the motion of the H atom from one isomer to the other.

The description of the vibrations low in energy in the CNH potential well would also require normal coordinates but the latter are different from the normal coordinates for the HCN potential since the two sets of coordinates are not defined with respect to the same geometry. In the same manner, higher in energy but below the isomerization barrier, the valence coordinates linking the nitrogen atom to the carbon atom and to the H atom would be more adapted. In other words, we would use valence coordinates that are different from those depicted in Fig. 5.1 since the latter link the carbon atom to the nitrogen atom and to the H atom. Again, only Jacobi coordinates allow us to adequately describe the isomerization from one well to the other. The different sets of coordinates for HCN/CNH are summarized on Fig. 5.3.

Another example may be helpful. The description of the vibrations of NOCl in its electronic ground state is similar to the HCN case. However, high in energy, the molecule does not isomerize but dissociates into NO + Cl. In the same manner, the first electronic excited singlet state, S_1, is purely dissociative, i.e. after absorption of light in the 650–420 nm domain, the transition from the electronic ground state to S_1 leads again to NO + Cl. The study of the vibrations of NOCl may be achieved in normal coordinates, low in energy, or in valence coordinates, higher in energy, but a set of Jacobi coordinates as shown in Fig. 5.4 fits much better the motion of dissociation since it describes the motion of Cl with respect to the center-of-mass of NO and then to NO globally and to the nitrogen atom only. After the breaking of the NCl bond, referring the motion of Cl with respect to the nitrogen atom is artificial and will lead to more numerical correlation in the Hamiltonian operator.

Another example of valence coordinates is depicted in Fig. 5.5 for water. As aforementioned the valence coordinates are well-adapted to describe the vibrations of the system especially above the low-amplitude region.

A third example of Jacobi coordinates is given in Fig. 5.6 for the $H + H_2$ scattering. We denote the three H atoms $H^{(a)}$, $H^{(b)}$ and $H^{(c)}$ to distinguish them. The three internal coordinates are the two lengths of the vectors and the angle between the two vectors.

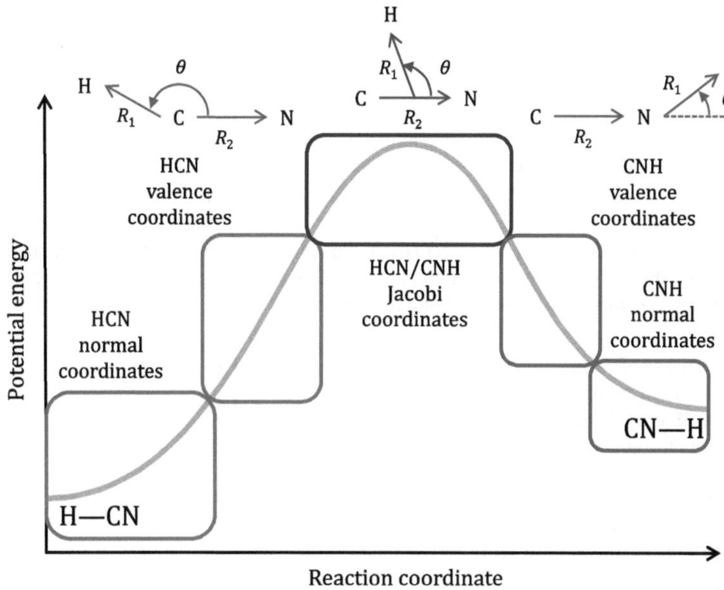

Fig. 5.3 Different sets of coordinates for HCN/CNH: low in energy the normal coordinates are optimal. They are different for HCN and for CNH. Higher in energy, in the local-mode regime, the valence coordinates are better adapted. The valence coordinates are different in the two potential wells. For the description of the isomerization, the Jacobi coordinates can be used. In particular, θ, provides a reaction coordinate for the isomerization process

Fig. 5.4 Jacobi coordinates for NOCl dissociation: $\vec{R}_1$ ($\boldsymbol{R}_1$ in the text) is the vector joining the N atom to the O atom, $\vec{R}_2$ ($\boldsymbol{R}_2$ in the text) is the vector joining the center of mass of NO to the Cl atom

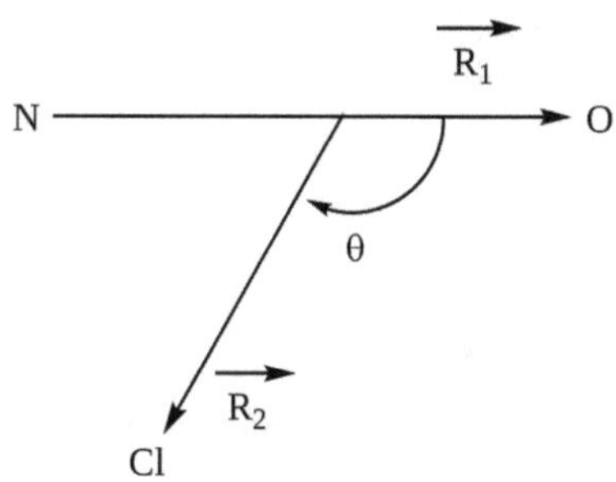

The beginning of the collision $H^{(a)} + H^{(b)}H^{(c)}$ is perfectly described by the set of Jacobi coordinates of Fig. 5.6a since the corresponding coordinates are adapted to the description of the motion of $H^{(a)}$ with respect to $H^{(b)}H^{(c)}$. Indeed, vector $\boldsymbol{R}_1$ describes the rotation and the vibration of the $H^{(b)}H^{(c)}$ molecule and $\boldsymbol{R}_2$ parameterizes the relative position of $H^{(a)}$ with respect to the $H^{(b)}H^{(c)}$ molecule. R_2 provides a single "reaction" coordinate. However, if $H^{(a)} + H^{(b)}H^{(c)}$ leads to $H^{(a)}H^{(b)} + H^{(c)}$ the first set of coordinates will yield artificial correlation in the corresponding exit

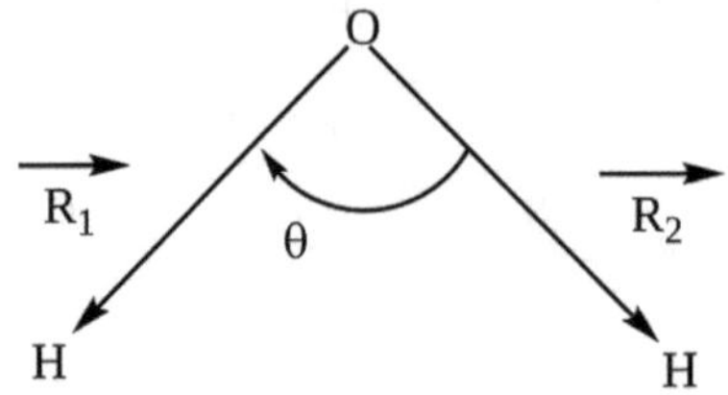

Fig. 5.5 Valence coordinates for water. $\vec{R}_1$ ($\boldsymbol{R}_1$ in the text) and $\vec{R}_2$ ($\boldsymbol{R}_2$ in the text) are the valence vectors joining the oxygen atom to the two H atoms. The three internal coordinates are the angle between the two vectors, θ, and the two bond lengths R_1 and R_2. $\theta \in [0, \pi]$

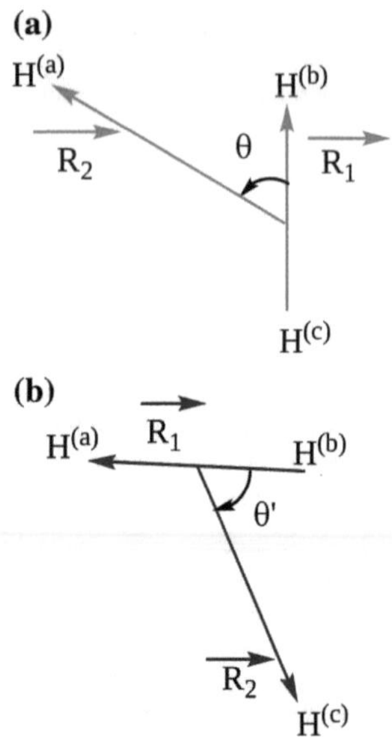

Fig. 5.6 Two sets of Jacobi coordinates for describing the collision $H + H_2$. In (**a**) $\vec{R}_2$ starts from the center of mass of $H^{(b)}H^{(c)}$ (at the middle of $H^{(b)}H^{(c)}$); in (**b**) $\vec{R}_2$ starts from the center of mass of $H^{(a)}H^{(b)}$ (at the middle of $H^{(b)}H^{(c)}$). The first set of coordinates, **a** is well-adapted for a inelastic scattering $H^{(a)} + H^{(b)}H^{(c)} \rightarrow H^{(a)} + H^{(b)}H^{(c)}$. However, if a reactive collision leads to $H^{(a)}H^{(b)} + H^{(c)}$, the second set of Jacobi coordinates, **b** is better-adapted to describe the corresponding exit channel

channel. The set of coordinates of Fig. 5.6b would be better to describe this exit channel. If the collision leads to $H^{(a)}H^{(c)} + H^{(b)}$, the best set of Jacobi coordinates would be the third one, i.e. with $\boldsymbol{R}_1$ joining $H^{(a)}$ to $H^{(c)}$ (not shown). We see that no single set of Jacobi coordinates can describe all the different channels without strong artificial correlation. Several solutions have been proposed to overcome this difficulty such as using a set of "hyperspherical" coordinates that treat all the different channels in a more symmetric way and partly reduce artificial correlation [8]. In practice, the resolution of the Schrödinger equation for the nuclei and the dynamics is often performed in the set of Jacobi coordinates of Fig. 5.6a since, for such small systems, modern computers allow one to converge the cross sections even with strong couplings in the exit channels.

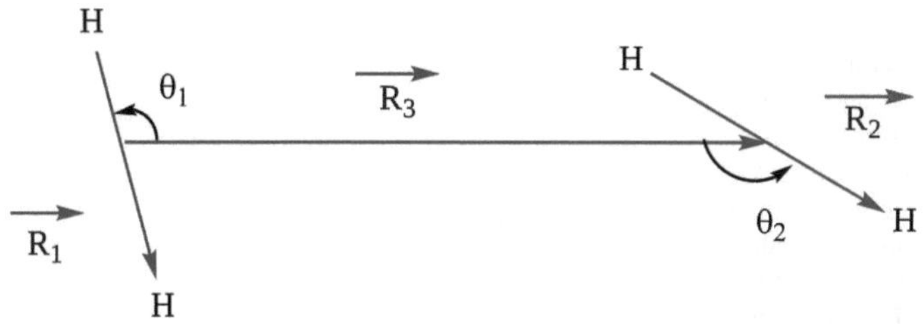

Fig. 5.7 Jacobi coordinates for the inelastic $H_2 + H_2$ scattering. $\vec{R}_3$ is the vector from the center of mass of the first H_2 molecule to the center of mass of the second H_2 molecule. The six internal coordinates are $R_1, R_2, R_3, \theta_1, \theta_2$ and the dihedral angle φ (not shown on the Figure) between the $(\vec{R}_3, \vec{R}_1)$ and $(\vec{R}_3, \vec{R}_2)$ half-planes. θ_1 and $\theta_2 \in [0, \pi]$ and $\varphi \in [0, 2\pi[$

5.3 Examples of Internal Coordinates for Tetra-atomic Systems

For a tetra-atomic system, the six internal coordinates could be the three lengths of the vectors, the two angles between one vector and the other two such as θ_1 and θ_2 in Figs. 5.7 and 5.9. The sixth coordinate could be a dihedral angle between the two half-planes $(\boldsymbol{R}_3, \boldsymbol{R}_1)$ and $(\boldsymbol{R}_3, \boldsymbol{R}_2)$. For non-reactive tetra-atomic AB + CD collisions, a very natural set of coordinates [9, 10] are the Jacobi ones depicted in Fig. 5.7 for $H_2 + H_2$. The vectors $\boldsymbol{R}_1$ and $\boldsymbol{R}_2$ are adapted to describe the rotation and the vibration of the two diatomic molecules and the third vector, $\boldsymbol{R}_3$, parameterizes the relative position of one molecule with respect to the other. When the molecules are separated, the Hamiltonian operator becomes separable. The correlation appears only when they enter into collision and the coordinate $\boldsymbol{R}_3$ provides a single "reaction" coordinate. The situation is more intricate when the collision becomes reactive [11–14]: for instance, $OH + H_2 \rightarrow H_2O + H$. Then, several dissociation channels are accessible as in the case of $H + H_2$. To overcome this difficulty, it has been proposed to use several sets of coordinates depending on the region of space: $\boldsymbol{R}_1$, $\boldsymbol{R}_2$ and $\boldsymbol{R}_3$ of Fig. 5.8a for the reactant region of space and $\boldsymbol{R}_1$, $\boldsymbol{R}_4$ and $\boldsymbol{R}_5$ of Fig. 5.8a for the product region of space, with A = O, C = H, B = H, and D = H [15, 16]. However, since it is rather difficult to switch from one set of coordinates to the other in the resolution of the Schrödinger equation for the nuclei, the dynamics is often performed in the set of coordinates $\boldsymbol{R}_1$, $\boldsymbol{R}_2$ and $\boldsymbol{R}_3$ of Fig. 5.8a only.

Let us now conclude with the trans-cis isomerization of HONO. The system explores two minima with a transition state between them. However, contrarily to the isomerization of HCN, the Jacobi coordinates are not the right solution for the problem since the system moves from one isomer to the other through a twisting motion that is better described by the torsion angle between the three successive chemical bonds. In particular, the Jacobi vectors as in Fig. 5.7 are no longer adapted:

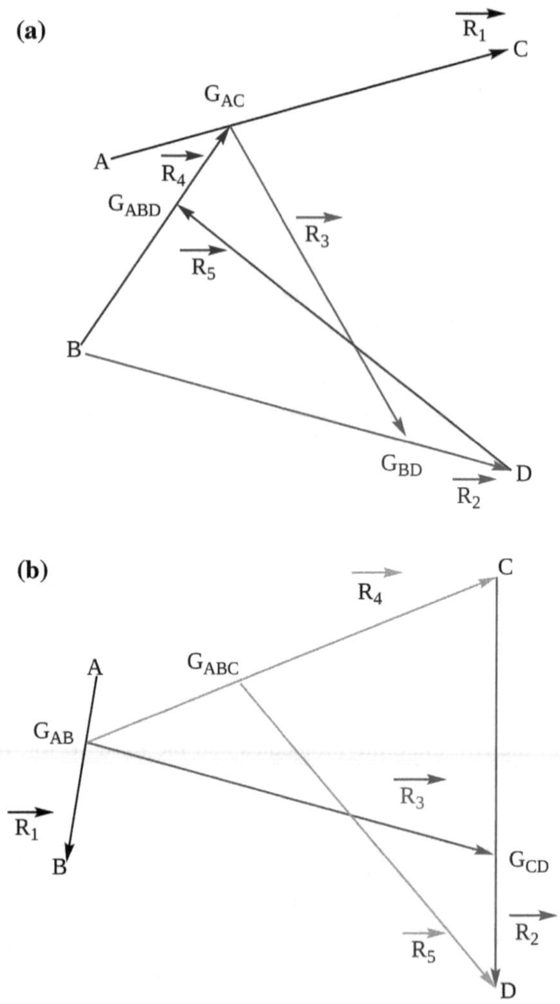

Fig. 5.8 Four sets of Jacobi vectors for a tetra-atomic system: $\vec{R}_1$, $\vec{R}_2$ and $\vec{R}_3$ or $\vec{R}_1$, $\vec{R}_4$ and $\vec{R}_5$ for both figures (**a**) and (**b**) (the vectors $\vec{R}_i$ appear in bold face $\boldsymbol{R}_i$ in the text). In the present Figure, A is heavier than C and D is heavier than B. **a** $\vec{R}_3$ is the vector from the center of mass of AC, G_{AC}, to the center of mass of BD, G_{BD}; $\vec{R}_4$ is the vector from B to the center of mass of AC; $\vec{R}_5$ is the vector from D to the center of mass of ABC, G_{ABD}. **b** $\vec{R}_3$ is the vector from the center of mass of AB, G_{AB}, to the center of mass of CD, G_{CD}; $\vec{R}_4$ is the vector from AB, G_{AB}, to the center of mass of CD, G_{CD}; $\vec{R}_5$ is the vector from the center of mass of ABC, G_{ABC}, to the D atom. The directions and the order of the vectors can be changed

the corresponding dihedral angle would not provide the angle between the three successive chemical bonds. It becomes necessary to switch to a set of valence coordinates as depicted in Fig. 5.9. Indeed, the valence coordinates describe more correctly the chemical bonds in the system. The dihedral angle, φ, is now the angle between the $(\boldsymbol{R}_3, \boldsymbol{R}_1)$ and $(\boldsymbol{R}_3, \boldsymbol{R}_2)$ half-planes. The dihedral angle is the required angle since it is very similar to the reaction coordinate leading from one isomer to the other.

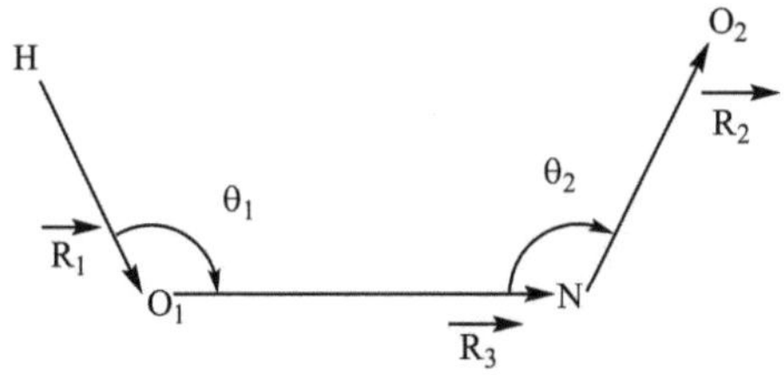

Fig. 5.9 Valence coordinates for the cis-trans isomerization of HONO. $\vec{R}_3$ is the vector from the central O_1 atom to the N atom, $\vec{R}_1$ is the OH vector and $\vec{R}_2$ is the final NO_2 vector. As for H_2 + H_2, the six internal coordinates are R_1, R_2, R_3, θ_1, θ_2, and the dihedral angle φ (not shown on the Figure) between the $(\vec{R}_3, \vec{R}_1)$ and $(\vec{R}_3, \vec{R}_2)$ half-planes. θ_1 and $\theta_2 \in [0, \pi]$ and $\varphi \in [0, 2\pi[$

5.4 Vector Parametrization of N-atom Molecular Systems

In the previous sections, several Jacobi or valence "vectors" have been introduced. Let us be more general and present the so-called "vector parametrization" of a molecular system that will be very useful to obtain kinetic energy operators in Chap. 6. As explained in Sect. 3.3.1, the configuration of a molecular system with N atoms can be described by $(N-1)$ relative position vectors, $\boldsymbol{R}_1$, $\boldsymbol{R}_2$, ..., $\mathbf{R}_{N-1}$. Adding the total LF center-of-mass vector, $\boldsymbol{R}^{CM}_{LF}$, these N vectors can be connected to the N LF-position vectors $\boldsymbol{r}^1_{LF}, \boldsymbol{r}^2_{LF}, \ldots, \boldsymbol{r}^N_{LF}$ by the following matrix relation[3]

$$\begin{bmatrix} \boldsymbol{R}_1 \\ \boldsymbol{R}_2 \\ \vdots \\ \boldsymbol{R}_{N-1} \\ \boldsymbol{R}^{CM}_{LF} \end{bmatrix} = \begin{bmatrix} & \boldsymbol{A} & \\ \frac{m}{M} & \frac{m_2}{M} \cdots & \frac{m_N}{M} \end{bmatrix} \begin{bmatrix} \boldsymbol{r}^1_{LF} \\ \boldsymbol{r}^2_{LF} \\ \vdots \\ \boldsymbol{r}^N_{LF} \end{bmatrix}, \tag{5.12}$$

where m_α $(\alpha = 1, \ldots, N)$ are the masses of the atoms, $M = \sum_{\alpha=1}^{N} m_\alpha$ and $\boldsymbol{A} = \left[A^i_\alpha\right]$ $(i = 1, \ldots, N-1, \alpha = 1, \ldots, N)$ is an $(N-1) \times N$ mass-dependent constant matrix satisfying the following constraints:

$$\sum_{\alpha=1}^{N} A^i_\alpha = 0 \ (i = 1, \ldots, N-1) \tag{5.13}$$

required to ensure the translational invariance of the relative position vectors.

To illustrate such a vector parametrization, we consider examples of valence and Jacobi vectors introduced before. It must be clear that other sets of vectors could be introduced, although Jacobi and valence vectors are the most frequently used. The valence vectors are sets of vectors that join one atom to another such as $\boldsymbol{R}_1$ and $\boldsymbol{R}_2$ for HCN and H_2O in Figs. 5.1 and 5.5 or $\boldsymbol{R}_1$, $\boldsymbol{R}_2$ and $\boldsymbol{R}_3$ for HONO in Fig. 5.9.

[3]Note that $\boldsymbol{r}$ does not refer to electronic coordinates here as in Sect. 3.

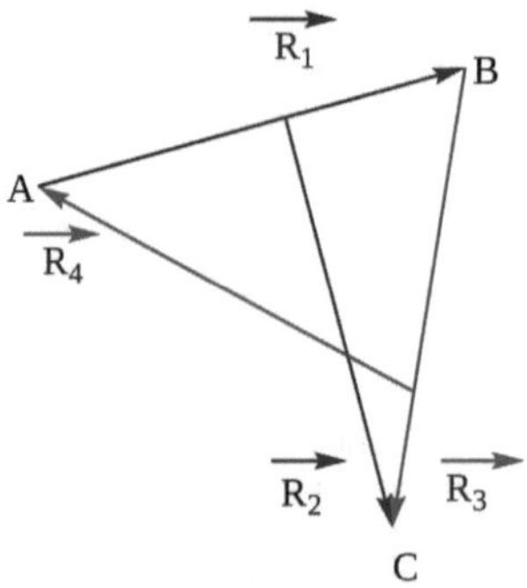

Fig. 5.10 Two sets of Jacobi vectors for a tri-atomic system (in *red* and *blue*). $\vec{R}_2$ is the vector from the center of mass of AB to the C atom. $\vec{R}_4$ is the vector from the center of mass of BC to the A atom. In the present picture, B is heavier than A and C is heavier than C. The corresponding three internal coordinates would be the angles between the two vectors in *red* or in *blue* and the lengths of the vectors

The definition of the Jacobi vectors is less straightforward and several "clustering" schemes are possible. Let us just consider the tri- and tetra-atomic cases to understand the schemes for constructing Jacobi vectors.

In Fig. 5.10, two sets of Jacobi vectors are depicted for a tri-atomic system: one Jacobi vector joins one atom to another and the second vector joins the center of mass of the first two atoms to the third one. The directions of the vectors can be changed. The corresponding three (= $3N$-6) internal coordinates could be the angle between the two vectors and the lengths of the two vectors. Figure 5.8 presents four sets of Jacobi vectors but for a tetra-atomic system. The first Jacobi vector, $\boldsymbol{R}_1$, joins one atom to the other. Then only two possibilities may occur: if the second vector, $\boldsymbol{R}_2$, joins the two remaining atoms then the third one, $\boldsymbol{R}_3$, must join the center of mass of the first diatomic system to the second one such as $\boldsymbol{R}_1$, $\boldsymbol{R}_2$, and $\mathbf{R}_3$ in Fig. 5.8a, b. Instead the second vector can join the third atom to the center of mass of the first diatomic system. Then the third Jacobi vector must be the vector joining the last atom to the center of mass of the triatomic system such as $\boldsymbol{R}_1$, $\boldsymbol{R}_4$, and $\boldsymbol{R}_5$ in Fig. 5.8a, b. Of course, these schemes can be generalized to any number of atoms [17].

We will see in Chap. 6 that Jacobi vectors, contrarily to valence vectors, lead to a simpler expression of the kinetic energy. Let us consider two examples of matrix $\boldsymbol{A}$ in Eq. (5.12). First, let us consider $H_2 + H_2$ parametrized by the set of three Jacobi vectors of Fig. 5.7, which is similar to the vectors $\boldsymbol{R}_1$, $\boldsymbol{R}_2$, and $\boldsymbol{R}_3$ of Fig. 5.8b. Let $\boldsymbol{r}_A$, $\boldsymbol{r}_B$, $\boldsymbol{r}_C$, and $\boldsymbol{r}_D$ be the position vectors from the center of mass G of the atoms A, B, C, and D. Then, we have

$$
\begin{aligned}
\boldsymbol{R}_1 &= \boldsymbol{r}_B - \boldsymbol{r}_A \\
\boldsymbol{R}_2 &= \boldsymbol{r}_D - \boldsymbol{r}_C \\
\boldsymbol{R}_3 &= \frac{m_C}{m_C + m_D}\boldsymbol{r}_C + \frac{m_D}{m_C + m_D}\boldsymbol{r}_D - \frac{m_A}{m_A + m_B}\mathbf{r}_A - \frac{m_B}{m_A + m_B}\boldsymbol{r}_B\,.
\end{aligned}
\tag{5.14}
$$

As regards valence coordinates, we choose as an example the HONO molecule as depicted in Fig. 5.9. We can use the following definitions (see Fig. 5.9):

$$\begin{aligned} \boldsymbol{R}_1 &= \boldsymbol{r}_{O_1} - \boldsymbol{r}_H \\ \boldsymbol{R}_2 &= \boldsymbol{r}_{O_2} - \boldsymbol{r}_N \\ \boldsymbol{R}_3 &= \boldsymbol{r}_N - \boldsymbol{r}_{O_1} \, . \end{aligned} \tag{5.15}$$

This results in six internal coordinates, R_1, R_2, R_3, θ_1, θ_2, and the dihedral angle φ. These coordinates are well-adapted to describe the cis/trans isomerization of the system as well as the other vibrations in each of the two potential wells corresponding to the two isomers.

5.5 Examples of BF Frames

In Sect. 3.3.3, the BF frame has been introduced in a general context. The explicit choice of the BF frame for a given molecular system is a difficult task. In principle, the BF frame, as the internal coordinates, must lead to moderate correlation in the Hamiltonian operator. More precisely, the BF frame must reduce numerically the "Coriolis coupling", i.e. the coupling between the Euler angles and the internal degrees of freedom. The minimization of the Coriolis coupling can be obtained systematically by the so-called "Eckart conditions" [18, 19]. However, the Eckart conditions are valid around one very specific geometry and the molecular system can display motions of large amplitude and explore large domains of space. In practice, as for the choice of the internal coordinates, one tries to find a BF frame that yields moderate Coriolis coupling. Two examples are provided in Fig. 5.11. Generally speaking, if z^{BF} is parallel to one of the $N-1$ vectors introduced in the previous section, it is often advantageous to take the vector with the largest associated reduced mass, since it can be proved that the Coriolis coupling is proportional to the inverse of this reduced mass.

In each case, the origin of the BF frame is the center of mass of the system, G. For HONO (see Fig. 5.11a), z^{BF} is parallel to $\boldsymbol{R}_3$, i.e. to the central ON bond. The (x^{BF}, z^{BF}), with $x^{BF} > 0$, half-plane contains $\boldsymbol{R}_2$ and thus the NO double bond. The atoms that play a role in the definition of the BF frame, i.e. the nitrogen atom and the oxygen atoms, are relatively heavy. Choosing $\boldsymbol{R}_1$ to define z^{BF} would result in a strong Coriolis coupling since the hydrogen atom is lighter. In other words, the rotation of the system is mainly carried by the ONO structure and it is thus logical to define the BF frame with respect to this subsystem. For the $H + H_2$ collision, see Fig. 5.11b, z^{BF} is parallel to $\boldsymbol{R}_2$, the Jacobi vector with the largest reduced mass and $\boldsymbol{R}_1$ is parallel to the (x^{BF}, z^{BF}), with $x^{BF} > 0$, half-plane.

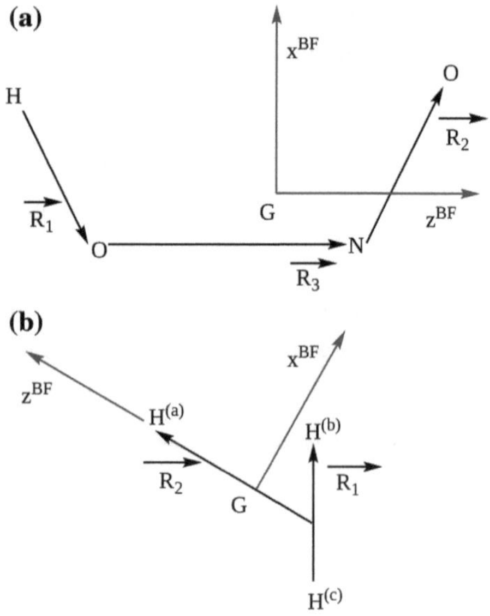

Fig. 5.11 Two possible definitions of the body-fixed frame for the isomerization of HONO (**a**) and for the $H + H_2$ collision (**b**). For HONO, the z^{BF} axis is parallel to the central ON bond and $\vec{R}_2$ lies parallel to the (x^{BF}, z^{BF}), with $x^{BF} > 0$, half-plane. For $H + H_2$, the z^{BF} axis is parallel to $\vec{R}_2$ and $\vec{R}_2$ lies parallel to the (x^{BF}, z^{BF}), with $x^{BF} > 0$, half-plane

References

1. Joubert-Doriol L, Lasorne B, Gatti F, Schröder M, Vendrell O, Meyer H-D (2012) Suitable coordinates for quantum dynamics: applications using the multiconfiguration time-dependent Hartree (MCTDH) algorithm. Comput Theor Chem 990:75
2. Wilson E, Decius J, Cross P (1955) Molecular vibrations. McGraw-Hill, New York
3. Atkins PW (1983) Molecular quantum mechanics, 2nd edn. OUP, Oxford, U.K
4. Herzberg GH (1945) Infrared and Raman Spectra. Van Nostrand Reihnold, Toronto
5. Herzberg G (1992) Molecular spectra and molecular structure. Krieger Pub Co
6. Miller WH, Handy NC, Adams JE (1980) Reaction path Hamiltonian for polyatomic molecules. J Chem Phys 72:99
7. Halonen L, Carrington T Jr (1988) Fermi resonances and local modes in water, hydrogen sulfide, and hydrogen selenide. J Chem Phys 88:4171
8. Launay JM, Dourneuf ML (1989) Hyperspherical close-coupling calculation of integral cross sections for the reaction $H+H_2 \rightarrow H_2+H$. Chem Phys Lett 163:178
9. Goldfield EM, Gray SK (2007) Quantum dynamics of chemical reactions. Adv Chem Phys 136:1
10. Gatti F, Otto F, Sukiasyan S, Meyer H-D (2005) Rotational excitation cross sections of para-H_2 + para-H_2 collisions. A full-dimensional wave packet propagation study using an exact form of the kinetic energy. J Chem Phys 123:174311
11. Clary DC (1991) Quantum reactive scattering of four-atom reactions with nonlinear geometry: $OH+H_2 \rightarrow H_2O+H$. J Chem Phys 95:7298
12. Zhang DH, Light JC (1996) A six-dimensional quantum study for atom-triatom reactions: the $H+H_2 \rightarrow H_2+OH$ reaction. J Chem Phys 104:4544
13. Zhang DH, Light JC (1996) Quantum state-to-state reaction probabilities for the $H+H_2O \rightarrow H_2$ +OH reaction in six dimensions. J Chem Phys 105:1291
14. Goldfield EM, Gray SK (2002) A quantum dynamics study of $H_2+OH \rightarrow H_2O+H$ employing Wu-Schatz-Lendvay-Fang-Harding potential function and a four-atom implementation of the real wave packet method. J Chem Phys 117:1604

15. Cvitas MT, Althorpe SC (2009) State-to-state reactive scattering using reactant-product decoupling. Phys Scr 80:048115
16. Cvitas MT, Althorpe SC (2011) State-to-state reactive scattering in six dimensions using reactant-product decoupling: $OH + H_2 \rightarrow H_2O + H$ ($J = 0$). J Chem Phys 134:024309
17. Hirschfelder JO (1969) Coordinates which diagonalize the kinetic energy of relative motion. Int J Quantum Chem 3S:17
18. Eckart C (1935) Some studies concerning rotating axes and polyatomic molecules. Phys Rev 47:552
19. Louck JD, Galbraith HW (1976) Eckart vectors, Eckart frames, and polyatomic molecules. Rev Mod Phys 48:69 (Coordinates)

Chapter 6
The Kinetic Energy Operator in Curvilinear Coordinates

As seen in the previous chapter, generalized curvilinear coordinates well-adapted to the description of a molecular process play an important role in molecular quantum dynamics. The present chapter is intended as a rather in-depth introduction to the classical kinetic energy and the quantum kinetic energy operator for the nuclei, denoted T and $\hat{T}$, respectively, in the following. The positions of the particles will be described by means of generalized curvilinear coordinates. By curvilinear coordinates we refer to coordinates that cannot be written as linear combinations of Cartesian coordinates with fixed coefficients. In the following sections, we follow the rather pedantic notation of the "upper" and "lower" indices as used in tensor analysis. Later on, however, this notational convention will be relaxed to avoid cumbersome expressions.

In Sect. 6.1, we recall the basic expressions, both *Lagrangian and Hamiltonian*, of the kinetic energy in *generalized curvilinear coordinates*. Section 6.2 is devoted to the *non-relativistic quantum kinetic energy operators* (KEO) and the various ways of writing them, with some emphasis on the less common topics of *normalization conventions* of the wave functions (Sect. 6.2.2) and a discussion of the *quasi-momentum operators* (Sect. 6.2.3). In the next four Sects. (6.3, 6.4, 6.5 and 6.6) applications to molecular physics are given, in particular to *polyspherical coordinates*, which play a central role in the present book. Finally, in the complements, some aspects of the *adjoint operators* are further developed and simple illustrations in the case of polar coordinates are given.

6.1 Classical Kinetic Energy

6.1.1 Classical Lagrangian Kinetic Energy

For a molecular system made of N atoms, the classical kinetic energy for the nuclei in Cartesian coordinates reads

F. Gatti et al., *Applications of Quantum Dynamics in Chemistry*,
Lecture Notes in Chemistry 98, DOI 10.1007/978-3-319-53923-2_6

$$T = \frac{1}{2}\sum_{\alpha=1}^{N} \tilde{m}_\alpha((\dot{\tilde{X}}^\alpha)^2 + (\dot{\tilde{Y}}^\alpha)^2 + (\dot{\tilde{Z}}^\alpha)^2) , \tag{6.1}$$

or

$$T = \frac{1}{2}\sum_{\mu=1}^{3N} m_\mu(\dot{X}^\mu)^2 , \tag{6.2}$$

where $\tilde{m}_\alpha$ and $\tilde{X}^\alpha$, $\tilde{Y}^\alpha$ and $\tilde{Z}^\alpha$ ($\alpha = 1, \ldots, N$) are, respectively, the mass and the Laboratory-Fixed (LF) Cartesian coordinates of the α-th atom, and μ and X^μ are defined in Eq. (5.3) of Sect. 5.1, i.e.

$$\begin{aligned} X^\mu &= \tilde{X}^\alpha \,; \text{if } \mu = 3\alpha - 2 , \\ X^\mu &= \tilde{Y}^\alpha \,; \text{if } \mu = 3\alpha - 1 , \\ X^\mu &= \tilde{Z}^\alpha \,; \text{if } \mu = 3\alpha . \end{aligned} \tag{6.3}$$

and

$$m_\mu = m_{\mu+1} = m_{\mu+2} = \tilde{m}_\alpha \tag{6.4}$$

Let us now express these Cartesian coordinates in terms of $3N$ *generalized curvilinear coordinates* $\boldsymbol{q} = (q^1, \ldots, q^i, \ldots, q^{3N})$ such that

$$X^\mu = X^\mu(q^1, \ldots, q^i, \ldots, q^{3N}) \;\; (\mu = 1, \ldots, 3N) , \tag{6.5}$$

in short,

$$\boldsymbol{X} = \boldsymbol{X}(\boldsymbol{q}) , \tag{6.6}$$

together with the inverse relations

$$q^i = q^i(X^1, \ldots, X^\mu, \ldots, X^{3N}) \;\; (i = 1, \ldots, 3N) , \tag{6.7}$$

in short,

$$\boldsymbol{q} = \boldsymbol{q}(\boldsymbol{X}) . \tag{6.8}$$

The $(3N \times 3N)$ *Jacobian matrix* corresponding to the coordinate change $\boldsymbol{X} \rightarrow \boldsymbol{q}$ is

$$\boldsymbol{J}_{X\rightarrow q} = [\frac{\partial X^\mu}{\partial q^i}] \;\; (\mu, i = 1, \ldots, 3N) , \tag{6.9}$$

μ and i being, respectively, the *row* and *column* indices and let

$$J_{X\rightarrow q} = |\,\mathrm{Det}\boldsymbol{J}_{X\rightarrow q}\,| , \tag{6.10}$$

denote the absolute value of its determinant. Since the coordinate change is assumed to be invertible, $J_{X\to q} \neq 0$.

Of course, the Jacobian matrix corresponding to the *inverse coordinate transformation* $\boldsymbol{q} \to \boldsymbol{X}$ is

$$\boldsymbol{J}_{q\to X} = [\frac{\partial q^i}{\partial X^\mu}] \ (i, \mu = 1, \ldots, 3N) , \tag{6.11}$$

μ and i being, respectively, the *column* and *row* indices. It is a well-known property of the Jacobian matrices that

$$\boldsymbol{J}_{q\to X} = \boldsymbol{J}^{-1}_{X\to q} , \tag{6.12}$$

and

$$J_{q\to X} = |\,\mathrm{Det}\boldsymbol{J}_{q\to X}\,| = \frac{1}{|\,\mathrm{Det}\boldsymbol{J}_{X\to q}\,|} = J^{-1}_{X\to q} . \tag{6.13}$$

Now, using the *chain rule* (see complement 3.6.3 for a quick reminder concerning the chain rule) for the time derivatives, the expressions of the Cartesian velocities $\dot{\boldsymbol{X}} = (\dot{X}^1, \ldots, \dot{X}^\mu, \ldots, \dot{X}^{3N})$ in terms of the generalized curvilinear velocities $\dot{\boldsymbol{q}} = (\dot{q}^1, \ldots, \dot{q}^i, \ldots, \dot{q}^{3N})$, read

$$\dot{X}^\mu = \sum_i^{3N} \frac{\partial X^\mu}{\partial q^i} \dot{q}^i \ \ (\mu = 1, \ldots, 3N) , \tag{6.14}$$

or in matrix form

$$\dot{\boldsymbol{X}}^T = \boldsymbol{J} \cdot \dot{\boldsymbol{q}}^T . \tag{6.15}$$

Inserting Eq. (6.14) into Eq. (6.2) yields the classical *Lagrangian* expression of the kinetic energy in terms of the *generalized velocities*

$$T(\boldsymbol{q}, \dot{\boldsymbol{q}}) = \frac{1}{2} \sum_{i,j=1}^{3N} g_{ij}(\boldsymbol{q}) \dot{q}^i \dot{q}^j , \tag{6.16}$$

where

$$g_{ij}(\boldsymbol{q}) = \sum_{\mu=1}^{3N} m_\mu (\frac{\partial X^\mu}{\partial q^i})(\frac{\partial X^\mu}{\partial q^j}) \ (i, j = 1, \ldots, 3N) , \tag{6.17}$$

are the *covariant* components of the *metric tensor*. In matrix form

$$\boldsymbol{g}(\boldsymbol{q}) = \left[g_{ij}\right] = \boldsymbol{J}^T_{X\to q} \cdot \boldsymbol{m} \cdot \boldsymbol{J}_{X\to q} , \tag{6.18}$$

with

$$\boldsymbol{m} = \begin{bmatrix} \ddots & & \\ & m_\mu & \\ & & \ddots \end{bmatrix} \tag{6.19}$$

and, in addition,

$$g(\boldsymbol{q}) = |\text{Det}\boldsymbol{g}| = (\Pi_\mu^{3N} m_\mu)\, J_{X\to q}^2 \,. \tag{6.20}$$

In matrix form, the Lagrangian kinetic energy reads

$$T(\boldsymbol{q}, \dot{\boldsymbol{q}}) = \frac{1}{2}\dot{\boldsymbol{q}} \cdot \boldsymbol{g} \cdot \dot{\boldsymbol{q}}^T \,. \tag{6.21}$$

In order to avoid any mistake or misunderstanding, it is important to point out that *mass-weighted Cartesian coordinates*, defined as

$$x^\mu = \sqrt{m_\mu}\, X^\mu \; (\mu = 1, \ldots, 3N) \tag{6.22}$$

are often used to simplify the notation (see Eq. (5.5) in Sect. 5.1). The coordinate changes and their inverses then read, respectively,

$$\boldsymbol{x} = \boldsymbol{x}(\boldsymbol{q}) \,, \tag{6.23}$$

and

$$\boldsymbol{q} = \boldsymbol{q}(\boldsymbol{x}) \,. \tag{6.24}$$

The corresponding $3N \times 3N$ Jacobian matrices and determinants are

$$[(\boldsymbol{J}_{x\to q})_i^\mu] = [\frac{\partial x^\mu}{\partial q^i}] \,, \tag{6.25}$$

$$J_{x\to q} = |\text{Det}\; J_{x\to q}| \,, \tag{6.26}$$

$$[(\boldsymbol{J}_{q\to x})_\mu^i] = [\frac{\partial q^i}{\partial x^\mu}] \,, \tag{6.27}$$

$$J_{q\to x} = |\text{Det}\; J_{q\to x}| \,, \tag{6.28}$$

together with the corresponding well-known properties

$$\boldsymbol{J}_{x\to q} = \boldsymbol{J}_{q\to x}^{-1} \,, \tag{6.29}$$

and

$$J_{x\to q} = J_{q\to x}^{-1} \,. \tag{6.30}$$

The $(3N \times 3N)$ matrix of the covariant components of the metric tensor can now be written as

$$\boldsymbol{g}(\boldsymbol{q}) = \left[g_{ij}\right] = \boldsymbol{J}^T_{x\rightarrow q} \cdot \boldsymbol{J}_{x\rightarrow q}\,, \tag{6.31}$$

and

$$g(\boldsymbol{q}) = |\mathrm{Det}\boldsymbol{g}| = J^2_{x\rightarrow q}\,. \tag{6.32}$$

After this little digression on notational problems, we now focus our attention on the *Lagrangian function* defined as

$$L = T - V\,, \tag{6.33}$$

where V is the interaction potential. In terms of generalized curvilinear coordinates, we obtain for the Lagrangian function

$$L(\boldsymbol{q}, \dot{\boldsymbol{q}}) = T(\boldsymbol{q}, \dot{\boldsymbol{q}}) - V(\boldsymbol{q})\,, \tag{6.34}$$

where

$$V(\boldsymbol{q}) = V(\boldsymbol{X}(\boldsymbol{q}))\,. \tag{6.35}$$

6.1.2 Classical Hamiltonian Kinetic Energy

The momenta, P_i, conjugate to the generalized coordinates, q^i $(i = 1, \ldots, 3N)$ are by definition

$$P_i = \frac{\partial L(\boldsymbol{q}, \dot{\boldsymbol{q}})}{\partial \dot{q}^i} = \frac{\partial T(\boldsymbol{q}, \dot{\boldsymbol{q}})}{\partial \dot{q}^i} \quad (i = 1, \ldots, 3N)\,, \tag{6.36}$$

since the potential function does not depend on the *generalized velocities*. Inserting into Eq. (6.36) the expression of the kinetic energy as given in Eq. (6.16), yields

$$P_i = \sum_{j=1}^{3N} g_{ij}(\boldsymbol{q})\dot{q}^j \quad (i = 1, \ldots, 3N)\,, \tag{6.37}$$

which, substituted in its turn into Eq. (6.16), leads to the following invariant form of the kinetic energy

$$T(\dot{\boldsymbol{q}}, \boldsymbol{P}) = \frac{1}{2}\sum_{i=1}^{3N} P_i \dot{q}^i\,. \tag{6.38}$$

Now, since the Jacobian matrices $\boldsymbol{J}_{X\rightarrow q}$ and $\boldsymbol{J}_{x\rightarrow q}$ are invertible and in view of Eqs. (6.18) and (6.29), it is possible to invert $\boldsymbol{g} = \left[g_{ij}(\boldsymbol{g})\right]$ as well. Let us define

$$\boldsymbol{G}(\boldsymbol{q}) = \left[G^{ij}(\boldsymbol{q})\right] = \boldsymbol{g}^{-1}(\boldsymbol{q}), \tag{6.39}$$

and invert Eq. (6.37) to obtain

$$\dot{q}^i = \sum_{j=1}^{3N} G^{ij} P_j \ (i = 1, \ldots, 3N). \tag{6.40}$$

On substitution into (6.38), we obtain the *Hamiltonian kinetic energy*, i.e. the kinetic energy expressed in terms of the *generalized coordinates* $\boldsymbol{q}$ and their *conjugate momenta* $\boldsymbol{P}$

$$T(\boldsymbol{q}, \boldsymbol{P}) = \frac{1}{2} \sum_{i,j}^{3N} G^{ij}(\boldsymbol{q}) P_i P_j, \tag{6.41}$$

or, in matrix form,

$$T(\boldsymbol{q}, \boldsymbol{P}) = \frac{1}{2} \boldsymbol{P} \cdot \boldsymbol{G} \cdot \boldsymbol{P}^T. \tag{6.42}$$

The G^{ij} $(i, j = 1, \ldots, 3N)$ are known as the *contravariant* components of the *metric tensor*.

From Eq. (6.18) and the usual rules of matrix inversion

$$\boldsymbol{G}(\boldsymbol{q}) = \boldsymbol{g}^{-1}(\boldsymbol{q}) = \boldsymbol{J}_{X\to q}^{-1} \cdot \boldsymbol{m}^{-1} \cdot (\boldsymbol{J}_{X\to q}^{-1})^T, \tag{6.43}$$

or still, in view of Eq. (6.31) when mass-weighted Cartesian coordinates are used

$$\boldsymbol{G} = \boldsymbol{g}^{-1} = \boldsymbol{J}_{x\to q}^{-1} \cdot (\boldsymbol{J}_{x\to q}^{-1})^T. \tag{6.44}$$

Taking the square root of the absolute value of the determinants of the matrices of Eqs. (6.43) and (6.44) leads to

$$\sqrt{G} = \sqrt{|\mathrm{Det}\boldsymbol{G}|} = \frac{1}{\sqrt{g}} = \frac{1}{J_{x\to q}} = \frac{1}{\Pi_{\mu}^{3N} m_{\mu} J_{X\to q}}. \tag{6.45}$$

So, when starting from the coordinate changes

$$\boldsymbol{X} = \boldsymbol{X}(\boldsymbol{q}) \text{ or } \boldsymbol{x} = \boldsymbol{x}(\boldsymbol{q}) \tag{6.46}$$

named $\boldsymbol{x}(\boldsymbol{q})$ *approach*, the inversion of a $3N \times 3N$ matrix seems unavoidable, i.e. direct inversion of $\boldsymbol{g}$ or $\boldsymbol{J}_{X\to q}$ or $\boldsymbol{J}_{x\to q}$.

However, when starting from the inverse relations

$$\boldsymbol{q} = \boldsymbol{q}(\boldsymbol{X}) \text{ or } \boldsymbol{q} = \boldsymbol{q}(\boldsymbol{x}), \tag{6.47}$$

named $\boldsymbol{q}(\boldsymbol{x})$ *approach* and in view of Eqs. (6.12) and (6.29), Eqs. (6.43) and (6.44) can be rewritten as, respectively,

$$\boldsymbol{G} = \left[G^{ij}\right] = \boldsymbol{g}^{-1} = (\boldsymbol{J}_{q\to X})^T \cdot \boldsymbol{m}^{-1} \cdot (\boldsymbol{J}^T_{q\to X}), \tag{6.48}$$

with

$$G^{ij} = \sum_{\mu=1}^{3N} \frac{1}{m_\mu}(\frac{\partial q^i}{\partial X^\mu})(\frac{\partial q^j}{\partial X^\mu}) \; (i, j = 1, \ldots, 3N), \tag{6.49}$$

or still

$$\boldsymbol{G} = \left[G^{ij}\right] = \boldsymbol{g}^{-1} = \boldsymbol{J}_{q\to x} \cdot \boldsymbol{J}^T_{q\to x}, \tag{6.50}$$

with

$$G^{ij} = \sum_{\mu=1}^{3N}(\frac{\partial q^i}{\partial x^\mu})(\frac{\partial q^j}{\partial x^\mu}) \; (i, j = 1, \ldots, 3N), \tag{6.51}$$

without having to resort to the inversion of a $3N \times 3N$ *matrix, inversion which is often the bottleneck for obtaining the contravariant components of the metric tensor, required for the classical Hamiltonian expression of the kinetic energy and, more importantly in our case, also required for the expression in curvilinear coordinates of the quantum kinetic energy operator* (see next section).

A word of caution: the Eqs. (6.48)–(6.51) give the impression that the explicit knowledge of $\boldsymbol{q}(\boldsymbol{X})$ or $\boldsymbol{q}(\boldsymbol{x})$ suffices to calculate G^{ij}. This is not entirely correct and there is a drawback. Indeed, $\frac{\partial q^i}{\partial x^\mu}$ and $\frac{\partial q^i}{\partial X^\mu}$ are functions of $\boldsymbol{x}$ and $\boldsymbol{X}$, respectively, whereas the contravariant components, G^{ij}, of the metric tensor are to be expressed as functions of $\boldsymbol{q}$. *In other words*, when the "$\boldsymbol{q}(\boldsymbol{x})$" *approach is used, matrix inversion is sidestepped but then the functions* $\boldsymbol{x}(\boldsymbol{q})$ or $\boldsymbol{X}(\boldsymbol{q})$ *must be known explicitly*. We refer to Sect. 6.7.3 for some simple examples of the metric tensors introduced above and their calculation in the $\boldsymbol{x}(\boldsymbol{q})$ and $\boldsymbol{q}(\boldsymbol{x})$ approaches, in the case of polar coordinates.

6.2 The Non Relativistic Quantum Kinetic Energy Operator

6.2.1 General Expressions

If one tries to derive the quantum kinetic energy operator from the expression of *twice*[1] the classical Hamiltonian kinetic energy (see previous section)

[1] It is more compact to write *twice* the kinetic energy.

$$2T(\boldsymbol{q},\boldsymbol{P}) = \sum_{i,j=1}^{3N} G^{ij}(\boldsymbol{q})P_iP_j\,, \tag{6.52}$$

by means of the usual correspondence rules between classical quantities and quantum operators, a well-known problem arises from the ambiguity related to the ordering of the operators (see Sect. 2.2.1).

However, in Cartesian coordinates, twice the kinetic energy reads (see Table 2.1)

$$2\hat{T} = -\hbar^2 \sum_{\mu=1}^{3N} \frac{1}{m_\mu}\frac{\partial^2}{\partial X^{\mu 2}}\,, \tag{6.53}$$

which, in mass-weighted Cartesian coordinates (see Eq. (6.22)), transforms into

$$2\hat{T} = -\hbar^2 \sum_{\mu=1}^{3N} \frac{\partial^2}{\partial x^{\mu 2}}\,. \tag{6.54}$$

A straightforward and safe way to obtain the quantum energy operator in generalized curvilinear coordinates, $\boldsymbol{q}$, is to apply the standard coordinate transformation $\boldsymbol{x} \rightarrow \boldsymbol{q}$ to the 3 *N-dimensional Laplacian operator*

$$\boldsymbol{\Delta} = \sum_{\mu=1}^{3N} \frac{\partial^2}{\partial x^{\mu 2}}\,. \tag{6.55}$$

The explicit expression of the Laplacian operator depends on the *normalization convention* used for the wavefunctions (see Sect. 6.2.2). If *the standard normalization condition* of a wavefunction $\Psi(\boldsymbol{x})$, given by

$$\int \Psi^\star(\boldsymbol{x})\Psi(\boldsymbol{x})d\boldsymbol{x} = 1\,, \tag{6.56}$$

with

$$d\boldsymbol{x} = dx^1dx^2\ldots dx^{3N}\,, \tag{6.57}$$

transforms, under the coordinate change $\boldsymbol{x} \rightarrow \boldsymbol{q}$, into

$$\int \Psi^\star(\boldsymbol{q})\Psi(\boldsymbol{q})d\tau = 1\,, \tag{6.58}$$

with

$$d\tau = J(\boldsymbol{q})dq^1\ldots dq^{3N}\,, \tag{6.59}$$

where (see Eqs. (6.25), (6.26) and (6.32))

$$J(\boldsymbol{q}) = J_{x\rightarrow q} = |\mathrm{Det}[\frac{\partial x^{\mu}}{\partial q^{i}}]| = \sqrt{g(\boldsymbol{q})}\,, \tag{6.60}$$

the *standard expression* of a $3N$ *Laplacian operator in curvilinear coordinates* yields the following *kinetic energy operator* [1–4]

$$2\hat{T} = -\hbar^{2} \sum_{i,j=1}^{3N} \frac{1}{\sqrt{g}} \frac{\partial}{\partial q^{i}} \sqrt{g}\, G^{ij} \frac{\partial}{\partial q^{j}}\,, \tag{6.61}$$

or, in view of Eq. (6.60),

$$2\hat{T} = -\hbar^{2} \sum_{i,j=1}^{3N} J^{-1} \frac{\partial}{\partial q^{i}} J\, G^{ij} \frac{\partial}{\partial q^{j}}\,, \tag{6.62}$$

where G^{ij} $(i, j = 1, \ldots, 3N)$ are the contravariant components of the metric tensor defined in the previous section.

Now, introducing the momentum operators (see Sect. 2.2.1)

$$\hat{P}_{i} = -i\hbar \frac{\partial}{\partial q^{i}}\,, (i = 1, \ldots, 3N)\,, \tag{6.63}$$

and their adjoints (Sects. 2.1.3 and 6.7.2)

$$\hat{P}_{i}^{\dagger} = -i\hbar J^{-1} \frac{\partial}{\partial q^{i}} J\,, (i = 1, \ldots, 3N)\,, \tag{6.64}$$

it is straightforward to rewrite Eq. (6.62) as

$$2\hat{T} = \sum_{i,j=1}^{3N} \hat{P}_{i}^{\dagger}\, G^{ij}\, \hat{P}_{j}\,, \tag{6.65}$$

which brings out the analogy with the classical Hamiltonian kinetic energy (see Eq. (6.52)). With

$$\hat{\boldsymbol{P}} = (\hat{P}_{1}, \ldots, \hat{P}_{i}, \ldots, \hat{P}_{3N})\,, \tag{6.66}$$

and

$$\hat{\boldsymbol{P}}^{\dagger} = (\hat{P}_{1}^{\dagger}, \ldots, \hat{P}_{i}^{\dagger}, \ldots, \hat{P}_{3N}^{\dagger})\,, \tag{6.67}$$

the kinetic energy operator reads in matrix form

$$2\hat{T} = \hat{\boldsymbol{P}}^{\dagger} \cdot \boldsymbol{G} \cdot \hat{\boldsymbol{P}}\,. \tag{6.68}$$

Equation (6.65) is the *key formula* of this chapter. Indeed, *it can be rewritten, developed and implemented in a variety of ways as well as easily adapted to many applications.*

(1) Of course, Eq. (6.65) can be used as it stands. So, for example, in the case of *polyspherical coordinates* (Sect. 6.4) the classical SF (Space-Fixed) Hamiltonian kinetic energy as given by Eq. (6.140) is straightforwardly turned into the corresponding kinetic energy operator, Eq. (6.141). These two equations are particular instances of Eqs. (6.52) and (6.65).
(2) The variety of ways of writing the momentum operators and their adjoints, the possibility of changing the normalization convention (Sect. 6.2.2) and introducing quasi momentum operators (Sect. 6.2.3), entail that Eq. (6.65) gives rise to a variety of *strictly equivalent ways* of writing the kinetic energy operators; for a rather extensive list of these operators see [3].
(3) A useful and common way of rewriting Eq. (6.65) goes as follows. Since (see Eqs. (6.190), (6.193) and (Sect. 6.7.2))

$$\begin{aligned}\hat{P}_i^\dagger \Psi &= J^{-1}\hat{P}_i J\Psi \\ &= \hat{P}_i\Psi + (J^{-1}\hat{P}_i J)\Psi\,,\end{aligned} \tag{6.69}$$

the *adjoint operator* may be written as

$$\hat{P}_i^\dagger = \hat{P}_i + \Lambda_i\,, \tag{6.70}$$

where

$$\Lambda_i = (J^{-1}\hat{P}_i J) = (\hat{P}_i \ln J) \tag{6.71}$$

is a *purely multiplicative operator*, i.e. a number. In the remainder of this section as well as in Sects. 6.3.1 and 6.3.2, in order to avoid any ambiguity in the writing of the operators, the following *parentheses convention* is used: *if a differential operator appears in an expression within parentheses, it does not operate beyond the parentheses.*[2]

In view of Eq. (6.70) and taking this convention into account, it is straightforward to develop Eq. (6.65) as

$$2\hat{T} = \sum_{i,j=1}^{3N} G^{ij}\hat{P}_i\hat{P}_j + + \sum_{j=1}^{3N}(\sum_{i=1}^{3N}\left[\Lambda_i + \hat{P}_i\right]G^{ij})\hat{P}_j\,, \tag{6.72}$$

and reverting to partial derivatives

---

[2]Incidentally, we will also tacitly use the convention that operators act beyond square brackets [ ] and curly brackets { }.

$$2\hat{T}\left(\boldsymbol{q},\frac{\partial}{\partial\boldsymbol{q}}\right)=-\hbar^2\left[\sum_{i,j=1}^{3N}G^{ij}(\boldsymbol{q})\frac{\partial^2}{\partial q^i\partial q^j}+\sum_{j=1}^{3N}\sum_{i=1}^{3N}G^{ij}(\boldsymbol{q})\left(\frac{\partial\ln J(\boldsymbol{q})}{\partial q^i}+\frac{\partial G^{ij}(\boldsymbol{q})}{\partial q^i}\right)\frac{\partial}{\partial q^j}\right]. \tag{6.73}$$

It is interesting to notice that this last expression turns out to have the following structure

$$2\hat{T}\left(\boldsymbol{q},\frac{\partial}{\partial\boldsymbol{q}}\right)=-\hbar^2\left[\sum_{i,j=1}^{3N}f^{ij}(\boldsymbol{q})\frac{\partial^2}{\partial q^i\partial q^j}+\sum_{j=1}^{3N}f^{j}(\boldsymbol{q})\frac{\partial}{\partial q^j}\right]. \tag{6.74}$$

Of course,

$$f^{ij}(\boldsymbol{q})=G^{ij}(\boldsymbol{q}), \tag{6.75}$$

and

$$f^{j}(\boldsymbol{q})=\sum_{i=1}^{3N}G^{ij}(\boldsymbol{q})\left(\frac{\partial\ln J(\boldsymbol{q})}{\partial q^i}+\frac{\partial G^{ij}(\boldsymbol{q})}{\partial q^i}\right). \tag{6.76}$$

It is worth noticing that a straightforward "brute force" application of the "chain rule" leads to an expression of $\hat{T}$ with the structure highlighted in Eq. (6.74). This approach can be regarded as a "$\boldsymbol{q}(\boldsymbol{x})$" approach, which implies that both $\boldsymbol{q}(\boldsymbol{x})$ and $\boldsymbol{x}(\boldsymbol{q})$ must be available (See Sect. 6.7.1).

It is also important to mention a *numerical approach*. Indeed, the analytical calculation of the functions $f^{ij}(\boldsymbol{q})=G^{ij}(\boldsymbol{q})$ and $f^{j}(\boldsymbol{q})$ given by Eqs. (6.75) and (6.76), often turns out to be awfully tedious and complex. Therefore, *numerical algorithms* have also been developed for computing these functions *without resorting to analytical expressions* [5–11]. So, for instance, with the help of the TNUM *algorithm* [9], the kinetic energy operator $\hat{T}(\boldsymbol{q},\frac{\partial}{\partial\boldsymbol{q}})$, can be calculated for a given value of the $3N$ generalized coordinates $\boldsymbol{q}$. This calculation only requires the *exact numerical* values of $\boldsymbol{X}(\boldsymbol{q})$ and their derivatives, so that the functions $f^{ij}(\boldsymbol{q})$ and $f^{j}(\boldsymbol{q})$ can be evaluated *numerically* and *exactly* for a given value of $\boldsymbol{q}$, e.g. at a grid point.

6.2.2 Change of Normalization Convention *

So far, all operators, $\hat{A}$, in particular the kinetic energy operator, $\hat{T}$, are designed for acting on wavefunctions $\Psi(\boldsymbol{q})$ that are normalized with the help of the Euclidean volume element given in Eq. (6.59) (see also Eq. (6.58)). The Euclidean volume element $d\tau^{(E)}=J(\boldsymbol{q})dq^1\ldots dq^{3N}$, however, is not always adapted and another volume element is often introduced

$$d\tau^{(\rho)}=\rho(\boldsymbol{q})dq^1\ldots dq^{3N}, \tag{6.77}$$

where $\rho(\boldsymbol{q})$ is a *real (positive) weightfunction*. In this case, the wavefunction Ψ and any operator, $\hat{A}$, must be replaced by, respectively, [3]

$$\Psi \to \Psi^{(\rho)} = J^{\frac{1}{2}}\rho^{-\frac{1}{2}}\Psi\,, \tag{6.78}$$

$$\hat{A} \to \hat{A}^{(\rho)} = J^{\frac{1}{2}}\rho^{-\frac{1}{2}}\hat{A}\rho^{\frac{1}{2}}J^{-\frac{1}{2}}\,, \tag{6.79}$$

in order to preserve

- the *normalization condition*

$$\int \Psi^{(\rho)\star}(\boldsymbol{q})\Psi^{(\rho)}(\boldsymbol{q})d\tau^{(\rho)} = \int \Psi^{\star}(\boldsymbol{q})\Psi(\boldsymbol{q})d\tau^{(E)} = 1\,, \tag{6.80}$$

- the *values of the matrix elements*

$$A_{12} = \int \Psi_1^{(\rho)\star}(\boldsymbol{q})(\hat{A}^{(\rho)}\Psi_2^{(\rho)}(\boldsymbol{q}))d\tau^{(\rho)} = \int \Psi_1^{\star}(\boldsymbol{q})(\hat{A}\Psi_2(\boldsymbol{q}))d\tau^{(E)}\,. \tag{6.81}$$

Of course, if $\Psi_1 = \Psi_2 = \Psi$, Eq. (6.81) reduces to preserving the mean value of the operator $\hat{A}$

$$\langle\hat{A}^{(\rho)}\rangle = \int \Psi^{(\rho)\star}(\boldsymbol{q})(\hat{A}^{(\rho)}\Psi^{(\rho)}(\boldsymbol{q}))d\tau^{(\rho)} = \int \Psi^{\star}(\boldsymbol{q})(\hat{A}\Psi(\boldsymbol{q}))d\tau^{(E)} = \langle\hat{A}\rangle\,. \tag{6.82}$$

If, in addition, $\hat{A}$ is the identity operator, Eq. (6.81) reduces to the normalization condition (6.80).

As far as the kinetic energy operator is concerned, it can be shown, straightforwardly but slightly tediously, that

$$\begin{aligned} 2\hat{T}^{(\rho)} &= 2J^{\frac{1}{2}}\rho^{-\frac{1}{2}}\hat{T}\rho^{\frac{1}{2}}J^{-\frac{1}{2}} \\ &= \sum_{i,j}^{3N} G^{ij}\hat{P}_i\hat{P}_j + \sum_{j}^{3N}\left(\sum_{i}^{3N}\left[\Lambda_i^{(\rho)} + \hat{P}_i\right]G^{ij}\right)\hat{P}_j + 2\hat{V}^{(\rho)}\,, \end{aligned} \tag{6.83}$$

where

$$\Lambda_i^{(\rho)} = (\hat{P}_i \ln\rho)\,, \tag{6.84}$$

and

$$\hat{V}^{(\rho)} = (J^{\frac{1}{2}}\rho^{-\frac{1}{2}}\hat{T}\rho^{\frac{1}{2}}J^{-\frac{1}{2}})\,. \tag{6.85}$$

This latter expression, though intrinsically of kinetic energy origin, is a *purely multiplicative extrapotential term*. It is interesting to note that, when an Euclidean normalization convention is used, i.e. $\rho = J$, we obtain $\hat{V}^{(J)} = 0$, $\Lambda_i^{(J)} = \Lambda_i$ and $\hat{T}^{(J)} = \hat{T}$, as expected.

Reverting to partial derivatives, Eq. (6.83) becomes

$$2\hat{T}^{(\rho)}(\boldsymbol{q}, \frac{\partial}{\partial \boldsymbol{q}}) = -\hbar^2 \left[\sum_{i,j}^{3N} G^{ij}(\boldsymbol{q}) \frac{\partial^2}{\partial q^i \partial q^j} + \sum_{j}^{3N} \sum_{i}^{3N} \left(G^{ij}(\boldsymbol{q}) \left(\frac{\partial}{\partial q^i} \ln \rho(\boldsymbol{q}) \right) + \frac{\partial G^{ij}}{\partial q^i} \right) \frac{\partial}{\partial q^i} \right] + 2\hat{V}^{(\rho)}(\boldsymbol{q}) . \tag{6.86}$$

Here again, a numerical algorithm can be used to calculate $G^{ij}(\boldsymbol{q})$, $f^j(\boldsymbol{q}) = \sum_i^{3N} \left(G^{ij}(\boldsymbol{q})(\frac{\partial}{\partial q^i} \ln \rho(\boldsymbol{q})) + \frac{\partial G^{ij}}{\partial q^i} \right)$ and $\hat{V}^{(\rho)}(\boldsymbol{q})$.

6.2.3 *Quasi-Momentum Operators*

As is often the case in quantum mechanics, the physically relevant momentum operators are not of the type $\hat{P}_i = -i\hbar \frac{\partial}{\partial q^i}$ but some linear combinations of such partial derivatives (e.g. the components of the total angular momentum are linear combinations of the partial derivatives with respect to the Eulerian angles, times $-i\hbar$: see Eq. (6.114)). We call them *quasi-momentum operators*[3] and write them as

$$\hat{\mathcal{P}}_K = \sum_{i=1}^{3N} B_K^i(\boldsymbol{q}) \hat{P}_i \quad (K = 1, \ldots, n \geq 3N) , \tag{6.87}$$

where

$$\boldsymbol{B} = \left[B_K^i(\boldsymbol{q}) \right] , \tag{6.88}$$

is a *real rectangular* (square, if $K = 3N$) matrix depending on the generalized curvilinear coordinates $\boldsymbol{q}$ only, i and K being, respectively, the row and column indices.

Unlike the conjugate momentum operators $\hat{P}_i$, which commute since they are essentially partial derivatives, the *quasi-momentum operators do not commute*, their commutation rules being either standard (as for angular momentum operators) or unusual [3, 12, 13].

Moreover, the fact that the number of quasi-momentum operators may be larger than the number of "true" momentum operators is not uncommon (see for instance Eq. (6.144)) but it entails that matrix $\boldsymbol{B}$ may be rectangular and thus noninvertible, so that some caution will be required in the derivation of the quantum mechanical expressions.

---

[3] To avoid all misunderstanding, it is worth noting that the usual angular momentum operators are particular cases of the quasi-momentum operators as defined here.

To determine the expression of the quantum kinetic energy operators in terms of quasi-momentum operators as defined by Eqs. (6.87) and (6.88), it is important to notice that, alternatively to the usual classical Hamiltonian kinetic energy in terms of "true" conjugate momenta (see Eq. (6.41))

$$T(\boldsymbol{q},\boldsymbol{P}) = \frac{1}{2}\sum_{i,j=1}^{3N} P_i G^{ij}(\boldsymbol{q}) P_j \,, \tag{6.89}$$

the expression of the *same* classical kinetic energy in terms of *classical quasi-momenta* is often known from classical mechanics theory

$$2T(\boldsymbol{q},\mathcal{P}) = \sum_{K,L=1}^{n} \mathcal{P}_K\, \mathcal{G}^{KL}\, \mathcal{P}_L \,, \tag{6.90}$$

where

$$\mathcal{P}_K = \sum_{i=1}^{3N} B^i_K(\boldsymbol{q}) P_i \,, \tag{6.91}$$

is the classical quasi-momentum corresponding to Eq. (6.87).

Since the two expressions, Eqs. (6.89) and (6.90) of the classical kinetic energy *must be equal*, after substitution of Eq. (6.91) into Eq. (6.90) and comparing with Eq. (6.89) one obtains

$$G^{ij} = \sum_{K,L=1}^{n} B^i_K \mathcal{G}^{KL} B^j_L \; (i, j = 1, \ldots, 3N) \,. \tag{6.92}$$

Now, the adjoint of $\hat{\mathcal{P}}_K$ is given by [3]

$$\hat{\mathcal{P}}^\dagger_K = \sum_{i=1}^{3N} \hat{P}^\dagger_i B^i_K(\boldsymbol{q}) \,, \tag{6.93}$$

so that the quantum counterpart of the classical expression of the kinetic energy Eq. (6.90), is given by

$$2\hat{T} = \sum_{K,L=1}^{n} \hat{\mathcal{P}}^\dagger_K\, \mathcal{G}^{KL}\, \hat{\mathcal{P}}_L \,. \tag{6.94}$$

Indeed, inserting Eqs. (6.87) and (6.93) into Eq. (6.94) yields

$$2\hat{T} = \sum_{i,j=1}^{3N} \hat{P}_i^\dagger \, G^{ij} \, \hat{P}_j \,. \tag{6.95}$$

Thus, both expressions (6.94) and (6.95) can be used to obtain the kinetic energy operator in generalized curvilinear coordinates and, as will be seen in Sects. 6.3, 6.4 and 6.5, *using quasi-momentum operators improves physical insight and greatly simplifies* the ways of obtaining the quantum expressions of the kinetic energy operators.

It is also interesting to consider some particular cases:

(a) If $\boldsymbol{B} = \left[B_K^i(\boldsymbol{q})\right]$ $(i, K = 1, \ldots, 3N)$ is a *square invertible* matrix, and

$$\boldsymbol{A} = \left[A_i^K(\boldsymbol{q})\right] = \left[B_K^i(\boldsymbol{q})\right]^{-1} \,, \tag{6.96}$$

Eq. (6.91) can be inverted

$$\hat{P}_i = \sum_{K=1}^{3N} A_i^K(\boldsymbol{q}) \hat{\mathcal{P}}_K \; (i = 1, \ldots, 3N) \,, \tag{6.97}$$

so that

$$\mathcal{G}^{KL} = \sum_{i,j=1}^{3N} A_i^K G^{ij} A_j^L \; (K, L = 1, \ldots, 3N) \,. \tag{6.98}$$

(b) If

$$\left[B_K^i(\boldsymbol{q})\right] = \left[\frac{\partial q^i}{\partial Q^K}\right] , \tag{6.99}$$

i.e. an ordinary Jacobian matrix of the coordinate transformation $q^i \to Q^K$. In view of Eqs. (6.91) and (6.87), $\mathcal{P}_K$ and $\hat{\mathcal{P}}_K$ reduce to "true" momenta and momentum operators conjugate to the generalized coordinates $\{Q^K, K = 1, \ldots, 3N\}$.

6.3 Applications in Molecular Physics

In Chap. 3, several molecular frames have been introduced to separate the translation, overall rotation and internal motions of a molecular system. In the present section, we will neglect the coupling between electrons and nuclei in the BF frame as well as the mass polarization terms (Sect. 3.3.3), which results in *the separation of nuclear and electronic kinetic energy operators* (see Eq. (3.129)). In addition, *the nuclear center of mass of the molecule* (NCM) *will be identified with the total center of mass of the molecule* (CM) *and the* NSF *frame with the* SF *frame* (Sect. 3.3).

It is important to note that in practice, we use the *masses of the atoms instead of the masses of the nuclei* in the final expressions of $\hat{T}$. Using atomic masses rather than nuclear masses in the expression of the kinetic energy operator is somewhat related to the effect of second-order non-adiabatic corrections (diagonal terms); see for example Ref. [14]. In short, considering atomic masses means that electrons are implicitly assumed to follow *instantaneously* the motion of the nuclei. As a consequence, the effective mass of a nucleus dressed with electrons that behave adiabatically is the *atomic mass*. This approximation has a small effect, except when dealing with highly-accurate rovibrational spectra of light diatoms such as, for example, H_2 and LiH. A more accurate description is then achieved when introducing a coordinate-dependent reduced mass that makes it possible to account for possible electron redistribution depending on the distance between the two nuclei.

6.3.1 General Case

We restrict ourselves to the kinetic energy of N nuclei and focus our attention on nuclear coordinate changes which particularize the coordinate change given by Eq. (6.5) as follows:

$$\begin{aligned} X^{\mu}_{LF} &= X^{\mu}_{LF}(q^1, \ldots, q^{3N-6}, \alpha, \beta, \gamma, X^{CM}_{LF}, Y^{CM}_{LF}, Z^{CM}_{LF}) \\ &= X^{\mu}_{LF}(\boldsymbol{q}, \boldsymbol{\Theta}, \boldsymbol{R}^{CM}_{LF}) \ (\mu = 1, \ldots, 3N), \end{aligned} \tag{6.100}$$

where (see Fig. 6.1)

(i) $\boldsymbol{q} = (q^1, \ldots, q^{3N-6})$[4] are the *internal or shape coordinates*;
(ii) $\boldsymbol{\Theta} = (\alpha, \beta, \gamma)$ are the three Euler angles that parametrize the orientation of the BF frame with respect to the SF frame;
(iii) $\boldsymbol{R}^{CM} = (X^{CM}_{LF}, Y^{CM}_{LF}, Z^{CM}_{LF})$ are the LF Cartesian coordinates of the center of mass.

The coordinate change (6.100) can be written more explicitly as (see Fig. 6.2)

$$\begin{aligned} \begin{bmatrix} X^{\mu=3\alpha-2}_{LF}(\boldsymbol{q}, \boldsymbol{\Theta}, \boldsymbol{R}^{CM}_{LF}) \\ X^{\mu=3\alpha-1}_{LF}(\boldsymbol{q}, \boldsymbol{\Theta}, \boldsymbol{R}^{CM}_{LF}) \\ X^{\mu=3\alpha}_{LF}(\boldsymbol{q}, \boldsymbol{\Theta}, \boldsymbol{R}^{CM}_{LF}) \end{bmatrix} &= \begin{bmatrix} \tilde{X}^{\alpha}_{LF}(\boldsymbol{q}, \boldsymbol{\Theta}, \boldsymbol{R}^{CM}_{LF}) \\ \tilde{Y}^{\alpha}_{LF}(\boldsymbol{q}, \boldsymbol{\Theta}, \boldsymbol{R}^{CM}_{LF}) \\ \tilde{Z}^{\alpha}_{LF}(\boldsymbol{q}, \boldsymbol{\Theta}, \boldsymbol{R}^{CM}_{LF}) \end{bmatrix} \\ &= \begin{bmatrix} X^{CM}_{LF} \\ Y^{CM}_{LF} \\ Z^{CM}_{LF} \end{bmatrix} + \boldsymbol{C}^T(\boldsymbol{\Theta}) \begin{bmatrix} \tilde{X}^{\alpha}_{BF}(\boldsymbol{q}) \\ \tilde{Y}^{\alpha}_{BF}(\boldsymbol{q}) \\ \tilde{Z}^{\alpha}_{BF}(\boldsymbol{q}) \end{bmatrix} \quad (\alpha = 1, \ldots, N), \end{aligned} \tag{6.101}$$

[4]In the present section $\boldsymbol{q}$ denotes the $3N - 6$ internal coordinates only and not the $3N$ generalized coordinates.

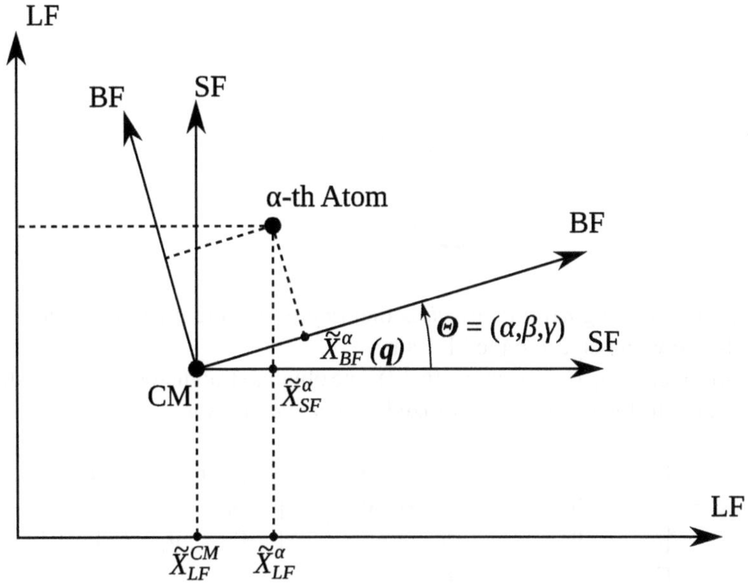

Fig. 6.1 Schematic reprsentation in 2D of an N-atom system in three reference frames. The dashed lines emphasize the Cartesian coordinates in these three frames **(i)** LF frame **(ii)** SF frame **(iii)** BF frame. $\boldsymbol{\Theta} = (\alpha, \beta, \gamma)$ are the three Euler angles that orient the BF frame with respect to SF frame. CM denotes the molecular center of mass (see Eq. (6.101))

where $\boldsymbol{C}(\boldsymbol{\Theta})$ is the (Euler) *rotation matrix* defined in Sect. 3.3.3 (Eqs. (3.114) and (3.115)).[5]

In Eq. (6.101), the $3N$ Cartesian BF coordinates are expressed as functions of the $(3N-6)$ internal coordinates

$$(\tilde{X}^{\alpha}_{BF}(\boldsymbol{q}), \tilde{Y}^{\alpha}_{BF}(\boldsymbol{q}), \tilde{Z}^{\alpha}_{BF}(\boldsymbol{q}); \alpha = 1, \ldots, N)\,, \tag{6.102}$$

so that the molecular system, viewed in the BF frame, is subjected to six constraints whereas it is *physically* not constrained at all, which implies that the $3N$ functions of Eq. (6.102) given above *completely* determine the BF frame. Of the six constraints, three are the *center-of-mass constraints*

[5]The transpose $\boldsymbol{C}^T(\boldsymbol{\Theta})$ is used here because it corresponds to a BF $\rightarrow$ SF transformation, whereas Eqs. (3.114) and (3.115) corresponds to a (N)SF $\rightarrow$ BF transformation.

$$\sum_{\alpha=1}^{N} \tilde{m}_\alpha \tilde{X}_{BF}^{\alpha} = 0$$
$$\sum_{\alpha=1}^{N} \tilde{m}_\alpha \tilde{Y}_{BF}^{\alpha} = 0$$
$$\sum_{\alpha=1}^{N} \tilde{m}_\alpha \tilde{Z}_{BF}^{\alpha} = 0\,, \tag{6.103}$$

the three others being *axial constraints* that *implicitly* determine the orientation of the BF frame with respect to the SF frame.

By introducing Eq. (6.101) in Eq. (6.17), one obtains straight forwardly but slightly tediously, the following nicely factorized "*covariant*" tensor,

$$\tilde{g} = \begin{bmatrix} \mathbf{1}_{(3N-6)\times(3N-6)} & \mathbf{0} & \mathbf{0} \\ \mathbf{0} & \boldsymbol{\Omega}_{3\times3}(\boldsymbol{\Theta}) & \mathbf{0} \\ \mathbf{0} & \mathbf{0} & \mathbf{1}_{3\times3} \end{bmatrix} \times \begin{bmatrix} \boldsymbol{S}(\boldsymbol{q}) & \boldsymbol{C}^T(\boldsymbol{q}) & \mathbf{0} \\ \boldsymbol{C}(\boldsymbol{q}) & \boldsymbol{I}(\boldsymbol{q}) & \mathbf{0} \\ \mathbf{0} & \mathbf{0} & M\,\mathbf{1}_{3\times3} \end{bmatrix}$$
$$\times \begin{bmatrix} \mathbf{1}_{(3N-6)\times(3N-6)} & \mathbf{0} & \mathbf{0} \\ \mathbf{0} & \boldsymbol{\Omega}^T_{3\times3}(\boldsymbol{\Theta}) & \mathbf{0} \\ \mathbf{0} & \mathbf{0} & \mathbf{1}_{3\times3} \end{bmatrix}, \tag{6.104}$$

where $\boldsymbol{S}(\boldsymbol{q})$, $\boldsymbol{C}(\boldsymbol{q})$ and $\boldsymbol{I}(\boldsymbol{q})$ are respectively, the *internal deformation* matrix, the *Coriolis* matrix, and the *inertia* tensor (for explicit expressions see Ref. [3]). M is the total mass of the molecular system. $\boldsymbol{\Omega}_{3\times3}(\boldsymbol{\Theta})$ is the 3×3 matrix relating the time derivatives of the Euler angles to the BF components of the angular velocities ω_x, ω_y and ω_z (of the BF frame),[6] i.e.

$$\omega^T = \boldsymbol{\Omega}^T(\boldsymbol{\Theta}) \cdot \dot{\boldsymbol{\Theta}}^T\,, \tag{6.105}$$

$$\begin{bmatrix} \omega_x \\ \omega_y \\ \omega_z \end{bmatrix} = \begin{bmatrix} -\sin\beta\cos\gamma & \sin\gamma & 0 \\ \sin\beta\sin\gamma & \cos\gamma & 0 \\ \cos\beta & 0 & 1 \end{bmatrix} \begin{bmatrix} \dot{\alpha} \\ \dot{\beta} \\ \dot{\gamma} \end{bmatrix}. \tag{6.106}$$

Leaving out the total center of mass, we obtain from Eq. (6.21)[7] the following *classical (Lagrangian) kinetic energy*

$$T(\boldsymbol{q}, \dot{\boldsymbol{q}}, \boldsymbol{\Theta}, \dot{\boldsymbol{\Theta}}) = \frac{1}{2}\left[\dot{\boldsymbol{q}}\ \dot{\boldsymbol{\Theta}}\right] \cdot g(\boldsymbol{q}, \boldsymbol{\Theta}) \cdot \begin{bmatrix} \dot{\boldsymbol{q}}^T \\ \dot{\boldsymbol{\Theta}}^T \end{bmatrix}, \tag{6.107}$$

with

[6] Introducing the (quasi) angular velocities arises quite naturally from the factorization of $\boldsymbol{g}$.

[7] In this equation $\boldsymbol{q}$ denotes the $3N$ generalized coordinates while in the present section $\boldsymbol{q}$ denotes the $(3N-6)$ internal coordinates.

$$g(\boldsymbol{q}, \boldsymbol{\Theta}) = \begin{bmatrix} \mathbf{1}_{(3N-6)\times(3N-6)} & \mathbf{0} \\ \mathbf{0} & \boldsymbol{\Omega}(\boldsymbol{\Theta}) \end{bmatrix} \times \begin{bmatrix} \boldsymbol{S}(\boldsymbol{q}) & \boldsymbol{C}^T(\boldsymbol{q}) \\ \boldsymbol{C}(\boldsymbol{q}) & \boldsymbol{I}(\boldsymbol{q}) \end{bmatrix} \times \begin{bmatrix} \mathbf{1}_{(3N-6)\times(3N-6)} & \mathbf{0} \\ \mathbf{0} & \boldsymbol{\Omega}^T(\boldsymbol{\Theta}) \end{bmatrix}, \tag{6.108}$$

or still, in view of Eq. (6.106)

$$T(\boldsymbol{q}, \dot{\boldsymbol{q}}, \omega) = \frac{1}{2}\left[\dot{\boldsymbol{q}}\ \omega\right] \cdot \begin{bmatrix} \boldsymbol{S}(\boldsymbol{q}) & \boldsymbol{C}^T(\boldsymbol{q}) \\ \boldsymbol{C}(\boldsymbol{q}) & \boldsymbol{I}(\boldsymbol{q}) \end{bmatrix} \cdot \begin{bmatrix} \dot{\boldsymbol{q}}^T \\ \omega \end{bmatrix}. \tag{6.109}$$

The kinetic energy can thus be split as follows

$$T^{Deformation}(\boldsymbol{q}, \dot{\boldsymbol{q}}) = \frac{1}{2}\dot{\boldsymbol{q}} \cdot \boldsymbol{S}(\boldsymbol{q}) \cdot \dot{\boldsymbol{q}}^T, \tag{6.110}$$

$$T^{Rotation}(\boldsymbol{q}, \omega) = \frac{1}{2}\omega \cdot \boldsymbol{I}(\boldsymbol{q}) \cdot \omega^T, \tag{6.111}$$

$$T^{Coriolis}(\boldsymbol{q}, \dot{\boldsymbol{q}}, \omega) = \frac{1}{2}(\dot{\boldsymbol{q}} \cdot \boldsymbol{C}^T \cdot \omega^T + \omega \cdot \boldsymbol{C} \cdot \dot{\boldsymbol{q}}^T). \tag{6.112}$$

6.3.2 Nuclear Angular Momentum

Let us now introduce the *total angular momentum for the nuclei*, $\boldsymbol{J}$, which is a quasi-momentum conjugate to the *angular velocities* defined as follows

$$\boldsymbol{J}^T = \boldsymbol{\Omega}^{-1}(\boldsymbol{\Theta})\boldsymbol{P}_{\Theta}^T, \tag{6.113}$$

or

$$\begin{bmatrix} J_x \\ J_y \\ J_z \end{bmatrix} = \begin{bmatrix} -\frac{\cos\gamma}{\sin\beta} & \sin\gamma & \cot\beta\cos\gamma \\ \frac{\sin\gamma}{\sin\beta} & \cos\gamma & -\cot\beta\sin\gamma \\ 0 & 0 & 1 \end{bmatrix} \begin{bmatrix} P_\alpha \\ P_\beta \\ P_\gamma \end{bmatrix}, \tag{6.114}$$

where J_x, J_y and J_z are the BF components of the (nuclear) total angular momentum and P_α, P_β and P_γ the momenta conjugate to the Euler angles.

The Hamiltonian kinetic energy is given by (see Eq. (6.90))

$$T(\boldsymbol{q}, \boldsymbol{P}_q, \boldsymbol{J}) = \frac{1}{2}\left[\boldsymbol{P}_q\ \boldsymbol{J}\right] \cdot \mathcal{G}(\boldsymbol{q}) \cdot \begin{bmatrix} \boldsymbol{P}_q^T \\ \boldsymbol{J}^T \end{bmatrix}, \tag{6.115}$$

where $\boldsymbol{P}_q$ are the momenta conjuguate to the internal coordinates $\boldsymbol{q}$ and

$$\mathcal{G}(\boldsymbol{q}) = \begin{bmatrix} \boldsymbol{S}(\boldsymbol{q}) & \boldsymbol{C}^T(\boldsymbol{q}) \\ \boldsymbol{C}(\boldsymbol{q}) & \boldsymbol{I}(\boldsymbol{q}) \end{bmatrix}^{-1} = \begin{bmatrix} \boldsymbol{\Sigma}(\boldsymbol{q}) & \boldsymbol{\Gamma}^T(\boldsymbol{q}) \\ \boldsymbol{\Gamma}(\boldsymbol{q}) & \boldsymbol{\mu}(\boldsymbol{q}) \end{bmatrix} . \tag{6.116}$$

The corresponding quantum kinetic energy operator (KEO) is given by (see Eq. (6.94))

$$2\hat{T}(\boldsymbol{q}, \boldsymbol{\Theta}) = \left[\hat{\boldsymbol{P}}_q^{\dagger}, \hat{\boldsymbol{J}}^{\dagger}\right] \begin{bmatrix} \boldsymbol{\Sigma}(\boldsymbol{q}) & \boldsymbol{\Gamma}^T(\boldsymbol{q}) \\ \boldsymbol{\Gamma}(\boldsymbol{q}) & \boldsymbol{\mu}(\boldsymbol{q}) \end{bmatrix} \begin{bmatrix} \hat{\boldsymbol{P}}_q^T \\ \hat{\boldsymbol{J}}^T \end{bmatrix} , \tag{6.117}$$

where

$$\hat{P}_{q^i} = -i\hbar \frac{\partial}{\partial q^i} = \hat{P}_i , \tag{6.118}$$

and

$$\begin{bmatrix} \hat{J}_x \\ \hat{J}_y \\ \hat{J}_z \end{bmatrix} = -i\hbar \boldsymbol{\Omega}^{-1}(\boldsymbol{\Theta}) \begin{bmatrix} \frac{\partial}{\partial \alpha} \\ \frac{\partial}{\partial \beta} \\ \frac{\partial}{\partial \gamma} \end{bmatrix} , \tag{6.119}$$

$\hat{J}_x$, $\hat{J}_y$ and $\hat{J}_z$ are the BF components of the (nuclear) *total angular momentum operator* and they satisfy the *anomalous commutation rules*:

$$\begin{aligned} \left[\hat{J}_x, \hat{J}_y\right] &= -i\hbar \hat{J}_z \\ \left[\hat{J}_y, \hat{J}_z\right] &= -i\hbar \hat{J}_x \\ \left[\hat{J}_z, \hat{J}_x\right] &= -i\hbar \hat{J}_y . \end{aligned} \tag{6.120}$$

For further details concerning anomalous and unusual commutation rules see [12, 13, 15].

As in the classical case (see Eqs. (6.110–6.112)) the quantum kinetic energy operator (Eq. (6.117)) can be expressed as a sum of three terms, the *deformation part*, $\hat{T}_{Def}$, the *Coriolis part*, $\hat{T}_{Cor}$, and the *rotational part*, $\hat{T}_{Rot}$:

$$\hat{T}(\boldsymbol{q}, \boldsymbol{\Theta}) = \hat{T}_{Def} + \hat{T}_{Cor} + \hat{T}_{Rot} , \tag{6.121}$$

where:

$$\hat{T}_{Def}(\boldsymbol{q}) = \frac{1}{2} \hat{\boldsymbol{P}}_q^{\dagger} \cdot \boldsymbol{\Sigma}(\boldsymbol{q}) \cdot \hat{\boldsymbol{P}}_q^T , \tag{6.122}$$

$$\hat{T}_{Cor}(\boldsymbol{q}, \boldsymbol{\Theta}) = \frac{1}{2} \left[\hat{\boldsymbol{J}}^{\dagger} \cdot \boldsymbol{\Gamma}(\boldsymbol{q}) \cdot \hat{\boldsymbol{P}}_q^T + \hat{\boldsymbol{P}}_q^{\dagger} \cdot \boldsymbol{\Gamma}(\boldsymbol{q}) \hat{\boldsymbol{J}}^T\right] , \tag{6.123}$$

$$\hat{T}_{Rot}(\boldsymbol{q}, \boldsymbol{\Theta}) = \frac{1}{2}\hat{\boldsymbol{J}}^{\dagger} \cdot \boldsymbol{\mu}(\boldsymbol{q}) \cdot \hat{\boldsymbol{J}}^{T} . \tag{6.124}$$

If the volume element used in normalizing the wavefunctions is of the form

$$d\tau^{\rho} = \rho(\boldsymbol{q}) d\boldsymbol{q} \sin\beta d\gamma d\alpha , \tag{6.125}$$

where $\rho(\boldsymbol{q})$ is a weight function depending on the internal coordinates only, it has been shown [3] that

$$\begin{aligned} \hat{J}_x^{\dagger} &= \hat{J}_x \\ \hat{J}_y^{\dagger} &= \hat{J}_y \\ \hat{J}_z^{\dagger} &= \hat{J}_z . \end{aligned} \tag{6.126}$$

Reverting to partial derivatives, the quantum KEO can then be expressed, rather straightforwardly, as

$$\hat{T}^{(\rho)}(\boldsymbol{q}, \boldsymbol{\Theta}) = \hat{T}^{(\rho)}_{Def} + \hat{T}^{(\rho)}_{Cor} + \hat{T}^{(\rho)}_{Rot} , \tag{6.127}$$

where

$$\hat{T}^{(\rho)}_{Def}(\boldsymbol{q}) = -\frac{\hbar^2}{2} \sum_{i,j=1}^{3N-6} \frac{1}{\rho(\boldsymbol{q})} \frac{\partial}{\partial q^i} \rho(\boldsymbol{q}) \Sigma^{ij}(\boldsymbol{q}) \frac{\partial}{\partial q^j} + \hat{V}_{extra}(\boldsymbol{q}) , \tag{6.128}$$

or still

$$\begin{aligned} \hat{T}^{(\rho)}_{Def}(\boldsymbol{q}) = &-\frac{\hbar^2}{2} \left[\sum_{i,j=1}^{3N-6} \Sigma^{ij}(\boldsymbol{q}) \frac{\partial^2}{\partial q^i \partial q^j} + \sum_{i,j=1}^{3N-6} \left(\frac{\partial}{\partial q^i} \Sigma^{ij}(\boldsymbol{q}) + \Sigma^{ij}(\boldsymbol{q}) \frac{\partial}{\partial q^i} \ln(\rho(\boldsymbol{q})) \right) \frac{\partial}{\partial q^j} \right] \\ &+ \hat{V}_{extra}(\boldsymbol{q}) , \end{aligned} \tag{6.129}$$

$$\hat{T}^{(\rho)}_{Cor}(\boldsymbol{q}, \boldsymbol{\Theta}) = -i\hbar \sum_{\alpha'=(x,y,z)} J_{\alpha'} \left[\sum_{i=1}^{3N-6} \frac{1}{2} \left(\frac{\partial}{\partial q^i} \Gamma^{\alpha' j}(\boldsymbol{q}) + \Gamma^{\alpha' j}(\boldsymbol{q}) \frac{\partial}{\partial q^i} \ln(\rho(\boldsymbol{q})) \right) + \Gamma^{\alpha' j}(\boldsymbol{q}) \frac{\partial}{\partial q^j} \right] , \tag{6.130}$$

$$\hat{T}^{(\rho)}_{Rot}(\boldsymbol{q}, \boldsymbol{\Theta}) = \sum_{\alpha',\beta'=x,y,z} \mu^{\alpha'\beta'}(\boldsymbol{q}) \hat{J}_{\alpha'} \hat{J}_{\beta'} , \tag{6.131}$$

$$\hat{V}^{(\rho)}_{extra}(\boldsymbol{q}) = (\tilde{J}^{\frac{1}{2}} \rho^{-\frac{1}{2}} \hat{T} \rho^{\frac{1}{2}} \tilde{J}^{-\frac{1}{2}}) . \tag{6.132}$$

To avoid any misinterpretation of the above formulae, we emphasize again that differential operators do not act beyond parentheses.

In the present section, we have introduced the total nuclear angular momentum, $\hat{\boldsymbol{J}}$, sometimes called the total nuclear orbital angular momentum. The latter must be clearly distinguished from the total angular momentum of the molecule, although we will assume that $\hat{\boldsymbol{J}}$ can be taken as the total angular momentum for practical applications in this book. Besides the total nuclear orbital momentum, there exist in a molecule [16]:

- the total nuclear spin angular momentum, $\hat{\boldsymbol{I}}$,
- the total electron spin angular momentum, $\hat{\boldsymbol{S}}$,
- the total electronic orbital angular momentum, $\hat{\boldsymbol{L}}$.

It is the sum of these three angular momenta plus the total nuclear orbital momentum that gives the total angular momentum of the molecule, $\hat{\boldsymbol{J}}_T$, which is a constant of the motion for an isolated system. The nuclear spin angular momenta play a very important role in specific fields, for instance in nuclear magnetic resonance or when the presence of spin nuclear isomers imposes restrictions on the possible rotational states of isomers such as H_2 (the elective occupation of rotational states that stems from the presence of the nuclear spins and the Pauli principle is called "nuclear statistics"). However, the coupling of the nuclear spin angular momentum with the other angular momenta is often very small (this coupling gives rise to the "nuclear hyperfine effects") and the role of the nuclear spin angular momentum can be separated from the rest.

The total electron spin and the total electronic orbital angular momentum differ from one electronic state to another for a given molecule: see the example of H_2 in Sects. 3.5.4 and 3.5.5 for the different spin states. But for a singlet state, this angular momentum is equal to zero. For a closed-shell electronic state, i.e. an electronic state without unpaired electrons, the total electronic orbital angular momentum and the electron spin angular momentum are equal to zero (singlet state). Therefore, for many molecules in the electronic ground state, we can consider $\hat{\boldsymbol{J}}$ only and take the corresponding quantum number, J, as a constant of the motion. However, many free radicals in their electronic ground states, and also many excited electronic states of molecules with closed-shell ground states, have electronic structures in which both electronic orbital and electronic spin angular momentum is present. It then may be necessary to replace $\hat{\boldsymbol{J}}$ by $\hat{\boldsymbol{J}}_T - \hat{\boldsymbol{L}} - \hat{\boldsymbol{S}}$ in $\hat{T}$ and to add to the molecular Hamiltonian the terms corresponding to the spin-orbit coupling for instance. There are cases where the rotational, vibrational, spin of electrons, and electronic contributions can all be intrinsically mixed in molecular eigenstates, e.g. the Renner-Teller effect that may appear in linear molecules [17–20]. But we will not consider such cases here for the sake of simplicity and assume that J is a good quantum number.

6.4 Polyspherical Coordinates

From now on, we do not take into account the "parentheses convention" any more, i.e. the operators may operate beyond parentheses.

Depending on the set of coordinates chosen to describe the dynamics of the system, the KEO can be derived *analytically* using the previous equations, i.e. by using the "chain rule" as illustrated in Sect. 6.7.1.

However, for an important family of coordinates, called "polyspherical coordinates", the KEO can be obtained using existing general expressions [21–27]. Since all the curvilinear coordinates used in the present book correspond to polyspherical coordinates, we briefly describe them here.

As explained in Sect. 5.4, a molecular system with N atoms can be described by $(N-1)$ relative position vectors, $\boldsymbol{R}_1$, $\boldsymbol{R}_2$, ..., $\boldsymbol{R}_{N-1}$. The transformation between the LF-position vectors of the atoms to the relative position vectors (plus the vector from the origin of the frame to the center of mass of the system) is given by Eq. (5.12). In view of Eq. (6.11), it is easy to obtain the Jacobian matrix:

$$\boldsymbol{J}_{R\to r_{LF}} = \left[\begin{array}{c} \boldsymbol{A} \\ \hline \frac{m_1}{M}\ \frac{m_2}{M} \cdots \frac{m_N}{M} \end{array}\right], \tag{6.133}$$

so that Eq. (6.48) particularizes into

$$\boldsymbol{G}_{R\to r_{LF}} = \boldsymbol{J}_{R\to r_{LF}} \cdot \boldsymbol{m}^{-1} \cdot \boldsymbol{J}^T_{R\to r_{LF}} = \begin{bmatrix} \boldsymbol{A}\cdot\boldsymbol{m}^{-1}\cdot\mathbf{A}^T & \mathbf{0} \\ \mathbf{0} & \frac{1}{M} \end{bmatrix}, \tag{6.134}$$

where the constraints (5.13) have been taken into account. If, instead of LF position vectors, SF position vectors $\boldsymbol{r}^1_{SF}, \boldsymbol{r}^2_{SF}, \ldots, \boldsymbol{r}^N_{SF}$ are used, we see from Eq. (6.11) that $\boldsymbol{R}^{CM}_{SF} = \mathbf{0}$ and, of course, in that case, leaving out the center of mass,

$$\boldsymbol{G}_{R\to r_{SF}} = \boldsymbol{A}\cdot\boldsymbol{m}^{-1}\cdot\boldsymbol{A}^T\,. \tag{6.135}$$

Let us now use the following notation: $\boldsymbol{A}\cdot\boldsymbol{m}^{-1}\cdot\boldsymbol{A}^T = \boldsymbol{M}$.

The classical SF Hamiltonian kinetic energy, relative to the center of mass, is simply given by

$$2T = \sum_{i,j=1,\ldots,N-1}^{\lambda=x,y,z} P_{i\,\lambda^{SF}} M^{ij} P_{j\,\lambda^{SF}}\,, \tag{6.136}$$

where $\boldsymbol{P}_i$ is the momentum associated with $\boldsymbol{R}^i_{SF}$.

It can be proven that for some families of vectors such as Jacobi vectors, this matrix $\boldsymbol{M}$ is diagonal. For other sets of vectors such as valence vectors, this matrix is not diagonal and the kinetic energy will have more terms. This is the reason why the *Jacobi coordinates* are said to be *orthogonal coordinates* and the *valence coordinates* are said to be *non-orthogonal coordinates*. The number of terms in the kinetic energy is rather limited for orthogonal coordinates even for large systems. On the other hand, for non-orthogonal coordinates, the number of terms may increase dramatically with the number of atoms.

The exact quantum counterpart of Eq. (6.136) is given by

$$2\hat{T} = \sum_{i,j=1,\ldots,N-1}^{\lambda=x,y,z} \hat{P}^{\dagger}_{i\,\lambda^{SF}} M^{ij} \hat{P}_{j\,\lambda^{SF}} , \tag{6.137}$$

with

$$\hat{P}_{i\,\lambda^{SF}} = -i\hbar \frac{\partial}{\partial R_{i\,\lambda^{SF}}} , \tag{6.138}$$

and

$$\hat{P}^{\dagger}_{i\,\lambda^{SF}} = \hat{P}_{i\,\lambda^{SF}} + \Lambda_i , \tag{6.139}$$

with Λ_i given by Eq. (6.71).

As already explained, in order to separate out overall rotation from internal deformation, *a body fixed (BF) frame with origin at the nuclear center of mass* is introduced. The classical kinetic energy can be recast as follows,

$$2T^{SF} = \sum_{i,j=1,\ldots,N-1}^{\lambda=x,y,z} P_{i\,\lambda^{BF}} M^{ij} P_{j\,\lambda^{BF}} , \tag{6.140}$$

where $P_{i\,\lambda^{BF}}$ are the Cartesian components of the *SF frame* (and not BF) *conjugate momenta in the BF frame* or, in other words, *the projections of the SF conjugate momenta onto the BF frame axes*. The exact quantum counterpart of this equation is given by

$$2\hat{T} = \sum_{i,j=1,\ldots,N-1}^{\lambda=x,y,z} \hat{P}^{\dagger}_{i\,\lambda^{BF}} M^{ij} \hat{P}_{j\,\lambda^{BF}} . \tag{6.141}$$

In addition, we have for each relative vector

$$\hat{\boldsymbol{P}}_i = \hat{P}_{R\,i}\boldsymbol{e}_i - \frac{\boldsymbol{e}_i \times \hat{\boldsymbol{L}}_i}{R_i} . \tag{6.142}$$

where $\boldsymbol{e}_{R_i} = \boldsymbol{R}_i/R_i$ denotes a *unit vector*, $R_i = ||\boldsymbol{R}_i||$ a *vector length*, $\hat{P}_{R\,i} = \boldsymbol{e}_{R_i} \cdot \hat{\boldsymbol{P}}_i = -i\hbar\frac{\partial}{\partial R_i}$ a *radial momentum*, and $\hat{\boldsymbol{L}}_i$ the *angular momentum associated* with $\boldsymbol{R}_i$.

Let us now define the *polyspherical coordinates*. First, we choose a particular definition of the BF frame: the BF frame is oriented such that z^{BF} is parallel to the relative vector $\boldsymbol{R}_{N-1}$ and such that $\boldsymbol{R}_{N-2}$, x^{BF} and z^{BF} lie in the same half-plane. Let us now parameterize the relative vectors by spherical coordinates. The spherical coordinates in the BF frame are defined as (see Fig. 6.2[8])

[8]Reprinted with permission from [27]. Copyright 2009. Licence number 3816530113208. Licence date Feb 26, 2016. Elsevier.

Fig. 6.2 Definition of the polyspherical coordinates: the $N-1$ vector lengths $R_i \in [0, \infty[$, $N-2$ (BF) planar angles $\theta_i^{BF} \in [0, \pi]$, and $N-3$ (BF) dihedral angles $\varphi_i^{BF} \in [0, 2\pi[$

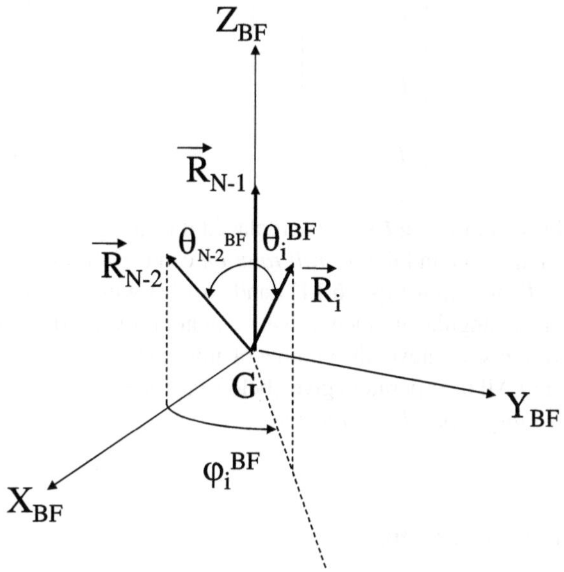

$$
\begin{aligned}
R_{i\,x^{BF}} &= R_i \sin\theta_i^{BF} \cos\varphi_i^{BF} \\
R_{i\,y^{BF}} &= R_i \sin\theta_i^{BF} \sin\varphi_i^{BF} \\
R_{i\,z^{BF}} &= R_i \cos\theta_i^{BF} \,,
\end{aligned}
\tag{6.143}
$$

with $0 \leq \theta_i^{BF} \leq \pi$ and $0 \leq \varphi_i^{BF} < 2\pi$. With our definition of the BF frame, $\theta_{N-1}^{BF}, \varphi_{N-1}^{BF}, \varphi_{N-2}^{BF}$ are equal to zero.[9]

Now, the expressions of BF projections of the angular momenta of Eq. (6.142) can be obtained with few calculations. For instance, the angular momenta $\boldsymbol{L}_i$ ($i = 1, ..., N-3$) associated to the rotating vectors $\boldsymbol{R}_i$ ($i = 1, ..., N-3$) not involved in the definition of the BF frame. They are characterized by the usual formula [12]

$$
\begin{bmatrix} \hat{L}_{i\,x^{BF}} \\ \hat{L}_{i\,y^{BF}} \\ \hat{L}_{i\,z^{BF}} \end{bmatrix} = \begin{bmatrix} -\sin\varphi_i^{BF} & -\cos\varphi_i^{BF}\cot\theta_i^{BF} \\ \cos\varphi_i^{BF} & -\sin\varphi_i^{BF}\cot\theta_i^{BF} \\ 0 & 1 \end{bmatrix} \begin{bmatrix} \frac{\hbar}{i}\frac{\partial}{\partial\theta_i^{BF}} \\ \frac{\hbar}{i}\frac{\partial}{\partial\varphi_i^{BF}} \end{bmatrix}.
\tag{6.144}
$$

The expressions for $\hat{\boldsymbol{L}}_{N-1}$ and $\hat{\boldsymbol{L}}_{N-2}$ are a little more complicated. The expressions for $\hat{\boldsymbol{L}}_{N-2}$ are

[9] In fact, φ_{N-1}^{BF} is not defined and we can use $\varphi_{N-1}^{BF} = 0$ in all the equations.

$$\begin{bmatrix} \hat{L}_{N-2x^{BF}} \\ \\ \hat{L}_{N-2y^{BF}} \\ \\ \hat{L}_{N-2z^{BF}} \end{bmatrix} = \begin{bmatrix} -\cot\theta^{BF}_{N-2}(\hat{J}_{z^{BF}} - \sum_{i=1}^{N-3}\hat{L}_{iz^{BF}}) \\ \\ -i\hbar\frac{\partial}{\partial\theta^{BF}_{N-2}} \\ \\ \hat{J}_{z^{BF}} - \sum_{i=1}^{N-3}\hat{L}_{iz^{BF}} \end{bmatrix} . \quad (6.145)$$

Instead of using $\hat{\boldsymbol{L}}_{N-1}$, we replaced it by $\hat{\boldsymbol{J}}$ - $\sum_{i=1,\ldots,N-2}\hat{\boldsymbol{L}}_i$ so that the *total angular momentum* and the *overall rotation* can be introduced simply.

Combining Eqs. (6.141) and (6.142) with the expression of the BF projections of the angular momenta gives a general expression that can be obtained for many systems, whatever the number of atoms and the nature of the vectors (Jacobi, valence, etc.) All the operators given below can be *obtained from this general expression*, i.e. *without using the chain rule.*

6.5 Examples

Let us give several examples of nuclear kinetic energy operators. Let us first consider NOCl in terms of the Jacobi coordinates of Fig. 5.4. If we express the classical kinetic energy in terms of the conjugate momenta associated with the two Jacobi vectors such as in Sect. 5.4, we get

$$2T^{SF} = \sum_{i,j=1}^{2} \boldsymbol{P}_i^{SF}\, M^{ij}\, \boldsymbol{P}_j^{SF} , \quad (6.146)$$

with the matrix $\boldsymbol{M}$, which reads

$$\boldsymbol{M} = \begin{bmatrix} \frac{1}{m_N} + \frac{1}{m_O} & 0 \\ 0 & \frac{1}{(m_N+m_O)} + \frac{1}{m_{Cl}} \end{bmatrix} . \quad (6.147)$$

The corresponding kinetic energy operator reads:

$$2\hat{T} = \sum_{i=1}^{2} \hat{\boldsymbol{P}}_i^{\dagger}\, M^{ii}\, \hat{\boldsymbol{P}}_i = \sum_{i=1}^{2} \frac{\hat{\boldsymbol{P}}_i^{\dagger} \cdot \hat{\boldsymbol{P}}_i}{\mu_{R_i}} , \quad (6.148)$$

with $\mu_{R_i} = 1/M^{ii}$, the *reduced masses* associated with the two vectors. In view of Eq. (6.147), the reduced masses are the following:

$$\mu_{R_1} = \frac{m_O m_N}{m_O + m_N},$$
$$\mu_{R_2} = \frac{(m_O + m_N)m_{Cl}}{m_O + m_N + m_{Cl}}. \tag{6.149}$$

To obtain the kinetic energy operator explicitly in terms of the internal coordinates, it could be possible to start with the Cartesian components of the momenta associated with the two vectors in Eq. (6.148) and to apply equations such as Eq. (6.73), i.e. to make a coordinate transformation from the Cartesian coordinates to the internal ones. The explicit coordinate transformation using Eq. (6.73) is rather tedious and, as explained in the previous section, since the set of coordinates is a particular case of polyspherical coordinates, the *KEO can be obtained without any calculation*. The KEO of NOCl for $J = 0$ in terms of the Jacobi coordinates of Fig. 5.4 reads

$$\begin{aligned}\hat{T}_{J=0} &= \frac{1}{2\mu_{R_1}}\hat{P}_{R_1}^{\dagger}\hat{P}_{R_1} + \frac{1}{2\mu_{R_2}}\hat{P}_{R_2}^{\dagger}\hat{P}_{R_2} \\ &+ \left(\frac{1}{2\mu_{R_1}R_1^2} + \frac{1}{2\mu_{R_2}R_2^2}\right)\hat{P}_{\theta}^{\dagger}\hat{P}_{\theta} \\ &= -\frac{\hbar^2}{2\mu_{R_1}}\frac{1}{R_1}\frac{\partial^2}{\partial R_1^2}R_1 - \frac{\hbar^2}{2\mu_{R_2}}\frac{1}{R_2}\frac{\partial^2}{\partial R_2^2}R_2 \\ &-\hbar^2\left(\frac{1}{2\mu_{R_1}R_1^2} + \frac{1}{2\mu_{R_2}R_2^2}\right)\frac{1}{\sin\theta}\frac{\partial}{\partial\theta}\sin\theta\frac{\partial}{\partial\theta}.\end{aligned} \tag{6.150}$$

This operator must be used with the following *Euclidean volume element*:

$$dV = \sin\theta\, R_1^2\, R_2^2\, d\theta\, dR_1\, dR_2\,. \tag{6.151}$$

Using Eq. (6.190), we obtain

$$\begin{aligned}\hat{P}_{R_1}^{\dagger} &= \hat{P}_{R_1} - \frac{2i\hbar}{R_1}, \\ \hat{P}_{R_2}^{\dagger} &= \hat{P}_{R_2} - \frac{2i\hbar}{R_2}, \\ \hat{P}_{\theta}^{\dagger} &= \hat{P}_{\theta} - i\hbar\cot\theta\,.\end{aligned} \tag{6.152}$$

For practical reasons, it is more convenient to work with the *non-Euclidean volume element*

$$dV' = \sin\theta\, d\theta\, dR_1\, dR_2\,. \tag{6.153}$$

since it makes the conjugate momenta $\hat{P}_{R_1}$ and $\hat{P}_{R_2}$ *Hermitian*. Changing the volume element comes down to changing the "*convention of normalization*" of the wavefunction [3, 27] (see also Sect. 6.6). In other words, the wavefunction, Ψ, and the kinetic energy operator are replaced by

$$\Psi' = R_1\, R_2\, \Psi\,, \tag{6.154}$$

and

$$\hat{T}' = R_1\, R_2\, \hat{T}\, R_1^{-1}\, R_2^{-1}\,. \tag{6.155}$$

It is rather straightforward to show that

$$\begin{aligned}
\hat{T}'_{J=0} &= \frac{1}{2\mu_{R_1}}\hat{P}^2_{R_1} + \frac{1}{2\mu_{R_2}}\hat{P}^2_{R_2} \\
&\quad -\hbar^2\left(\frac{1}{2\mu_{R_1}R_1^2} + \frac{1}{2\mu_{R_2}R_2^2}\right)\frac{1}{\sin\theta}\frac{\partial}{\partial\theta}\sin\theta\frac{\partial}{\partial\theta} \\
&= -\frac{\hbar^2}{2\mu_{R_1}}\frac{\partial^2}{\partial R_1^2} - \frac{\hbar^2}{2\mu_{R_2}}\frac{\partial^2}{\partial R_2^2} \\
&\quad -\hbar^2\left(\frac{1}{2\mu_{R_1}R_1^2} + \frac{1}{2\mu_{R_2}R_2^2}\right)\frac{1}{\sin\theta}\frac{\partial}{\partial\theta}\sin\theta\frac{\partial}{\partial\theta}\,,
\end{aligned} \tag{6.156}$$

which is to be used with the volume element of Eq. (6.153). We arrive at the important conclusion that the KEO of Eq. (6.156) *has no cross terms*, i.e. no terms proportional to $\frac{\partial^2}{\partial R_1 R_2}$ for instance. This is due to the fact that we are using Jacobi coordinates. Even for much larger systems, the KEO in Jacobi coordinates will not include cross terms involving the conjugate momenta of the lengths of vectors. However, *cross terms mixing the angles* will crop up for systems with more than three atoms ever for Jacobi coordinates.

If $J = 0$, the kinetic energy operator for the $H+H_2$ system is identical to the one of Eq. (6.156) except that the reduced masses are different:

$$\begin{aligned}
\mu_{R_1} &= \frac{m_H m_H}{m_H + m_H} = \frac{m_H}{2}\,, \\
\mu_{R_2} &= \frac{(m_H + m_H)m_H}{m_H + m_H + m_H} = \frac{2}{3}m_H\,.
\end{aligned} \tag{6.157}$$

Now, to calculate the *cross sections* of the collision $H + H_2$, it becomes necessary to include the cases with $J > 0$. We use the definition of the BF frame of Fig. 5.11b and the coordinates of Fig. 5.6a. We recall that for any angular momentum $\hat{L}_+ = \hat{L}_x + i\hat{L}_y$ and $\hat{L}_- = \hat{L}_x - i\hat{L}_y$. In the following equations, x and y will refer to the corresponding BF axes. The KEO reads

$$\begin{aligned}
\hat{T}^J &= -\frac{\hbar^2}{2\mu_{R_1}}\frac{\partial^2}{\partial R_1^2} - \frac{\hbar^2}{2\mu_{R_2}}\frac{\partial^2}{\partial R_2^2} + \frac{\hat{j}^2}{2\mu_{R_1}R_1^2} + \frac{\hat{j}^2}{2\mu_{R_2}R_2^2} \\
&\quad + \frac{1}{2\mu_{R_2}R_2^2}\left[\hat{\boldsymbol{J}}^2 + 2\frac{\partial^2}{\partial\gamma^2} - \hat{J}_+\hat{j}_- - \hat{J}_-\hat{j}_+\right],
\end{aligned} \tag{6.158}$$

with

$$\hat{j}^2 = -\hbar^2 \left(\frac{1}{\sin\theta} \frac{\partial}{\partial\theta} \sin\theta \frac{\partial}{\partial\theta} + \frac{1}{\sin^2\theta} \frac{\partial^2}{\partial\gamma^2} \right), \tag{6.159}$$

and

$$\hat{j}_{\pm} = \mp i\hbar \frac{\partial}{\partial\theta} + i\hbar \frac{\partial}{\partial\gamma} \cot\theta \,. \tag{6.160}$$

In addition, we have

$$\hat{J}_{\pm} = i\hbar \exp \mp i\gamma \left[-\cot\beta \frac{\partial}{\partial\gamma} + \frac{1}{\sin\beta} \frac{\partial}{\partial\alpha} \mp i \frac{\partial}{\partial\beta} \right], \tag{6.161}$$

and

$$\hat{\boldsymbol{J}}^2 = \hbar^2 \left[-\frac{\partial^2}{\partial\beta^2} - \cot\beta \frac{\partial}{\partial\beta} - \frac{1}{\sin^2\beta} (\frac{\partial^2}{\partial\alpha^2} + \frac{\partial^2}{\partial\gamma^2} - 2\cos\beta \frac{\partial^2}{\partial\alpha\partial\gamma}) \right]. \tag{6.162}$$

We also add that

$$\hat{J}_z = -i\hbar \frac{\partial}{\partial\gamma}, \tag{6.163}$$

and thus

$$\hat{J}_z^2 = -\hbar^2 \frac{\partial^2}{\partial\gamma^2}, \tag{6.164}$$

where z denotes the z^{BF} axis. If $J = 0$, the KEO of Eq. (6.162) reduces to the KEO of Eq. (6.156) with the reduced masses of the $H+H_2$ system (see Eq. (6.157)).

Let us now consider a tri-atomic molecule in *valence coordinates: the water molecule* with the coordinates of Fig. 5.5. If we express the classical kinetic energy in terms of the conjugate momenta associated with the two valence vectors, we get

$$2T^{SF} = \sum_{i,j=1}^{2} \boldsymbol{P}_i^{SF} \, M^{ij} \, \boldsymbol{P}_j^{SF}, \tag{6.165}$$

with the matrix $\boldsymbol{M}$, which reads

$$\boldsymbol{M} = \begin{bmatrix} \frac{1}{m_H} + \frac{1}{m_O} & \frac{1}{m_O} \\ \frac{1}{m_O} & \frac{1}{m_H} + \frac{1}{m_O} \end{bmatrix}. \tag{6.166}$$

As for Jacobi vectors, we can define a reduced mass to each vector $\mu_{R_i} = 1/M^{ii}$. Here, $\mu_{R_1} = \mu_{R_2} = \frac{m_O m_H}{m_O + m_H}$. The corresponding kinetic energy operator is given by

$$2\hat{T} = \sum_{i,j=1}^{2} \hat{\boldsymbol{P}}_i^\dagger \, M^{ii} \, \hat{\boldsymbol{P}}_j \ , \tag{6.167}$$

and, if we explicitly express the operator in terms of the coordinates, we obtain for $J = 0$

$$\begin{aligned}\hat{T}_{J=0} = & \frac{1}{2\mu_{R_1}} \hat{P}_{R_1}^\dagger \hat{P}_{R_1} + \frac{1}{2\mu_{R_2}} \hat{P}_{R_2}^\dagger \hat{P}_{R_2} + (\frac{1}{2\mu_{R_1} R_1^2} + \frac{1}{2\mu_{R_2} R_2^2}) \hat{P}_\theta^\dagger \hat{P}_\theta \\ & - \frac{(\hat{P}_{R_1}^\dagger \sin\theta \hat{P}_\theta + \hat{P}_\theta^\dagger \sin\theta \hat{P}_{R_1})}{2m_O R_2} - \frac{(\hat{P}_{R_2}^\dagger \sin\theta \hat{P}_\theta + \hat{P}_\theta^\dagger \sin\theta \hat{P}_{R_2})}{2m_O R_1} \\ & + \frac{(\hat{P}_{R_1}^\dagger \hat{P}_{R_2} + \hat{P}_{R_2}^\dagger \hat{P}_{R_1}) \cos\theta}{2m_O} - \frac{\hat{P}_\theta^\dagger \cos\theta \hat{P}_\theta}{2m_O R_1 R_2} \ .\end{aligned} \tag{6.168}$$

This kinetic energy operator must be used with the volume element given by Eq. (6.151) and the adjoints of the conjugate momenta are given by Eq. (6.152). Comparing Eqs. (6.156) and (6.168), we see that the kinetic energy operator for H_2O has new terms compared to the kinetic energy operator of NOCl. All these new terms are proportional to $1/m_O$ and thus result from the non-diagonal elements in the matrix of Eq. (6.166). This is the reason why *valence coordinates* are said to be *non-orthogonal* and lead to a more complicated kinetic energy. On the other hand, *Jacobi coordinates* are said to be *orthogonal* because the matrix of Eq. (6.147) is diagonal.

Now, if we want to use the volume element of Eq. (6.153), we have to transform the kinetic energy operator of Eq. (6.168) according to Eq. (6.155). This yields

$$\begin{aligned}\hat{T}'_{J=0} = & \frac{1}{2\mu_{R_1}} \hat{P}_{R_1}^2 + \frac{1}{2\mu_{R_2}} \hat{P}_{R_2}^2 + (\frac{1}{2\mu_{R_1} R_1^2} + \frac{1}{2\mu_{R_2} R_2^2}) \hat{P}_\theta^\dagger \hat{P}_\theta \\ & - \frac{(\hat{P}_{R_1} \sin\theta \hat{P}_\theta + \hat{P}_\theta^\dagger \sin\theta \hat{P}_{R_1})}{2m_O R_2} - \frac{(\hat{P}_{R_2} \sin\theta \hat{P}_\theta + \hat{P}_\theta^\dagger \sin\theta \hat{P}_{R_2})}{2m_O R_1} \\ & + \frac{\hat{P}_{R_1} \hat{P}_{R_2} \cos\theta}{m_O} - \frac{\hat{P}_\theta^\dagger \cos\theta \hat{P}_\theta}{2m_O R_1 R_2} - \frac{\hbar^2 \cos\theta}{m_O R_1 R_2} \ .\end{aligned} \tag{6.169}$$

The last term in the previous equation, $-\frac{\hbar^2 \cos\theta}{m_O R_1 R_2}$, is called *"extrapotential"* term (see Eq. (6.85), Sect. 6.3.2) since it is a purely multiplicative operator. It crops up only because we use a non-Euclidean convention of normalization (see Eq. (6.153)). Finally, reverting to partial derivatives, the kinetic energy operator of water in valence coordinates of Eq. (6.169) can be recast as

$$
\begin{aligned}
\hat{T}'_{J=0} &= -\frac{\hbar^2}{2\mu_{R_1}}\frac{\partial^2}{\partial R_1^2} - \frac{\hbar^2}{2\mu_{R_2}}\frac{\partial^2}{\partial R_2^2} - \hbar^2\left(\frac{1}{2\mu_{R_1}R_1^2} + \frac{1}{2\mu_{R_2}R_2^2}\right)\frac{1}{\sin\theta}\frac{\partial}{\partial\theta}\sin\theta\frac{\partial}{\partial\theta} \\
&- \frac{\hbar^2}{m_O}\cos\theta\frac{\partial^2}{\partial R_2\partial R_1} + \frac{\hbar^2}{m_O R_2}\frac{\partial^2}{\partial R_1\partial\theta}\sin\theta + \frac{\hbar^2}{m_O R_1}\frac{\partial^2}{\partial R_2\partial\theta}\sin\theta \\
&+ \frac{\hbar^2}{2m_O R_1 R_2}[(\frac{1}{\sin\theta}\frac{\partial}{\partial\theta}\sin\theta\frac{\partial}{\partial\theta})\cos\theta + \cos\theta(\frac{1}{\sin\theta}\frac{\partial}{\partial\theta}\sin\theta\frac{\partial}{\partial\theta}\sin\theta)] - \frac{\hbar^2\cos\theta}{m_O R_1 R_2}\,.
\end{aligned}
\tag{6.170}
$$

For *very large systems*, the number of terms may increase dramatically. In this context, it is worth noting that the program TANA developed by M. Ndong, D. Lauvergnat and co-workers provides automatically the explicit expression of the KEO in polyspherical coordinates whatever the number of atoms [28, 29].

6.6 The Problem of Singularities in the KEO *

If curvilinear coordinates are used, some operators may become singular for specific geometries when the Euclidean volume element is equal to zero. This is the case for operators such as

$$
\begin{aligned}
\hat{j}^2 &= -\hbar^2\left(\frac{1}{\sin\theta}\frac{\partial}{\partial\theta}\sin\theta\frac{\partial}{\partial\theta} + \frac{1}{\sin^2\theta}\frac{\partial^2}{\partial\phi^2}\right), \\
\hat{j}_\pm &= \hbar e^{\pm i\phi}\left(\pm\frac{\partial}{\partial\theta} + i\cot\theta\frac{\partial}{\partial\phi}\right),
\end{aligned}
\tag{6.171}
$$

that appear in the kinetic energy operator of Eq. (6.158) (see Eqs. (6.159) and (6.160), $\theta \in [0, \pi]$ and $\phi \in [0, 2\pi]$). These operators are singular for $\theta = 0$ or π since $\sin\theta$ tends to zero when θ tends to zero or π. These "singularities" appear because the Euclidean volume element, $dV = \sin\theta d\theta d\phi$ vanishes for the corresponding geometries.

The singularity appears only in coordinate space, all matrix elements are non-singular when an appropriate basis set, e.g. the *spherical harmonics* $Y_{lm}(\theta, \phi)$, are used. The spherical harmonics are given by

$$
Y_{lm}(\theta, \phi) = \tilde{P}_l^m(\cos\theta)\frac{e^{im\phi}}{\sqrt{2\pi}}, \tag{6.172}
$$

with

$$
\tilde{P}_l^m(\cos\theta) = (-1)^m\sqrt{\frac{(2l+1)(l-m)!}{2(l+m)!}}P_l^m(\cos\theta), \tag{6.173}
$$

the normalized associated Legendre functions. In fact,

$$\hat{j}^2 Y_{lm}(\theta,\phi) = \hbar^2 l(l+1) Y_{lm}(\theta,\phi)$$
$$\hat{j}_{\pm} Y_{lm}(\theta,\phi) = \hbar\sqrt{l(l+1) - m(m\pm 1)}\, Y_{l,m\pm 1}(\theta,\phi)\,. \qquad (6.174)$$

In particular, after integration over the functions, $\frac{e^{im\phi}}{\sqrt{2\pi}}$, we see that the operator j^2 is diagonal in this basis set of ϕ and that the corresponding matrix element reads

$$\hat{j}^2 \delta_{mn} = -\hbar^2 \left(\frac{1}{\sin\theta}\frac{\partial}{\partial\theta}\sin\theta\frac{\partial}{\partial\theta} - \frac{m^2}{\sin^2\theta} \right) \delta_{mn}\,, \qquad (6.175)$$

and

$$l(l+1) f(\cos\theta) = \left(\frac{1}{\sin\theta}\frac{\partial}{\partial\theta}\sin\theta\frac{\partial}{\partial\theta} - \frac{m^2}{\sin^2\theta} \right) f(\cos\theta)\,, \qquad (6.176)$$

corresponds to the general Legendre equation with $y = f(x)$ and $x = \cos\theta$. The associated Legendre polynomials, $P_l^m(x)$, are the canonical solutions of this equation. Two points must be emphasized at this level: (i) the basis set of spherical harmonics of Eq. (6.172) is not a direct product basis, because $\tilde{P}_l^m(\cos\theta)$ depends on both l and m. In other words, θ and ϕ are intrinsically linked together in the basis set. (ii) The choice of this direct product basis guarantees that the functions have the correct asymptotic behavior when θ tends to zero or π, i.e. that the matrix elements are non-singular. If the chosen basis functions do not have such an asymptotic behavior, the convergence of the matrix elements with respect to the number of basis functions will be very poor and the basis set is said to be incomplete for the corresponding mathematical problem.

On the other hand, if for physical reasons (e.g. because of potential barriers) the geometries corresponding to $|\sin\theta| \ll 1$ can not be reached, a direct product basis set such as

$$\tilde{P}_l(\cos\theta)\frac{e^{im\phi}}{\sqrt{2\pi}}\,, \qquad (6.177)$$

with $\tilde{P}_l(\cos\theta) = \tilde{P}_l^0(\cos\theta)$ being normalized Legendre polynomials that are restricted to an interval smaller than $[0,\pi]$, can be employed. Now, if we use $x = \cos\theta$, the volume element becomes $dV = dx d\varphi$ and the well-known particle-in-a-box functions (i.e. the $\sin\frac{n\pi x}{L}$ functions with L the interval) can also be used instead of the Legendre polynomials for x. Such systems, where potential barriers exclude the singular regions, are called "semi-rigid". Note that, the H+H_2 system mentioned in Sect. 6.5 is not semi-rigid since the H_2 molecule rotates freely before collision. Thus, the geometries near zero or π can be reached. For this system, it is then necessary to use a non direct product basis set such as the spherical harmonics.

6.7 Complements to Chapter 6

6.7.1 Chain Rule and Laplacian Operator

To apply the chain rule to the Laplacian in Cartesian coordinates

$$\Delta = \sum_{\alpha} \frac{\partial^2}{\partial x^\alpha \partial x^\alpha}, \tag{6.178}$$

we start from

$$\frac{\partial}{\partial x^\alpha} = \sum_i \frac{\partial q^i}{\partial x^\alpha} \frac{\partial}{\partial q^i}, \tag{6.179}$$

and

$$\begin{aligned} \frac{\partial}{\partial x^\alpha} \frac{\partial}{\partial x^\alpha} &= \frac{\partial}{\partial x^\alpha} \sum_i \frac{\partial q^i}{\partial x^\alpha} \frac{\partial}{\partial q^i} \\ &= \sum_i \left(\frac{\partial^2 q^i}{\partial x^\alpha \partial x^\alpha}\right) \frac{\partial}{\partial q^i} + \sum_{i,j} \left(\frac{\partial q^i}{\partial x^\alpha}\right)\left(\frac{\partial q^j}{\partial x^\alpha}\right) \frac{\partial^2}{\partial q^i \partial q^j}, \end{aligned} \tag{6.180}$$

so that

$$\Delta = \sum_{i,j} \sum_{\alpha} \left(\frac{\partial q^i}{\partial x^\alpha}\right)\left(\frac{\partial q^j}{\partial x^\alpha}\right) \frac{\partial^2}{\partial q^i \partial q^j} + \sum_i \sum_{\alpha} \left(\frac{\partial^2 q^i}{\partial x^\alpha \partial x^\alpha}\right) \frac{\partial}{\partial q^i}. \tag{6.181}$$

Comparing Eq. (6.181) with Eq. (6.75), we have

$$f^{ij}(\boldsymbol{q}) = G^{ij}(\boldsymbol{q}) = \sum_{\alpha} \left(\frac{\partial q^i}{\partial x^\alpha}\right)\left(\frac{\partial q^j}{\partial x^\alpha}\right), \tag{6.182}$$

which is the expression of the *contravariant components* of the metric tensor in the $\boldsymbol{q}(\boldsymbol{x})$ approach and

$$f^i(\boldsymbol{q}) = \sum_{\alpha} \frac{\partial^2 q^i}{\partial x^\alpha \partial x^\alpha}. \tag{6.183}$$

Obviously, $\boldsymbol{x}(\boldsymbol{q})$ has to be used in Eqs. (6.182) and (6.183).

6.7.2 Adjoint Operator $\hat{P}_i^\dagger$ of the Momentum Operator $\hat{P}_i = -i\hbar\frac{\partial}{\partial q^i}$

The adjoint of an operator $\hat{A}$ is defined as follows (see Eq. (2.14), Sect. 2.1.3). For all functions Ψ and Φ:

$$\langle \hat{A}\Psi|\Phi\rangle = \int (\hat{A}\Psi)^\star \Phi dV = \int \Psi^\star(\hat{A}^\dagger\Phi)dV$$
$$= \langle\Psi|\hat{A}^\dagger|\Phi\rangle\,, \tag{6.184}$$

where $dV = d\tau^{(E)} = dx^1 \dots dx^{3N} = J_{x\to q}\, dq^1 \cdots dq^{3N}$ is the volume element in Cartesian space. The parentheses in Eq. (6.184) indicate that $\hat{A}$ *does not operate beyond the parentheses.*

For simplicity, $J_{x\to q}$ will be denoted J and $dq^1 \dots dq^{3N}$ by $d\boldsymbol{q}$, so that Eq. (6.184) reads

$$\int (\hat{A}\Psi)^\star \Phi J d\boldsymbol{q} = \int \Psi^\star(\hat{A}^\dagger\Phi) J d\boldsymbol{q}\,, \tag{6.185}$$

where Ψ, Φ and J are functions of the $3N$ curvilinear coordinates $\boldsymbol{q} = (q^1 \dots q^{3N})$. In addition, the product $J\Psi^\star\Phi$ is assumed to vanish on the boundaries of the range of variation of the coordinates $\boldsymbol{q}$. Particularizing to the momentum operator $\hat{P}_i = -i\hbar\frac{\partial}{\partial q^i}$, we obtain successively

$$\langle \hat{P}_i\Psi|\Phi\rangle = \int (-i\hbar\frac{\partial}{\partial q^i}\Psi)^\star \Phi J d\boldsymbol{q}\,. \tag{6.186}$$

Integrating by parts yields

$$\int (i\hbar\frac{\partial}{\partial q^i}\Psi^\star)\Phi J d\boldsymbol{q} = i\hbar\left[\Psi^\star\Phi J\right]_{-\infty}^{\infty} - i\hbar\int \Psi^\star(\frac{\partial}{\partial q^i}J\Phi)d\boldsymbol{q}\,. \tag{6.187}$$

Since $J\Psi^\star\Phi$ vanishes on the boundaries and inserting J and J^{-1}, we obtain

$$\int (-i\hbar\frac{\partial}{\partial q^i}\Psi)^\star \Phi J d\boldsymbol{q} = \int \Psi^\star(J^{-1}(-i\hbar)\frac{\partial}{\partial q^i}J\Phi)J d\boldsymbol{q}\,, \tag{6.188}$$

or still, reverting to the momentum operator $\hat{P}_i = -i\hbar\frac{\partial}{\partial q^i}$,

$$\int (\hat{P}_i\Psi)^\star \Phi J d\boldsymbol{q} = \int \Psi^\star(J^{-1}\hat{P}_i J\Phi)J d\boldsymbol{q}\,. \tag{6.189}$$

Comparing Eqs. (6.189) and (6.184) obviously leads to the following expression for the adjoint of the momentum operator in curvilinear coordinates[10]

$$\hat{P}_i^\dagger = J^{-1}\,\hat{P}_i\,J\ (i = 1, \ldots, 3N)\,. \tag{6.190}$$

Now,

$$\hat{P}_i^\dagger \Psi = J^{-1}\,\hat{P}_i\,J\Psi = \hat{P}_i\Psi + (J^{-1}\,\hat{P}_i\,J)\Psi\,, \tag{6.191}$$

or still

$$\hat{P}_i^\dagger = \hat{P}_i + \Lambda_i\,, \tag{6.192}$$

where

$$\Lambda_i = (J^{-1}\,\hat{P}_i\,J) = (\hat{P}_i \ln J)\,. \tag{6.193}$$

Parenthesis convention: When an operator appears in an expression in parentheses, it is always assumed not to operate beyond the parentheses. Thus, $J^{-1}\,\hat{P}_i\,J \neq (J^{-1}\,\hat{P}_i\,J)$. The left hand side is an operator while the right hand side is a number.

6.7.3 Metric Tensor for Polar Coordinates

$x(q)$ approach

$$\begin{aligned} x(r,\theta) &= r\cos\theta \\ y(r,\theta) &= r\sin\theta\,. \end{aligned} \tag{6.194}$$

The Jacobian matrix corresponding to the coordinate change from the Cartesian coordinates to the polar coordinates reads

$$\boldsymbol{J} = \begin{bmatrix} \frac{\partial x}{\partial r} & \frac{\partial x}{\partial \theta} \\ \frac{\partial y}{\partial r} & \frac{\partial y}{\partial \theta} \end{bmatrix} = \begin{bmatrix} \cos\theta & -r\sin\theta \\ \sin\theta & r\cos\theta \end{bmatrix}. \tag{6.195}$$

The absolute value of the Jacobian determinant is

$$J = |\text{Det}\ \boldsymbol{J}| = r\,. \tag{6.196}$$

The matrix of the *covariant* components of the metric tensor reads

[10]When Cartesian coordinates are used $J = 1$ and the well-known result $\hat{P}_i^\dagger = \hat{P}_i$ is obtained.

$$\begin{aligned} \boldsymbol{g} &= \begin{bmatrix} g_{rr} & g_{r\theta} \\ g_{\theta r} & g_{\theta\theta} \end{bmatrix} = \boldsymbol{J}^T \cdot \boldsymbol{J} = \begin{bmatrix} \cos\theta & \sin\theta \\ -r\sin\theta & r\cos\theta \end{bmatrix} \begin{bmatrix} \cos\theta & -r\sin\theta \\ \sin\theta & r\cos\theta \end{bmatrix} \\ &= \begin{bmatrix} 1 & 0 \\ 0 & r^2 \end{bmatrix} . \end{aligned} \tag{6.197}$$

We see that Det$\boldsymbol{g} = r^2 = J^2$. The matrix of the *contravariant* components of the metric tensor reads

$$\boldsymbol{G} = \begin{bmatrix} g^{rr} & g^{r\theta} \\ g^{\theta r} & g^{\theta\theta} \end{bmatrix} = \boldsymbol{g}^{-1} = \begin{bmatrix} 1 & 0 \\ 0 & \frac{1}{r^2} \end{bmatrix}, \tag{6.198}$$

or still, since

$$\boldsymbol{J}^{-1} = \frac{1}{r} \begin{bmatrix} r\cos\theta & r\sin\theta \\ -\sin\theta & \cos\theta \end{bmatrix}, \tag{6.199}$$

$$\begin{aligned} \boldsymbol{G} = \boldsymbol{J}^{-1} \cdot (\boldsymbol{J}^{-1})^T &= \begin{bmatrix} \cos\theta & \sin\theta \\ -\frac{\sin\theta}{r} & \frac{\cos\theta}{r} \end{bmatrix} \begin{bmatrix} \cos\theta & -\frac{\sin\theta}{r} \\ \sin\theta & \frac{\cos\theta}{r} \end{bmatrix} \\ &= \begin{bmatrix} 1 & 0 \\ 0 & \frac{1}{r^2} \end{bmatrix}, \end{aligned} \tag{6.200}$$

$q(x)$ approach

$$\begin{aligned} r &= \sqrt{x^2+y^2} \\ \theta &= \arctan\frac{y}{x}, \end{aligned} \tag{6.201}$$

$$\boldsymbol{J}^{-1} = \begin{bmatrix} \frac{\partial r}{\partial x} & \frac{\partial r}{\partial y} \\ \frac{\partial \theta}{\partial x} & \frac{\partial \theta}{\partial y} \end{bmatrix} = \begin{bmatrix} \frac{x}{(x^2+y^2)^{\frac{1}{2}}} & \frac{y}{(x^2+y^2)^{\frac{1}{2}}} \\ -\frac{y}{x^2+y^2} & \frac{x}{x^2+y^2} \ . \end{bmatrix},$$

which is obviously a function of the Cartesian coordinates (x, y). *Replacing* by $x = r\cos\theta$ and $y = r\sin\theta$ *i.e. reverting to* $\boldsymbol{x}(\boldsymbol{q})$, gives

$$\boldsymbol{J}^{-1} = \begin{bmatrix} \frac{r\cos\theta}{r} & \frac{r\sin\theta}{r} \\ -\frac{r\sin\theta}{r^2} & -\frac{r\cos\theta}{r^2} \end{bmatrix} = \begin{bmatrix} \cos\theta & \sin\theta \\ -\frac{\sin\theta}{r} & \frac{\cos\theta}{r} \end{bmatrix},$$

and

$$|\text{Det}\,\boldsymbol{J}^{-1}| = \frac{1}{r} = J^{-1}. \tag{6.202}$$

Thus,

$$\boldsymbol{G} = \begin{bmatrix} G^{rr} & G^{r\theta} \\ G^{\theta r} & G^{\theta\theta} \end{bmatrix} = \begin{bmatrix} 1 & 0 \\ 0 & \frac{1}{r^2} \end{bmatrix},$$

and

$$|\mathrm{Det}\boldsymbol{G}| = \frac{1}{r^2} = J^{-2}\,. \tag{6.203}$$

When the metric tensors are diagonal, the curvilinear coordinates are called *orthogonal coordinates*.

6.7.4 Computation of the Laplacian Operator in Polar Coordinates

In the case of the polar coordinates of Sect. 6.7.3, the expression (6.62) particularizes as follows

$$\begin{aligned}\Delta = -2T/\hbar^2 = J^{-1}[&\frac{\partial}{\partial r}JG^{rr}\frac{\partial}{\partial r} + \frac{\partial}{\partial r}JG^{r\theta}\frac{\partial}{\partial \theta} \\ &+ \frac{\partial}{\partial \theta}JG^{\theta r}\frac{\partial}{\partial r} + \frac{\partial}{\partial \theta}JG^{\theta\theta}\frac{\partial}{\partial \theta}]\,,\end{aligned} \tag{6.204}$$

with

$$\begin{aligned}J &= r \\ G^{rr} &= 1 \\ G^{\theta\theta} &= \frac{1}{r^2} \\ G^{r\theta} &= G^{\theta r} = 0\,,\end{aligned} \tag{6.205}$$

so that we obtain the well-known result

$$\begin{aligned}\Delta = -2T/\hbar^2 &= \frac{1}{r}\left[\frac{\partial}{\partial r}r\frac{\partial}{\partial r} + \frac{\partial}{\partial \theta}\frac{1}{r}\frac{\partial}{\partial \theta}\right] \\ &= \frac{1}{r}\frac{\partial}{\partial r}r\frac{\partial}{\partial r} + \frac{1}{r^2}\frac{\partial^2}{\partial \theta^2} \\ &= \frac{1}{r}\frac{\partial}{\partial r} + \frac{\partial^2}{\partial r^2} + \frac{1}{r^2}\frac{\partial^2}{\partial \theta^2}\,.\end{aligned} \tag{6.206}$$

Or still, Eq. (6.73) particularizes as follows

$$\Delta = -2T/\hbar^2 = G^{rr}\frac{\partial^2}{\partial r^2} + G^{r\theta}\frac{\partial^2}{\partial r\partial\theta} + G^{\theta r}\frac{\partial^2}{\partial\theta\partial r} + G^{\theta\theta}\frac{\partial^2}{\partial\theta^2}$$
$$+ (G^{rr}\frac{\partial \ln J}{\partial r} + \frac{\partial}{\partial r}G^{rr})\frac{\partial}{\partial r}$$
$$+ (G^{r\theta}\frac{\partial \ln J}{\partial r} + \frac{\partial}{\partial r}G^{r\theta})\frac{\partial}{\partial\theta}$$
$$+ (G^{\theta r}\frac{\partial \ln J}{\partial\theta} + \frac{\partial}{\partial\theta}G^{r\theta})\frac{\partial}{\partial r}$$
$$+ (G^{\theta\theta}\frac{\partial \ln J}{\partial\theta} + \frac{\partial}{\partial\theta}G^{\theta\theta})\frac{\partial}{\partial\theta}\,. \tag{6.207}$$

Using Eq. (6.205), we obtain of course the same result

$$\Delta = -2T/\hbar^2 = \frac{\partial^2}{\partial r^2} + \frac{1}{r^2}\frac{\partial^2}{\partial\theta^2} + (\frac{\partial \ln r}{\partial r})\frac{\partial}{\partial r}$$
$$= \frac{1}{r}\frac{\partial}{\partial r} + \frac{\partial^2}{\partial r^2} + \frac{1}{r^2}\frac{\partial^2}{\partial\theta^2}\,. \tag{6.208}$$

If we change the normalization convention and use the volume element $d\tau^{(\rho)} = \rho(r,\theta)drd\theta$ instead of $d\tau^{(E)} = Jdrd\theta = rdrd\theta$, we obtain from Eq. (6.86) and using Eq. (6.205)

$$\Delta^{(\rho)} = -2T^{(\rho)}/\hbar^2 = \frac{\partial^2}{\partial r^2} + \frac{1}{r^2}\frac{\partial^2}{\partial\theta^2} + (\frac{\partial \ln \rho}{\partial r})\frac{\partial}{\partial r}$$
$$+ (r^{1/2}\rho^{-1/2}\Delta\rho^{1/2}r^{-1/2})\,. \tag{6.209}$$

In $(r^{1/2}\rho^{-1/2}\Delta\rho^{1/2}r^{-1/2})$, the parentheses indicate that the derivative operators do not operate beyond them. In other words, $(r^{1/2}\rho^{-1/2}\Delta\rho^{1/2}r^{-1/2})$ is purely multiplicative.

For instance, for $\rho(r,\theta) = 1$, $\ln\rho = 0$ and

$$\Delta^{(\rho=1)} = \frac{\partial^2}{\partial r^2} + \frac{1}{r^2}\frac{\partial^2}{\partial\theta^2} + \frac{1}{4r^2}\,. \tag{6.210}$$

Indeed, it is easy to check that

$$(r^{1/2}[\frac{1}{r}\frac{\partial}{\partial r} + \frac{\partial^2}{\partial r^2} + \frac{1}{r^2}\frac{\partial^2}{\partial\theta^2}]r^{-1/2}) = \frac{1}{4r^2}\,, \tag{6.211}$$

which is a multiplicative operator.

References

1. Podolsky B (1928) Quantum-mechanically correct form of hamiltonian function for conservative systems. Phys Rev 32:812
2. Brillouin L (1938) Les tenseurs en mécanique et en électricité. Masson, Paris
3. Nauts A, Chapuisat X (1985) Momentum, quasi-momentum and hamiltonian operators in terms of arbitrary curvilinear coordinates, with special emphasis on molecular hamiltonians. Mol Phys 55:1287
4. Kemble EC (2005) The fundamental principles of quantum mechanics. Dover Publications, New York
5. Meyer R (1979) Flexible models for intramolecular motion, a versatile treatment and its application to glyoxal. J Mol Spec 76:266
6. Harthcock MA, Laane J (1982) Calculation of kinetic energy terms for the vibrational Hamiltonian: Application to large-amplitude vibrations using one-, two-, and three-dimensional models. J Mol Spec 91:300
7. Nino A, Munoz-Caro C (1994) Computation of kinetic constants for large range internal motions in molecules. Comput Chem 18:27
8. Senent M (1998) Determination of the kinetic energy parameters of non-rigid molecules. Chem Phys Lett 296:299
9. Lauvergnat D, Nauts A (2002) Exact numerical computation of a kinetic energy operator in curvilinear coordinates. J Chem Phys 116:8560
10. Matyus E, Simunek J, Csaszar AG (2009) On the variational computation of a large number of vibrational energy levels and wave functions for medium-sized molecules. J Chem Phys 131:074106
11. Strobusch D, Scheurer C (2011) The hierarchical expansion of the kinetic energy operator in curvilinear coordinates extended to the vibrational configuration interaction method. J Chem Phys 135:144101
12. Gatti F, Nauts A (2003) Vector parametrization, partial angular momenta and unusual commutation relations in molecular physics. Chem Phys 295:167–174
13. Nauts A, Gatti F (2010) Unusual commutation relations in physics. Am J Phys 78:1365
14. Kutzelnigg W (2007) Which masses are vibrating or rotating in a molecule? Mol Phys 105:2627
15. Van Vleck JH (1951) The coupling of angular momentum vectors in molecules. Rev Mod Phys 23:213
16. Bunker PR, Jensen P (1998) Molecular symmetry and spectroscopy, 2nd edn. NRC Research Press, Ottawa
17. Renner R (1934) Z Phys 92:172
18. Poluyanov LV, Domcke W (2009) Spin-orbit vibronic coupling in Jahn-Teller and Renner systems. In: Köppel H, Yarkony DR, Barentzen H (eds) Springer series in chemical physics: the Jahn-Teller effect, fundamentals and implications for physics and chemistry, vol 97. Springer, pp 77
19. Jutier L, Léonard C, Gatti F (2009) Renner-Teller effect in linear tetra-atomic molecules II. Rovibronic levels analysis of the $X_2\pi_u$ electronic state of $HCCH^+$. J Chem Phys 130:134301
20. Jutier L, Léonard C, Gatti F (2009) Renner-Teller effect in linear tetra-atomic molecules: I. Variational method including couplings between all degrees of freedom on six-dimensional potential energy surface of $HCCH^+$. J Chem Phys 130:134302
21. Gatti F, Iung C, Menou M, Justum Y, Nauts A, Chapuisat X (1998) Vector parametrization of the n-atom problem in quantum mechanics. I. Jacobi vectors. J Chem Phys 108:8804
22. Gatti F, Iung C, Menou M, Chapuisat X (1998) Vector parametrization of the n-atom problem in quantum mechanics. II. Coupled-angular-momentun spectral representations for four atom systems. J Chem Phys 108:8821
23. Gatti F (1999) Vector parametrization of the n-atom problem in quantum mechanics. III. Separation into two sub-systems. J Chem Phys 111:7225
24. Gatti F, Munoz C, Iung C (2001) A general expression of the exact kinetic energy operator in polyspherical coordinates. J Chem Phys 114:8275

25. Gatti F, Iung C (2003) Exact and constrained kinetic energy operators in polyspherical coordinates. J Theor Comp Chem 2:507
26. Iung C, Gatti F (2006) Polyspherical parametrization of a n-atom system: principles and applications. Int J Quant Chem 106:130
27. Gatti F, Iung C (2009) Exact and constrained kinetic energy operators for polyatomic molecules: the polyspherical approach. Phys Rep 484:1
28. Ndong M, Joubert Doriol L, Meyer HD, Nauts A, Gatti F, Lauvergnat D (2012) Automatic computer procedure for generating exact and analytical kinetic energy operators based on the polyspherical approach. J Chem Phys 136:034107
29. Ndong M, Nauts A, Joubert-Doriol L, Meyer H-D, Gatti F, Lauvergnat D (2013) Automatic computer procedure for generating exact and analytical kinetic energy operators based on the polyspherical approach: general formulation and removal of singularities. J Chem Phys 139:204107

Chapter 7
Group Theory and Molecular Symmetry

In various chapters of this book, we have mentioned how a group-theoretical approach could be applied to *molecular symmetry* and help in the context of vibrational and vibronic problems. In the present chapter, we shall give further information on three examples addressed elsewhere: the harmonic vibrations of water (see Sect. 9.3 in Chap. 9); the vibronic Hamiltonian of the butratriene cation (see Sect. 12.2 in Chap. 12); the umbrella inversion of twice-deuterated ammonia (see Sects. 13.3–13.5 in Chap. 13).

Note that applications of group theory to the treatment of angular momenta (rotations and spins of electrons and nuclei) and of their various compositions require an involved formalism that is beyond the scope of the present book. Our discussion will thus be limited to vibronic variables. In practice, we shall consider transformations of nuclear displacements and atomic orbitals to define symmetry-adapted combinations (vibrational modes and molecular orbitals) to be used ultimately for building vibrational and electronic wavefunctions belonging to well-defined symmetry species (because the invariances of the molecular Hamiltonian are transferred to the Hessian and Fock matrices). The positions of the electrons and the nuclei of a molecular system will be represented either in a space-fixed frame or in a body-fixed frame (assuming coincidence between the molecular and the nuclear centres of mass). They will be denoted (X, Y, Z) or (x, y, z), respectively, throughout the present chapter for notational simplicity.

7.1 Introduction

Group theory is a powerful branch of mathematics with many applications in physics and chemistry. *It is often used* as a practical "tool" *to predict—before any formal derivation or numerical calculation—that the value of an observable should be zero for symmetry reasons*, which is related to the structural properties of the object under

F. Gatti et al., *Applications of Quantum Dynamics in Chemistry*,
Lecture Notes in Chemistry 98, DOI 10.1007/978-3-319-53923-2_7

study and subsequent invariance laws. This is how and why, for example, selection rules in molecular spectroscopy are able to tell us in advance that a given transition between two levels is allowed or forbidden and should thus be observed or not in the experimental spectrum. However, if a transition is not symmetry-forbidden, predicting its magnitude still requires solving the relevant equations derived from the fundamental laws of physics or making relevant approximations to determine whether there are some physical reasons that should make the intensity large or small.

A group-theoretical analysis thus provides formal relationships that reflect invariance properties under particular transformations affecting the space and spin variables of the electrons and nuclei in a molecular system. *In the practical case of a molecule with several identical nuclei (or in situations of coplanar or collinear arrangements where there are intrinsic symmetries with respect to the three-dimensional space), the "shape" of the system (determined from the relative positions of the nuclei) occurs to be invariant under some geometrical transformations (symmetry operations, such as axial rotations, planar reflections, permutations of identical nuclei, etc.).* If so, the geometry of the molecule, in other words its structure, is said to have non-trivial symmetry. In addition, identical fermions or bosons are characterised by wavefunctions that are antisymmetric or symmetric, respectively, with respect to the permutations of both their space and spin coordinates (note that the density of probability of presence—the square-modulus of the wavefunction—is symmetric in both cases, as expected for indistinguishable particles).

The set of such transformations forms an algebraic structure known in mathematics as a symmetry group. This set is closed under composition (sequential action of two transformations), has a neutral element (the identity), and each transformation has an inverse (reversing the effect of the transformation). Representation theory is then used to determine what are known as the irreducible representations of the symmetry group (also called symmetry species) that are typical of the group structure. Each is characterised by a specific behaviour, called a character, under all transformations (for example odd or even). The corresponding list of such characters is called the *character table* of the symmetry group. Molecular properties that depend on the geometry can thus be rationalised according to the irreducible representations to which they belong. From this knowledge, it is then possible to understand why a physical quantity is zero at some geometry in much the same way as why a one-dimensional odd function is zero at the origin and gives rise to a vanishing integral over a symmetric domain or why an even function corresponds to a horizontal tangent at the origin, etc.

When used in chemistry, *group theory is often derived from the geometrical shape of the equilibrium structure of the molecule under study* (for example, the shape of the methane molecule is that of a regular tetrahedron). In other words, because identical nuclei are indistinguishable, an internuclear arrangement that exhibits some symmetry is either geometrically-invariant or turned into its enantiomeric partner under a set of rotations and/or reflections (possibly combined together). In this, *the geometry of this arrangement is identified to a "solid" object of the same shape* (i.e., a sphere, a cylinder, a cone, or some type of polyhedron), irrespectively of the labelling pattern of identical nuclei. In mathematical terms, this implies that the *point group* (the set of such symmetry operations) that characterises this geometric structure can be used to determine how the properties of the molecule are affected under these operations. We will refer to this treatment as the "standard" point-group approach hereafter.

From a more rigorous perspective, group theory applied to a molecular system should reflect the invariances of the molecular Hamiltonian with respect to transformations acting on the spin and space variables of the particles. Essentially, the molecular Hamiltonian (see Chap. 3) is unchanged if the positions and spins of indistinguishable particles are permuted or if the position of each particle is inverted in space through the centre of mass of the molecule. This description is often referred to as *the molecular-symmetry approach.*

Point-group symmetry thus concerns geometrical invariance while molecular symmetry deals with energetic invariance. Using one for the other could seem paradoxical at first glance. However, it occurs that both approaches are intimately related. This is the reason why they are often mixed up, as *they provide equivalent results in most application cases where only small-amplitude deformations (vibrations) from a single equilibrium structure are to be considered.* In contrast, *the standard point-group approach becomes inadequate in situations where a large-amplitude deformation (contortion) connecting two distinct equilibrium structures is to be considered* (this will be illustrated in the last section of this chapter).

Let us now clarify some of the fundations of the molecular-symmetry approach. As mentioned in Chap. 3, a molecular system is a collection of electrons and nuclei, in other words a chemical formula that uniquely determines the general expression of the molecular Hamiltonian according to the number of electrons and the numbers and natures of the various nuclei (in this, different isotopes of the same element are distinguishable, as they have different masses). A given chemical formula often corresponds to several possible isomers (in a general sense: local minima of the potential energy surface or asymptotic minima when considering fragmented species). Each isomer is characterised by its structural formula determined by the relative arrangement of the nuclei. They all belong to the same energy landscape (for a given electronic state) and may or may not be interconverted among each other under feasible transformations according to the energy available in the system and the relative barrier heights between minima. In addition, considering all possible ways of labelling the nuclei may yield a set of identical conformers that can be interconverted through barriers between multiple but equivalent wells. Such dynamical processes are not chemical reactions as such (because the reactant and the product are identical species), but they may have spectroscopic consequences (for example, vibrational tunnelling splitting). In addition to this permutational invariance, we must consider the consequences of the chirality of the three-dimensional space (a Cartesian frame can be right-handed or left-handed). For example, a pair of R and S enantiomers such as CHFClBr corresponds to two structures that are energetically degenerate but geometrically different despite their apparent equivalence. They are mirror images of each other and cannot be interconverted under any rigid-body rotation. The symmetry operation that relates them is called *parity trans-*

formation[1] or *space inversion* and denoted E^*. It turns the laboratory- or space-fixed Cartesian coordinates of *all* the electrons *and* nuclei in the molecular system into their negatives, $(X, Y, Z) \rightarrow (-X, -Y, -Z)$, and thus inverses left/right chirality. Such considerations imply that there is a more global symmetry in the system than simply indicated by the geometrical shape of a particular isomer.[2]

More specifically, the full set of variables entering the definition of the molecular Hamiltonian is made of the laboratory-fixed space coordinates and spin variables of all electrons and nuclei. Note that identical particles are labelled arbitrarily but unambiguously in the expression of the Hamiltonian. Indistinguishability is further ensured when considering the eigenstates of the Hamiltonian, which are made symmetric or antisymmetric according to the fermionic (half-integer spin) or bosonic (integer spin) natures of the particles. The full symmetry group of the Hamiltonian of the isolated molecular system should thus reflect all possible invariances: the position of the frame origin (space uniformity), the orientation of the frame axes (space isotropy), the space and spin permutations of identical particles (indistinguishability), and the space inversion of all particles through the frame origin (parity conservation). Electronic permutations (spin and space) can be treated separately, when generating electronic wavefunctions so as to be antisymmetric with respect to such operations. Further removing translational and rotational invariances yields the so-called *complete nuclear-permutation inversion (CNPI) group*. It is made of *all permutations of identical nuclei* together with the *space inversion of all electrons and nuclei*; such operations could be defined with respect to a laboratory-fixed frame but are more conveniently applied in a space-fixed frame so as to separate overall translations from the onset (the treatment of overall rotations in this context is more subtle and will not be addressed in the present chapter; see Ref. [1] for further details). As already pointed out, electrons are fermions, and it is well-established that the electronic wavefunction must be made antisymmetric with respect to their permutations (see Sect. 3.5 in Chap. 3). Not doing so yields a significant error on the total energy. Nuclei are either bosons or fermions depending on their spins. In contrast with electrons, making the nuclear wavefunction symmetric or antisymmetric, according to the case, has a small effect on the total energy within the range of vibrational transitions (infrared). As a consequence, they are often omitted from a description in terms of CNPI group, unless specifically required (for example, when nuclear-spin statistics must be accounted for in the selection rules of rotational spectra).

In practice, the operations of the CNPI group (space and spin permutations of identical nuclei with and without space inversion) are transformations that fall into

---

[1]Charge conjugation (C), parity transformation (P), and time reversal (T) form together the famous CPT symmetry under which all physics laws are assumed to be invariant.

[2]These aspects are sometimes explained using the concept of version. The various versions of a structural formula are produced upon permuting the labels of identical nuclei in all possible ways. Versions are considered distinct if they cannot be interconverted under a mere rigid-body rotation. In some cases, the set of permutational versions may get doubled upon further considering the image of each under space inversion (two enantiomers of the same isomer are considered as two distinct versions). As a result, a degenerate set of distinct versions corresponds to equivalent but distinct multiple wells in the energy landscape.

three types: (i) transformations that correspond to mere rigid-body rotations, hence with no activation barrier; (ii) feasible transformations between permutationally-equivalent isomers or between enantiomers (i.e., which can be achieved dynamically along a continuous deformation going across a superable activation barrier according to the energy given to the system and the time required for observing its effect); (iii) unfeasible transformations between permutationally-equivalent isomers or between enantiomers (i.e., of no observable consequence because the barrier is not superable in any reasonable time[3]). Discarding unfeasible transformations from the CNPI group yields the so-called molecular-symmetry group, which is an invariance subgroup of the full symmetry group of the molecular Hamiltonian with respect to the spin and space variables of the particles, irrespectively of any specific internuclear arrangement. *For a single isomer, the molecular-symmetry group is isomorphic to the corresponding point group.* contrast, when a process connects several isomers, each characterised by its own point group, the molecular-symmetry group is larger than any of them.

For a more comprehensive overview, we propose a selection of seminal references about the foundations of group theory and its applications to molecular physics and chemistry [1–8]. In what follows, we will address three examples that illustrate how group theory can help in the context of molecular quantum dynamics for classifying vibrational and electronic states. Two application cases (the water molecule and the butatriene cation) will be treated within the framework of point-group theory, adapted to small geometrical deformations around a single minimum (vibrations), while the third one (the twice-deuterated ammonia molecule) will make use of molecular-symmetry (nuclear-permutation inversion) group theory to consider large-amplitude deformations (contortions) connecting two equivalent minima through a barrier (symmetric double well). The latter formalism is less widely-used in chemistry than the former but is, in fact, a generalisation of it, as already mentioned above (this aspect is carefully explained in the comprehensive book of Bunker and Jensen [1]). For clarification, a comparison between both approaches will be made first in the case of the water molecule.

7.2 The Harmonic Vibrations of Water

In this section, we start with a standard point-group treatment where the character table can be used quite readily to determine the symmetry properties of the harmonic vibrations of water (see Sect. 9.3 in Chap. 9). We then make a more rigorous connection with molecular symmetry to illustrate how both approaches are related. Finally, we show how the ground-state dipole moment behaves to first order as a function

[3]From a quantum-mechanical perspective, an unfeasible transformation means that tunnelling is too weak to be detected experimentally. When so, the unsuperable barrier between two equivalent wells can be considered as infinite and this structural degenerescence can be omitted from the description.

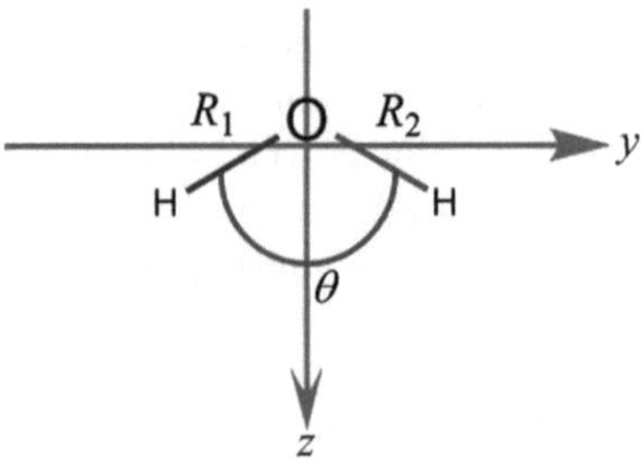

Fig. 7.1 Water molecule: body-fixed axes (Mulliken's convention) and curvilinear coordinates (here, the x-axis is perpendicular to the representation plane and points towards the opposite side of the reader sight; the signs of the axes are such that the H-nucleus with label 2 (corresponding to the OH bond length R_2 on the *right-hand* side) is on the positive side of the y-axis and the O-nucleus on the negative side of the z-axis)

of the nuclear coordinates to derive vibrational selection rules and how the potential energy surface can be expanded as a quadratic form of the nuclear coordinates.

7.2.1 Standard Point-Group Treatment

The equilibrium geometry of the water molecule (H_2O) in its electronic ground state is V-shaped and corresponds to an isosceles triangular prism when viewed as a solid object; see Fig. 7.1 (the molecule is planar and both H-nuclei are at the same distance from the O-nucleus, which implies that they are permutationally-equivalent: permutating them yields the same nuclear positions, up to a rigid-body rotation). This structure thus exhibits three symmetry elements: a twofold rotation axis (bisecting the H–O–H sector) and two mirror planes containing this axis (the molecular plane and the orthogonal plane intersecting it along the rotation axis). They all intersect through a common point at least (here an axis), hence the name point group, which is the molecular centre of mass (assumed here to coincidence with the nuclear centre of mass). Each symmetry element generates symmetry operations defined such that they leave the shape of the molecule unchanged, irrespectively of the labelling of identical nuclei: a half-turn rotation around the axis (denoted C_2) and two reflections with respect to each mirror plane (both denoted σ_v). Note that, as usual, symmetry elements and symmetry operations are referred to with the same symbols.

A symmetry-adapted body-fixed Cartesian frame must be such that the three directions are invariant under the symmetry operations, which means that the signed axes are transformed into plus or minus themselves under the various rotations and reflections (see Table 7.1).[4] For symmetry reasons, these three axes are parallel to

[4]In this chapter, we will limit our discussion to non-degenerate cases, whereby body-fixed axes are turned into themselves or their negatives. However, degenerate cases may involve transformations that mix pairs of equivalent axes together.

Table 7.1 Behaviours of the body-fixed axes under the C_{2v} point-group operations (in the standard point-group treatment, the right-handed frame becomes left-handed under reflections)

E	$C_2(z)$	$\sigma_v(yz)$	$\sigma_v(xz)$
x	$-x$	$-x$	x
y	$-y$	y	$-y$
z	z	z	z

the eigenvectors of the inertia tensor of the molecule and, as such, are also called principal axes of inertia. Using Mulliken's convention[5] [4, 5] for labelling them yields z as the rotation axis, x as the axis perpendicular to the molecular plane, and y as the remaining axis orthogonal to the other two; see Fig. 7.1. In other words, C_2 corresponds to z and both σ_v to (xz) and (yz), the latter being the molecular plane. Further, the sign of each axis must be specified according to some body-fixed constraints (for example, on Fig. 7.1, the H-nucleus with label 2 corresponds to the positive side of the y-axis).

The three aforementioned symmetry operations, $C_2(z)$, $\sigma_v(yz)$, and $\sigma_v(xz)$, along with the identity, E, form a point group named C_{2v}. These four distinct operations commute and the group is thus said to be Abelian. As a consequence, it can be represented with four non-degenerate (one-dimensional) irreducible representations, which are denoted A_1, A_2, B_1, and B_2. A_1 is the totally-symmetric irreducible representation. A_2 is even with respect to the rotation and odd with respect to both reflections. B_1 and B_2 are both odd with respect to $C_2(z)$; B_1 is odd with respect to $\sigma_v(yz)$ and even with respect to $\sigma_v(xz)$, and the reverse for B_2. According to their definitions, the z-axis behaves as A_1, the x-axis as B_1, and the y-axis as B_2 (see Table 7.1).

The definitions of the irreducible representations of C_{2v} are summarised in its character table (see Table 7.2). As usual, the last columns indicate the behaviours of the first- and second-order monomials of x, y, and z, along with those of R_x, R_y, and R_z, which denote the infinitesimal generators of rigid-body rotations around the body-fixed axes (from a practical point of view, they can be pictured as circular arrows around the corresponding axes). The properties of x, y, and z indicate how body-fixed coordinates and the body-fixed components of vectors behave (for example, the infinitesimal generators of rigid-body translations, T_x, T_y, and T_z, or the components of the dipole moment, μ_x, μ_y, and μ_z). The properties of xx, xy, etc. indicate how the body-fixed components of symmetric second-order tensors behave (for example, the components of the polarizability tensor, α_{xx}, α_{xy}, etc.). Finally, the properties of R_x, R_y, and R_z indicate how the body-fixed components of pseudovectors behave.[6]

[5] Note that alternative conventions are sometimes found in the literature. The x- and y-axes are often swapped, for example. Or, in Ref. [1], the x-axis is defined as the rotation axis while the y-axis is perpendicular to the molecular plane.

[6] Pseudovectors (or axial vectors), as opposed to true vectors (or polar vectors) actually are antisymmetric second-order tensors for which the three independent components are gathered

Table 7.2 Character table of the C_{2v} point group

C_{2v}	E	$C_2(z)$	$\sigma_v(yz)$	$\sigma_v(xz)$		
A_1	1	1	1	1	z	x^2, y^2, z^2
A_2	1	1	−1	−1	R_z	xy
B_1	1	−1	−1	1	x, R_y	xz
B_2	1	−1	1	−1	y, R_x	yz

Table 7.3 Direct-product table of the C_{2v} point group

$\otimes$	A_1	A_2	B_1	B_2
A_1	A_1	A_2	B_1	B_2
A_2	A_2	A_1	B_2	B_1
B_1	B_1	B_2	A_1	A_2
B_2	B_2	B_1	A_2	A_1

Another useful tool is the direct-product table of all irreducible representations (see Table 7.3). Each entry is obtained by first multiplying the characters of both irreducible representations and then determining to what belongs the result. One shows easily that $A_2 \otimes B_{1/2} = B_{2/1}$ and $B_1 \otimes B_2 = A_2$.

The description of molecular vibrations within the framework of point-group theory consists in determining the list of irreducible representations (their natures and numbers of occurrence) that characterise the harmonic normal modes (see Sect. 5.1 in Chap. 5) of a given equilibrium structure. Such irreducible representations correspond to the point group of this geometry. This treatment is based on the infinitesimal nuclear displacements from the equilibrium position of each nucleus. They are combined collectively and classified into three types: rigid-body translations (all nuclei shift together along an axis[7]), rigid-body rotations (all nuclei rotate together around an axis), and vibrations (small-amplitude motions that change the shape or size of the molecule[8]). For an N-atom system with $3N$ nuclear coordinates, there are three translations, three rotations, and $3N - 6$ vibrations (except for a diatom or if the equilibrium geometry is collinear, in which cases there are two rotations and $3N - 5$ vibrations). The fact that vibrations can be classified according to their

(Footnote 6 continued)
as vectors. They usually are the results of vector products (cross products) of two true vectors. If an operation turns true vectors into their negatives, it leaves pseudovectors unchanged: $\mathbf{u} \times \mathbf{v} \rightarrow (-\mathbf{u}) \times (-\mathbf{v}) = \mathbf{u} \times \mathbf{v}$. Angular momenta are typical examples of pseudovectors.

[7]In fact, an infinitesimal translation expressed in terms of body-fixed coordinates reflects the constraint that the body-fixed frame keeps being centred at the centre of mass when the nuclei move. It thus corresponds to displacements multiplied by masses rather than a collective translation that would shift the centre of mass with respect to its original position in the laboratory-fixed frame.

[8]The treatment of large-amplitude deformations (contortions) able to connect two equivalent isomers (for example in the case of a symmetric double well) will be treated in the last section of this chapter.

symmetries is a result of the harmonic approximation, whereby the eigenvectors of the mass-weighted Hessian matrix (second derivatives of the potential energy with respect to mass-weighted nuclear coordinates) at the equilibrium geometry, called the normal modes, belong to specific irreducible representations to preserve the totally-symmetric property of the potential energy function (see for example the book of Wilson, Decius, and Cross for a comprehensive overview of this aspect [6]).

The first step in this treatment consists in determining the behaviours of all infinitesimal displacements of the nuclei along the body-fixed axes from the equilibrium geometry. Here, there are nine linearly-independent variations to consider (three per atom): Δx_{O}, Δy_{O}, Δz_{O}, Δx_{H1}, Δy_{H1}, Δz_{H1}, Δx_{H2}, Δy_{H2}, and Δz_{H2} (here, 1 and 2 are the labels of the two H-nuclei on the left- and right-hand sides of Fig. 7.1, respectively). The total representation thus generated, Γ_{tot}, is nine-dimensional and reducible (in mathematical terms, it is a direct sum of nine one-dimensional irreducible representations, some of which occurring several times). Γ_{tot} contains the three-dimensional representation made of the three infinitesimal generators of rigid-body translations, Γ_{trans}, the three-dimensional representation made of the three infinitesimal generators of rigid-body rotations, Γ_{rot}, and the three-dimensional representation made of the three infinitesimal generators of vibrations, Γ_{vib}, which reads $\Gamma_{\mathrm{tot}} = \Gamma_{\mathrm{trans}} \oplus \Gamma_{\mathrm{rot}} \oplus \Gamma_{\mathrm{vib}}$.

As already mentioned the irreducible representations of T_x, T_y, and T_z behave as x, y, and z, and can be found in the character table: B_1, B_2, and A_1, respectively. The same applies to those of R_x, R_y, and R_z: B_2, B_1, and A_2, respectively. Thus,

$$\Gamma_{\mathrm{trans}} = A_1 \oplus B_1 \oplus B_2\,, \tag{7.1}$$

$$\Gamma_{\mathrm{rot}} = A_2 \oplus B_1 \oplus B_2\,. \tag{7.2}$$

Γ_{vib} will be obtained by subtracting them from Γ_{tot}, which is thus to be determined. To this end, one must now examine in detail how each infinitesimal displacement behaves under all operations. The equilibrium position of the O-nucleus does not move, such that Δx_{O}, Δy_{O}, and Δz_{O} simply behave as x, y, and z, respectively. However, the equilibrium positions of both H-nuclei are permuted under $C_2(z)$ and $\sigma_{\mathrm{v}}(xz)$, which must be accounted for, in addition to how the three Cartesian coordinates behave. This is summarised in Table 7.4.

When an infinitesimal displacement is unchanged, it accounts for a character 1. When it is turned into its negative, it accounts for a character -1. When it is changed for another, the corresponding character is 0. Summing all characters gives 9 for E, -1 for $C_2(z)$, 3 for $\sigma_{\mathrm{v}}(yz)$, and 1 for $\sigma_{\mathrm{v}}(xz)$.

Sets of characters of different irreducible representations can be viewed as orthogonal vectors. Their square-lengths are equal to the number of operations in the group, here four. The decomposition of a reducible representation is thus achieved from an orthogonal-projection technique where the result is to be divided by four. Hence, A_1 occurs $\frac{9-1+3+1}{4} = 3$ times, A_2 $\frac{9-1-3-1}{4} = 1$ time, B_1 $\frac{9+1-3+1}{4} = 2$ times, and B_2 $\frac{9+1+3-1}{4} = 3$ times. This yields $\Gamma_{\mathrm{tot}} = 3A_1 \oplus A_2 \oplus 2B_1 \oplus 3B_2$.

Table 7.4 Behaviours of the nine infinitesimal displacements of the nuclei under the C_{2v} point-group operations

E	$C_2(z)$	$\sigma_v(yz)$	$\sigma_v(xz)$
Δx_O	$-\Delta x_O$	$-\Delta x_O$	Δx_O
Δy_O	$-\Delta y_O$	Δy_O	$-\Delta y_O$
Δz_O	Δz_O	Δz_O	Δz_O
Δx_{H1}	$-\Delta x_{H2}$	$-\Delta x_{H1}$	Δx_{H2}
Δy_{H1}	$-\Delta y_{H2}$	Δy_{H1}	$-\Delta y_{H2}$
Δz_{H1}	Δz_{H2}	Δz_{H1}	Δz_{H2}
Δx_{H2}	$-\Delta x_{H1}$	$-\Delta x_{H2}$	Δx_{H1}
Δy_{H2}	$-\Delta y_{H1}$	Δy_{H2}	$-\Delta y_{H1}$
Δz_{H2}	Δz_{H1}	Δz_{H2}	Δz_{H1}

From the reductions of Γ_{trans} and Γ_{rot} derived above and considering the following decomposition,

$$\Gamma_{\text{tot}} = (\Gamma_{\text{trans}} \oplus \Gamma_{\text{rot}}) \oplus \Gamma_{\text{vib}} = (A_1 \oplus A_2 \oplus 2B_1 \oplus 2B_2) \oplus \Gamma_{\text{vib}}\,, \tag{7.3}$$

we thus get

$$\Gamma_{\text{vib}} = 2A_1 \oplus B_2\,. \tag{7.4}$$

This reduction of Γ_{vib} means that the three-dimensional space that characterises the possible vibrations of the water molecule can be described in terms of three linearly-independent degrees of freedom: a symmetric-stretching-type mode, ν_1, and a bending-type mode, ν_2, that belong to A_1 (they are both totally symmetric and preserve the C_{2v} point group) and an antisymmetric-stretching-type mode, ν_3, that belongs to B_2 (it breaks the equivalence between both H-nuclei and lowers the symmetry to the C_s subgroup made of E and $\sigma_v(yz)$ only). The corresponding normal modes are plotted on Fig. 7.2 (see also Fig. 9.5 in Chap. 9).

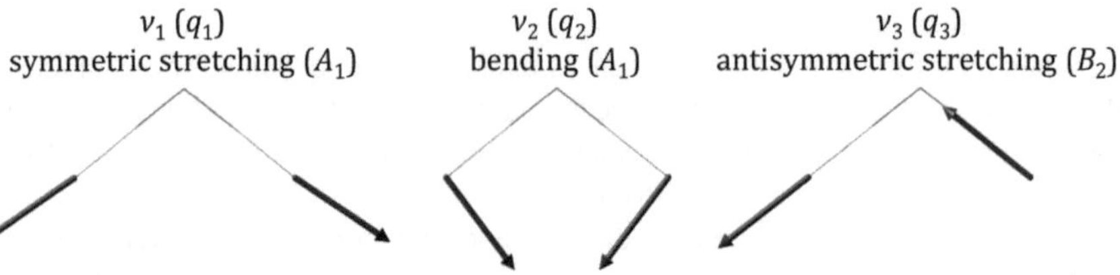

Fig. 7.2 The three normal modes of vibration of the water molecule. Normal coordinates (denoted q_1, q_2, and q_3) are the amplitudes of the collective nuclear displacements along the directions of the normal modes (labelled ν_1, ν_2, and ν_3)

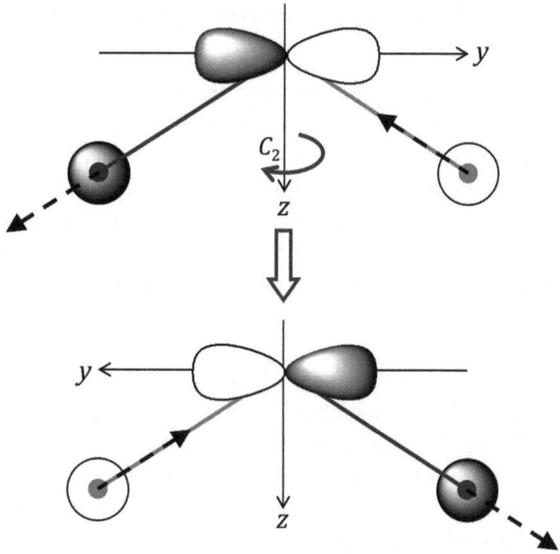

Fig. 7.3 Simultaneous action of $C_2(z)$ on the molecule, the B_2 vibrational mode, the y-axis, and a b_2 molecular orbital, all considered as solid objects irrespectively of the labels of the H-nuclei (standard point-group treatment)

Using the standard approach of point-group theory (see Fig. 7.3), one can determine the symmetry of each vibrational mode upon applying the symmetry operations to the molecule and the arrows representing the nuclear displacements as if they moved together as a solid object. For example, applying the rotation $C_2(z)$ to the B_2 vibrational mode results in a permutation of the two arrows such that each arrow is turned into its negative. The same approach can be used for a b_2 molecular orbital considered as a solid object: it behaves as if it were attached to the y-axis and rotated with it to be turned into its negative (note that orbitals are usually labelled from their irreducible representations using non-capitalised letters).

Note that vibrational modes and molecular orbitals can be defined more easily in terms of symmetry-adapted linear combinations of nuclear displacements and atomic orbitals:

- A_1: $\{\Delta z_{\mathrm{O}}, \Delta y_{\mathrm{H1}} - \Delta y_{\mathrm{H2}}, \Delta z_{\mathrm{H1}} + \Delta z_{\mathrm{H2}}\}$; a_1: $\{2s_{\mathrm{O}}, 2p_{z\mathrm{O}}, 1s_{\mathrm{H1}} + 1s_{\mathrm{H2}}\}$;
- A_2: $\{\Delta x_{\mathrm{H1}} - \Delta x_{\mathrm{H2}}\}$;
- B_1: $\{\Delta x_{\mathrm{O}}, \Delta x_{\mathrm{H1}} + \Delta x_{\mathrm{H2}}\}$; b_1: $\{2p_{x\mathrm{O}}\}$;
- B_2: $\{\Delta y_{\mathrm{O}}, \Delta y_{\mathrm{H1}} + \Delta y_{\mathrm{H2}}, \Delta z_{\mathrm{H1}} - \Delta z_{\mathrm{H2}}\}$; b_2: $\{2p_{y\mathrm{O}}, 1s_{\mathrm{H1}} - 1s_{\mathrm{H2}}\}$.

The relative weights of contributions belonging to the same irreducible representation in a vibrational mode or in a molecular orbital are to be determined from a calculation, except in some trivial cases: for example, $\Delta x_{\mathrm{H1}} - \Delta x_{\mathrm{H2}}$ is the only A_2 displacement and obviously corresponds to R_z, the infinitesimal generator of rotation around the z-axis.

7.2.2 Rigorous Molecular-Symmetry Treatment *

When examined more closely, the standard approach to point-group theory used above raises three paradoxes: (i) point-group symmetry operations concern the shape of the molecule viewed as a solid object rather than the invariances of the molecular Hamiltonian and yet brings the same information for a rigid non-linear molecule; (ii) reflections seem to turn the original right-handed body-fixed frame into a left-handed image; (iii) when rotating the arrows representing a vibrational mode or a molecular orbital together with the molecule (see Fig. 7.3), the comparison is made in the observer-fixed frame but there actually is no change within the body-fixed frame.

These issues disappear when using a rigorous formulation in terms of molecular-symmetry groups such as exposed in Ref. [1]. Let us illustrate this on the water molecule. For this system, the indistinguishability of the two H-nuclei and the parity transformation yield four *permutation-inversion operations*: the identity, E, the space inversion, E^*, the permutation of both H-nuclei, denoted (12), and the composition of E^* and (12), denoted $(12)^*$. In this simple case, there are no unfeasible transformations to be removed from the CNPI group, which can thus be identified to the molecular-symmetry group of water. It is isomorphic to $\mathcal{C}_{2v}$ and, as such, called $\mathcal{C}_{2v}(\mathrm{M})$. Its character table is provided in the last section of the present chapter. Note that nuclear and electronic spins will be omitted here for simplicity.

The effects of the four operations on the space-fixed Cartesian coordinates of the three nuclei and of an electron are given in Table 7.5. These are not only valid for a $\mathcal{C}_{2v}$ geometry but can be applied to any distorted arrangement of the nuclei. We recall here that (12) *permutes the nuclei but does not act on the electrons*. However,

Table 7.5 Effects of the $\mathcal{C}_{2v}(\mathrm{M})$ molecular-symmetry operations on the space-fixed Cartesian coordinates of the three nuclei and of an electron

E	(12)	E^*	$(12)^*$
X_O	X_O	$-X_O$	$-X_O$
Y_O	Y_O	$-Y_O$	$-Y_O$
Z_O	Z_O	$-Z_O$	$-Z_O$
X_{H1}	X_{H2}	$-X_{H1}$	$-X_{H2}$
Y_{H1}	Y_{H2}	$-Y_{H1}$	$-Y_{H2}$
Z_{H1}	Z_{H2}	$-Z_{H1}$	$-Z_{H2}$
X_{H2}	X_{H1}	$-X_{H2}$	$-X_{H1}$
Y_{H2}	Y_{H1}	$-Y_{H2}$	$-Y_{H1}$
Z_{H2}	Z_{H1}	$-Z_{H2}$	$-Z_{H1}$
X_e	X_e	$-X_e$	$-X_e$
Y_e	Y_e	$-Y_e$	$-Y_e$
Z_e	Z_e	$-Z_e$	$-Z_e$

Table 7.6 Effects of the C_{2v} point-group operations on the body-fixed Cartesian coordinates of the three nuclei and of an electron (rigourous molecular-symmetry treatment)

E	$C_2(z)$	$\sigma_v(yz)$	$\sigma_v(xz)$
$x_O = x_O^{eq} + \Delta x_O$	$x_O = x_O^{eq} - \Delta x_O$	$x_O = x_O^{eq} - \Delta x_O$	$x_O = x_O^{eq} + \Delta x_O$
$y_O = y_O^{eq} + \Delta y_O$	$y_O = y_O^{eq} - \Delta y_O$	$y_O = y_O^{eq} + \Delta y_O$	$y_O = y_O^{eq} - \Delta y_O$
$z_O = z_O^{eq} + \Delta z_O$	$z_O = z_O^{eq} + \Delta z_O$	$z_O = z_O^{eq} + \Delta z_O$	$z_O = z_O^{eq} + \Delta z_O$
$x_{H1} = x_{H1}^{eq} + \Delta x_{H1}$	$x_{H1} = x_{H1}^{eq} - \Delta x_{H2}$	$x_{H1} = x_{H1}^{eq} - \Delta x_{H1}$	$x_{H1} = x_{H1}^{eq} + \Delta x_{H2}$
$y_{H1} = y_{H1}^{eq} + \Delta y_{H1}$	$y_{H1} = y_{H1}^{eq} - \Delta y_{H2}$	$y_{H1} = y_{H1}^{eq} + \Delta y_{H1}$	$y_{H1} = y_{H1}^{eq} - \Delta y_{H2}$
$z_{H1} = z_{H1}^{eq} + \Delta z_{H1}$	$z_{H1} = z_{H1}^{eq} + \Delta z_{H2}$	$z_{H1} = z_{H1}^{eq} + \Delta z_{H1}$	$z_{H1} = z_{H1}^{eq} + \Delta z_{H2}$
$x_{H2} = x_{H2}^{eq} + \Delta x_{H2}$	$x_{H2} = x_{H2}^{eq} - \Delta x_{H1}$	$x_{H2} = x_{H2}^{eq} - \Delta x_{H2}$	$x_{H2} = x_{H2}^{eq} + \Delta x_{H1}$
$y_{H2} = y_{H2}^{eq} + \Delta y_{H2}$	$y_{H2} = y_{H2}^{eq} - \Delta y_{H1}$	$y_{H2} = y_{H2}^{eq} + \Delta y_{H2}$	$y_{H2} = y_{H2}^{eq} - \Delta y_{H1}$
$z_{H2} = z_{H2}^{eq} + \Delta z_{H2}$	$z_{H2} = z_{H2}^{eq} + \Delta z_{H1}$	$z_{H2} = z_{H2}^{eq} + \Delta z_{H2}$	$z_{H2} = z_{H2}^{eq} + \Delta z_{H1}$
x_e	$-x_e$	$-x_e$	x_e
y_e	$-y_e$	y_e	$-y_e$
z_e	z_e	z_e	z_e

E^* inverses through space both the nuclei and the electrons. As a consequence, both E^* and $(12)^*$ turn space-fixed electronic coordinates into their negatives, whereas E and (12) leave them unchanged.

Let us now clarify the connection between both approaches. The molecular-symmetry operations of C_{2v}(M) are defined in the space-fixed frame (see Table 7.5). In contrast, the point-group operations of C_{2v} are defined in the body-fixed frame and act on vibronic coordinates: body-fixed electronic coordinates (according to the transformations given in Table 7.1) and body-fixed components of nuclear displacements from their equilibrium positions (see Table 7.4). At this stage, it must be stressed that *point-group operations* are defined, rigorously speaking, such that they *keep the equilibrium positions of the nuclei unchanged* and thus *do not move the body-fixed axes*. What they really change are body-fixed coordinates and body-fixed vector components. This is summarised in Table 7.6.

At this stage, we understand better the actual meaning of Table 7.1. It does not really give the results of the point-group operations acting on the body-fixed axes but rather on the body-fixed components of vectors affected by such operations. The behaviours of body-fixed electronic coordinates, (x_e, y_e, z_e), are identical to those of (x, y, z) (see Table 7.1) and a comparison of Table 7.6 with Table 7.4 shows that the effect on nuclear displacements is identical to what we derived using the standard approach.

Moving the equilibrium positions of the nuclei with respect to the space-fixed frame is further achieved under rigid-body rotations around the body-fixed axes that change the values of the Euler angles. The body-fixed frame is thus rotated according to the new space-fixed positions of labelled nuclei. Applying point-group operations and rigid-body rotations sequentially yields the effect of the permutation-inversion operations of the molecular-symmetry group. In other words, the operations

Table 7.7 Effects of the molecular rotation-group operations on the body-fixed axes with respect to the original ones (in this rigorous picture, all transformed frames now stay right-handed as expected)

R_0	$R_\pi(z)$	$R_\pi(x)$	$R_\pi(y)$
x	$-x$	x	$-x$
y	$-y$	$-y$	y
z	z	$-z$	$-z$

of $\mathcal{C}_{2v}$(M) are compositions of point-group operations of $\mathcal{C}_{2v}$ together with rigid-body rotations around the body-fixed axes: (12) has the same effect as $C_2(z)$ followed by a rigid-body rotation around z through π turning (x, y, z) into $(-x, -y, z)$; E^* has the same effect as the reflection $\sigma_v(yz)$ followed by a rigid-body rotation around x through π turning (x, y, z) into $(x, -y, -z)$; $(12)^*$ has the same effect as the reflection $\sigma_v(xz)$ followed by a rotation around y through π turning (x, y, z) into $(-x, y, -z)$. In summary,

$$E = R_0 E \,, \tag{7.5}$$

$$(12) = R_\pi(z) C_2(z) \,, \tag{7.6}$$

$$E^* = R_\pi(x) \sigma_v(yz) \,, \tag{7.7}$$

$$(12)^* = R_\pi(y) \sigma_v(xz) \,, \tag{7.8}$$

where R_0 is the identity rotation[9] and $R_\pi(\xi)$ is a rotation through π around ξ (where $\xi = x, y, z$); see Table 7.7. We now see why the standard interpretation of point-group operations as involving rigid-body rotations is somewhat misleading but works in practice, as the rigid-body rotations are accounted for implicitly. This is illustrated on Fig. 7.4, to be compared with Fig. 7.3.

Defining how these various operations act on electronic coordinates requires some care. Let us examine the example shown on Fig. 7.4. The nuclear permutation (12) does not involve E^* and thus leaves the space-fixed electronic coordinates, (X_e, Y_e, Y_e), unchanged (see Table 7.5). However, the body-fixed electronic coordinates, (x_e, y_e, z_e), behave as (x, y, z) (see Tables 7.1 and 7.2). They are thus turned into $(-x_e, -y_e, z_e)$ under $C_2(z)$ (see Table 7.6). This is compensated by $R_\pi(z)$ that

[9]Following Ref. [1], a conceptual distinction must be made between the *molecular point group* of a rigid non-linear molecule and the *point group* of a solid object of the same shape. Using the properties of the point group in place of the molecular point group (what we have called the standard approach above) is less rigorous but works in practice because both yield the same results as long as vibronic coordinates are concerned. The molecular-symmetry group is made of operations composed of operations belonging to: (i) the molecular point group; (ii) the molecular rotation group, $\mathcal{K}$(mol); and (iii) the nuclear-spin-permutation group. For the water molecule, $\mathcal{K}$(mol) is made of R_0, $R_\pi(x)$, $R_\pi(y)$, and $R_\pi(z)$ whereby the three axes are distinguishable, as expected for an antisymmetric top molecule (it thus identifies to the point group $\mathcal{D}_2$). Because the molecular point group concerns vibronic coordinates in the body-fixed frame, it is sometimes called *molecular vibronic group*, although the latter concept is more general and also applies to the case of non-rigid molecules (see last section).

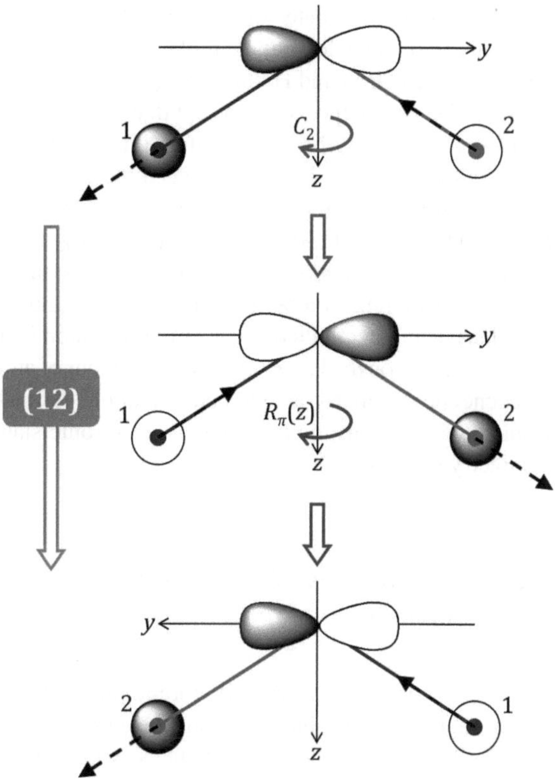

Fig. 7.4 Action of $(12) = R_\pi(z)C_2(z)$ on the molecule, the B_2 vibrational mode, the body-fixed frame, and a b_2 molecular orbital, represented in the space-fixed frame (let us recall here that the y-axis is body-fixed with respect to the nuclei in such a way that the H-nucleus with label 2 is on the positive side)

turns the body-fixed axes, (x, y, z), into $(-x, -y, z)$ in the space-fixed frame (see Table 7.7), consistently with the fact that (12) has no effect on the space-fixed electronic coordinates.

In practice, when considering electrons, it is more practical to apply group theory to their representation in terms of atomic orbitals rather than in terms of their spatial coordinates. From the transformations of body-fixed electronic coordinates under point-group operations, one readily deduces the behaviours of their various monomials,[10] hence of atomic orbitals or symmetry-adapted combinations of them. An atomic orbital centred on O (for example, $2p_{y\mathrm{O}}$ on Fig. 7.4) behaves as y and thus belongs to B_2. For an atomic orbital centred on one of the H-nuclei, one must consider an additional aspect: if the point-group operation under consideration changes

[10]Such monomials appear in the expressions of the real and imaginary parts of the spherical harmonics. When considered as origin-centred (or centred on a nucleus that does not move), s-orbitals behave as pure numbers and thus belong to the totally-symmetric irreducible representation, (p_x, p_y, p_z)-orbitals behave as (x, y, z), and d-orbitals behave as linear combinations of second-order monomials of x, y, and z (for example, $d_{x^2-y^2}$-orbitals behave as $x^2 - y^2$). For f-orbitals and beyond, the behaviours of higher-order monomials are obtained upon further considering the direct product table to generate them (see Table 7.3).

the body-fixed coordinates of the electron it then swaps the distances between the electron and both nuclei. For example, $1s_{\mathrm{H1}}$ is expressed as a function of the distance between the electron and H number 1 (i.e., $1s_{\mathrm{H1}}$ is an atomic orbital centred on H1); $1s_{\mathrm{H2}}$ has the same expression, except that it is centred on H2. $C_2(z)$ turns (x_e, y_e, z_e) into $(-x_e, -y_e, z_e)$, which means that it rotates the electron around z, from one side to the other side. The distance between the electron and H number 2 is now equal to what it was between the electron and H number 1 before the rotation. In practical terms, this means that $C_2(z)$ swaps $1s_{\mathrm{H1}}$ and $1s_{\mathrm{H2}}$, such as shown on Fig. 7.4. In brief, this means that atomic orbitals behave much as nuclear displacements with respect to nuclear labelling: if a point-group operation moves a nuclear displacement from a nucleus to another nucleus it also moves an atomic orbital centred on the former to make it centred on the latter. Once the behaviour of atomic orbitals is determined, one can generate symmetry-adapted combinations for building molecular orbitals, electronic configurations, and, ultimately, electronic states that belong to well-defined symmetry species.

7.2.3 Ground-State Dipole Moment and Potential Energy Surface

Let us now consider the infrared vibrational spectroscopy of water. First-order transition intensities are obtained from the matrix elements of the permanent dipole moment between two vibrational states (with final integration performed over the internal nuclear coordinates). The permanent dipole moment, $\boldsymbol{\mu}(\boldsymbol{q})$, is first obtained upon integrating the total dipole moment over the electronic coordinates for the electronic ground state, as mentioned in Sect. 3.4 of Chap. 3.[11] If overall rotations are not considered, $\boldsymbol{q}$ is the collection of internal nuclear coordinates. Here, $\boldsymbol{q} = (q_1, q_2, q_3)$ where, within the harmonic approximation, q_1, q_2, and q_3 are the normal coordinates[12] of water (see Fig. 7.2).

The origin, $\boldsymbol{q} = \mathbf{0}$, corresponds to the C_{2v} equilibrium geometry. A normal coordinate belongs to the same irreducible representation as the normal mode that it describes: q_1 and q_2 are A_1 while q_3 is B_2. Each body-fixed Cartesian component, $\mu_\xi(\mathbf{0})$ (where $\xi = x, y, z$), belongs to the same irreducible representation as the ξ-axis (first-order monomials in the character table). For $\mu_\xi(\mathbf{0})$ to be a finite number, it must transform as A_1 (a constant is not affected by any symmetry operation and is thus totally-symmetric). As a consequence, only $\mu_z(\mathbf{0})$ is not zero, as expected intuitively for a C_{2v} geometry (partial charges are equally distributed on both H-nuclei).

---

[11]Thus, in this section $\boldsymbol{\mu}(\boldsymbol{q})$ corresponds to $\boldsymbol{\mu}_{00}(\boldsymbol{q})$ in Sect. 3.4 of Chap. 3.

[12]The terms *normal mode* and *normal coordinate* are often used in place of one another in the literature. Strictly speaking, a normal mode is the set of rectilinear directions followed collectively by the nuclei for a given vibration and the normal coordinate is the amplitude of the corresponding displacement. For a one-dimensional motion along an axis, the mode would be the direction of the axis and the coordinate would be the position along this axis.

In contrast, $\mu_x(\mathbf{0}) = \mu_y(\mathbf{0})$ because they belong to B_1 and B_2, respectively, which are not totally-symmetric.

This approach can be extended in the vicinity of the C_{2v} point. Both q_1 and q_2 are A_1, which implies that distortions along them from the origin preserve C_{2v}. If so, $\mu_z(q_1, q_2, 0)$ is still A_1, which means that there is no restriction for its finite value to stay finite and to vary with q_1 and q_2. In contrast, q_3 is B_2, hence not totally-symmetric. Any linear function of q_3 is thus odd. As a consequence, in order for $\mu_z(0, 0, q_3)$ to stay A_1, it can only be an even function of q_3 and must be at least quadratic ($B_2 \otimes B_2 = A_1$). In contrast, as $\mu_y(\boldsymbol{q})$ is B_2 at the origin, it can be an odd function of q_3: it is zero at the origin but can vary linearly with q_3 to first order. This is consistent with the fact that the new point group obtained for $q_3 \neq 0$ is C_s, in which B_2 correlates with the totally-symmetric irreducible representation of this subgroup: $\mu_y(0, 0, q_3)$ being now totally-symmetric can take a finite value. However, the x-component is zero to all orders because there is no monomial of q_1, q_2, and q_3 that could belong to A_1 (all powers of A_1 and B_2 are either A_1 or B_2, hence, never B_1). This reflects the fact that the molecular plane is always preserved for a triatom (the smallest possible group of any triatom is C_s, and cannot be the trivial group C_1 made of the identity only, for which there would no longer be any symmetry restriction). In practice, expanding $\mu_\xi(\boldsymbol{q})$ around the origin involves only monomials of the coordinates that belong to the irreducible representation of ξ determined at the origin. One can thus determine the expression of the first-order expansion of each component,

$$\mu_x(\boldsymbol{q}) = 0\,, \tag{7.9}$$

$$\mu_y(\boldsymbol{q}) \approx \left[\partial_{q_3}\mu_y\right]_{\mathbf{0}} q_3\,, \tag{7.10}$$

$$\mu_z(\boldsymbol{q}) \approx \mu_z(\mathbf{0}) + \left[\partial_{q_1}\mu_z\right]_{\mathbf{0}} q_1 + \left[\partial_{q_2}\mu_z\right]_{\mathbf{0}} q_2\,. \tag{7.11}$$

Now, to determine the intensity of a $(\nu_j)_0^1$ transition (where $j = 1, 2, 3$), as explained in Sect. 3.4 of Chap. 3, we must evaluate an integral over q_1, q_2, and q_3 involving the vibrational ground-state (an even function with respect to all internal coordinates; approximately a three-dimensional Gaussian function within the harmonic approximation) and the first excited state with respect to mode ν_j (with a node for this mode, thus an odd wavefunction with respect to q_j but even for the other two coordinates). For the integral over q_j not to vanish, it thus must involve an odd term in one of the components of the dipole moment to compensate the odd product of both vibrational wavefunctions, $\Psi_1(\boldsymbol{q})$ and $\Psi_0(\boldsymbol{q})$. As a consequence, $(\nu_1)_0^1$ and $(\nu_2)_0^1$ transitions involving A_1 modes are induced to first order by variations of $\mu_z(\boldsymbol{q})$ from $\mu_z(\mathbf{0})$ with q_1 and q_2, respectively (i.e., by $\left[\partial_{q_1}\mu_z\right]_{\mathbf{0}} q_1$ and $\left[\partial_{q_2}\mu_z\right]_{\mathbf{0}} q_2$), while the $(\nu_3)_0^1$ transition involving the B_2 mode is induced to first order by a variation of $\mu_y(\boldsymbol{q})$ from $\mu_y(\mathbf{0})$ with q_3 (i.e., $\left[\partial_{q_3}\mu_y\right]_{\mathbf{0}} q_3$). All $(\nu_j)_0^1$ transitions are thus active in the infrared but they are not polarised along the same body-fixed axes.

Let us now turn to the expression of the ground-state potential energy function of water, $V(\mathbf{q})$. This quantity is totally-symmetric with respect to infinitesimal variations of $\mathbf{q}$ from $\mathbf{0}$ (this reflects the property that the electronic Hamiltonian operator is totally-symmetric with respect to all symmetry operations). In practice, this means that expanding $V(\mathbf{q})$ around the origin $\mathbf{q} = \mathbf{0}$ should only involve A_1 terms. In particular, there can only be even powers of the B_2 normal coordinate q_3, hence no gradient component for this coordinate. However, there may be some gradient along the totally-symmetric coordinates q_1 and q_2 (in other words, the gradient operator is totally-symmetric, the potential energy function is totally-symmetric, hence the gradient of the potential energy is totally-symmetric and each component along q_j transforms as q_j). Again, this is quite intuitive: there is no physical reason for the force driving the nuclear motion to make any distinction between both equivalent H-nuclei (a non-zero component along q_3, the normal coordinate associated to the antisymmetric stretching mode, would induce the lengthening of one bond and the shortening of the other).

As $\mathbf{q} = \mathbf{0}$ was chosen as the equilibrium geometry (a minimum, but this would also be true for any type of stationary point such as a saddle point), there is no gradient by definition. The expression of $V(\mathbf{q})$ around this point involves second-order terms and higher. Restricting it to a quadratic expansion (harmonic approximation) reads

$$V(\mathbf{q}) \approx V(\mathbf{0}) + \frac{1}{2}\left[\partial^2_{q_1} V\right]_{\mathbf{0}} q_1^2 + \frac{1}{2}\left[\partial^2_{q_2} V\right]_{\mathbf{0}} q_2^2 + \frac{1}{2}\left[\partial^2_{q_3} V\right]_{\mathbf{0}} q_3^2 + \left[\partial_{q_1}\partial_{q_2} V\right]_{\mathbf{0}} q_1 q_2 \,. \tag{7.12}$$

There is no cross term between A_1 and B_2 coordinates because $V(\mathbf{q})$ transforms as A_1. This implies that the Hessian matrix (second derivatives) is block-diagonal with respect to the various irreducible representations of the point group corresponding to the origin. In fact, if q_1, q_2, and q_3 are normal coordinates, this further implies that the corresponding Hessian matrix has been diagonalised, such that the symmetry-allowed cross term $\left[\partial_{q_1}\partial_{q_2} V\right]_{\mathbf{0}}$ is also zero by construction. Note that third-order anharmonicity along a given coordinate (a cubic term) could only be due to q_1 or q_2 (totally-symmetric), whereas q_3 (non-totally-symmetric) can only bring even powers, hence fourth-order anharmonicity (a quartic term) or higher.

Finally, in this example, defining symmetry-adapted curvilinear coordinates from normal coordinates is quite obvious. From the internal valence coordinates depicted on Fig. 7.1 (bond lengths and planar angle), one can identify $R_+ = \frac{R_1+R_2}{2}$ and θ as A_1 coordinates (symmetric stretching and bending) and $R_- = \frac{R_1-R_2}{2}$ as a B_2 coordinate (antisymmetric stretching). Let us set (R_+^0, θ^0, R_-^0) as a point corresponding to any $\mathcal{C}_{2v}$ geometry (not necessarily the equilibrium geometry). If so, $R_-^0 = 0$ by construction. Again, expanding $V(R_+, \theta, R_-)$ around $(R_+^0, \theta^0, 0)$ should only involve A_1 terms. In particular, there can only be even powers of R_- (which is consistent with not having any gradient component along this non-totally-symmetric coordinate). To second order, this reads

$$\begin{aligned} V(R_+, \theta, R_-) \approx\; & V(R_+^0, \theta^0, 0) \\ & + \left[\partial_{R_+} V\right]_{(R_+^0, \theta^0, 0)} (R_+ - R_+^0) \\ & + \left[\partial_\theta V\right]_{(R_+^0, \theta^0, 0)} (\theta - \theta^0) \end{aligned} \tag{7.13}$$

$$
\begin{aligned}
&+ \frac{1}{2} \left[\partial^2_{R_+} V\right]_{(R^0_+,\theta^0,0)} (R_+ - R^0_+)^2 \\
&+ \frac{1}{2} \left[\partial^2_{\theta} V\right]_{(R^0_+,\theta^0,0)} (\theta - \theta^0)^2 \\
&+ \frac{1}{2} \left[\partial^2_{R_-} V\right]_{(R^0_+,\theta^0,0)} R^2_- \\
&+ \left[\partial_{R_+} \partial_\theta V\right]_{(R^0_+,\theta^0,0)} (R_+ - R^0_+)(\theta - \theta^0) \,,
\end{aligned}
$$

where there is no cross term involving R_- in the second-order terms.

7.3 The Vibronic Hamiltonian of the Butratriene Cation

Regarding the electronic states and their energies as functions of the nuclear coordinates (i.e., the potential energies for the nuclei), group theory is not limited to the description of the ground state, as was the case in the previous example. It can also be used in a non-adiabatic context where several electronic states and their couplings must be considered. If the vibronic Hamiltonian is represented in a basis set of electronic states that have a well-defined symmetry at some chosen reference geometry, the rules of point-group symmetry can be applied to the entries of the matrix representation of the electronic Hamiltonian, which also belong to the irreducible representations of the point group at the origin. This is useful to determine the behaviours of such matrix elements when considering small-amplitude deformations from this point. Here, we will provide further details on the construction of the vibronic Hamiltonian model of the butatriene cation, $C_4H_4^+$, already mentioned in Sect. 4.4 of Chap. 4. Note that another, but similar, type of use of point groups concerns the behaviour of the transition dipole between two electronic states of a neutral system (see, e.g., an application to the photoabsorption spectrum of pyrazine in Ref. [9]).

The vibronic Hamiltonian model of the butatriene cation presented in Ref. [10] was built to calculate a photoelectron spectrum. The initial state of the molecule, before photoionisation, was assumed to be the vibrational ground state of the neutral molecule centred at the equilibrium geometry of the electronic ground state. This Franck-Condon point, which will serve as our reference in the following, corresponds to a planar geometry of $\mathcal{D}_{2h}$ symmetry (see Fig. 7.5). This point group is associated

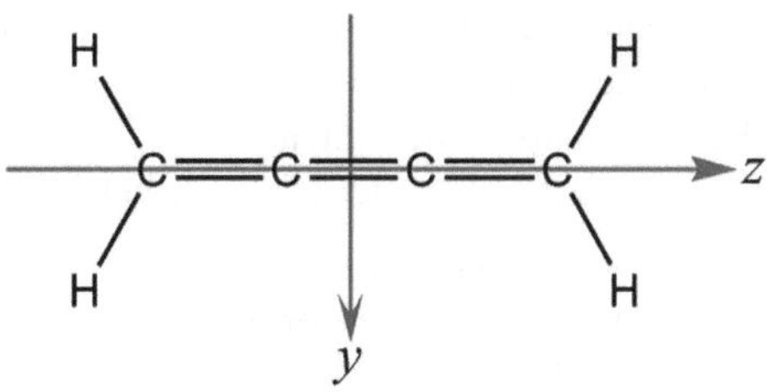

Fig. 7.5 Butatriene molecule: body-fixed axes (Mulliken's convention; here, the x-axis is perpendicular to the representation plane and points towards the reader sight)

Table 7.8 Character table of the $\mathcal{D}_{2h}$ point group

$\mathcal{D}_{2h}$	E	$C_2(z)$	$C_2(y)$	$C_2(x)$	i	$\sigma_v(xy)$	$\sigma_v(xz)$	$\sigma_v(yz)$		
A_g	1	1	1	1	1	1	1	1		x^2, y^2, z^2
B_{1g}	1	1	−1	−1	1	1	−1	−1	R_z	xy
B_{2g}	1	−1	1	−1	1	−1	1	−1	R_y	xz
B_{3g}	1	−1	−1	1	1	−1	−1	1	R_x	yz
A_u	1	1	1	1	−1	−1	−1	−1		
B_{1u}	1	1	−1	−1	−1	−1	1	1	z	
B_{2u}	1	−1	1	−1	−1	1	−1	1	y	
B_{3u}	1	−1	−1	1	−1	1	1	−1	x	

to the shape of a rectangular prism (all orthogonal faces with three different edge lengths). It can also be viewed as the merging of two $\mathcal{C}_{2v}$ V-shaped objects within the same plane and along their common rotation axis to form a sort of X-shaped object. There are now seven symmetry elements: an inversion centre (at the geometric centre of the molecule, which coincides with the centre of mass), three twofold rotation axes and three mirror planes, each orthogonal to one of the rotation axes. Again, they all intersect through a common point: the centre of mass.

The corresponding symmetry operations are the inversion of all nuclei through the inversion centre (denoted i), three half-turn rotation around the axes (all denoted C_2) and three reflections with respect to each mirror plane (all denoted σ_v).

As for water, there is some ambiguity for labelling the axes of the symmetry-adapted body-fixed frame. Using Mulliken's convention[13] yields z as the rotation axis going through the greatest number of nuclei (here, the four C-nuclei), x as the axis perpendicular to the molecular plane, and y as the remaining axis orthogonal to the other two. The molecular plane is thus (yz); see Fig. 7.5. Again, these three axes are the principal axes of inertia for symmetry reasons.

The eight symmetry operations of $\mathcal{D}_{2h}$ (the seven aforementioned ones, along with the identity, E) all commute and this point group is thus Abelian and made of eight non-degenerate irreducible representations given in the following character table (see Table 7.8), along with the behaviours of R_x, R_y, and R_z as well as those of the first- and second-order monomials of x, y, and z. The integer labels in B_{1u}, B_{2u}, and B_{3u} correspond to z, y, and x, respectively. The g or u labels indicate even or odd (*gerade* or *ungerade* in German) with respect to i. The direct-product table of all irreducible representations is also provided (see Table 7.9).

Again, the 24-dimensional reducible representation generated by the infinitesimal displacements of the eight nuclei, Γ_{tot}, can be decomposed into $\Gamma_{\text{tot}} = \Gamma_{\text{trans}} \oplus \Gamma_{\text{rot}} \oplus \Gamma_{\text{vib}}$. From the character table, we have

[13] As in the case of $\mathcal{C}_{2v}$, an alternative convention is sometimes found in the literature, whereby x and y are swapped. In fact, there are six possible conventions here because there are three axes of same degree.

Table 7.9 Direct-product table of the $\mathcal{D}_{2h}$ point group

$\otimes$	A_g	B_{1g}	B_{2g}	B_{3g}	A_u	B_{1u}	B_{2u}	B_{3u}
A_g	A_g	B_{1g}	B_{2g}	B_{3g}	A_u	B_{1u}	B_{2u}	B_{3u}
B_{1g}	B_{1g}	A_g	B_{3g}	B_{2g}	B_{1u}	A_u	B_{3u}	B_{2u}
B_{2g}	B_{2g}	B_{3g}	A_g	B_{1g}	B_{2u}	B_{3u}	A_u	B_{1u}
B_{3g}	B_{3g}	B_{2g}	B_{1g}	A_g	B_{3u}	B_{2u}	B_{1u}	A_u
A_u	A_u	B_{1u}	B_{2u}	B_{3u}	A_g	B_{1g}	B_{2g}	B_{3g}
B_{1u}	B_{1u}	A_u	B_{3u}	B_{2u}	B_{1g}	A_g	B_{3g}	B_{2g}
B_{2u}	B_{2u}	B_{3u}	A_u	B_{1u}	B_{2g}	B_{3g}	A_g	B_{1g}
B_{3u}	B_{3u}	B_{2u}	B_{1u}	A_u	B_{3g}	B_{2g}	B_{1g}	A_g

$$\Gamma_{\text{trans}} = B_{1u} \oplus B_{2u} \oplus B_{3u} \,, \tag{7.14}$$

$$\Gamma_{\text{rot}} = B_{1g} \oplus B_{2g} \oplus B_{3g} \,. \tag{7.15}$$

The reduction of Γ_{tot} is made easier upon noticing that only four operations leave some of the nuclear displacements on their original nuclei, namely, $E, C_2(z), \sigma_v(xz)$, and $\sigma_v(yz)$. From Table 7.10, we get (24, −4, 4, 8) as the corresponding characters of Γ_{tot}. Note that there is no actual need here to specify the labels of the nuclei. Using the same orthogonal-projection technique as for the water molecule (where we now must divide by 8, the total number of symmetry operations), we obtain that A_g and B_{1u} $(1, 1, 1, 1)$ occur $\frac{24-4+4+8}{8} = 4$ times, B_{1g} and A_u $(1, 1, -1, -1)$ occur $\frac{24-4-4-8}{8} = 1$ time, B_{2g} and B_{3u} $(1, -1, 1, -1)$ occur $\frac{24+4+4-8}{8} = 3$ times, and B_{3g} and B_{2u} $(1, -1, -1, 1)$ occur $\frac{24+4-4+8}{8} = 4$ times. In other words,

$$\Gamma_{\text{tot}} = 4A_g \oplus B_{1g} \oplus 3B_{2g} \oplus 4B_{3g} \oplus A_u \oplus 4B_{1u} \oplus 4B_{2u} \oplus 3B_{3u} \,. \tag{7.16}$$

As a result, the 18 vibrational modes of butatriene can be classified according to their irreducible representations as

$$\Gamma_{\text{vib}} = 4A_g \oplus 2B_{2g} \oplus 3B_{3g} \oplus A_u \oplus 3B_{1u} \oplus 3B_{2u} \oplus 2B_{3u} \,. \tag{7.17}$$

As for water, the vibrational subspace is spanned by infinitesimal displacements of the nuclei along collective directions called normal modes (see Sect. 5.1 in Chap. 5).

They are obtained as the eigenvectors of the Hessian matrix of the ground-state potential energy at the equilibrium geometry. Their respective symmetries are consistent with the reduction of Γ_{vib} into its irreducible representations. The descriptions, labels, and irreducible representations of the 18 normal modes are given on Figs. 7.6, 7.7 and 7.8 (see also Ref. [10]). In what follows, $\mathbf{q}$ will denote the collection of

Table 7.10 Characters of the 24 infinitesimal nuclear displacements and of the total representation under the four $\mathcal{D}_{2h}$ point-group operations that leave some of the nuclear displacements on their original nuclei

	E	$C_2(z)$	$\sigma_v(xz)$	$\sigma_v(yz)$
Δx_{C1}	1	−1	1	−1
Δy_{C1}	1	−1	−1	1
Δz_{C1}	1	1	1	1
Δx_{C2}	1	−1	1	−1
Δy_{C2}	1	−1	−1	1
Δz_{C2}	1	1	1	1
Δx_{C3}	1	−1	1	−1
Δy_{C3}	1	−1	−1	1
Δz_{C3}	1	1	1	1
Δx_{C4}	1	−1	1	−1
Δy_{C4}	1	−1	−1	1
Δz_{C4}	1	1	1	1
Δx_{H1}	1	0	0	−1
Δy_{H1}	1	0	0	1
Δz_{H1}	1	0	0	1
Δx_{H2}	1	0	0	−1
Δy_{H2}	1	0	0	1
Δz_{H2}	1	0	0	1
Δx_{H3}	1	0	0	−1
Δy_{H3}	1	0	0	1
Δz_{H3}	1	0	0	1
Δx_{H4}	1	0	0	−1
Δy_{H4}	1	0	0	1
Δz_{H4}	1	0	0	1
Γ_{tot}	24	−4	4	8

18 dimensionless (frequency-mass-weighted) normal coordinates[14] ($\mathbf{q} = \mathbf{0}$ corresponding to the ground-state equilibrium geometry). Setting $V(\mathbf{0}) = 0$ as the energy origin, the harmonic approximation of the ground-state potential energy surface of the neutral molecule thus reads

$$V(\mathbf{q}) \approx \sum_{j=1}^{18} \frac{\hbar\omega_j}{2} q_j^2 \,, \tag{7.18}$$

[14]Dimensionless normal coordinates q_j are scaled (frequency-mass-weighted) upon multiplying lengths by the factor $\sqrt{\frac{\mu_j \omega_j}{\hbar}}$, where μ_j is the reduced mass of mode ν_j and $\hbar\omega_j$ is the corresponding quantum of vibrational energy.

Symmetry	Label	Description	Displacement
A_g	8	C–C stretching (terminal)	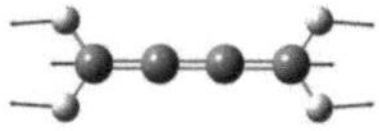
	12	CH_2 scissoring	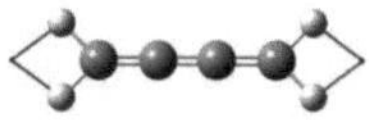
	14	C–C stretching (central)	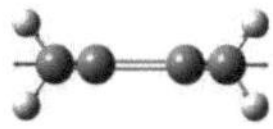
	15	C–H stretching	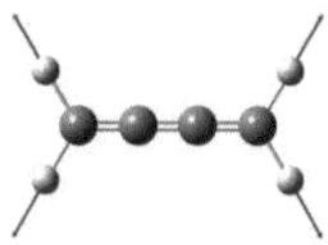
A_u	5	twisting (torsion)	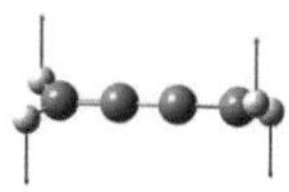
B_{2g}	4	skeletal bending (out of plane)	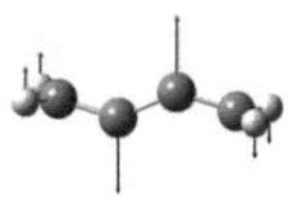
	7	CH_2 wagging	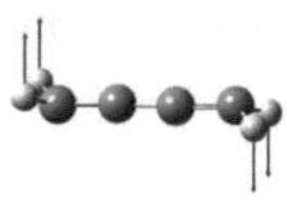

Fig. 7.6 A_g, A_u, and B_{2g} neutral-butatriene normal modes (geometries displaced from the equilibrium geometry along the normal modes—*arrows*—are depicted in the last column for the sake of clarity)

B_{3g}	3	skeletal bending (in plane)	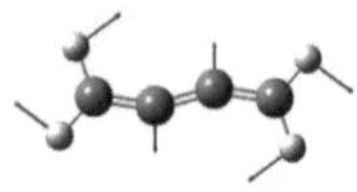
	9	CH_2 rocking	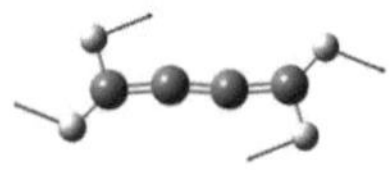
	18	C–H stretching	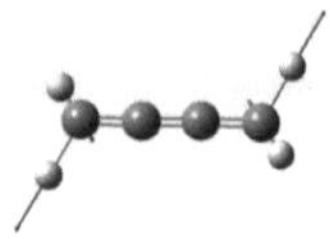
B_{1u}	11	CH_2 scissoring	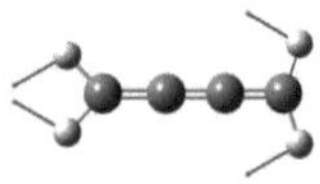
	13	C–C stretching	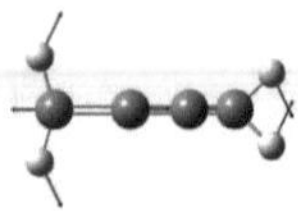
	16	C–H stretching	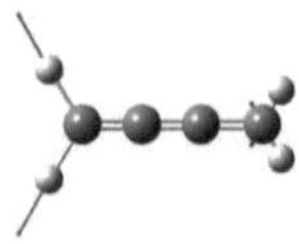

Fig. 7.7 B_{3g} and B_{1u} neutral-butatriene normal modes (geometries displaced from the equilibrium geometry along the normal modes—*arrows*—are depicted in the last column for the sake of clarity)

where ω_j are the harmonic vibrational (angular) frequencies of the normal modes. Note that a quadratic expansion with no symmetry at all would yield $\frac{18\times19}{2} = 171$ independent parameters (second derivatives in the Hessian matrix). Considering symmetry yields a block-diagonal Hessian matrix (according to the irreducible representations involved in Γ_{vib}) with $\frac{4\times5}{2} + 0 + \frac{2\times3}{2} + \frac{3\times4}{2} + \frac{1\times2}{2} + \frac{3\times4}{2} + \frac{3\times4}{2} + \frac{2\times3}{2} = 35$ terms. Further diagonalising the Hessian removes the cross terms to produce a sum made of 18 quadratic terms (again, this final reduction is not due to symmetry but to the definition of normal modes as the eigenvectors of the Hessian matrix).

B_{2u}	1	skeletal bending (in plane)	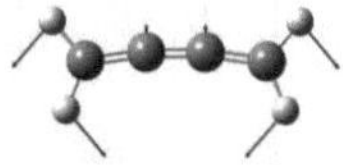
	10	CH_2 rocking	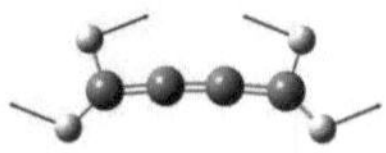
	17	C–H stretching	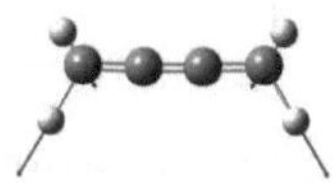
B_{3u}	2	skeletal bending (out of plane)	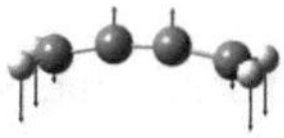
	6	CH_2 wagging	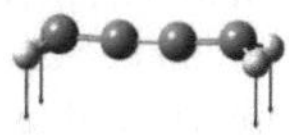

Fig. 7.8 B_{2u} and B_{3u} neutral-butatriene normal modes (geometries displaced from the equilibrium geometry along the normal modes—*arrows*—are depicted in the last column for the sake of clarity)

We now need to determine the irreducible representations of the electronic states in the context of point-group theory. Rather than considering transformations acting on electronic coordinates, one can directly consider the behaviours of atomic or molecular orbitals under symmetry operations, as already pointed out in the case of the water molecule. In practice, the first step for determining the irreducible representation of an electronic state consists in determining the most representative electronic configuration for the state of interest (note that, for symmetry reasons, other configurations belong to the same irreducible representation as the dominant one). Molecular orbitals can be vacant or occupied once or twice. When a non-degenerate orbital is twice-occupied, it accounts for a totally symmetric contribution (the same applies for a fully-occupied set of degenerate orbitals when there are degenerate irreducible representations in the point group). Such so-called closed shells can thus be ignored. As a result, only singly-occupied orbitals (or non-fully

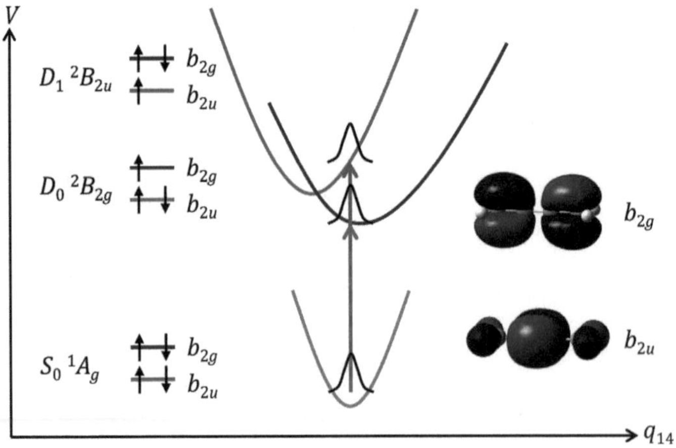

Fig. 7.9 Singly-occupied molecular orbitals in the dominant configurations of the first two *diabatic* electronic states of the butatriene cation and cuts of the potential energy surfaces along coordinate q_{14} (mode ν_{14} is A_g and describes the stretching of the central C–C bond)

occupied degenerate sets) are determining. Let us consider as an example an open-shell electronic configuration: [closed shells]a^1b^1. If orbitals a and b belong to Γ_a and Γ_b, respectively, the irreducible representation of the electronic configuration is $\Gamma_a \otimes \Gamma_b$. The determination of Γ_a and Γ_b is then obtained upon determining how orbitals a and b transform under the various symmetry operations of the point group. Again, this can be achieved upon considering a standard point-group approach where orbitals are viewed as solid objects that move and are transformed into themselves or their negatives[15] under symmetry operations.

Let us now turn to the butatriene cation. As mentioned in Sect. 4.4 of Chap. 4, the first two adiabatic electronic states are denoted $X^2B_{2g}(D_0)$ and $A^2B_{2u}(D_1)$. Their symmetry species are given with respect to the ground-state equilibrium geometry (Franck-Condon point) where their dominant configurations are [closed shells]$b_{2u}^2b_{2g}^1$ and [closed shells]$b_{2u}^1b_{2g}^2$, respectively. In other words, as a good approximation, state $X^2B_{2g}(D_0)$ is obtained from the ground-state neutral molecule upon removing an electron from the highest-occupied molecular orbital (b_{2g}), whereas $A^2B_{2u}(D_1)$ is obtained upon removing an electron from the highest-minus-one-occupied molecular orbital (b_{2u}). These molecular orbitals are depicted on Fig. 7.9. Their symmetry species can be determined quite easily upon noticing that molecular orbitals b_{2u} and b_{2g} are similar to molecule-centred atomic orbitals $p_y(B_{2u})$ and $d_{xz}(B_{2g})$, respectively.

These two states are known to be well-separated in energy from the higher-excited states of the cation. However, they interact strongly together, as there is a conical

[15] As already pointed out, we have limited our discussion to non-degenerate cases, whereby body-fixed axes are turned into themselves or their negatives and the same for orbitals. Degenerate cases may involve transformations that mix pairs of equivalent orbitals together.

intersection near the Franck-Condon point, with consequences on the photoelectron spectrum. The two (quasi)diabatic states underlying the vibronic-coupling Hamiltonian model proposed in Ref. [10] are chosen such that states 1 and 2 coincide with $X^2B_{2g}(D_0)$ and $A^2B_{2u}(D_1)$ at the Franck-Condon point ($\mathbf{q} = \mathbf{0}$) and keep belonging to the B_{2g} and B_{2u} irreducible representations, respectively, for other geometries $\mathbf{q} \neq \mathbf{0}$ as long as $\mathcal{D}_{2\mathrm{h}}$ symmetry is preserved. For $\alpha, \beta = 1, 2$, the entries of the diabatic Hamiltonian matrix, $\mathbf{H}^{\mathrm{dia}}(\mathbf{q})$, read

$$H^{\mathrm{dia}}_{\alpha\beta}(\mathbf{q}) = \langle \Psi^{\mathrm{dia}}_{\alpha}; \mathbf{q} | H^{el}(\mathbf{q}) | \Psi^{\mathrm{dia}}_{\beta}; \mathbf{q} \rangle \,. \tag{7.19}$$

As $H^{el}(\mathbf{q})$ is A_g (totally-symmetric), $H^{\mathrm{dia}}_{\alpha\beta}(\mathbf{q})$ transforms as $\Gamma_\alpha \otimes A_g \otimes \Gamma_\beta = \Gamma_\alpha \otimes \Gamma_\beta$. The diagonal entries are totally symmetric terms, $B_{2g} \otimes B_{2g} = B_{2u} \otimes B_{2u} = A_g$, while the off-diagonal entries belong to $B_{2g} \otimes B_{2u} = A_u$. Expanding this matrix as a quadratic form in terms of $\mathbf{q}$ around the origin $\mathbf{q} = \mathbf{0}$ involves first- and second-order monomials of normal coordinates that can be classified into four sets, as follows, where we use the mode labelling given in Figs. 7.6, 7.7 and 7.8.

G_1 is the set made of the four normal coordinates q_j that have A_g symmetry:

- A_g: $j = 8, 12, 14, 15 \to 4\, q_j$.

G_2 is the set of pairs of normal coordinates (q_j, q_k) the products of which, $q_j q_k$, have A_g symmetry ($\Gamma_j \otimes \Gamma_k = A_g$). From the direct-product table (Table 7.9), this is achieved only if both modes belong to the same irreducible representation. This amounts for a total of $\frac{4\times5}{2} + 0 + \frac{2\times3}{2} + \frac{3\times4}{2} + \frac{1\times2}{2} + \frac{3\times4}{2} + \frac{3\times4}{2} + \frac{2\times3}{2} = 35$ unordered pairs as indicated in the following list,

- (A_g, A_g): $(j, k) = (8, 8), (8, 12), (8, 14), (8, 15),$
  $(12, 12), (12, 14), (12, 15), (14, 14), (14, 15), (15, 15) \to 10\, q_j q_k$;
- (B_{2g}, B_{2g}): $(j, k) = (4, 4), (4, 7), (7, 7) \to 3\, q_j q_k$;
- (B_{3g}, B_{3g}): $(j, k) = (3, 3), (3, 9), (3, 18),$
  $(9, 9), (9, 18), (18, 18) \to 6\, q_j q_k$;
- (A_u, A_u): $(j, k) = (5, 5) \to 1\, q_j q_k$;
- (B_{1u}, B_{1u}): $(j, k) = (11, 11), (11, 13), (11, 16), (13, 13),$
  $(13, 16), (16, 16) \to 6\, q_j q_k$;
- (B_{2u}, B_{2u}): $(j, k) = (1, 1), (1, 10), (1, 17),$
  $(10, 10), (10, 17), (17, 17) \to 6\, q_j q_k$;
- (B_{3u}, B_{3u}): $(j, k) = (2, 2), (2, 6), (6, 6) \to 3\, q_j q_k$.

G_3 is the set made of the sole normal coordinates q_j that has A_u symmetry:

- A_u: $j = 5 \to 1\, q_j$.

G_4 is the set of pairs of normal coordinates (q_j, q_k) the products of which, $q_j q_k$, have A_u symmetry ($\Gamma_j \otimes \Gamma_k = A_u$). From the direct-product table (Table 7.9), this is achieved with $A_u \otimes A_g = A_u$, $B_{1u} \otimes B_{1g} = A_u$, $B_{2u} \otimes B_{2g} = A_u$, and $B_{3u} \otimes B_{3g} = A_u$. This amounts for a total of $1 \times 4 + 0 + 3 \times 2 + 2 \times 3 = 16$ unordered pairs as indicated in the following list,

- (A_u, A_g): $(j, k) = (5, 8), (5, 12), (5, 14), (5, 15) \rightarrow 4\, q_j q_k$;
- (B_{2u}, B_{2g}): $(j, k) = (1, 4), (10, 4), (17, 4),$
  $(1, 7), (10, 7), (17, 7) \rightarrow 6\, q_j q_k$;
- (B_{3u}, B_{3g}): $(j, k) = (2, 3), (2, 9), (2, 18),$
  $(6, 3), (6, 9), (6, 18) \rightarrow 6\, q_j q_k$.

Using this convention, the diabatic Hamiltonian matrix (see also Sect. 4.4 of Chap. 4) reads

$$\begin{aligned}\mathbf{H}^{\text{dia}}(\mathbf{q}) &= \sum_{j=1}^{18} \frac{\hbar\omega_j}{2} q_j^2 \mathbf{1} + \begin{bmatrix} \varepsilon_1 & 0 \\ 0 & \varepsilon_2 \end{bmatrix} \\ &+ \sum_{j \in G_1} \begin{bmatrix} \kappa_j^{(1)} & 0 \\ 0 & \kappa_j^{(2)} \end{bmatrix} q_j + \sum_{(j,k) \in G_2} \begin{bmatrix} \gamma_{jk}^{(1)} & 0 \\ 0 & \gamma_{jk}^{(2)} \end{bmatrix} q_j q_k \\ &+ \sum_{j \in G_3} \begin{bmatrix} 0 & \lambda_j \\ \lambda_j & 0 \end{bmatrix} q_j + \sum_{(j,k) \in G_4} \begin{bmatrix} 0 & \mu_{jk} \\ \mu_{jk} & 0 \end{bmatrix} q_j q_k \,.\end{aligned} \tag{7.20}$$

where ω_j are the harmonic vibrational (angular) frequencies of the neutral ground state (see above). We recall that q_j, here, are dimensionless (frequency-mass-weighted) normal coordinates. Modes belonging to G_1 induce a first-order energy shift (due to $\kappa_j^{(1)} \neq \kappa_j^{(2)}$) and are called tuning modes. The one belonging to G_3 (here, the single A_u mode, ν_5, spanned by coordinate q_5) induces a first-order interstate coupling (due to $\lambda_j \neq 0$) and is called coupling mode. G_2-modes further induce "second-order tuning" ($\gamma_{jk}^{(1)} \neq \gamma_{jk}^{(2)}$) and G_4-modes "second-order coupling" ($\mu_{jk} \neq 0$).

This matrix involves 18 ω_j-parameters, 2 ε_α-parameters, $2 \times 4 = 8$ $\kappa_j^{(\alpha)}$-parameters, $2 \times 35 = 70$ $\gamma_{jk}^{(\alpha)}$-parameters, 1 λ_j-parameter, and 16 μ_{jk}-parameters (where $\alpha = 1$ or 2). The total number of non-zero parameters is thus 115. In contrast, not using point-group symmetry would have yielded $18 + 2 \times (1 + 18 + \frac{18 \times 19}{2}) + 18 + \frac{18 \times 19}{2} = 587$ parameters to be determined. In other words, 472 parameters vanish for symmetry reasons. This illustrates how group theory is helpful in order to determine in advance the parameters that are zero, before any numerical treatment. In practice, this can turn out to be crucial to save time and/or memory but also to avoid numerical instabilities when optimising the values of the parameters (those that should be zero would introduce "noise" and disturb the convergence of the fitting procedure).

7.4 The Umbrella Inversion of Twice-Deuterated Ammonia

As mentioned in the previous sections, a description of the internuclear motion in terms of normal modes and their behaviours in terms of irreducible representations of the point group determined at some equilibrium geometry is sufficient for characterising small-amplitude deformations (vibrations) from this reference point. In this context, the equilibrium geometry corresponds to a single minimum at the bottom of a given potential-energy well in the energy landscape. Point-group theory proves useful if this structure shows non-trivial symmetry (when the point group is larger than C_1).

As already pointed out, in the case of water, there is a single permutational isomer for each internuclear arrangement because permuting the position of both H-nuclei can be achieved dynamically via a barrierless rigid-body rotation. In particular, the ground-state equilibrium geometry corresponds to a unique well within the energy landscape. Using either C_{2v} (point group) or C_{2v}(M) (molecular-symmetry group) thus yields vibronic descriptions that are equivalent. In contrast, *in the case of ammonia, the ground-state equilibrium geometry has C_{3v} symmetry and corresponds to two distinct permutational isomers* (they would be enantiomers if the three H-nuclei were distinguishable). The deformation between both distinct versions (the umbrella inversion, characterised by a double well in the energy landscape) is feasible dynamically, as it gives rise to significant tunnelling with observable consequences on the infrared spectrum. In this context, this large-amplitude deformation is sometimes called a contortion, as opposed to a vibration. We will now illustrate this on NHD_2, which also yields significant tunnelling, as illustrated in Refs. [11, 12] (this isotopologue of ammonia is somewhat simpler because its treatment does not involve non-degenerate irreducible representations; see below).

The equilibrium geometry of NHD_2 is similar to that of NH_3: a trigonal pyramid ("umbrella"); see Fig. 7.10 (see also Fig. 13.1 in Chap. 13). Because of isotopic substitution, the corresponding point group is not C_{3v} as in ammonia but C_s. As already mentioned, the umbrella inversion is a feasible contortion. It is characterised by a symmetric double well with a small transition barrier of about 1800 cm^{-1} (here,

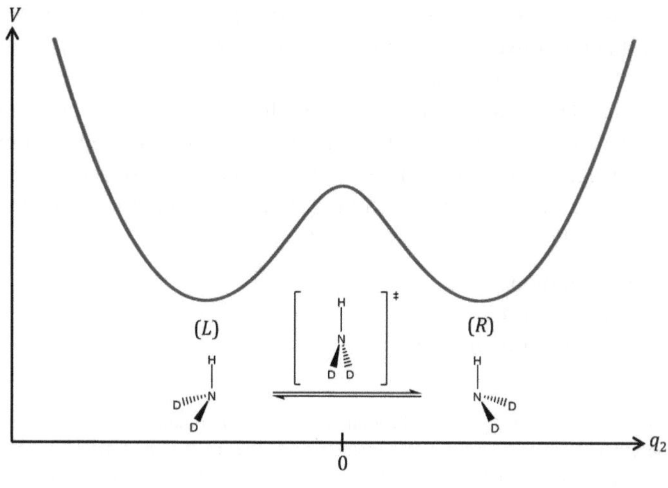

Fig. 7.10 Double well of the umbrella inversion of twice-deutared ammonia

Table 7.11 Character table of the $\mathcal{C}_{2v}$(M) and $\mathcal{G}_4$ molecular-symmetry groups and equivalence with the $\mathcal{C}_{2v}$ point group

$\mathcal{C}_{2v}$(M)/$\mathcal{G}_4$	E	(12)	E^*	(12)*		
$\mathcal{C}_{2v}$	E	$C_2(z)$	$\sigma_v(yz)$	$\sigma_v(xz)$		
A_1/A^+	1	1	1	1	z	x^2, y^2, z^2
A_2/A^-	1	1	-1	-1	R_z	xy
B_1/B^-	1	-1	-1	1	x, R_y	xz
B_2/B^+	1	-1	1	-1	y, R_x	yz

small means that tunnelling splitting is observed experimentally; see Refs. [11, 12]). The transition structure, midway between both minima, is trigonal and flat (point group $\mathcal{C}_{2v}$; $\mathcal{D}_{3h}$ for ammonia).

The relevant *molecular-symmetry group* that connects continuously[16] the two equivalent but distinct $\mathcal{C}_s$ minima through the $\mathcal{C}_{2v}$ transition structure is denoted $\mathcal{G}_4$, also known as $\mathcal{MS}_4$. It is made of the identity, E, the parity transformation, E^*, the permutation of both D-nuclei, denoted (12), and the composition of E^* and (12), denoted (12)*. The corresponding character table is given in Table 7.11. In $\mathcal{G}_4$, A/B denote whether irreducible representations are even/odd with respect to (12) and $+/-$ whether they are even/odd with respect to E^*. These operations are the same as those that define $\mathcal{C}_{2v}$(M) (the aforementioned molecular-symmetry group of water) and both molecular-symmetry groups are thus isomorphic. As a consequence, $\mathcal{G}_4$ also is isomorphic to the point group $\mathcal{C}_{2v}$ such that the labels of both types of irreducible representations can be used for convenience, according to the one-to-one correspondence indicated in the character table. As is often the case, the molecular-symmetry group corresponding to a feasible deformation involving *a symmetric double well* is isomorphic to the point group of the transition structure (here, $\mathcal{C}_{2v}$), which is more symmetrical than both equivalent minima (here, $\mathcal{C}_s$).

Using Mulliken's convention as in the case of the water molecule, the body-fixed axes (x, y, z) of the transition structure of NHD_2 are such that z is the rotation axis parallel to NH and (yz) is the molecular plane. These axes can be kept to describe the two equivalent $\mathcal{C}_s$ minima on both sides of the $\mathcal{C}_{2v}$ transition structure, for example if both D-nuclei wagging together across the (yz) plane are described symmetrically with respect to (xz) with NH fixed along z. Following the nomenclature given in Ref. [13], the six vibrational modes of both $\mathcal{C}_s$ minima are:

- ν_1: NH stretching mode (A');
- ν_2: NHD_2 pyramidalisation mode (A');
- ν_{3a}: symmetric ND stretching mode (A');

[16] The molecular-symmetry group is valid for any relative arrangement of the nuclei, as opposed to the point group that characterises a specific arrangement. This is why the former can be considered as "containing" all possible subgroups that can be connected dynamically during the course of a reactive process. Molecular-symmetry groups are sometimes called dynamical groups for this reason.

- ν_{3b}: antisymmetric ND stretching mode (A'');
- ν_{4a}: symmetric HND bending mode (A');
- ν_{4b}: antisymmetric HND bending mode (A'').

At the $\mathcal{C}_{2v}$ transition structure, they become:

- ν_1: NH stretching mode (A_1);
- ν_2: umbrella-inversion mode (B_1);
- ν_{3a}: symmetric ND stretching mode (A_1);
- ν_{3b}: antisymmetric ND stretching mode (B_2);
- ν_{4a}: symmetric HND bending mode (A_1);
- ν_{4b}: antisymmetric HND bending mode (B_2).

In other words, mode ν_2 can be spanned by a large-amplitude coordinate, q_2, that connects both minima through the transition structure (see Fig. 7.10). This umbrella-inversion coordinate is B_1 in $\mathcal{C}_{2v}$(M) (i.e., B^- in $\mathcal{G}_4$). The remaining small-amplitude coordinates are essentially vibrations that can be described in terms of normal modes that barely change along this reaction path.

The irreducible representation of mode ν_2 treated as a contortion can be determined in $\mathcal{C}_{2v}$(M). The left (L), $q_2 < 0$, and right (R), $q_2 > 0$, minima can be viewed as resulting from nuclear displacements with respect to the transition structure, $q_2 = 0$. If the transition barrier were too high for tunnelling to be observed, we would have degenerate (L) and (R) vibrational states for each single well (mode ν_2 would thus be treated as a vibration around a $\mathcal{C}_s$ minimum). Such local states can be used as zero-order states to build the first vibrational-contortional states of NHD_2. To this end, we must consider plus and minus combinations. For example, the first two vibrational states read approximately

$$\Psi_0^s(\mathbf{q}) \approx \frac{\Psi_0^L(\mathbf{q}) + \Psi_0^R(\mathbf{q})}{\sqrt{2}}, \tag{7.21}$$

$$\Psi_0^a(\mathbf{q}) \approx \frac{\Psi_0^L(\mathbf{q}) - \Psi_0^R(\mathbf{q})}{\sqrt{2}}. \tag{7.22}$$

The symmetric and antisymmetric wavefunctions, $\Psi_0^s(\mathbf{q})$ and $\Psi_0^a(\mathbf{q})$, correspond to the vibrational energy levels labelled $(0)^s$ and $(0)^a$, respectively (see Table 13.1 in Chap. 13). The fact that q_2 is B_1 implies that either (12) or E^* turn $\Psi_0^L(\mathbf{q})$ into $\Psi_0^R(\mathbf{q})$ and $\Psi_0^R(\mathbf{q})$ into $\Psi_0^L(\mathbf{q})$ because such operations swap the (L) and (R) minima that correspond to opposite values of q_2 ($q_2 = 0$ being the position of the barrier that corresponds to the $\mathcal{C}_{2v}$ transition structure). As shown in Table 7.12, $\Psi_0^s(\mathbf{q})$ is A_1/A^+ while $\Psi_0^a(\mathbf{q})$ is B_1/B^-. From a more intuitive perspective, $\Psi_0^s(\mathbf{q})$ has no node along q_2 and $\Psi_0^a(\mathbf{q})$ has one node at $q_2 = 0$ (see Fig. 13.1 in Chap. 13). When the number of quanta increases in mode ν_2, it is possible to consider symmetric and antisymmetric combinations of zero-order states. Plus and minus signs depend on the phase convention chosen for them. However, there is no ambiguity when considering the number of nodes: an even number corresponds to a symmetric state (A_1/A^+)

Table 7.12 Behaviours of the first two (L) and (R) zero-order vibrational states and their symmetric and antisymmetric combinations

E	(12)	E^*	(12)*
$\Psi_0^L(\mathbf{q})$	$\Psi_0^R(\mathbf{q})$	$\Psi_0^R(\mathbf{q})$	$\Psi_0^L(\mathbf{q})$
$\Psi_0^R(\mathbf{q})$	$\Psi_0^L(\mathbf{q})$	$\Psi_0^L(\mathbf{q})$	$\Psi_0^R(\mathbf{q})$
$\Psi_0^s(\mathbf{q})$	$\Psi_0^s(\mathbf{q})$	$\Psi_0^s(\mathbf{q})$	$\Psi_0^s(\mathbf{q})$
$\Psi_0^a(\mathbf{q})$	$-\Psi_0^a(\mathbf{q})$	$-\Psi_0^a(\mathbf{q})$	$\Psi_0^a(\mathbf{q})$

while and odd number to an antisymmetric state (B_1/B^-). As a consequence, the levels ordered by increasing energy are:

- no node: $(0)^s$ (A_1/A^+);
- one node: $(0)^a$ (B_1/B^-);
- two nodes: $(2^1)^s$ (A_1/A^+);
- three nodes: $(2^1)^a$ (B_1/B^-);
- four nodes: $(2^2)^s$ (A_1/A^+);
- five nodes: $(2^2)^a$ (B_1/B^-);
- etc.

Above the barrier, the number of quanta in mode ν_2 based on (L) and (R) zero-order vibrational states gradually becomes meaningless and should be changed for the global number of nodes of the wavefunctions that are delocalised over the double well as if it were a single well.

Other states occur as tunnelling doublets in the vibrational spectrum (see Table 13.1 in Chap. 13). They involve $(0)^s$ and $(0)^a$ with extra vibrational quanta in the other modes. For example, ν_1, ν_{3a}, and ν_{4a} are A_1/A^+, so that $(1^1)^s$, $(3_a^1)^s$, and $(4_a^1)^s$ are A_1/A^+, whereas $(1^1)^a$, $(3_a^1)^a$, and $(4_a^1)^a$ are B_1/B^-. In contrast, ν_{3b} and ν_{4b} are B_2/B^+, so that $(3_b^1)^s$ and $(4_b^1)^s$ are B_2/B^+, whereas $(3_b^1)^a$ and $(4_b^1)^a$ are A_2/A^- (because $B_1 \otimes B_2 = A_2$, i.e., $B^- \otimes B^+ = A^-$).

References

1. Bunker PR, Jensen P (1998) Molecular symmetry and spectroscopy, 2nd edn. NRC Research Press, Ottawa
2. Weyl H (1950) The theory of groups and quantum mechanics. Dover Publications, New York
3. Herzberg G (1992) Molecular spectra and molecular structure. Krieger Pub Co
4. Mulliken RS (1955) Report on notation for the spectra of polyatomic molecules. J Chem Phys 23:1997
5. Mulliken RS (1956) Erratum: report on notation for the spectra of polyatomic molecules. J Chem Phys 24:1118
6. Wilson E, Decius J, Cross P (1955) Molecular vibrations. McGraw-Hill, New York
7. Cotton FA (1990) Chemical applications of group theory, 3rd edn. Wiley, New York
8. Wales DJ (2003) Energy landscapes—applications to clusters. Biomolecules and glasses. Cambridge University Press, Cambridge, UK

9. Raab A, Worth G, Meyer H-D, Cederbaum LS (1999) Molecular dynamics of pyrazine after excitation to the S_2 electronic state using a realistic 24-mode model Hamiltonian. J Chem Phys 110:936
10. Cattarius C, Worth GA, Meyer H-D, Cederbaum LS (2001) All mode dynamics at the conical intersection of an octa-atomic molecule: multi-configuration time-dependent Hartree (MCTDH) investigation on the butatriene cation. J Chem Phys 115:2088
11. Marquardt R, Sanrey M, Gatti F, Quere FL (2010) Full-dimensional quantum dynamics of vibrationally highly excited NHD_2. J Chem Phys 133:174302
12. Sala M, Guérin S, Gatti F, Marquardt R, Meyer H-D (2012) Laser induced enhancement of tunneling in NHD_2. J Chem Phys 136:194308
13. Snels M, Hollenstein H, Quack M (2003) The NH and ND stretching fundamentals of $^{14}ND_2H$. J Chem Phys 119:7893

Chapter 8
Introduction to Numerical Methods

The present chapter is dedicated to the numerical methods for solving the time-dependent Schrödinger equation for the nuclei.[1]

We first introduce *a variational principle for the time-dependent Schrödinger equation* (Sect. 8.1.1) and then discuss *primitive basis sets* used to build matrix-elements associated to operators (Sects. 8.1.2 and 8.1.3). The primitive basis set can be a *Discrete variable representation (DVR)*, adapted to the potential since in the DVR representation any multiplicative operator is represented by a diagonal matrix. On the other hand, a *Finite Basis Representation (FBR)* is adapted to derivative operators which appear in the kinetic energy since the action of derivative operators on FBR basis functions is analytical.

Several *numerical time propagation algorithms* (Split Operator, Lanczos, etc.) can be used to propagate a wavepacket expanded in a fixed primitive basis set (so called *standard method*), are described in Sect. 8.1.6. Since the size of a primitive product basis set increases exponentially with the number of degrees of freedom, *we introduce intermediate time-dependent basis functions that lead to a much more compact description of the systems*: after having sketched the Time-Dependent Hartree (TDH) approach in Sect. 8.2, we present in detail the *MultiConfiguration Time-Dependent Hartree (MCTDH)* method in Sect. 8.3. More details about the use of the propagation algorithms in MCTDH are given in Sect. 8.3.8. We explain briefly that MCTDH can also be utilized to solve the time-independent Schrödinger equation (see Sect. 8.4). Next, we present in Sect. 8.5 complex absorbing potentials that allow reflections to be prevented that otherwise occur at the boundaries of the domain of definition of the coordinates. Then, in Sect. 8.6, we discuss the flux analysis algorithm, which can be used to extract reaction probabilities or (integral) cross sections from a propagated

---

[1] A large part of this chapter is taken form the MCTDH lecture notes of H.-D. Meyer written by Daniel Pelaez-Ruiz. The authors sincerely thank Dr. Pelaez for letting them use his script.

F. Gatti et al., *Applications of Quantum Dynamics in Chemistry*,
Lecture Notes in Chemistry 98, DOI 10.1007/978-3-319-53923-2_8

wavepacket- Finally, we discuss in Sect. 8.7 how a compact and efficient representation of operators, in particular of the PES, can be achieved. In the present section, atomic units are used, in particular, $\hbar = 1$.

8.1 The Standard Method

8.1.1 Variational Principle

In a mathematically-rigorous formulation of quantum mechanics, the possible quantum states are represented by state vectors residing in an infinite Hilbert space [1]. However, for numerical resolutions of the Schrödinger equation, it is necessary to truncate the basis set to some finite dimension [2]. Now, given a particular mathematical expression of the wavefunction depending on several parameters (see Eqs. (8.7), (8.96), and (8.151) for the standard, TDH and MCTDH expressions, respectively), the approximate solution of the time-dependent Schrödinger equation (TDSE) can be obtained from a variational principle. This guarantees that the parameters in the wavefunction are chosen such that the wavefunction is as close as possible to the exact solution of the time-dependent Schrödinger equation. Here, we use the so-called Dirac-Frenkel [3, 4] principle that reads (see Appendix A of Ref. [5] for a discussion on different variational principles for the time-dependent Schrödinger equation):

$$\langle \delta\Psi | H - i\frac{\partial}{\partial t} | \Psi \rangle = 0\,, \tag{8.1}$$

where $\delta\Psi$ denotes a variation of the wavefunction generated by varying the parameters, $\delta\Psi = \sum \partial\Psi/\partial\lambda_k \cdot \delta\lambda_k$, where λ_k formally denote the parameters of the model wavefunction.

The equations of motion derived from the Dirac-Frenkel variational principle conserve both norm and mean energy, if the Hamiltonian is Hermitian and time-independent, $\frac{\partial}{\partial t}H = 0$, and if the model wavefunction Ψ *itself is contained in the space of the allowed variations:* $\Psi \in \{\delta\Psi\}$.

In order to prove this important result, let us first derive an expression for the time derivative of the expectation value of the nth moment of the Hamiltonian:

$$\begin{aligned}\frac{d}{dt}\langle \Psi \mid H^n \mid \Psi \rangle &= \langle \dot{\Psi} \mid H^n \mid \Psi \rangle + \langle \Psi \mid H^n \mid \dot{\Psi} \rangle \\ &= 2\,\mathrm{Re}\,\langle \Psi \mid H^n \mid \dot{\Psi} \rangle = 2\,\mathrm{Re}\,\langle H^n\Psi \mid \frac{\partial}{\partial t} \mid \Psi \rangle\,, \end{aligned} \tag{8.2}$$

and since $2\,\mathrm{Re}(a) = -2\,\mathrm{Im}(-ia)$, we obtain

$$\frac{d}{dt}\langle \Psi \mid H^n \mid \Psi \rangle = -2\,\mathrm{Im}\,\langle H^n \Psi \mid -i\,\frac{\partial}{\partial t} \mid \Psi \rangle \,. \tag{8.3}$$

Now, since $\langle \Psi \mid H^{n+1} \mid \Psi \rangle$ is real and thus $\mathrm{Im}\langle H^n \Psi \mid H \mid \Psi \rangle = 0$, this yields

$$\frac{d}{dt}\langle \Psi \mid H^n \mid \Psi \rangle = -2\,\mathrm{Im}\,\langle H^n \Psi \mid H - i\,\frac{\partial}{\partial t} \mid \Psi \rangle \,. \tag{8.4}$$

For $n = 0$, Eq. (8.4) proves the conservation of the norm:

$$\frac{d}{dt}\langle \Psi \mid 1 \mid \Psi \rangle = -2\,\mathrm{Im}\,\langle \Psi \mid H - i\,\frac{\partial}{\partial t} \mid \Psi \rangle = 0\,, \tag{8.5}$$

as Eq. (8.1) ensures that $\langle \Psi \mid H - i\,\frac{\partial}{\partial t} \mid \Psi \rangle$ vanishes because $\delta\Psi = \Psi$ is an allowed variation from the assumption $\Psi \in \{\delta\Psi\}$. This assumption is valid if the model ansatz for Ψ contains a parameter which controls its length, e.g. if there are linear parameters. This is true for both TDH and MCTDH.

The case $n = 1$ demonstrates the conservation of the energy:

$$\begin{aligned}
\frac{d}{dt}\langle \Psi \mid H \mid \Psi \rangle &= -2\,\mathrm{Im}\,\langle H\Psi \mid H - i\,\frac{\partial}{\partial t} \mid \Psi \rangle \\
&= -2\,\mathrm{Im}\Big[\Big\langle \Big(H - i\,\frac{\partial}{\partial t}\Big)\Psi \Big| \Big(H - i\,\frac{\partial}{\partial t}\Big)\Psi \Big\rangle + \Big\langle i\,\frac{\partial}{\partial t}\Psi \Big| H - i\,\frac{\partial}{\partial t} \Big| \Psi \Big\rangle\Big] \\
&= 2\,\mathrm{Re}\Big\langle \frac{\partial}{\partial t}\Psi \Big| H - i\,\frac{\partial}{\partial t} \Big| \Psi \Big\rangle \\
&= 0\,,
\end{aligned} \tag{8.6}$$

since $\delta\Psi = \frac{\partial}{\partial t}\Psi$ is always an allowed variation, $\dot{\Psi} \in \{\delta\Psi\}$, because $\dot{\Psi} = \sum \dot{\lambda}_k\, \partial\Psi/\partial\lambda_k$.

All approaches we will discuss, such as the standard, TDH, and the MCTDH methods, conserve norm and mean energy. If in an actual calculation these quantities are not conserved, this must be due to numerical errors, in particular to the use of finite integrator step sizes. If norm and energy conservation is insufficient, one has to increase the integrator accuracy.

8.1.2 *The Standard Method in the Finite Basis Set Representation (FBR)*

The most direct way to solve the TDSE is to expand the wavefunction (WF) into a product of time-independent (TI) basis set. This is what is called the standard method:

$$\Psi(q_1, q_2, \ldots, q_f, t) = \sum_{j_1 \cdots j_f}^{N_1 \cdots N_f} C_{j_1 \cdots j_f}(t)\, \chi_{j_1}^{(1)}(q_1) \cdots \chi_{j_f}^{(f)}(q_f), \tag{8.7}$$

where the χ_j are orthonormal primitive basis functions. The basis set must be numerically complete for the problem. The choice of the basis set is dictated by the physics of the system, the volume element, the boundary conditions and also the possible presence of singularities in the kinetic energy operator as explained in Sect. 6.6. For numerical applications, the number of primitive functions is finite and N denotes the number of these functions. The corresponding basis set is called a *Finite Basis Representation* or FBR [6–14]. Several examples are listed below:

- The particle-in-a-box eigenfunctions as a basis: the box boundaries are x_0 and x_{N+1}, and $L = x_{N+1} - x_0$ denotes the length of the box. The basis functions are thus

$$\chi_j(x) = \begin{cases} \sqrt{2/L}\, \sin\left(j\pi(x - x_0)/L\right) & \text{for } x_0 \le x \le x_{N+1} \\ 0 & \text{elsewhere} \end{cases} . \tag{8.8}$$

  Here, j starts from 1. These functions are equal to zero at x_0 and x_{N+1}. They are very often used for degrees of freedom with a constant volume element $dV = dx$ and if the wavefunction vanishes at finite boundaries.
- The Hermite functions that are the solutions of the one-dimensional harmonic oscillator problem that can be considered as a zero-order Hamiltonian for many systems:

$$\chi_j(x) = \left(2^j j!\right)^{-1/2} (m\omega/\pi)^{1/4} H_j\left(\sqrt{m\omega}\,(x - x_0)\right) e^{-m\omega(x-x_0)^2/2}, \tag{8.9}$$

  where $j = 0, \ldots, N-1$, x_0 is the equilibrium position, ω the oscillator frequency, and m its mass. Here again, $dV = dx$. These functions are typically used for vibrational modes, for instance when normal coordinates or lengths of chemical bonds are utilized as coordinates.
- Normalized Legendre polynomials that are eigenfunctions of $j^2 = -\frac{1}{\sin\theta}\frac{\partial}{\partial\theta}\sin\theta\frac{\partial}{\partial\theta}$ with the volume element $dV = \sin\theta\, d\theta$:

$$\chi_{l+1}(\theta) = \sqrt{\frac{2l+1}{2}}\, P_l(\cos\theta), \tag{8.10}$$

  l starts from zero and $\theta \in [0, \pi]$. They are often used for planar angles (i.e. the polar θ angles in spherical coordinates).
- Periodic exponential functions (Fourier):

$$\chi_j(x) = L^{-1/2} \exp\left(2i\pi j(x - x_0)/L\right), \tag{8.11}$$

  with $-n \leqslant j \leqslant n$ and $L = x_N - x_0$. The wavefunctions to be represented satisfy periodic boundary conditions, $\psi(x_0) = \psi(x_N)$. Here, $N = 2n + 1$ and $dV = dx$.

These functions are typically used for dihedral angles (i.e. the azimuthal φ angles in spherical coordinates with $\varphi \in [0, 2\pi[)$ or for the coordinates that describe the translations of an atom or a molecule on a periodic surface of a solid.

If the kinetic energy operator displays some singularities as explained in Sect. 6.6, it may be necessary to use non-direct product basis sets, i.e. functions that depend on more than one degree of freedom. The two most important examples are listed below.

- The two-dimensional spherical harmonics:

$$Y_{jm}(\theta, \varphi) = (-1)^m \sqrt{\frac{2j+1}{4\pi} \frac{(j-m)!}{(j+m)!}}\, P_j^m(\cos\theta)\, e^{im\varphi}\ for m \geq 0. \tag{8.12}$$

The relation

$$Y_{jm}(\theta, \varphi) = (-1)^m\ Y_{j-m}(\theta, \varphi)$$

provides spherical harmonics for negative m.
The matrix elements of the angular momentum operators $j^2 = -\left(\frac{1}{\sin\theta}\frac{\partial}{\partial\theta}\sin\theta\frac{\partial}{\partial\theta} + \frac{1}{\sin^2\theta}\frac{\partial^2}{\partial\varphi^2}\right)$, $J_\pm = e^{\pm i\varphi}\left(\pm\frac{\partial}{\partial\theta} + i\cot\theta\frac{\partial}{\partial\varphi}\right)$, and $J_z = \frac{1}{i}\frac{\partial}{\partial\varphi}$ are then given by simple formulas (see Eq. (6.173)) and the corresponding matrix elements are non-singular. Here, $dV = \sin\theta\, d\theta\, d\varphi$ and $|m| \leq j$.
- Three-dimensional rotational Wigner functions $\overline{D}^j_{m,k}(\alpha, \beta, \gamma)$ that are defined as:

$$\overline{D}^j_{m,k}(\alpha, \beta, \gamma) = \sqrt{\frac{2j+1}{8\pi^2}}\, D^j_{m,k}(\alpha, \beta, \gamma)\,, \tag{8.13}$$

with

$$D^j_{m,k}(\alpha, \beta, \gamma) = e^{-im\alpha} d^j_{m,k}(\beta) e^{-ik\gamma}\,. \tag{8.14}$$

They are used to describe the three-dimensional rotation of a body: α, β, γ are the three Euler angles as shown on Fig. 3.4 of Sect. 3.3.3. $d^j_{m,k}(\beta)$ is a Jacobi function [15]. The matrix elements of the projections of the total angular momentum onto the axes of the SF and BF frame in this basis set are non-singular and analytical [15]. Here, $dV = \sin\beta\, d\alpha\, d\beta\, d\gamma$ and $|m|, |k| \leq j$.

In the Heidelberg MCTDH package [16], the wavefunction is not expanded in an FBR but in the corresponding Discrete Variable Representation (DVR) as explained in Sect. 8.1.3. More precisely, the wavefunction and the matrix of the potential operator (or of any multiplicative operator) are expressed on a grid representation, i.e. in the DVR. The corresponding FBR basis is only used to generate matrix representations of derivative operators.

As regards the FBR, we arrive at the following important results: *in the FBR, the action of the derivative operators is simple and analytical.* For instance, the operator $\frac{\partial^2}{\partial x^2}$ is diagonal in the basis set of particle-in-a-box eigenfunctions, $\sqrt{2/L}\ \sin(j\pi\,(x - x_0)/L)$. In the same manner, $j^2 = -\frac{1}{\sin\theta}\frac{\partial}{\partial\theta}\sin\theta\frac{\partial}{\partial\theta}$ is diagonal in the basis set or Legendre polynomials, $\sqrt{\frac{2l+1}{2}}\,P_l(\cos\theta)$ and $j^2 = -\left(\frac{1}{\sin\theta}\frac{\partial}{\partial\theta}\sin\theta\frac{\partial}{\partial\theta} + \frac{1}{\sin^2\theta}\frac{\partial^2}{\partial\varphi^2}\right)$ is diagonal in the basis set of spherical harmonics.

Let us go back to the resolution of the time-dependent Schrödinger equation with the standard approach. The parameters to be optimized by the variational principle of Eq. (8.1) are the coefficients C. To derive the equations of motion for the coefficients C, we employ the Dirac-Frenkel variational principle. This approach is similar to a full Configuration Interaction (CI) method in quantum chemistry where the primitive functions are the atomic orbitals. Since the objects to be varied here are just numbers, the variation is a partial differentiation:

$$\delta\Psi = \sum_{l_1\cdots l_f} \frac{\partial\Psi}{\partial C_{l_1\cdots l_f}}\,\delta C_{l_1\cdots l_f} = \sum_{l_1\cdots l_f} \chi_{l_1}^{(1)}(q_1)\cdots\chi_{l_f}^{(f)}(q_f)\,\delta C_{l_1\cdots l_f}\,, \tag{8.15}$$

and

$$\dot{\Psi} = \sum_{j_1\cdots j_f} \dot{C}_{j_1\cdots j_f}\,\chi_{j_1}^{(1)}\cdots\chi_{j_f}^{(f)}\,. \tag{8.16}$$

Because the variations are independent one may set

$$\delta C_{l_1\cdots l_f} = \begin{cases} 1 & \text{for } l_1\cdots l_f = l_1^{(0)}\cdots l_f^{(0)} \\ 0 & \text{else}\,. \end{cases}$$

From $\langle\delta\Psi|H - i\frac{\partial}{\partial t}|\Psi\rangle = 0$, and replacing $l_\kappa^{(0)}$ with l_κ, we obtain

$$\begin{aligned} \langle\chi_{l_1}\cdots\chi_{l_f}|\sum_{j_1\cdots j_f} C_{j_1\cdots j_f} H\chi_{j_1}\cdots\chi_{j_f}\rangle &= \\ i\langle\chi_{l_1}\cdots\chi_{l_f}|\sum_{j_1\cdots j_f} \dot{C}_{j_1\cdots j_f}\,\chi_{j_1}\cdots\chi_{j_f}\rangle&. \end{aligned} \tag{8.17}$$

or

$$\sum_{j_1\cdots j_f} \langle\chi_{l_1}\cdots\chi_{l_f}|H|\chi_{j_1}\cdots\chi_{j_f}\rangle\, C_{j_1\cdots j_f} = i\dot{C}_{l_1\cdots l_f}\,, \tag{8.18}$$

Defining composite indices $J = (j_1, \ldots, j_f)$ and configurations $\chi_J = \prod_{\kappa=1}^{f}\chi_{j_\kappa}$, one arrives at the compact expression

$$i\dot{C}_L = \sum_J \langle\chi_L|H|\chi_J\rangle\, C_J\,. \tag{8.19}$$

This is a simple first-order differential equation with constant coefficients. It has the formal solution (for time-independent Hamiltonians):

$$\boldsymbol{C}(t) = e^{-i\boldsymbol{H}t}\,\boldsymbol{C}(0)\,, \tag{8.20}$$

where the bold faces shall indicate the vector and matrix form of coefficients and Hamiltonian, respectively. This system of differential equations is difficult to solve, because the number of coupled equations, its dimension, is large.

In general one needs at least 10 basis functions per degree of freedom. Hence, there are about 10^f coupled equations to be solved. Consider a molecule with 6 atoms, then there are $f = 3N - 6 = 12$ degrees of freedom, and 10^{12} coupled equations. This is not doable. In general only up to 4 atom systems (6D) may be treated by the standard method with today's computers. One hence has to resort to cleverer, but also more approximate methods. But before this, we will introduce the Discrete Variable Representation.

8.1.3 Discrete Variable Representation (DVR)

As aforementioned, a function on a computer, such as the wavefunctions Ψ or $H\Psi$, has to be represented by a *finite* set of numbers, i.e. by a discrete vector. For the sake of simplicity, we concentrate here on one-dimensional systems. We have introduced in the previous section the FBRs. In an FBR approach, the wavefunction is expressed as

$$\Psi = \sum_{j=1}^{N} a_j \phi_j \quad \text{with } a_j = \langle \phi_j | \Psi \rangle \,, \tag{8.21}$$

where the ϕ_j functions correspond to the functions introduced above (Hermite, Legendre functions, etc.). In a vector notation, we obtain

$$\Psi \to \boldsymbol{a} = (a_1, a_2, \ldots, a_N)^T \,. \tag{8.22}$$

However, it is often an advantage to express the wavefunction on a grid of points

$$x_\alpha, \quad \alpha = 1, \ldots, N \,. \tag{8.23}$$

Thus we have

$$\Psi(x) \to (\Psi(x_1), \ldots, \Psi(x_N))^T = (\Psi_1, \ldots, \Psi_N)^T = \{\Psi_\alpha\} \,. \tag{8.24}$$

The great advantage of grid methods is that the application of the, in general complicated, potential operator is very simple [6–10, 12–14, 17, 18]

$$(V\Psi)_\alpha = (V\Psi)(x_\alpha) = V(x_\alpha)\Psi(x_\alpha) \,. \tag{8.25}$$

In other words, on the grid representation, the potential and any multiplicative operator is diagonal. For evaluating matrix-elements by quadrature over the grid, one needs weights in addition. Within a quadrature approximation, we have

$$\langle \Psi | \Phi \rangle \approx \sum_{\alpha=1}^{N} w_\alpha \Psi^*(x_\alpha) \Phi(x_\alpha) \,. \tag{8.26}$$

In general, it is not obvious how to find optimal points and weights to be used in the previous equation.

In addition, a difficult problem for a grid representation are the differential operators, because there is no differentiable function any more. If one interpolates the points *locally*, one arrives at the finite-difference formulas, e.g.

$$\Psi''(x_\alpha) \approx \frac{1}{h^2} \left(\Psi(x_{\alpha+1}) - 2\Psi(x_\alpha) + \Psi(x_{\alpha-1}) \right) \,, \tag{8.27}$$

(local quadratic interpolation on an equidistant grid, where h is the grid spacing). Unfortunately, finite differences are not very accurate and this is the reason why one switches back to the FBR to evaluate the matrix elements for differential operators (see below).

But what exactly is a DVR? A DVR, as a basis representation, is a global approximation of high accuracy. There is a unitary transformation from the FBR to the DVR. Thus, the two representations are mathematically equivalent. We refer the reader to Refs. [2, 5, 8, 14] for a comprehensive discussion of the DVR theory. Here, we just mention that one way to arrive at a DVR is to diagonalize the matrix representation of the coordinate operator in the FBR. Indeed, we want to obtain a new representation that makes multiplicative operators diagonal. Thus, we build the matrix [5, 19]

$$Q_{jk} = \langle \varphi_j | x | \varphi_k \rangle \,, \tag{8.28}$$

where φ_j are the FBR functions. The diagonalization provides the eigenvector matrix, $\boldsymbol{U}$, where:

$$\boldsymbol{Q} = \boldsymbol{U} \; \boldsymbol{X} \; \boldsymbol{U}^T \,, \tag{8.29}$$

and an eigenvalue matrix:

$$\boldsymbol{X}_{\alpha\beta} = x_\alpha \delta_{\alpha\beta} \,, \tag{8.30}$$

where the eigenvalues, x_α, are taken as grid points. Here, two possibilities may occur:

(i) $\boldsymbol{Q}$ is tri-diagonal, then the weights are given as

$$w_\alpha^{1/2} = \frac{U_{k,\alpha}}{\varphi_k^*(x_\alpha)} \,, \tag{8.31}$$

where the right-hand side can be shown [20] to be independent of k. We know not only the points but also the weights as we have a quadrature rule (see Eq. (8.26)) to be used in Eq. (8.26) to calculate matrix-elements and the DVR is said to be a proper DVR [5];

(ii) $\boldsymbol{Q}$ is not tri-diagonal and we cannot define weights in a systematic way and the DVR is said to be an improper DVR.

We will assume in the following that our DVR is a proper DVR and for the DVRs associated with the FBRs introduced above (sine, Legendre, Hermite, etc., it can be shown that the corresponding DVRs are proper[2]). We just emphasize that it is not always the case. Now, if we have a proper DVR then we have a quadrature rule, and the matrix elements

$$\langle\varphi_j|\varphi_k\rangle = \sum_{\alpha=1}^{N} w_\alpha\, \varphi_j^*(x_\alpha)\, \varphi_k(x_\alpha) = \delta_{jk}\,, \tag{8.32}$$

$$\langle\varphi_j|x|\varphi_k\rangle = \sum_{\alpha=1}^{N} w_\alpha\, \varphi_j^*(x_\alpha)\, x_\alpha\, \varphi_k(x_\alpha) = Q_{jk}\,, \tag{8.33}$$

are exactly computed by quadrature for $j, k \leq N$. Equation (8.32) is called *discrete orthonormality*. From the unitary condition $\boldsymbol{U}^\dagger \boldsymbol{U} = \mathbf{1}$ follows *discrete completeness*:

$$\sum_{j=1}^{N} (w_\alpha w_\beta)^{\frac{1}{2}}\, \varphi_j^*(x_\alpha)\, \varphi_j(x_\beta) = \delta_{\alpha\beta}\,. \tag{8.34}$$

Next we introduce DVR-functions defined as

$$\chi_\alpha(x) = \sum_{j=1}^{N} \varphi_j(x)\, U_{j\alpha}\,. \tag{8.35}$$

The DVR functions are orthonormal

$$\langle\chi_\alpha|\chi_\beta\rangle = \delta_{\alpha\beta}\,, \tag{8.36}$$

and they behave like δ-functions on the grid

$$\chi_\alpha(x_\beta) = w_\alpha^{-1/2}\, \delta_{\alpha\beta}\,. \tag{8.37}$$

The DVR functions are very localized around one of the grid points, x_α, see Fig. 8.1 and below for examples. Potential matrix elements are now simple

[2]The quadrature rule defined through a proper DVR, i.e. defined by Eqs. (8.30) and (8.31), is of Gaussian type, i.e. it yields an exact result for polynomials of degree $2n - 1$ or less.

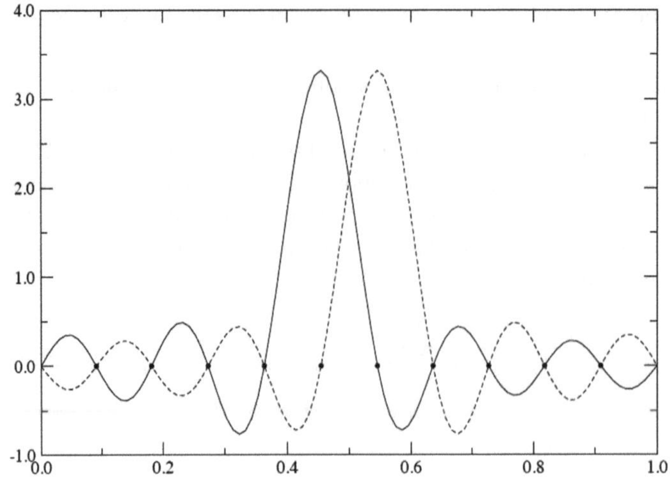

Fig. 8.1 First two sine DVR functions (*solid* and *dashed lines*) centered at two consecutive DVR points. Note that the functions are strictly zero at all DVR points (*black dots*) but the one that labels the function

$$\begin{aligned}\langle\chi_\alpha|V|\chi_\beta\rangle &= \sum_{\gamma=1}^{N} w_\gamma\, \chi_\alpha^*(x_\gamma)\, V(x_\gamma)\, \chi_\beta(x_\gamma) \\ &= \sum_{\gamma=1}^{N} w_\gamma\, w_\alpha^{-1/2}\, w_\beta^{-1/2}\, \delta_{\alpha\gamma}\, \delta_{\beta\gamma}\, V(x_\gamma) \\ &= V(x_\alpha)\, \delta_{\alpha\beta}\,. \end{aligned} \tag{8.38}$$

We arrive at the very important result: *in the DVR, the potential (and any multiplicative operator) is assumed to be diagonal. This is the DVR approximation.* It is an approximation because the matrix element is done by quadrature, not exactly: the integral is replaced by a discrete sum. Similarly, the equation

$$\langle\chi_\alpha|\Psi\rangle = \sum_{\gamma=1}^{N} w_\gamma\, \chi_\alpha^*(x_\gamma)\, \Psi(x_\gamma) = w_\alpha^{-1/2}\, \Psi(x_\alpha)\,, \tag{8.39}$$

connects grid and basis set representations. However, Eq. (8.39) is strictly valid only if Ψ lies entirely in the space spanned by the basis.

We now represent the wavefunction (WF) by its values at the grid points times square roots of weights

$$\Psi(x) \to \boldsymbol{\Psi} = \left(w_1^{1/2}\Psi(x_1), \ldots, w_N^{1/2}\Psi(x_N)\right)^T, \tag{8.40}$$

which is both, a grid and a spectral representation (see Eq. 8.39, pseudo-spectral methods). Integrals are now simple

$$\langle \Psi | \Phi \rangle = \sum_{\alpha=1}^{N} w_\alpha \, \Psi_\alpha^*(x_\alpha) \, \Phi(x_\alpha) = \sum_{\alpha=1}^{N} \Psi_\alpha^* \, \Phi_\alpha = \boldsymbol{\Psi}^* \cdot \boldsymbol{\Phi} \, . \tag{8.41}$$

In fact, one almost never needs the weights, as they are built into the WF. Weights are required only for plotting a WF or generating an initial WF from an analytic expression.

To derive the kinetic energy operator for the DVR-grid representations, we start considering its basis set representation (finite basis representation, FBR).

$$T_{jk}^{\mathrm{FBR}} = \langle \varphi_j | T | \varphi_k \rangle \, , \tag{8.42}$$

where we assume that the matrix elements can be obtained analytically.

The DVR-representation is then given by *a unitary transformation*

$$T_{\alpha\beta}^{\mathrm{DVR}} = \langle \chi_\alpha | T | \chi_\beta \rangle = \left(\boldsymbol{U}^\dagger \, \boldsymbol{T}^{\mathrm{FBR}} \, \boldsymbol{U} \right)_{\alpha\beta} . \tag{8.43}$$

Remarks:

(i) a DVR must be consistent with the volume element to be used: $dr, r^2 dr, sin\theta \, d\theta$, etc.
(ii) a DVR must be consistent with the applied boundary conditions.
(iii) the potential should be smoother than the WF to ensure that the DVR error is small, (no hard walls). The variational property is destroyed because the potential matrix elements are not evaluated exactly, i.e. computed eigenvalues are not necessarily upper bounds to the exact ones.
(iv) for smooth potentials and not too few grid points, the DVR error (Cf. Eq. (8.38)) is in general smaller or of the same order as the basis set truncation error.

Let us now consider two examples.

The Sine-DVR

In most cases one cannot derive the DVR grid points, weights, etc. analytically, because this requires the diagonalization of a matrix. For the sine-DVR, however, one can do everything analytically. The underlying basis functions are the "particle in a box" functions (see Fig. 8.2 for the first two functions).

$$\varphi_j(x) = \begin{cases} \sqrt{2/L} \, \sin(j\pi(x - x_0)/L) \text{ for } x_0 \le x \le L \\ 0 \text{ else} \, , \end{cases} \tag{8.44}$$

with $L = x_{n+1} - x_0$. The following matrix elements can be obtained exactly.

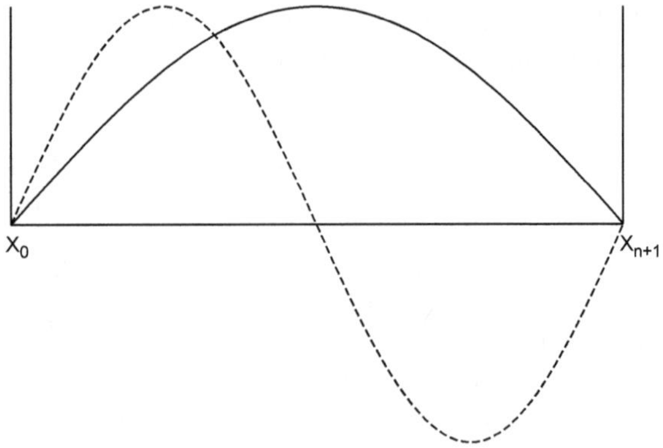

Fig. 8.2 First two particle-in-a-box functions

$$
\begin{aligned}
\langle\varphi_j|\varphi_k\rangle &= \delta_{jk}\,, \\
\langle\varphi_j|\frac{\partial}{\partial x}|\varphi_k\rangle &= \mathrm{mod}(j-k,2)\frac{4}{L}\frac{jk}{j^2-k^2} \text{ for } j \neq k\,, \\
\langle\varphi_j|\frac{\partial^2}{\partial x^2}|\varphi_k\rangle &= -\left(\frac{j\pi}{L}\right)^2 \delta_{jk}\,.
\end{aligned} \tag{8.45}
$$

The matrix $\langle\varphi_j|x|\varphi_k\rangle$ is not tri-diagonal. But after transforming the coordinate

$$f(x) = \cos(\pi(x - x_0)/L)\,, \tag{8.46}$$

one finds

$$F_{jk} = \langle\varphi_j|f(x)|\varphi_k\rangle = \frac{1}{2}(\delta_{j,k+1} + \delta_{j,k-1}) = \frac{1}{2}\begin{pmatrix} 0 & 1 & 0 & \cdots \\ 1 & 0 & 1 & \cdots \\ 0 & 1 & 0 & \cdots \\ \vdots & \vdots & \vdots & \ddots \end{pmatrix}\,. \tag{8.47}$$

This matrix is diagonalized analytically:

$$U_{j\alpha} = \sqrt{\frac{2}{n+1}}\sin\left(\frac{j\alpha\pi}{n+1}\right)\,, \tag{8.48}$$

with eigenvalues

$$f_\alpha = \cos\left(\frac{\alpha\pi}{n+1}\right)\,. \tag{8.49}$$

This yields the DVR grid-points

$$x_\alpha = f^{-1}(f_\alpha) = x_0 + \frac{L}{\pi}\arccos(f_\alpha) = x_0 + \alpha\frac{L}{n+1} = x_0 + \alpha\Delta x\,, \tag{8.50}$$

with $\alpha = 1, 2, \ldots, n$ and $\Delta x = \frac{L}{n+1}$. Note that x_0 and x_{n+1} do not belong to the grid. The wavefunction is vanishing there anyway. The weights are constant.

$$w_\alpha^{1/2} = U_{j\alpha}/\varphi_j(x_\alpha) = \sqrt{\frac{L}{n+1}} = \sqrt{\Delta x}\,, \tag{8.51}$$

hence $w_\alpha = \Delta x$, as it is to be expected for an evenly spaced grid.

The FBR derivative matrices are to be transformed to DVR. For the second derivative matrix, this can be done analytically (see Eq. (B.65) of Ref. [5]). The sum $\chi_\alpha(x) = \sum_{j=1}^n \varphi_j(x)U_{j\alpha}$ can also be done analytically and one obtains

$$\chi_\alpha(x) = \frac{1}{2\sqrt{L(n+1)}}\left\{\frac{\sin\left[\frac{\pi}{2}(2n+1)\frac{x-x_\alpha}{L}\right]}{\sin\left[\frac{\pi}{2}\frac{x-x_\alpha}{L}\right]} - \frac{\sin\left[\frac{\pi}{2}(2n+1)\frac{x+x_\alpha}{L}\right]}{\sin\left[\frac{\pi}{2}\frac{x+x_\alpha}{L}\right]}\right\}\,, \tag{8.52}$$

as the expression of the DVR-functions.

Two-Dimensional Spherical Harmonics DVR

If one uses spherical coordinates θ and ϕ, there will appear operators like

$$\begin{aligned} j^2 &= -\left(\frac{1}{\sin\theta}\frac{\partial}{\partial\theta}\sin\theta\frac{\partial}{\partial\theta} + \frac{1}{\sin^2\theta}\frac{\partial^2}{\partial\phi^2}\right)\,, \\ j_\pm &= e^{\pm i\varphi}\left(\pm\frac{\partial}{\partial\theta} + i\cot\theta\frac{\partial}{\partial\phi}\right)\,. \end{aligned} \tag{8.53}$$

These operators are *singular* for $\theta = 0$ or π. The singularity appears only in coordinate space, all matrix elements are non-singular when an appropriate basis set, e.g. the spherical harmonics Y_{lm}, are used. In fact,

$$\begin{aligned} j^2 Y_{lm} &= l(l+1)Y_{lm}\,, \\ j_\pm Y_{lm} &= \sqrt{l(l+1) - m(m\pm1)}\,Y_{l,m\pm1}\,. \end{aligned} \tag{8.54}$$

However, rather than a basis set, we would like to use a DVR. To this end, let us introduce the normalized associated Legendre functions

$$\tilde{P}_l^m(\cos\theta) = (-1)^m\sqrt{\frac{(2l+1)(l-m)!}{2(l+m)!}}\,P_l^m(\cos\theta)\,. \tag{8.55}$$

The spherical harmonics are then given by

$$Y_{lm}(\theta,\varphi) = \tilde{P}_l^m(\cos\theta)\frac{e^{im\phi}}{\sqrt{2\pi}}\,. \tag{8.56}$$

This is *not* a product basis, because $\tilde{P}_l^m$ depends on m. For $m = 0$, we obtain the usual Legendre functions $P_l^{m=0} = P_l$, which are polynomials in $x = \cos\theta$. The associated functions P_l^m are not polynomials as they contain the factor $(1-x^2)^{|m|/2}$. For the Legendre polynomials, there exist a Gauss-Legendre quadrature and hence a proper DVR for the coordinate θ.

Cory, Tromp, and Lemoine (see Ref. [21]) have noticed that the Gauss-Legendre quadrature can be used for the $m \neq 0$ case as well. They showed that

$$\langle \tilde{P}_l^m|\cos^k\theta|\tilde{P}_{l'}^m\rangle = \sum_{\alpha=1}^{n} w_\alpha \tilde{P}_l^m(\cos\theta_\alpha)\cos^k\theta_\alpha \tilde{P}_{l'}^m(\cos\theta_\alpha)\,, \tag{8.57}$$

is exact for $l + l' + k \leq 2n - 1$. We define $l_{\max} = n - 1$ and restrict l and l' to $0 \leq l \leq l_{\max}$, and we find that all matrix elements are exact by quadrature for $k = 0$ and 1. This looks like a proper DVR. However, for $m \neq 0$, there are fewer basis functions than grid points, because $l \geq |m|$ while we keep the grid points built from $m = 0$. Before we proceed, let us change the nomenclature and substitute l, m by j, k. Furthermore, we replace the angle ϕ by the discrete angular-momentum variable k via a Fourier transform.

$$\begin{aligned}
\psi(\theta,\phi) &= \frac{1}{\sqrt{2\pi}}\sum_k \psi(\theta,k)e^{ik\phi}\,,\\
\psi(\theta,k) &= \frac{1}{\sqrt{2\pi}}\int_0^{2\pi}\psi(\theta,\phi)e^{-ik\phi}\,\mathrm{d}\phi\,.
\end{aligned} \tag{8.58}$$

The associated operators now read

$$\begin{aligned}
j^2 &= -\left(\frac{1}{\sin\theta}\frac{\partial}{\partial\theta}\sin\theta\frac{\partial}{\partial\theta} - \frac{k^2}{\sin^2\theta}\right),\\
j_\pm &= \pm\frac{\partial}{\partial\theta} - k\cot\theta \text{ and } k \to k \pm 1\,,\\
j^2\tilde{P}_j^k &= j(j+1)\tilde{P}_j^k\,,\\
j_\pm\tilde{P}_j^k &= \sqrt{j(j+1) - k(k\pm 1)}\,\tilde{P}_j^{k\pm 1}\,.
\end{aligned} \tag{8.59}$$

To introduce a DVR, called extended Legendre DVR [22], we define transformation matrices $\boldsymbol{U}^k$ for each value of k individually:

$$U_{j\alpha}^k = w_\alpha^{1/2}\tilde{P}_j^k(\cos\theta_\alpha)\,. \tag{8.60}$$

The rows of $\boldsymbol{U}^k$ are orthonormal as long as $j \le j_{\max} \equiv n-1$. This follows immediately from Eq. (8.57). However, if one restricts j to $j \le j_{\max}$, there are only $j_{\max}+1-|k|$ rows but $n = j_{\max}+1$ columns. To make $\boldsymbol{U}^k$ square and unitary, we let j run from $|k|$ to $j_{\max}+|k|$ and successively Schmidt-orthogonalize the rows $j_{\max}+1, \ldots, j_{\max}+|k|$ to the lower ones. This ad hoc procedure for building unitary matrices works well, because only the high j-states, which should be only weakly populated, are modified.

Let us analyze what we have done. Schmidt-orthogonalization is equivalent to QR-decomposition, i.e. decomposition of a matrix in a unitary and a triangular one. In our case, it reads

$$w_\alpha^{1/2} \tilde{P}_j^k(\cos\theta_\alpha) = \sum_{j'} R^k_{jj'} U^k_{j'\alpha} , \tag{8.61}$$

with

$$\boldsymbol{R}^k = \begin{pmatrix} 1 & 0 & 0 & 0 \cdots 0 \cdots \\ 0 & 1 & 0 & 0 \cdots 0 \cdots \\ \vdots & & \ddots & & \vdots \\ 0 \cdots & & 0 & 1 & 0 & 0 \cdots \\ * & * & \cdots * & 1 & 0 \cdots \\ * & * & \cdots * & * & 1 \cdots \\ \vdots & \vdots & \vdots \ \vdots & \vdots & & \ddots \end{pmatrix} , \tag{8.62}$$

where the stars $*$ denotes (in general small) non-zero entries of R^k. The upper-left $(n-k) \times (n-k)$ corner of R^k is a unit matrix. Keeping $\tilde{P}_j^k(\cos\theta)$ as a basis, we define DVR-functions as:

$$\chi_\alpha^k(\theta) = \sum_j \tilde{P}_j^k(\cos\theta) U^k_{j\alpha} = w_\alpha^{1/2} \sum_{jj'} \tilde{P}_j^k(\cos\theta)(\boldsymbol{R}^{-1})^k_{jj'} \tilde{P}^k_{j'}(\cos\theta_\alpha) . \tag{8.63}$$

The χ's are orthonormal, as they are generated by a unitary transformation of the orthonormal $\tilde{P}$, but the χ's are not discrete orthonormal, and they do not obey the discrete δ-property. This introduces an additional error on top of the DVR-approximation when evaluating potential matrix elements. Note, however, that the kinetic energy operators are represented exactly. To arrive at working equations for the kinetic energy operators we define the tensors

$$\begin{aligned} j^2(\alpha,\beta,k) &= \sum_{j=|k|}^{j_{\max}+|k|} U^k_{j\alpha}\, j(j+1)\, U^k_{j\beta} , \\ j_+(\alpha,\beta,k) &= \sum_{j=\max(|k|,|k+1|)}^{j_{\max}+\min(|k|,|k+1|)} U^{k+1}_{j\alpha} \sqrt{j(j+1)-k(k+1)}\, U^k_{j\beta} , \\ j_-(\alpha,\beta,k) &= j_+(\beta,\alpha,k-1) , \end{aligned} \tag{8.64}$$

and the operations of the angular momentum operators in grid space is given through these tensors by

$$(j^2\psi)(\theta_\alpha, k) = \sum_{\beta=1}^{n} j^2(\alpha, \beta, k)\psi(\theta_\beta, k)\,,$$
$$(j_\pm\psi)(\theta_\alpha, k \pm 1) = \sum_{\beta=1}^{n} j_\pm(\alpha, \beta, k)\psi(\theta_\beta, k)\,. \quad (8.65)$$

8.1.4 The Primitive Basis Set: A Summary

In practice, one expresses the wavefunction in the DVR or the FBR. In the MCTDH package, the wavefunction is expressed in the DVR. In the later, all the multiplicative operators are diagonal. There is a unitary transformation between DVR and FBR. It is thus very easy to switch back and forth between the two representations. Starting from the DVR, one switches to the FBR to calculate the action of derivative operators which is analytical in the FBR. The procedure is summarized in Table 8.1.

8.1.5 Contracted Intermediate Basis Set Functions

The number of primitive functions increases drastically with the number of degrees of freedom due to the large correlation, or coupling, between the nuclear coordinates. The problem of correlation is well-known in quantum chemistry for the electrons but this correlation is much stronger for the nuclei due to their large masses. The large masses make the energy difference between the vibrational eigenstates smaller compared to the difference between the electronic eigenstates. In practice, one often has to deal with very highly excited states in molecular quantum dynamics and thus with a huge correlation. To try to reduce the size of the basis set, a contracted intermediate basis set is often used. The contracted functions, $\xi^{(i)}_{j_\kappa}(q_i)$ are written as linear combinations of the primitive functions:

$$\xi^{(i)}_{j_\kappa}(q_i) = \sum_{j_i=1}^{N_i} D^{j_\kappa}_{j_i} \chi^{(i)}_{j_i}(q_i)\,. \quad (8.66)$$

Now, the wavefunction can be expressed in the new intermediate basis set:

$$\Psi(q_1, q_2, \ldots, q_f, t) = \sum_{j_1 \cdots j_f}^{n_1 \cdots n_f} B_{j_1 \cdots j_f}(t)\, \xi^{(1)}_{j_1}(q_1) \cdots \xi^{(f)}_{j_f}(q_f)\,. \quad (8.67)$$

Table 8.1 Discrete Variable Representation (DVR) and Finite Basis Representation (FBR) representations. Example of the sine FBR and DVR

FBR	Unitary Transformation $\longleftrightarrow$	DVR
Delocalized along the range of definition		Localized around some grid points
Derivative operators (T)	Adapted to	Multiplicative operators (V)
	Example sine-DVR	
$\sqrt{2/L}\sin(j\pi(x-x_0)/L)$		DVR Functions obtained by diagonalization of $\langle n\vert x\vert m\rangle$ in the FBR The eigenvalues correspond to the grid points x_α
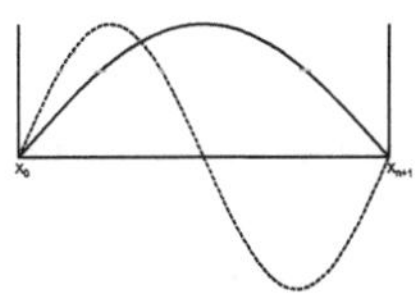		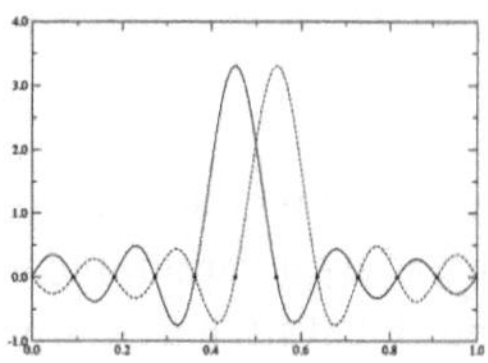
$-\frac{1}{2m}\frac{\partial^2}{\partial x^2}$, diagonal $(\frac{j^2\pi^2}{2mL^2})$		$V(x)$, diagonal $(\mathrm{V}(x_\alpha))$

We hope that the contracted intermediate basis functions have been defined in an intelligent way so that $n_1 \cdots n_f$ are smaller that $N_1 \cdots N_f$.

Many strategies have been defined to construct these intermediate functions. One simple way is to define a one-dimensional operator for each degree of freedom such as

$$-\frac{1}{2\mu}\frac{\partial^2}{\partial q^2} + V_{eff}(q)\,. \tag{8.68}$$

In other words, only the most important term is kept in the kinetic energy operator for the coordinate q (for angles, the expression can be more complicated). For the potential, one can fix all the other degrees of freedom to their values at a reference geometry or minimize the potential for each value of q with respect to the other degrees of freedom.

The one-dimensional operator can then be diagonalized in the FBR representation. Since the new functions take into account the form of the potential (in an approximate way), the eigenfunctions of high energy can be removed. A new basis set is obtained (using the eigenvectors) that is expressed in the original FBR. The position operator, q, can be diagonalized in this new, smaller, basis set of functions. New DVR functions and DVR grid points are extracted. Their number is smaller but they are better adapted to the potential. This approach allows one to easily obtain a smaller basis set and the new DVR is called a Potential-Optimized Discrete Variable Representation (PODVR) [19, 23].

More efficient multidimensional contraction schemes have been devised [24–31]. Contraction schemes based on the systematic use of a variational principle to optimize the contracted functions have also been used (Self-Consistent Field (SCF), Configuration-Interaction Self-Consistent Field (CI-SCF) [32–34], vibrational multiconfiguration self-consistent field [35–37] methods).

We will see in Sect. 8.3 that MCTDH also uses intermediate contracted functions that are obtained by a time-dependent vibrational multiconfiguration self-consistent field approach:

$$\varphi_{j_\kappa}^{(i)}(q_i, t) = \sum_{j_i=1}^{N_i} c_{j_i}^{(j_\kappa)}(t)\, \chi_{j_i}^{(i)}(q_i)\,, \tag{8.69}$$

where we allow the functions $\varphi_{j_\kappa}^{(i)}(q_i, t)$ to be time-dependent. These functions are optimized by a variational principle.

The global strategy is summarized in Fig. 8.3: the primitive basis set creates a huge primitive mathematical space represented by square (a) on Fig. 8.3. The contraction scheme extracts a much smaller active space (square (b) on Fig. 8.3) in which the Schrödinger equation for the nuclei is solved. In the MCTDH method, this active space changes with time.

8.1.6 Integrators for the Time-Dependent Schrödinger Equation

In this section we discuss three popular integrator algorithms designed for solving large sets of ordinary first order differential equations, in particular the time-dependent Schrödinger equation treated within the standard method. The time-dependent Schrödinger equation,

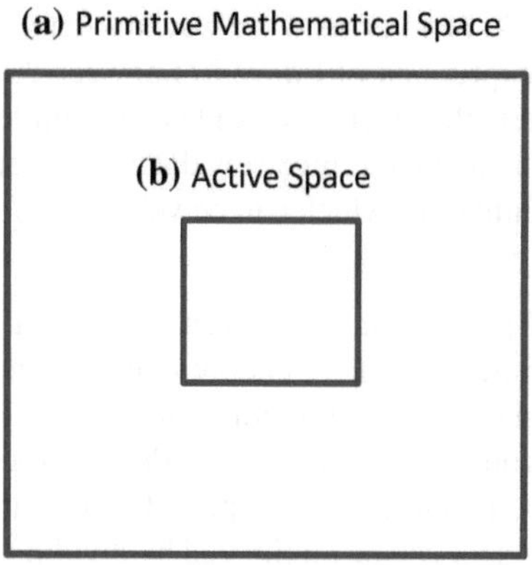

Fig. 8.3 In the numerical method used to solve the Schrödinger equation for the nuclei, the primitive functions build a mathematical space (**a**) that becomes very large with the number of degrees of freedom. In general an intermediate basis set is used: a basis set of contracted functions that builds a smaller active subspace (**b**) in which the Schrödinger equation is solved. In MCTDH, the active space changes with time

$$i\frac{d|\psi(t)\rangle}{dt} = H|\psi(t)\rangle \ , \tag{8.70}$$

has, for time-independent Hamiltonians, the formal solution

$$|\psi(t)\rangle = e^{-iHt}|\psi(0)\rangle \ . \tag{8.71}$$

Once the initial wavefunction $|\psi(0)\rangle$ is known, the behavior of the wavefunction is also known via evaluating the time-evolution operator e^{-iHt} operating on $|\psi(0)\rangle$. This procedure of evaluating the time-dependent wavefunction is also known as *propagation*.

Using the standard method of Sect. 8.1.2, the equations of motions are formally identical to Eq. (8.70). The differential equation to be solved is, according to Eq. (8.19),

$$i\,\dot{\boldsymbol{C}} = \boldsymbol{H}\ \boldsymbol{C}\,, \tag{8.72}$$

where $\boldsymbol{C}$ denotes the vector of coefficients, and $\boldsymbol{H}$ a matrix representation of the Hamiltonian in the primitive basis. A formal solution is given by Eq. (8.20). The first order linear differential equation (8.72), allows to derive more efficient integrator algorithms, which are specifically adapted to solving Eq. (8.72). In the following, we will, for the sake of simplicity, replace $\boldsymbol{C}$ with Ψ but remind the reader that, in numerical applications, Ψ will be a vector, not a function.

We refer the reader to Sect. 11.7 of Ref. [2] for a more exhaustive presentation of the different approaches and to Ref. [38] for a systematic comparison between the different propagation schemes. Here, we present only a few examples to explain how the problem can be solved.

A first possibility would be to expand the wavepacket as a linear combination of eigenstates and use the expression of Eq. (2.37) of Sect. 2. However, this approach requires to know accurately the eigenstates and is then limited to small systems. The advantage of a propagation method is precisely that we do not need to know explicitly all the eigenstates, the number of which can be very large for realistic problems.

Split Operator *

The split operator (SOP) propagator has been developed by Feit and Fleck [39, 40]. Its spirit is to separate the kinetic energy operator and the potential operator the exponential form and evaluate them individually, i.e. $e^{iHt} \sim e^{iTt}e^{iVt}$. However, separating the two operators in the exponent introduces errors to the original operator since T and V do not commute and we expect the error to be proportional to the commutator $[T, V]$. The error can be minimized by dividing a long propagation time t to multi-step short propagation times. For example, Trotter (1959) proposed

$$e^{-\beta(T+V)} = \left(e^{-\beta(T+V)/n}\right)^n = \lim_{n\to\infty} \left(e^{-\beta T/n}e^{-\beta V/n}\right)^n . \tag{8.73}$$

Replacing β by it, we immediately obtain the working equation for the propagation. Defining $\tau = t/n$, we would have the propagated wavefunction at time t as the following

$$|\psi(t)\rangle \simeq \left(e^{-iT\tau}e^{-iV\tau}\right)^n |\psi(0)\rangle . \tag{8.74}$$

In practice, we cannot take n to infinity, but we can estimate the error that is introduced to the propagation. The Baker-Campbell-Hausdorff formula reads

$$e^A e^B = e^{A+B+\frac{1}{2}[A,B]+\frac{1}{12}[A,[A,B]]+\frac{1}{12}[[A,B]B]+\cdots} . \tag{8.75}$$

Therefore, the effective time-evolution operator reads

$$e^{-iT\tau}e^{-iV\tau} = e^{-iH_{\text{eff}}\,\tau} = e^{-i(T+V)\tau-\frac{1}{2}\tau^2[T,V]+\tau^3\cdots} . \tag{8.76}$$

The first term in the exponent is the system Hamiltonian, and the rest are the errors introduced by the split operator (SOP) propagator. As we already mentioned, the larger the τ, the larger the error.

This method can be improved by symmetrization. If we separate the exponent according to $H = V/2 + T + V/2$, applying the Baker-Campbell-Hausdorff formula gives the effective time-evolution operator under symmetrization.

$$e^{-iV\tau/2}e^{-iT\tau}e^{-iV\tau/2} = e^{-iH_{\text{eff}}\,\tau} = e^{-iH\tau+\frac{i\tau^3}{24}[H,[T,V]]+\tau^4\cdots} . \tag{8.77}$$

In this case, the error starts with the third order of τ. We obtain one order more by symmetrization and we will always refer to the symmetrized form when mentioning the split operator propagator hereafter. The effective Hamiltonian reads

$$H_{\text{eff}} = H - \frac{\tau^2}{24}[H,[T,V]] + O\left(\tau^3\right) . \qquad (8.78)$$

It is Hermitian and hence secures norm conservation. However, the effective Hamiltonian does not commute with the original system Hamiltonian. The energy is not conserved. How to estimate the error term for the SOP propagator is discussed in the following three examples.

General One-Dimensional Problem

Given a kinetic energy operator $T = -\frac{1}{2m}\frac{\partial^2}{\partial x^2}$ and a general potential $V(x)$, the leading error term can be estimated by evaluating $[H,[T,V]]$. First we evaluate $[T,V]$.

$$[T,V] = TV - VT = -\frac{1}{2m}\left(V'' + 2V'\frac{d}{dx}\right) . \qquad (8.79)$$

Then we continue to evaluate the following commutators.

$$[V,[T,V]] = \frac{1}{m}(V')^2 ;$$
$$[T,[T,V]] = \frac{1}{4m^2}\left(V''\frac{d^2}{dx^2} + 2\frac{d}{dx}V''\frac{d}{dx} + \frac{d^2}{dx^2}V''\right) . \qquad (8.80)$$

The error term $[H,[T,V]]$ is just the sum of these two terms. From these equations follows that, when $\frac{1}{m} \to 0$, the propagation by SOP will be exact ($H_{\text{eff}} = H$). Thus the SOP exhibits some semi-classical behavior, although it is not exact for harmonic Hamiltonians. However, as $\frac{1}{m} \neq 0$, the propagation is not exact and $\langle E\rangle$ is not conserved ($[H_{\text{eff}}, H] \neq 0$).

Harmonic Oscillator

Given a harmonic oscillator potential $V(x) = \frac{1}{2}m\omega^2x^2$, based on Eqs. (8.80) and (8.77), the leading error term of the propagator reads

$$[H,[T,V]] = 2\omega^2\left(\frac{1}{2}m\omega^2x^2\right) + 2\omega^2\left(\frac{1}{2m}\frac{d^2}{dx^2}\right) = 2\omega^2\,(V - T) . \qquad (8.81)$$

In addition, we have $|\langle V - T\rangle| < \langle H\rangle = \left(n + \frac{1}{2}\right)\omega$. Thus, the SOP error term can be bound by:

$$\text{Err(SPO)} < \frac{\left(n + \frac{1}{2}\right)}{12}\omega^3\tau^3 . \qquad (8.82)$$

In contrast, the error introduced when using a Taylor expansion of the exponential to second order reads

$$\frac{\tau^3H^3}{3!} = \frac{\left(n + \frac{1}{2}\right)^3}{6}\omega^3\tau^3 , \qquad (8.83)$$

which is much larger than the SOP error, when $n > 1$.

The SOP propagator is usually combined with the Fast Fourier Transform (FFT) algorithm in computations. The idea is to exponentiate both potential and KEO in their own eigenspaces such that both representations are diagonal (see Ref. [6]). The procedure is as such: one first obtains $\left(e^{-iV(x)\tau/2}\psi(x,t)\right)$ in the x-space and then performs a Fourier transform to the p-space, multiplies with $e^{-i\frac{p^2}{2m}\tau}$ and then performs an inverse Fourier transform and multiplies with $e^{-iV(x)\tau/2}$, *i.e.*

$$\psi(x,t+\tau) = e^{-iV(x)\tau/2}F^{-1}\left(e^{-i\frac{p^2}{2m}\tau}F\left(e^{-iV(x)\tau/2}\psi(x,t)\right)\right) . \tag{8.84}$$

The propagator is stable, i.e. it conserves the norm, even in the region where the potential is high or the wavefunction is highly oscillating. On the other hand, SOP requires performing $e^{-iV\tau/2}$ and $e^{-iT\tau}$ exactly. The latter exponentiation becomes problematic if one uses generalized coordinates, because then T ceases to be diagonal in momentum representation.

Second Order Difference (SOD) *

Previously we illustrated the propagators that involve only single time steps, i.e. evaluating $|\psi(t+\tau)\rangle$ based only on $|\psi(t)\rangle$. There are also methods to propagate a wavefunction by making use of the wavefunction at several previous times. Second-order difference (SOD) is one of those multi-step integrators. Suppose we have the wavefunctions $|\psi(t-\tau)\rangle$,$|\psi(t)\rangle$, and $|\psi(t+\tau)\rangle$:

$$\begin{aligned} |\psi_+\rangle &:= |\psi(t+\tau)\rangle = e^{-iH\tau}|\psi(t)\rangle , \\ |\psi_0\rangle &:= |\psi(t)\rangle , \\ |\psi_-\rangle &:= |\psi(t-\tau)\rangle = e^{iH\tau}|\psi(t)\rangle . \end{aligned} \tag{8.85}$$

The wavefunction difference is given by

$$|\psi_+\rangle - |\psi_-\rangle = (e^{-iH\tau} - e^{iH\tau})|\psi_0\rangle = -2i\sin(H\tau)|\psi_0\rangle . \tag{8.86}$$

The exact propagation is given by multi-steps, i.e. $|\psi_+\rangle$ depends on $|\psi_-\rangle$ and $|\psi_0\rangle$.

$$|\psi_+\rangle = |\psi_-\rangle - 2i\sin(H\tau)|\psi_0\rangle . \tag{8.87}$$

However, when performing the propagation, one usually uses an approximated form by linearizing the sine function,

$$|\psi_+\rangle = |\psi_-\rangle - 2iH\tau|\psi_0\rangle . \tag{8.88}$$

This working equation, Eq. (8.88), can be viewed as using an effective Hamiltonian H_{eff} in Eq. (8.87). The effective Hamiltonian has the following expression:

$$\sin(H_{\text{eff}}\tau) = H\tau\,,$$
$$H_{\text{eff}} = \frac{1}{\tau}\arcsin(H\tau)\,. \tag{8.89}$$

Note that H_{eff} is Hermitian and the norm of the wavefunction is conserved only if the operator norm (Hilbert norm) of $H\tau$ satisfies $\|H\tau\|_H < 1$, because otherwise the arcsin function becomes complex and H_{eff} hence non-Hermitian. It can be shown that the norm of the wavefunction will explode suddenly when $\|H\tau\|_H > 1$, i.e. when the absolute largest eigenvalue of H is larger than $1/\tau$.

The Lanczos and the Lanczos-Arnoldi Integrators *
We briefly review the Short Iterative Lanczos (SIL) integration scheme [41], and discuss two different error estimates for the SIL method. We consider here the complex Lanczos, or Lanczos-Arnoldi, algorithm [42–44], which includes the Hermitian Lanczos as a special case.

In the Lanczos scheme the exact Hamiltonian H is approximated by the reduced Hamiltonian $H_L = P_L H P_L$, where P_L denotes the projector on the Krylov space spanned by the set of vectors $H^k\psi(t)$, $k = 0, \ldots, L$. The Lanczos-Arnoldi recursion,

$$\chi_{j+1}^{(0)} = H\psi_j\,, \tag{8.90}$$
$$\text{for } i = 0, \ldots, j: \quad \chi_{j+1}^{(i+1)} = \chi_{j+1}^{(i)} - \beta_{ij}\psi_i \quad \text{with} \quad \beta_{ij} = \langle \psi_i \mid \chi_{j+1}^{(i)} \rangle$$
$$\psi_{j+1} = \chi_{j+1}^{(j+1)}/\beta_{j+1,j} \quad \text{with} \quad \beta_{j+1,j} = \|\, \chi_{j+1}^{(j+1)} \,\|\,,$$

constructs an orthonormal basis $\psi_0, \ldots, \psi_L$, starting from a normalized state ψ_0, in which the reduced Hamiltonian H_L is a complex upper Hessenberg matrix:

$$\langle \psi_j \mid H_L \mid \psi_k \rangle = \begin{cases} \beta_{jk} & \text{for } j \leq k+1 \\ 0 & \text{else} \end{cases}\,, \quad j, k = 0, \ldots, L\,. \tag{8.91}$$

(If H is Hermitian, then matrix (8.91) becomes symmetric, i.e. tridiagonal.) The Lanczos-Arnoldi algorithm requires $L+1$ evaluations of $H \mid \psi\rangle$.

The SIL integrator approximates the propagated wavefunction $\psi(t+\tau)$ by

$$\mid \psi(t+\tau)\rangle = e^{-iH_L\tau} \mid \psi(t)\rangle = \|\, \psi(t) \,\| \sum_{k=0}^{L} a_k^{(L)} \mid \psi_k\rangle\,, \tag{8.92}$$

with

$$a_k^{(L)} = \sum_{j=0}^{L} T_{kj}\, e^{-i\lambda_j\tau} \left(T^{-1}\right)_{j0}\,, \tag{8.93}$$

and $\psi_0 = \psi(t)/\,\|\, \psi(t) \,\|$. The λ_j specify the eigenvalues and $\boldsymbol{T}$ the eigenvector matrix of the small, and hence easy to diagonalise, upper Hessenberg matrix (8.91).

The Lanczos recurrence in its standard form is in so far inefficient as the vector ψ_{L+1} as well as the matrix element $\beta_{L+1,L}$ are determined but never used. A slight modification of the SIL algorithm can circumvent this. It has been suggested [45] that one may increase the SIL order to $L' = L + 1$ by adding a further, approximate, column to the Lanczos matrix. To this end we set $\beta_{L'L'} = \beta_{LL}$, $\beta_{L,L'} = \beta_{L',L}$, and $\beta_{j,L'} = 0$ for $j < L$. The upper Hessenberg matrix is then an $L' \times L'$ matrix, and the summation in Eqs. (8.92) and (8.93) extends to L' rather than L. The wavefunction propagated with this modified algorithm is correct up to an order $\tau^{L'}$, while it is only accurate up to an order τ^{L} in the conventional Lanczos scheme.

For the conventional SIL integrator there exists a very convenient estimate for the difference $\Delta\psi$ of the propagated and the exact wavefunction [41], namely

$$\| \Delta\psi \| \approx \frac{\beta_{10} \ldots \beta_{L+1,L}}{(L+1)!} \tau^{L+1}, \tag{8.94}$$

which may be used to adjust either the step size τ or the order L. For the modified SIL integrator one could use the same formula but replace L by L'. However, $\beta_{L'+1,L'}$ is unknown, and instead of estimating it we prefer to use Eq. (8.94) as it is. Since for any one-step method, such as Lanczos, the order of the global discretisation error is a power one lower than that of the local error [46], the error estimate therefore controls the global rather than the local error.

Turning again to the conventional Lanczos integrator we note that the error estimate (8.94) is correct to an order of $L + 1$, yet the true error also contains contributions from higher orders of τ. These contributions depend not only on the subdiagonal but on all elements of matrix (8.91). Despite these higher contributions, the error formula (8.94) is remarkably accurate for small step sizes. However, it may grossly overestimate the error when the integrator is running at high order and thus taking large step sizes [47].

A more reliable error estimate for the (modified) Lanczos integrator is given by [47]

$$\| \Delta\psi \| = \| \boldsymbol{a}^{(L')} - \boldsymbol{a}^{(L'-1)} \| , \tag{8.95}$$

where a zero is appended to the shorter of the two vectors, $\boldsymbol{a}^{(L'-1)}$. This criterion thus takes as an error estimate the norm of the difference between the wavefunctions propagated by the SIL method of order L' and $L' - 1$. The accuracy of the propagation can be controlled by increasing the SIL order L' until the predicted error (8.95) becomes smaller than the prescribed error tolerance.

The improved error estimate is predominantly determined by the error of the solution obtained with an order of $L' - 1$. Hence the propagated wavefunction is one order in τ more accurate than the error estimate indicates. This is very much in the spirit of what has been discussed above, namely using Eq. (8.94) for the modified Lanczos integrator as well. Equation (8.95) seems to be rather costly because it requires the diagonalisation of matrix (8.91) after each SIL iteration. However, for large systems the evaluation of the matrix-vector product $H \mid \psi\rangle$ needs so much numerical effort that the effort for the error estimation remains negligible. For large step sizes the improved error measure (8.95) estimates the SIL integration error more accurately than the standard estimate (8.94), while for small step sizes both estimates are equally accurate [47].

In the MCTDH package, the SIL method is often used to solve the EOM of the A-coefficients in Eq. (8.206) because this is a linear differential equation.

8.2 The Time-Dependent Hartree (TDH) Approach

As aforementioned, the standard method of quantum dynamics scales exponentially with molecular size and thus renders simulations with more than four nuclei challenging, even with present-day computer hardware. One of the simplest alternative propagation methods is the TDH approach [48, 49]. The wavefunction is expressed as a simple product of one-dimensional functions but the latter are now time-dependent [5, 50]:

$$\Psi(q_1, q_2, \ldots, q_f, t) = a(t)\, \varphi_1(q_1, t) \cdots \varphi_f(q_f, t)\,. \tag{8.96}$$

The functions $\varphi_1(q_1, t)$, $\cdots$, $\varphi_f(q_f, t)$ are called *single-particle functions* (SPF), which are, of course, expressed in a time-independent basis set, the primitive basis:

$$\varphi_{j_\kappa}^{(\kappa)}(q_\kappa, t) = \sum_{i_\kappa=1}^{N_\kappa} c_{i_\kappa}^{(\kappa, j_\kappa)}(t)\, \chi_{i_\kappa}^{(\kappa)}(q_\kappa)\,, \tag{8.97}$$

where $\chi_{i_\kappa}^{(\kappa)}(q_\kappa)$ is a primitive basis function for the κ-th degree of freedom.

The representation (8.96) is not unique because

$$\varphi_1 \cdot \varphi_2 = (\frac{\varphi_1}{b}) \cdot (\varphi_2 \cdot b)\,, \tag{8.98}$$

holds for any complex constant $b \neq 0$. The additional factor $a(t)$ increases the redundancy, but because of this coefficient there is now a free factor for each function φ_κ. All SPFs are now treated on the same footing. To arrive at unique equations of motion one has to introduce constraints, which remove the non-uniqueness but do not narrow the variational space. If a function changes in time by a complex factor only, then this is equivalent to a time derivative which is in the direction of the function itself

$$\dot{\varphi} \propto \varphi\,. \tag{8.99}$$

To see this more explicitly, let us write

$$\begin{aligned} \varphi &= \alpha\tilde{\varphi} \quad \text{with } \|\tilde{\varphi}\| = 1\,, \\ \dot{\varphi} &= \dot{\alpha}\,\tilde{\varphi} + \alpha\,\dot{\tilde{\varphi}}\,, \end{aligned} \tag{8.100}$$

and

$$\langle\varphi|\dot{\varphi}\rangle = \alpha^*\dot{\alpha} + |\alpha|^2\langle\tilde{\varphi}|\dot{\tilde{\varphi}}\rangle . \tag{8.101}$$

This shows that we can prescribe $\langle\varphi|\dot{\varphi}\rangle$ any value, such a prescription will merely determine $\dot{\alpha}$. Hence, we propose the constraints:

$$i\langle\varphi_\kappa(t)|\dot{\varphi}_\kappa(t)\rangle = g_\kappa(t) , \tag{8.102}$$

for some set of functions $g_\kappa(t)$. Note again that this constraint, by construction, does not further restricts possible values for $\dot{\Psi}$.

Of course, the SPFs should be normalized. Their change in norm is given by

$$\begin{aligned}\frac{d}{dt}\|\varphi_\kappa\|^2 &= \frac{d}{dt}\langle\varphi_\kappa|\varphi_\kappa\rangle \\ &= \langle\dot{\varphi}_\kappa|\varphi_\kappa\rangle + \langle\varphi_\kappa|\dot{\varphi}_\kappa\rangle \\ &= 2Re\,\langle\varphi_\kappa|\dot{\varphi}_\kappa\rangle = 2Im\ g_\kappa ,\end{aligned} \tag{8.103}$$

which implies that the norm is conserved if the constraints g_κ are real.

8.2.1 TDH Equations *

In the TDH approach, the wavefunction is expressed as

$$\Psi(q_1, q_2, \ldots, q_f, t) = a(t)\prod_{\kappa=1}^{f}\varphi_\kappa(q_\kappa, t) = a(t)\Phi(t) , \tag{8.104}$$

with the constraints:

$$i\langle\varphi_\kappa(t)|\dot{\varphi}_\kappa(t)\rangle = g_\kappa(t) , \tag{8.105}$$

and with g_κ real, but otherwise arbitrary. Later, we will choose g_κ such that the EOM become as simple as possible. Without restriction we may choose the initial SPFs $\varphi_\kappa(t=0)$ to be normalized and Eq. (8.103) then tells us that they stay normalized for all times.

We are now ready to perform the variation.

$$\dot{\Psi} = \dot{a}(t)\prod_{\kappa=1}^{f}\varphi_\kappa(q_\kappa, t) + a(t)\sum_{\kappa=1}^{f}\dot{\varphi}_\kappa\prod_{\nu\neq\kappa}^{f}\varphi_\nu = \dot{a}(t)\ \Phi + a\sum_{\kappa=1}^{f}\dot{\varphi}_\kappa\ \Phi^{(\kappa)} , \tag{8.106}$$

and

$$\delta\Psi = (\delta a)\Phi + a\sum_{\kappa=1}^{f}(\delta\varphi_\kappa)\ \Phi^{(\kappa)} , \tag{8.107}$$

where we have used the definitions

$$\Phi = \prod_{\kappa=1}^{f} \varphi_\kappa \quad \text{and} \quad \Phi^{(\kappa)} = \prod_{\substack{\nu=1 \\ \nu \neq \kappa}}^{f} \varphi_\nu \,. \tag{8.108}$$

From the variational principle (VP) of Eq. (8.1) follows

$$\begin{aligned} &\big\langle \delta a\, \Phi \big| H \big| a\, \Phi \big\rangle - i \big\langle \delta a\, \Phi \big| \dot a\, \Phi + a \sum_\kappa \dot\varphi_\kappa\, \Phi^{(\kappa)} \big\rangle \\ &+ \sum_{\kappa=1}^{f} \big\{ \big\langle \delta\varphi_\kappa\, a\, \Phi^{(\kappa)} \big| H \big| a\, \Phi \big\rangle - i \big\{ \big\langle \delta\varphi_\kappa\, a\, \Phi^{(\kappa)} \big| \dot a\, \Phi + a \sum_{\kappa'} \dot\varphi_{\kappa'}\, \Phi^{(\kappa')} \big\rangle \big\} = 0 \,. \end{aligned} \tag{8.109}$$

Since δa and all $\delta\varphi_\kappa$ are independent of each other, each line has to vanish individually.

For δa:

$$(\delta a)^*\, a \big\langle \Phi \big| H \big| \Phi \big\rangle = i\, (\delta a)^*\, \dot a + i\, (\delta a)^*\, a \sum_\kappa \big\langle \Phi \big| \dot\varphi_\kappa\, \Phi^{(\kappa)} \big\rangle \,. \tag{8.110}$$

Since

$$i \big\langle \Phi \big| \dot\varphi_\kappa\, \Phi^{(\kappa)} \big\rangle = i \big\langle \varphi_1 \cdots \varphi_\kappa \cdots \varphi_f \big| \varphi_1 \cdots \dot\varphi_\kappa \cdots \varphi_f \big\rangle = i \big\langle \varphi_\kappa \big| \dot\varphi_\kappa \big\rangle = g_\kappa \tag{8.111}$$

follows

$$i\, \frac{\dot a}{a} = \big\langle \Phi \big| H \big| \Phi \big\rangle - \sum_\kappa g_\kappa \,, \tag{8.112}$$

or, introducing

$$E = \big\langle \Phi \big| H \big| \Phi \big\rangle = \frac{\big\langle \Psi \big| H \big| \Psi \big\rangle}{\big\langle \Psi \big| \Psi \big\rangle} \,, \tag{8.113}$$

it follows

$$i\, \dot a = (E - \sum_{\kappa=1}^{f} g_\kappa)\, a \,. \tag{8.114}$$

On the other hand, by varying a particular φ_κ, we obtain

$$\begin{aligned} \big\langle (\delta\varphi_\kappa)\, a\, \Phi^{(\kappa)} \big| H \big| a\, \Phi \big\rangle = i \big\langle (\delta\varphi_\kappa)\, a\, \Phi^{(\kappa)} \big| \dot a\, \Phi \big\rangle + \\ i \big\langle (\delta\varphi_\kappa)\, a\, \Phi^{(\kappa)} \big| a \sum_{\nu=1}^{f} \dot\varphi_\nu\, \Phi^{(\nu)} \big\rangle \,. \end{aligned} \tag{8.115}$$

Now

$$\begin{aligned}\left\langle(\delta\varphi_\kappa)\,a\,\Phi^{(\kappa)}\middle|H\middle|a\,\Phi\right\rangle &= |a|^2\left\langle(\delta\varphi_\kappa)\,\Phi^{(\kappa)}\middle|H\middle|\varphi_\kappa\,\Phi^{(\kappa)}\right\rangle \\ &= |a|^2\left\langle(\delta\varphi_\kappa)\middle|\left\langle\Phi^{(\kappa)}\middle|H\middle|\Phi^{(\kappa)}\right\rangle\middle|\varphi_\kappa\right\rangle \\ &= |a|^2\left\langle(\delta\varphi_\kappa)\middle|\mathcal{H}^{(\kappa)}\middle|\varphi_\kappa\right\rangle,\end{aligned} \tag{8.116}$$

with the definition

$$\mathcal{H}^{(\kappa)} = \left\langle\Phi^{(\kappa)}\middle|H\middle|\Phi^{(\kappa)}\right\rangle. \tag{8.117}$$

$\mathcal{H}^{(\kappa)}$ is called a *mean field*. Note that it is an operator on the κ-th degree of freedom. The second term of the Eq. (8.115) is transformed to

$$i\left\langle(\delta\varphi_\kappa)\,a\,\Phi^{(\kappa)}\middle|\dot a\,\Phi\right\rangle = i\dot a a^*\left\langle\delta\varphi_\kappa\middle|\varphi_\kappa\right\rangle = |a|^2\,(E-\sum_{\nu=1}^{f} g_\nu)\left\langle\delta\varphi_\kappa\middle|\varphi_\kappa\right\rangle, \tag{8.118}$$

and the third term of (8.115) to

$$\begin{aligned}i\left\langle(\delta\varphi_\kappa)\,a\,\Phi^{(\kappa)}\middle|a\sum_{\nu=1}^{f}\dot\varphi_\nu\,\Phi^{(\nu)}\right\rangle &= i|a|^2\left\langle\delta\varphi_\kappa\middle|\dot\varphi_\kappa\right\rangle + i|a|^2\sum_{\nu\neq\kappa}\left\langle\delta\varphi_\kappa\,\varphi_\nu\middle|\dot\varphi_\nu\,\varphi_\kappa\right\rangle \\ &= i|a|^2\left\langle\delta\varphi_\kappa\middle|\dot\varphi_\kappa\right\rangle + i|a|^2\left\langle\delta\varphi_\kappa\middle|\varphi_\kappa\right\rangle\cdot\sum_{\nu\neq\kappa}\left\langle\varphi_\nu\middle|\dot\varphi_\nu\right\rangle \\ &= i|a|^2\left\langle\delta\varphi_\kappa\middle|\dot\varphi_\kappa\right\rangle + |a|^2\left\langle\delta\varphi_\kappa\middle|\varphi_\kappa\right\rangle\sum_{\nu\neq\kappa} g_\nu\,.\end{aligned} \tag{8.119}$$

Using (8.116) = (8.118) + (8.119) and dividing by $|a|^2$ yields

$$\begin{aligned}&\left\langle(\delta\varphi_\kappa)\middle|\mathcal{H}^{(\kappa)}\middle|\varphi_\kappa\right\rangle = \\ &(E-\sum_{\nu=1}^{f} g_\nu)\left\langle\delta\varphi_\kappa\middle|\varphi_\kappa\right\rangle + i\left\langle\delta\varphi_\kappa\middle|\dot\varphi_\kappa\right\rangle + \sum_{\nu\neq\kappa} g_\nu\left\langle\delta\varphi_\kappa\middle|\varphi_\kappa\right\rangle,\end{aligned} \tag{8.120}$$

or

$$i\left\langle\delta\varphi_\kappa\middle|\dot\varphi_\kappa\right\rangle = \left\langle\delta\varphi_\kappa\middle|\mathcal{H}^{(\kappa)}\middle|\varphi_\kappa\right\rangle - (E-g_\kappa)\left\langle\delta\varphi_\kappa\middle|\varphi_\kappa\right\rangle. \tag{8.121}$$

Since $\delta\varphi_\kappa$ is arbitrary, we finally arrive at

$$\boxed{\begin{aligned} i\,\dot\varphi_\kappa &= (\mathcal{H}^{(\kappa)} - E + g_\kappa)\,\varphi_\kappa\,, \\ i\,\dot a &= (E - \sum_{\kappa=1}^{f} g_\kappa)\,a\,.\end{aligned}} \tag{8.122}$$

Everything may be time-dependent.

If we multiply the first of the EOM by $\langle\varphi_\kappa|$ we see that the constraint is obeyed,

$$i\,\langle\varphi_\kappa|\dot{\varphi}_\kappa\rangle = \langle\varphi_\kappa|\mathcal{H}^{(\kappa)}|\varphi_\kappa\rangle - E + g_\kappa = g_\kappa\,, \tag{8.123}$$

because

$$\langle\varphi_\kappa|\mathcal{H}^{(\kappa)}|\varphi_\kappa\rangle = \langle\varphi_\kappa\,\Phi^{(\kappa)}|H|\varphi_\kappa\,\Phi^{(\kappa)}\rangle = \langle\Phi|H|\Phi\rangle = E\,. \tag{8.124}$$

We now have to decide what to use for g_κ. Remember *any* function $g_\kappa(t)$ can be chosen as long as it is real. The simplest choice is $g_\kappa \equiv 0$. This yields:

$$\boxed{\begin{aligned} a(t) &= a(0)\cdot exp\big(-i\int_0^t E(t')\,dt'\big)\,,\\ i\,\dot{\varphi}_\kappa &= (\mathcal{H}^{(\kappa)} - E)\,\varphi_\kappa\\ &= \big(1 - |\varphi_\kappa\rangle\langle\varphi_\kappa|\big)\,\mathcal{H}^{(\kappa)}\,\varphi_\kappa\,. \end{aligned}} \tag{8.125}$$

The very last line is introduced because of its similarity with the MCTDH EOM. It holds because of (8.124).

For Hermitian time-independent Hamiltonians the Dirac-Frenkel variational principle ensures that the norm and the mean energy of the wavepacket (WP) are conserved. Hence $E(t)$ is real and time-independent. For Hermitian time-dependent Hamiltonians E will become time-dependent but stays real. For non-Hermitian Hamiltonians E will become both complex and time-dependent.

Hence for Hermitian Hamiltonians there are two other meaningful choices for g_κ, namely

$$g_\kappa = E\,, \tag{8.126}$$

and

$$g_\kappa = E/f\,. \tag{8.127}$$

Then

$$\begin{aligned} a(t) &= a(0)\cdot exp\big(i(f-1)\int_0^t E(t')\,dt'\big)\,,\\ i\dot{\varphi}_\kappa &= \mathcal{H}^{(\kappa)}\,\varphi_\kappa\,, \end{aligned} \tag{8.128}$$

and

$$\begin{aligned} a(t) &= a(0)\,,\\ i\dot{\varphi}_\kappa &= \big(\mathcal{H}^{(\kappa)} - \Big(\frac{f-1}{f}\Big)\,E(t)\big)\,\varphi_\kappa\,. \end{aligned} \tag{8.129}$$

Hence the various choices of g_κ merely shift phase-factors from a to φ_κ and vice versa.

The derivation of the EOM of the TDH method is now concluded. The TDH solution is approximate because of the very restricted form of the wavefunction. To investigate the quality of a TDH solution, we adopt the idea of an effective Hamiltonian:

$$i\dot{\Psi} = H_{\text{eff}}\,\Psi\,, \tag{8.130}$$

where Ψ denotes the TDH solution. Using the last set of EOM and remembering (since $\dot{a} = 0$ there)

$$\dot{\Psi} = a \sum_{\kappa} \dot{\varphi}_\kappa\,\Phi^{(\kappa)}\,, \tag{8.131}$$

one readily finds

$$H_{\text{eff}} = \Big(\sum_{\kappa=1}^{f} \mathcal{H}^{(\kappa)}\Big) - (f-1)\,E\,. \tag{8.132}$$

The TDH solution is the exact solution of the TDSE using H_{eff} as the Hamiltonian. To proceed, we split the Hamiltonian into separable and non-separable terms.

$$H = \sum_{\kappa=1}^{f} h^{(\kappa)} + V_{\text{corr}}\,, \tag{8.133}$$

where $h^{(\kappa)}$ operates only on the κ-th degree of freedom. V_{corr} denotes the correlated part of the potential. The separable potential terms as well as the kinetic terms, which for the sake of simplicity are assumed to consist of 1D-terms only, are included in $h^{(\kappa)}$.

The mean fields and the effective Hamiltonian can now be evaluated somewhat more explicitly

$$\begin{aligned}\mathcal{H}^{(\kappa)} &= \langle\Phi^{(\kappa)}|H|\Phi^{(\kappa)}\rangle \\ &= h^{(\kappa)}\,\langle\Phi^{(\kappa)}|\Phi^{(\kappa)}\rangle + \sum_{\nu\neq\kappa}\langle\Phi^{(\kappa)}|h^{(\nu)}|\Phi^{(\kappa)}\rangle + \langle\Phi^{(\kappa)}|V_{corr}|\Phi^{(\kappa)}\rangle\,,\end{aligned} \tag{8.134}$$

or more compactly

$$\mathcal{H}^{(\kappa)} = h^{(\kappa)} + \sum_{\nu\neq\kappa} E_{\text{uncorr}}^{(\nu)} + w^{(\kappa)}\,, \tag{8.135}$$

with

$$w^{(k)}(x_k) = \langle \varphi_1 \cdots \varphi_{k-1}\varphi_{k+1} \cdots \varphi_f \big| V_{corr}(x_1, \ldots, x_f) \big| \varphi_1 \cdots \varphi_{k-1}\varphi_{k+1} \cdots \varphi_f \rangle, \tag{8.136}$$

i.e. one averages the potential over the other degrees of freedom. In addition,

$$\begin{aligned} E^{(\nu)}_{\text{uncorr}} &= \langle \Phi^{(\kappa)} \big| h^{(\nu)} \big| \Phi^{(\kappa)} \rangle = \langle \varphi^{(\nu)} \big| h^{(\nu)} \big| \varphi^{(\nu)} \rangle, \quad (\nu \neq \kappa), \\ E_{\text{uncorr}} &= \langle \Phi \big| \sum_\nu h^{(\nu)} \big| \Phi \rangle = \sum_\nu E^{(\nu)}_{\text{uncorr}}, \end{aligned} \tag{8.137}$$

and

$$E_{\text{corr}} = \langle \Phi \big| V_{\text{corr}} \big| \Phi \rangle = \frac{\langle \Psi \big| V_{\text{corr}} \big| \Psi \rangle}{\langle \Psi \big| \Psi \rangle}. \tag{8.138}$$

Hence

$$E = \langle \Phi \big| H \big| \Phi \rangle = E_{\text{uncorr}} + E_{\text{corr}}. \tag{8.139}$$

Next we derive the equation for the mean fields

$$\mathcal{H}^{(\kappa)} = h^{(\kappa)} + w^{(\kappa)} + E_{\text{uncorr}} - E^{(\kappa)}_{\text{uncorr}}. \tag{8.140}$$

Thus,

$$\sum_{\kappa=1}^{f} \mathcal{H}^{(\kappa)} = \sum_{\kappa=1}^{f} \left(h^{(\kappa)} + w^{(\kappa)}\right) + (f-1)\, E_{\text{uncorr}}, \tag{8.141}$$

and

$$H_{\text{eff}} = \sum_{\kappa=1}^{f} \mathcal{H}^{(\kappa)} - (f-1)\, E = \sum_{\kappa=1}^{f} \left(h^{(\kappa)} + w^{(\kappa)}\right) - (f-1)\, E_{\text{corr}}, \tag{8.142}$$

and

$$\begin{aligned} H - H_{\text{eff}} &= V_{corr} - \sum_{\kappa=1}^{f} w^{(\kappa)} + (f-1)\, E_{\text{corr}} \\ &= (V_{corr} - E_{\text{corr}}) - \sum_{\kappa=1}^{f} (w^{(\kappa)} - E_{\text{corr}}). \end{aligned} \tag{8.143}$$

This makes it clear that TDH is exact, i.e. $H = H_{\text{eff}}$, if $V_{corr} \equiv 0$, in other words, if the Hamiltonian is separable.

8.2.2 TDH Equations: 2-Dimensional Case

To investigate the errors introduced by the TDH approximation let us assume that the Hamiltonian has the form:

$$H = -\frac{1}{2m}\frac{\partial^2}{\partial x_1^2} - \frac{1}{2m}\frac{\partial^2}{\partial x_2^2} + V_1(x_1) + V_2(x_2) + W_1(x_1)W(x_2)\,. \tag{8.144}$$

A two-dimensional potential energy can always be written in the form $V(x_1, x_2) = \sum_i W_{1,i}(x_1)W_{2,i}(x_2)$ but keeping only the terms $V_1(x_1)$, $V_2(x_2)$, and $W_1(x_1)\,W_2(x_2)$ is sufficient for the discussion that follows. Now, we obtain

$$E_{\text{corr}} = \langle\varphi_1|W_1|\varphi_1\rangle\langle\varphi_2|W_2|\varphi_2\rangle \equiv \langle W_1\rangle\langle W_2\rangle\,. \tag{8.145}$$

In addition

$$w^{(1)}(x_1) = W(x_1)\langle W_2\rangle\,, \tag{8.146}$$

and

$$w^{(2)}(x_2) = W(x_2)\langle W_1\rangle\,, \tag{8.147}$$

and

$$H_{\text{eff}} = -\frac{1}{2m}\frac{\partial^2}{\partial x_1^2} - \frac{1}{2m}\frac{\partial^2}{\partial x_2^2} + V_1(x_1) + V_2(x_2) + W(x_1)\langle W_2\rangle + W(x_2)\langle W_1\rangle - \langle W_1\rangle\langle W_2\rangle\,, \tag{8.148}$$

and

$$H - H_{\text{eff}} = W_1(x_1)\,W_2(x_2) - W_2(x_2)\,\langle W_1\rangle - W_1(x_1)\,\langle W_2\rangle + \langle W_1\rangle\,\langle W_2\rangle\,. \tag{8.149}$$

This yields

$$H - H_{\text{eff}} = \big(W_1(x_1) - \langle W_1\rangle\big)\big(W_2(x_2) - \langle W_2\rangle\big)\,. \tag{8.150}$$

Hence the TDH-error is small if $W_1(x_1)$ and $W_2(x_2)$ are almost constant over the width of the wavepacket. What is the physical meaning of $W_1(x_1)\,W_2(x_2)$? This term includes the correlation within the potential between the two coordinates x_1 and x_2. If $W_1(x_1)\,W_2(x_2) = 0$, the Hamiltonian is said to be separable and the variation of the potential along x_2 does not depend on x_1 and vice versa. If $\big(W_1(x_1) - \langle W_1\rangle\big)\big(W_2(x_2) - \langle W_2\rangle\big)$ is small, the form of the potential along x_2 smoothly depends on x_1 as in Fig. 8.4 for the NOCl potential (here $x_1 = r_d$ is the dissociative Jacobi coordinate and $x_2 = r_v$ is the vibrational Jacobi coordinate which connects N and O). If the shape of the potential along x_2 rapidly changes with x_1, $\big(W_1(x_1) - \langle W_1\rangle\big)\big(W_2(x_2) - \langle W_2\rangle\big)$ becomes large. This happens in Fig. 8.5 (here $x_1 = r_d$ and $x_2 = r_v$) for the reactive $H+H_2$ molecular system when the H-atom comes

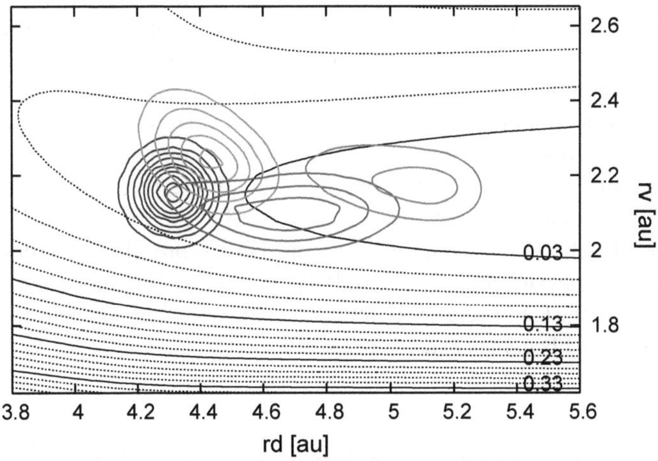

Fig. 8.4 A contour plot of the S_1 excited electronic state PES of NOCl as a function of two Jacobi coordinates (r_d, the distance between the center of mass of NO and Cl, r_v the NO bond length) is shown for $\theta = 127°$, θ being the angle between the two Jacobi vectors. Superimposed are contours of a numerically exact wavepacket (averaged over all angles) for $t = 0, 12, 24, 36$, and 48 fs describing the photodissociation into NO + Cl starting from a Franck-Condon transition. In a Franck-Condon transition, the vibrational ground state of the electronic ground state S_0 is vertically moved to the potential energy surface of the excited electronic state (here the S_1 state)

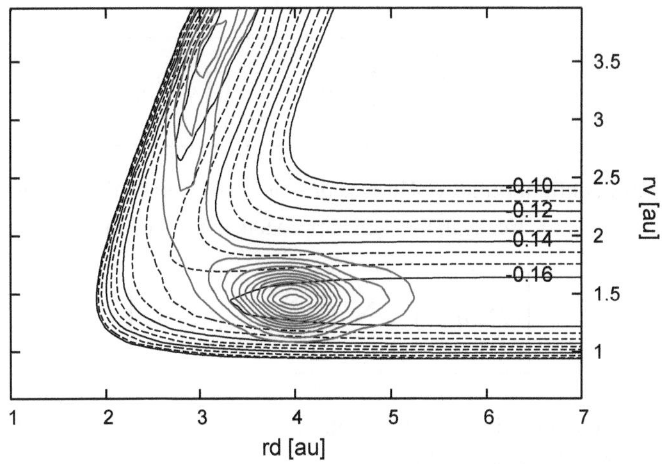

Fig. 8.5 A contour plot of the electronic ground state PES of $H+H_2$ as a function of two Jacobi coordinates (r_d, the distance between the center of mass of H_2 and H, r_v the H_2 bond length) is shown for $\theta = 0°$ (collinear geometry), θ being the angle between the two Jacobi vectors. Superimposed are contours of a wavepacket (averaged over all angles) at $t = 20$ fs describing the reactive $H+H_2$ collision

close to the H_2 molecule, i.e. for $r_d < 4.0$ a.u. and especially around the transition state ($r_d \approx 2.8$ a.u., $r_v \approx 1.8$ a.u.), and in the reactive exit channel ($r_v > 2.5$ a.u.).

TDH reduces an f-dimensional equation to a set of f one-dimensional equations. That is an enormous simplification. Assume we have 20 basis functions per DOF and 12 degrees of freedom. Then there are $20^{12} = 4 \cdot 10^{15}$ coupled differential equations to be solved for the standard method but only $12 \cdot 20 = 240$ equations for TDH. The first problem is undoable, the latter very simple. However, for many problems, for instance for the $H + H_2$ molecular system, the TDH approach cannot describe even qualitatively the physics of the problem [51], as several configurations are required. Note also that in the (semi-)classical limit, when the wavefunction converges towards a δ-function, TDH becomes exact (Eq. (8.150)).

8.3 The Multi-configuration Time-Dependent Hartree Method

8.3.1 MCTDH Fundamentals

To overcome the limitations of TDH, we turn to a multiconfigurational ansatz [52] and write the WF as [5, 53, 54]

$$\boxed{\Psi(q_1, \ldots, q_f, t) = \sum_{j_1}^{n_1} \cdots \sum_{j_f}^{n_f} A_{j_1 \ldots j_f}(t) \prod_{\kappa=1}^{f} \varphi_{j_\kappa}^{(\kappa)}(q_\kappa, t) \, .} \tag{8.151}$$

The number of configurations, or Hartree products, is given by the product $n_1 \ldots n_f$. The SPFs, $\varphi_{j_\kappa}^{(\kappa)}(q_\kappa, t)$, are, as in the TDH approach, expressed in a time-independent basis set:

$$\boxed{\varphi_{j_\kappa}^{(\kappa)}(q_\kappa, t) = \sum_{i_\kappa=1}^{N_\kappa} c_{i_\kappa}^{(\kappa, j_\kappa)}(t) \, \chi_{i_\kappa}^{(\kappa)}(q_\kappa) \, ,} \tag{8.152}$$

where $\chi_{i_\kappa}^{(\kappa)}(q_\kappa)$ is a primitive basis functions, in general a DVR function, that depends on coordinate q_κ. If one sets in Eq. (8.151) $n_1 = n_2 = \cdots = n_f = 1$, one returns to the TDH approach. On the other hand, if $n_1 = N_1, n_2 = N_2, \ldots, n_f = N_f$, MCTDH is equivalent to the standard approach since there is a unitary transformation between the primitive and SPF basis sets. Contrarily to the standard method, both the coefficients and the basis functions are time-dependent. Both are optimized using a variational principle. The SPFs adapt along the propagation of the wavepacket and since $n_1, n_2, \ldots, n_f$ are, in general, smaller than $N_1, N_2, \ldots, N_f$, the number of equations to be solved is smaller and it is possible to treat larger systems. We are performing a contraction of the basis set since we have extracted a much smaller active space from the original space built by the primitive functions. MCTDH can be compared to the MultiConfigurational Self-Consistent Field (MCSCF) methods used in quantum chemistry to solve the time-independent electronic Schrödinger equation [55, 56]. MCTDH is an MCSCF method but for the nuclear coordinates and time-dependent. If one uses MCTDH to solve the time-independent Schrödinger equation as discussed in Sect. 8.4, the algorithm will generate *vibrational orbitals* similar to the molecular orbitals for the electrons in quantum chemistry [57]. Since we are solving the time-dependent Schrödinger equation here, we call the contracted basis functions SPFs rather than orbitals. Note that, contrarily to the quantum chemistry methods, there is no exchange operator since we do not have to antisymmetrize the wavefunction. The wavefunction is antisymmetrized only if MCTDH is applied to a system of fermions as in the MCTDHF method [58]. In the same manner, if MCTDH is applied to a system of bosons, the wavefunction must be symmetrized leading to the MCTDHB method [59].

As in TDH, this ansatz is not unique. One may perform linear transformations among the SPFs (orbitals) and the inverse transformations on the coefficients (A-vector). Defining transformed SPFs and coefficients as

$$\begin{aligned}\tilde{\varphi}_{j_\kappa}^{(\kappa)} &= \sum_{l_\kappa} U_{j_\kappa l_\kappa}^{(\kappa)} \, \varphi_{l_\kappa}^{(\kappa)} , \\ \tilde{A}_{j_1 \ldots j_f} &= \sum_{l_1 \cdots l_f} A_{l_1 \ldots l_f} (U^{(1)})^{-1}_{l_1 j_1} \cdots (U^{(f)})^{-1}_{l_f j_f} ,\end{aligned} \tag{8.153}$$

where $\boldsymbol{U}^{(\kappa)}$ denotes an arbitrary regular matrix, one obtains an unchanged total wavefunction

$$\Psi = \sum_{j_1 \ldots j_f} \tilde{A}_{j_1 \ldots j_f} \, \tilde{\varphi}_{j_1} \cdots \tilde{\varphi}_{j_f} . \tag{8.154}$$

As in TDH, we need constraints to lift the ambiguity. As constraints we choose

$$i \langle \varphi_l^{(\kappa)} | \dot{\varphi}_j^{(\kappa)} \rangle = \langle \varphi_l^{(\kappa)} | g^{(\kappa)} | \varphi_j^{(\kappa)} \rangle , \tag{8.155}$$

with some arbitrary constraint operator $g^{(\kappa)}$. The operator $g^{(\kappa)}$ defines the transformation matrix $\boldsymbol{U}^{(\kappa)}$. In fact, after the equations of motion are derived, one can show

$$i \, \dot{\boldsymbol{U}}^{(\kappa)} = \boldsymbol{g}^{(\kappa)^T} \, \boldsymbol{U}^{(\kappa)} , \tag{8.156}$$

where

$$(\boldsymbol{g}^{(\kappa)})_{lj} = \langle \varphi_l^{(\kappa)} | g^{(\kappa)} | \varphi_j^{(\kappa)} \rangle , \tag{8.157}$$

A formal solution is hence

$$\boldsymbol{U}^{(\kappa)}(t) = \mathcal{T} \, \exp\big(-i \int_0^t \boldsymbol{g}^{(\kappa)^T}(t') \, dt' \big) , \tag{8.158}$$

where $\mathcal{T}$ is the time-ordering operator, and $\boldsymbol{U}^{(\kappa)}$ is the transformation matrix from the SPFs computed with $g^{(\kappa)} \equiv 0$ to those computed with $g^{(\kappa)}$.

It is of great advantage if the SPFs are orthonormal. This is not a restriction as one can always find a transformation $\boldsymbol{U}^{(\kappa)}$ that orthogonalizes the SPFs. The overlap matrix is given by (dropping κ for the sake of simplicity)

$$S_{lj} = \langle \varphi_l | \varphi_j \rangle \,, \tag{8.159}$$

and

$$\dot{S}_{lj} = \langle \varphi_l | \dot{\varphi}_j \rangle + \langle \dot{\varphi}_l | \varphi_j \rangle = -i\;(g_{lj} - g_{jl}^*) = -i\;(\boldsymbol{g} - \boldsymbol{g}^\dagger)_{lj} \,.$$

Hence,

$$\dot{\boldsymbol{S}} = 0 \quad \text{if} \;\; \boldsymbol{g} = \boldsymbol{g}^\dagger \,, \tag{8.160}$$

and thus we require Hermitian constraint operators. If the initial WF, $\Psi(0)$, is built on orthornomal SPFs

$$S_{lj}^{(\kappa)}(0) = \langle \varphi_l^{(\kappa)}(0) | \varphi_j^{(\kappa)}(0) \rangle = \delta_{lj} \,, \tag{8.161}$$

then it follows that the SPFs stay orthonormal at all times, because $\dot{\boldsymbol{S}}^{(\kappa)} = 0$ and, hence, $\boldsymbol{S}^{(\kappa)}(t) = \mathbf{1}$.

Before we derive the MCTDH equations of motion we have to introduce some notations:

- Composite indices:

  $J \equiv (j_1, \ldots, j_f)$,
  $A_J \equiv A_{j_1 \cdots j_f}$.

- Configuration or Hartree product:

  $\Phi_J \equiv \prod_{\kappa=1}^f \varphi_{j_\kappa}^{(\kappa)}$,

which allows us to write the MCTDH wavefunction in a compact form

$$\Psi = \sum_J A_J \Phi_J \,. \tag{8.162}$$

Next, we introduce single-hole functions. The WF Ψ lies, of course, in the space spanned by the Hartree product of SPFs and we can make use of completeness

$$\Psi = \sum_{l=1}^{n_\kappa} |\varphi_l^{(\kappa)}\rangle\langle\varphi_l^{(\kappa)}|\Psi\rangle_\kappa = \sum_{l=1}^{n_\kappa} \varphi_l^{(\kappa)}\;\Psi_l^{(\kappa)} \,. \tag{8.163}$$

To make this clear, we write the single-hole function $\Psi_l^{(\kappa)}$ for the first DOF $\kappa = 1$

$$\Psi_l^{(1)} = \langle \varphi_l^{(1)} | \Psi \rangle = \sum_{j_2=1}^{n_2} \cdots \sum_{j_f=1}^{n_f} A_{lj_2 \cdots j_f} \, \varphi_{j_2}^{(2)} \cdots \varphi_{j_f}^{(f)} . \tag{8.164}$$

For a general definition, we need an extended nomenclature:

- $J^\kappa \equiv (j_1, \ldots, j_{\kappa-1}, j_{\kappa+1}, \ldots, j_f)$,
- $J_l^\kappa \equiv (j_1, \ldots, j_{\kappa-1}, l, j_{\kappa+1}, \ldots, j_f)$,
- $\Phi_{J^\kappa} \equiv \prod_{\substack{\nu=1 \\ \nu \neq \kappa}}^{f} \varphi_{j_\nu}^{(\nu)}$.

Then

$$\Psi_l^{(\kappa)} = \sum_{J^\kappa} A_{J_l^\kappa} \Phi_{J^\kappa} . \tag{8.165}$$

The single-hole functions allow us to introduce mean fields

$$\langle H \rangle_{jl}^{(\kappa)} = \langle \Psi_j^{(\kappa)} | H | \Psi_l^{(\kappa)} \rangle . \tag{8.166}$$

Note that we have not only one mean field for each degree of freedom, but a matrix of mean fields. Next, we introduce the density matrix:

$$\rho_{kl}^{(\kappa)} = \langle \Psi_k^{(\kappa)} | \Psi_l^{(\kappa)} \rangle . \tag{8.167}$$

As the SPFs are orthonormal, the integrations are readily performed to yield

$$\rho_{kl}^{(\kappa)} = \sum_{J^\kappa} A_{J_k^\kappa}^* \, A_{J_l^\kappa} .$$

Note that

$$\langle \Psi | \Psi \rangle = \sum_J A_J^* \, A_J = \|A\|^2 , \tag{8.168}$$

again due to the orthonormality of the SPFs. Hence,

$$\mathrm{Tr}\,[\rho^{(\kappa)}] = \sum_{j=1}^{n_\kappa} \rho_{jj}^{(\kappa)} = \|\Psi\|^2 . \tag{8.169}$$

We are now ready to derive the MCTDH-EOM. But before doing so, let us make some remarks on densities.

*8.3.2 Remarks on Densities **

We show here that the MCTDH density matrix is the transposed of the well-known one-particle reduced density matrix.

The density matrix of a mixed state reads

$$\rho = \sum_n p_n \big|\Psi_n\big\rangle\big\langle\Psi_n\big| ,$$

where $p_n \geq 0$ denote statistical weights. The density matrix of a pure state is given by:

$$\rho = |\Psi\rangle\langle\Psi| .$$

A reduced density is obtained by tracing out unwanted DOFs

$$\rho_{\text{red}} = \underset{\text{unwanted dofs}}{\text{Trace}} \left\{|\Psi\rangle\langle\Psi|\right\} ,$$

and the trace of an operator is given by

$$\text{Trace}\left\{A\right\} = \sum_n \big\langle n\big|A\big|n\big\rangle ,$$

for *any* complete orthonormal basis $\big|n\big\rangle$.

Choosing $\big|q\big\rangle$ as the basis one obtains the one-particle reduced densities

$$\rho_{\text{red}}^{(\kappa)}(q_\kappa, q_\kappa') = \int \Psi(q_1 \cdots q_\kappa \cdots q_f)\, \Psi^*(q_1 \cdots q_\kappa' \cdots q_f)\, dq_1 \cdots dq_{\kappa-1} dq_{\kappa+1} \cdots dq_f ,$$

and

$$\begin{aligned}
\big\langle\varphi_j^{(\kappa)}\big|\rho_{\text{red}}^{(\kappa)}\big|\varphi_l^{(\kappa)}\big\rangle &= \int \varphi_j^{(\kappa)*}\; \Psi\, \Psi^*\; \varphi_l^{(\kappa)}\;\; dq_\kappa\, dq_\kappa'\, dq_1 \cdots dq_{\kappa-1} dq_{\kappa+1} \cdots dq_f \\
&= \int \Psi_j^{(\kappa)}\; \Psi_l^{(\kappa)*}\;\; dq_1 \cdots dq_{\kappa-1} dq_{\kappa+1} \cdots dq_f \\
&= \big\langle\Psi_l^{(\kappa)}\big|\Psi_j^{(\kappa)}\big\rangle = \rho_{lj}^{\text{MCTDH}} .
\end{aligned} \tag{8.170}$$

Hence

$$\left(\boldsymbol{\rho}_{\text{red}}^{(\kappa)}\right) = \left(\boldsymbol{\rho}_{\text{MCTDH}}^{(\kappa)}\right)^T , \tag{8.171}$$

and the operator $\rho_{\text{red}}^{(\kappa)}$ reads

$$\begin{aligned}
\rho_{\text{red}}^{(\kappa)} &= \sum_{jl} \big|\varphi_j^{(\kappa)}\big\rangle \left(\boldsymbol{\rho}_{\text{red}}^{(\kappa)}\right)_{jl} \big\langle\varphi_l^{(\kappa)}\big| \\
&= \sum_{jl} \big|\varphi_j^{(\kappa)}\big\rangle \left(\boldsymbol{\rho}_{\text{MCTDH}}^{(\kappa)}\right)_{lj} \big\langle\varphi_l^{(\kappa)}\big| \\
&= \sum_{jl} \big|\varphi_j^{(\kappa)}\big\rangle \left(\boldsymbol{\rho}_{\text{MCTDH}}^{(\kappa)}\right)_{jl}^* \big\langle\varphi_l^{(\kappa)}\big| .
\end{aligned} \tag{8.172}$$

Note that the transformation to natural orbitals reads:

$$|\varphi_k^{(\kappa,\mathrm{nat})}\rangle = \sum_j |\varphi_j^{(\kappa)}\rangle T_{jk}^* , \tag{8.173}$$

where $\boldsymbol{T}$ denotes the eigenvector matrix of $\boldsymbol{\rho}_{\mathrm{MCTDH}}^{(\kappa)}$

The diagonal values, $q_\kappa = q'_\kappa$, of the reduced density are given by

$$\rho^{(\kappa)}(q_\kappa, q_\kappa) \equiv \rho^{(\kappa)}(q_\kappa) = \int |\Psi(q_1 \cdots q_f)|^2 dq_1 \cdots dq_{\kappa-1} dq_{\kappa+1} \cdots dq_f . \tag{8.174}$$

These diagonal densities are frequently plotted as they conveniently highlight how a wavepacket moves in time along a coordinate κ.

8.3.3 MCTDH Equations of Motion (MCTDH-EOM) *

To derive the MCTDH-EOM, we first repeat the MCTDH ansatz for the wavefunction

$$\begin{aligned}\Psi(q_1, \ldots, q_f, t) &= \sum_{j_1}^{n_1} \cdots \sum_{j_f}^{n_f} A_{j_1 \ldots j_f}(t)\, \varphi_{j_1}^{(1)}(q_1, t) \cdots \varphi_{j_f}^{(f)}(q_f, t) \\ &= \sum_J A_J\, \Phi_J \\ &= \sum_{j=1}^{n_\kappa} \varphi_j^{(\kappa)}\, \Psi_j^{(\kappa)} ,\end{aligned} \tag{8.175}$$

where the last equation holds for any $\kappa = 1, \ldots, f$. The variations with respect to coefficients and SPFs yield configurations and single-hole functions, respectively,

$$\frac{\delta \Psi}{\delta A_J} = \Phi_J , \tag{8.176}$$

and

$$\frac{\delta \Psi}{\delta \varphi_j^{(\kappa)}} = \Psi_j^{(\kappa)} . \tag{8.177}$$

And the time differentiation is given by

$$\dot{\Psi} = \sum_J \dot{A}_J\, \Phi_J + \sum_{\kappa=1}^{f} \sum_{j=1}^{n_\kappa} \dot{\varphi}_j^{(\kappa)}\, \Psi_j^{(\kappa)} . \tag{8.178}$$

We first consider variations with respect to the coefficients only. For δA_J follows:

$$\begin{aligned}\langle\delta\Psi|H|\Psi\rangle &= \langle\Phi_J|H|\Psi\rangle \\ &= \sum_L \langle\Phi_J|H|\Phi_L\rangle\, A_L \qquad (8.179)\\ i\,\langle\delta\Psi|\dot{\Psi}\rangle &= i\langle\Phi_J|\dot{\Psi}\rangle \\ &= i\sum_L \langle\Phi_J|\dot{A}_L\Phi_L\rangle + i\sum_\kappa\sum_l \langle\Phi_J|\dot{\varphi}_l^{(\kappa)}\,\Psi_l^{(\kappa)}\rangle \\ &= i\dot{A}_J + i\sum_\kappa\sum_l \langle\varphi_{j_\kappa}^{(\kappa)}|\dot{\varphi}_l^{(\kappa)}\rangle\,\langle\Phi_{J^\kappa}|\Psi_l^{(\kappa)}\rangle \\ &= i\dot{A}_J + i\sum_\kappa\sum_l \left(-i\, g_{j_\kappa l}^{(\kappa)}\right) A_{J_l^\kappa}\,. \qquad (8.180)\end{aligned}$$

The Dirac-Frenkel VP sets Eqs. (8.179, 8.180) equal, and solving for $\dot{A}$ yields

$$g_{j_\kappa l}^{(\kappa)} \equiv \langle\varphi_j^{(\kappa)}|g^{(\kappa)}|\varphi_l^{(\kappa)}\rangle = i\,\langle\varphi_j^{(\kappa)}|\dot{\varphi}_l^{(\kappa)}\rangle, \qquad (8.181)$$

which holds because

$$\langle\Phi_{J^\kappa}|\Psi_l^{(\kappa)}\rangle = \sum_{L^\kappa} \langle\Phi_{J^\kappa}|A_{L_l^\kappa}\Phi_{L^\kappa}\rangle = A_{J_l^\kappa}, \qquad (8.182)$$

and

$$i\dot{A}_J = \sum_L \langle\Phi_J|H|\Phi_L\rangle\, A_L - i\sum_{\kappa=1}^{f}\sum_{l=1}^{n_\kappa} g_{j_\kappa l}^{(\kappa)}\, A_{J_l^\kappa}\,. \qquad (8.183)$$

Next we consider variations with respect to the SPFs. For $\delta\varphi_j^{(\kappa)}$ follows:

$$\begin{aligned}\langle\delta\Psi|H|\Psi\rangle &= \langle\Psi_j^{(\kappa)}|H|\Psi\rangle \\ &= \langle\Psi_j^{(\kappa)}|H|\sum_l \Psi_l^{(\kappa)}\varphi_l^{(\kappa)}\rangle \\ &= \sum_{l=1}^{n_\kappa} \langle H\rangle_{jl}^{(\kappa)}\,\varphi_l^{(\kappa)} \qquad (8.184)\end{aligned}$$

$$i\,\langle\delta\Psi|\dot{\Psi}\rangle = \underbrace{i\sum_L \langle\Psi_j^{(\kappa)}|\Phi_L\rangle\,\dot{A}_L}_{\text{part 1}} + \underbrace{i\sum_L \langle\Psi_j^{(\kappa)}|\sum_{\nu=1}^{f}\sum_{l=1}^{n_\nu} \dot{\varphi}_l^{(\nu)}\Psi_l^{(\nu)}\rangle}_{\text{part 2}}\,. \qquad (8.185)$$

For the sake of simplicity we set $g^{(\kappa)} \equiv 0$ in the following. Then Eq. (8.183) simplifies to

$$i\dot{A}_L = \langle\Phi_L|H|\Psi\rangle = \sum_K \langle\Phi_L|H|\Phi_K\rangle\, A_K\,, \tag{8.186}$$

and Eq. (8.185) part 1 reads

$$i\sum_L \langle\Psi_j^{(\kappa)}|\Phi_L\rangle \dot{A}_L = \sum_L \langle\Psi_j^{\kappa}|\Phi_L\rangle\,\langle\Phi_L|H|\Psi\rangle\,, \tag{8.187}$$

which, with

$$\Phi_L = \Phi_{L^\kappa}\,\varphi_{l_\kappa}^{(\kappa)}\,, \tag{8.188}$$

and

$$\Psi_j^{(\kappa)} = \sum_{J^\kappa} A_{J_j^\kappa}\,\Phi_{J^\kappa}\,, \tag{8.189}$$

can be turned into

$$(\text{part1}) = \sum_{L^\kappa, l_\kappa} A^*_{L_j^\kappa} |\varphi_{l_\kappa}^{(\kappa)}\rangle\langle\varphi_{l_\kappa}^{(\kappa)}\Phi_{L^\kappa}|H|\Psi\rangle = P^{(\kappa)}\langle\Psi_j^{(\kappa)}|H|\Psi\rangle\,, \tag{8.190}$$

where we have introduced the MCTDH projector

$$P^{(\kappa)} = \sum_{j=1}^{n_\kappa} |\varphi_j^{(\kappa)}\rangle\langle\varphi_j^{(\kappa)}|\,. \tag{8.191}$$

Hence for part 1 of Eq. (8.185) we arrive at

$$i\sum_L \langle\Psi_j^{(\kappa)}|\Phi_L\rangle\,\dot{A}_L = P^{(\kappa)}\langle\Psi_j^{(\kappa)}|H|\Psi\rangle = P^{(\kappa)}\sum_{l=1}^{n_\kappa}\langle H\rangle_{jl}^{(\kappa)}\,\varphi_l^{(\kappa)}\,. \tag{8.192}$$

Next, we turn to part 2 of Eq. (8.185)

$$i\langle\Psi_j^{(\kappa)}|\sum_{\nu=1}^{f}\sum_{l=1}^{n_\nu}\dot{\varphi}_l^{(\nu)}\,\Psi_l^{(\nu)}\rangle = i\langle\Psi_j^{(\kappa)}|\sum_l \dot{\varphi}_l^{(\kappa)}\Psi_l^{(\kappa)}\rangle = i\sum_l \rho_{jl}^{(\kappa)}\dot{\varphi}_l^{(\kappa)}\,. \tag{8.193}$$

Here we have used

$$\langle\varphi_j^{(\kappa)}|\dot{\varphi}_l^{(\kappa)}\rangle = 0\,, \tag{8.194}$$

which holds for any j and l because $g^{(\kappa)} \equiv 0$ is assumed. Only when $\nu = \kappa$ there is no SPF with which $\dot{\varphi}$ is to be overlapped.

Using again the Dirac-Frenkel VP and equate Eq. (8.184) with the sum of Eqs. (8.192) and (8.193) we have

$$\sum_{l=1}^{n_\kappa} \langle H \rangle_{jl}^{(\kappa)} \varphi_l^{(\kappa)} = P^{(\kappa)} \sum_{l=1}^{n_\kappa} \langle H \rangle_{jl}^{(\kappa)} \varphi_l^{(\kappa)} + i \sum_l \rho_{jl}^{(\kappa)} \dot{\varphi}_l^{(\kappa)} , \tag{8.195}$$

or

$$i\dot{\varphi}_j^{(\kappa)} = \sum_{k,l} \left(\rho^{(\kappa)^{-1}}\right)_{jk} \left(1 - P^{(\kappa)}\right) \langle H \rangle_{kl}^{(\kappa)} \varphi_l^{(\kappa)} . \tag{8.196}$$

Hence for $g^{(\kappa)} \equiv 0$ we have the following set of EOM:

$$i\dot{A}_J = \sum_L \langle \Phi_J | H | \Phi_L \rangle A_L , \tag{8.197}$$

$$i\dot{\varphi}_j^{(\kappa)} = \left(1 - P^{(\kappa)}\right) \sum_{k,l=1}^{n_\kappa} \left(\rho^{(\kappa)^{-1}}\right)_{jk} \langle H \rangle_{kl}^{(\kappa)} \varphi_l^{(\kappa)} . \tag{8.198}$$

Introducing vectors of SPFs

$$\boldsymbol{\varphi}^{(\kappa)} = \left(\varphi_1^{(\kappa)} \cdots \varphi_{n_\kappa}^{(\kappa)}\right)^T , \tag{8.199}$$

we can write the last equation more compactly

$$i\dot{\boldsymbol{\varphi}}^{(\kappa)} = \left(1 - P^{(\kappa)}\right) \boldsymbol{\rho}^{(\kappa)^{-1}} \langle \boldsymbol{H} \rangle^{(\kappa)} \boldsymbol{\varphi}^{(\kappa)} . \tag{8.200}$$

In full generality the EOM are derived in the complement, Sect. 8.8. They read

$$\begin{aligned} i\dot{A}_J &= \sum_L \langle \Phi_J | H | \Phi_L \rangle A_L - \sum_{\kappa=1}^{f} \sum_{l=1}^{n_\kappa} g_{j_\kappa l}^{(\kappa)} A_{J_l^\kappa} , \\ i\dot{\boldsymbol{\varphi}}^{(\kappa)} &= \left(g^{(\kappa)} \mathbf{1}\right) \boldsymbol{\varphi}^{(\kappa)} + \left(1 - P^{(\kappa)}\right) \left\{ \boldsymbol{\rho}^{(\kappa)^{-1}} \langle \boldsymbol{H} \rangle^{(\kappa)} - g^{(\kappa)} \mathbf{1} \right\} \boldsymbol{\varphi}^{(\kappa)} . \end{aligned} \tag{8.201}$$

The last equation may be written as

$$i\dot{\boldsymbol{\varphi}}^{(\kappa)} = P^{(\kappa)} g^{(\kappa)} \boldsymbol{\varphi}^{(\kappa)} + \left(1 - P^{(\kappa)}\right) \boldsymbol{\rho}^{(\kappa)^{-1}} \langle \boldsymbol{H} \rangle^{(\kappa)} \boldsymbol{\varphi}^{(\kappa)} , \tag{8.202}$$

or

$$i\dot{\boldsymbol{\varphi}}^{(\kappa)} = \left\{ \boldsymbol{g}^{(\kappa)^T} + \left(1 - P^{(\kappa)}\right) \boldsymbol{\rho}^{(\kappa)^{-1}} \langle \boldsymbol{H} \rangle^{(\kappa)} \right\} \boldsymbol{\varphi}^{(\kappa)} , \tag{8.203}$$

As

$$
\begin{aligned}
P^{(\kappa)}\, g^{(\kappa)}\, \varphi_j^{(\kappa)} &= \sum_l |\varphi_l^{(\kappa)}\rangle\langle\varphi_l^{(\kappa)}|g^{(\kappa)}|\varphi_j^{(\kappa)}\rangle \\
&= \sum_l |\varphi_l^{(\kappa)}\rangle\, g_{lj}^{(\kappa)} \\
&= \sum_l g_{lj}^{(\kappa)}\, \varphi_l^{(\kappa)} = \left(\boldsymbol{g}^T\, \boldsymbol{\varphi}\right)_j .
\end{aligned}
\tag{8.204}
$$

Defining

$$
H_g = H - \sum_\kappa g^{(\kappa)} , \tag{8.205}
$$

one arrives at the EOM

$$
\begin{aligned}
i\dot{A}_J &= \sum_L \langle\Phi_J|H_g|\Phi_L\rangle\, A_L , \\
i\dot{\boldsymbol{\varphi}}^{(\kappa)} &= \left\{g^{(\kappa)}\, \mathbf{1} + \left(1 - P^{(\kappa)}\right) \boldsymbol{\rho}^{(\kappa)^{-1}} \langle \boldsymbol{H}_g\rangle^{(\kappa)}\right\} \boldsymbol{\varphi}^{(\kappa)} .
\end{aligned}
\tag{8.206}
$$

To prove Eq. (8.206), we note:

$$
\langle\Phi_J|H|\Phi_L\rangle = \langle\Phi_J|H_g + \sum_\kappa g^{(\kappa)}|\Phi_L\rangle = \langle\Phi_J|H_g|\Phi_L\rangle + \sum_\kappa g_{j_\kappa l_\kappa}^{(\kappa)}\, \delta_{J^\kappa L^\kappa} , \tag{8.207}
$$

where the previous last term cancels the last term of the $i\dot{A}$ equation. And similarly

$$
\begin{aligned}
\langle H\rangle_{jl}^{(\kappa)} &= \langle H_g + \sum_\kappa g^{(\kappa)}\rangle_{jl}^{(\kappa)} \\
&= \langle\Psi_j^{(\kappa)}|g^{(\kappa)}|\Psi_l^{(\kappa)}\rangle + \sum_{\nu\neq\kappa} \langle\Psi_j^{(\kappa)}|g^{(\nu)}|\Psi_l^{(\kappa)}\rangle + \langle H_g\rangle_{jl}^{(\kappa)} \\
&= g^{(\kappa)}\, \rho_{jl}^{(\kappa)} + \epsilon_{jl} + \langle H_g\rangle_{jl}^{(\kappa)} ,
\end{aligned}
\tag{8.208}
$$

which defines the matrix ϵ_{jl}. The EOM for the SPF hence reads

$$
\begin{aligned}
i\dot{\boldsymbol{\varphi}}^{(\kappa)} &= g^{(\kappa)}\, \boldsymbol{\varphi}^{(\kappa)} + \left(1 - P^{(\kappa)}\right)\left\{\boldsymbol{\rho}^{-1}\left[\langle \boldsymbol{H}_g\rangle^{(\kappa)} + \boldsymbol{\epsilon} + g^{(\kappa)}\, \boldsymbol{\rho}\right] - g^{(\kappa)}\right\}\boldsymbol{\varphi}^{(\kappa)} \\
&= g^{(\kappa)}\, \boldsymbol{\varphi}^{(\kappa)} + \left(1 - P^{(\kappa)}\right) \boldsymbol{\rho}^{-1} \langle \boldsymbol{H}_g\rangle^{(\kappa)}\, \boldsymbol{\varphi}^{(\kappa)} ,
\end{aligned}
\tag{8.209}
$$

as $\boldsymbol{\epsilon}\, \boldsymbol{\varphi}^{(\kappa)}$ is annihilated by the projector $(1 - P^{(\kappa)})$. This finishes the proof.

The two most obvious choices for constraint operator are either

$$
g^{(\kappa)} \equiv 0 , \tag{8.210}
$$

or

$$g^{(\kappa)} = h^{(\kappa)} , \tag{8.211}$$

where $h^{(\kappa)}$ denotes an operator acting on the κ-th DOF only, such that

$$H = \sum_{\kappa} h^{(\kappa)} + H_R , \tag{8.212}$$

i.e. the $\sum h^{(\kappa)}$ term stands for the separable part of H, and H_R for the non-separable or residual part.

With the arguments just given (replacing $g^{(\kappa)}$ with $h^{(\kappa)}$ and H_g with H_R in Eqs. (8.206, 8.208)), we find for $g^{(\kappa)} \equiv 0$

$$\begin{aligned} i\dot{A}_J &= \sum_L \langle \Phi_J | H | \Phi_L \rangle A_L , \\ i\dot{\boldsymbol{\varphi}}^{(\kappa)} &= \left(1 - P^{(\kappa)}\right) \left\{ h^{(\kappa)}\, \mathbf{1} + \boldsymbol{\rho}^{(\kappa)^{-1}} \langle \boldsymbol{H_R} \rangle^{(\kappa)} \right\} \boldsymbol{\varphi}^{(\kappa)} , \end{aligned} \tag{8.213}$$

whereas for $g^{(\kappa)} = h^{(\kappa)}$ we arrive at

$$\begin{aligned} i\dot{A}_J &= \sum_L \langle \Phi_J | H_R | \Phi_L \rangle A_L , \\ i\dot{\boldsymbol{\varphi}}^{(\kappa)} &= \left\{ h^{(\kappa)}\, \mathbf{1} + \left(1 - P^{(\kappa)}\right) \boldsymbol{\rho}^{(\kappa)^{-1}} \langle \boldsymbol{H_R} \rangle^{(\kappa)} \right\} \boldsymbol{\varphi}^{(\kappa)} . \end{aligned} \tag{8.214}$$

In the MCTDH package one may switch between those two sets of EOM with the keywords `proj-h`, and `h-proj` with obvious meaning.

It is illustrative to study the separable case $H = \sum_\kappa h^{(\kappa)}$, i.e. $H_R \equiv 0$. For $g^{(\kappa)} \equiv 0$ one obtains:

$$\begin{aligned} i\dot{A}_J &= \sum_L \sum_\kappa \langle \Phi_J | h^{(\kappa)} | \Phi_L \rangle A_L \\ &= \sum_{\kappa=1}^{f} \sum_{l=1}^{n_\kappa} \langle \varphi_{j_\kappa}^{(\kappa)} | h^{(\kappa)} | \varphi_{l_\kappa}^{(\kappa)} \rangle A_{J_l^\kappa} , \end{aligned} \tag{8.215}$$

$$i\dot{\varphi}_j^{(\kappa)} = \left(1 - P^{(\kappa)}\right) h^{(\kappa)} \varphi_j^{(\kappa)} \tag{8.216}$$

and for $g^{(\kappa)} = h^{(\kappa)}$:

$$\begin{aligned} i\dot{A}_J &= 0 , \\ i\dot{\varphi}_j^{(\kappa)} &= h^{(\kappa)}\, \varphi_j^{(\kappa)} . \end{aligned} \tag{8.217}$$

This suggests that the choice $g^{(\kappa)} = h^{(\kappa)}$ is of advantage, at least if H_R is small compared to the separable part $\sum h^{(\kappa)}$. However, for the sake of simplicity, we will assume $g^{(\kappa)} = 0$ in the following, if not stated otherwise.

8.3.4 Examples

Let us go back to the discussion of the end of Sect. 8.2.1 and consider the Hamiltonian operator of the 2D model defined by Eq. (8.144). We saw that the quality of the TDH approximation was linked to the correlation between the two coordinates x_1 and x_2, i.e. to the term $W(x_1)W(x_2)$ (see Eq. (8.144)). In the same manner, the number of configurations in the MCTDH expansion will be linked to the correlation between the coordinates. If $(W_1 - \langle W_1 \rangle)(W_2 - \langle W_2 \rangle)$ remains small, the number of configurations will be small and even the TDH approximation (i.e. one single configuration only) can provide a qualitatively correct description of the physics. On the other hand, if terms like $W(x_1)W(x_2)$ become large, the number of configurations will increase accordingly.

For the dissociation of NOCl on the electronic state S_1, Fig. 8.4 displays a numerically exact propagation of the wavepacket with MCTDH. Since the wavepacket remains rather localized and the shape of the potential along r_v varies slowly with r_d, the correlation effects coming from terms of the form $(W_1 - \langle W_1 \rangle)(W_2 - \langle W_2 \rangle)$ should remain rather weak. The problem should be optimally suited for the Hartree-type description. Indeed, in Ref. [54] it has been shown that with one SPF in each degree of freedom (TDH approximation), a structureless photodissociation spectrum can already be obtained that correctly describes the center and the width of the exact spectrum. With two SPFs in each degree of freedom (8 configurations since there are three internal degrees of freedom), structures corresponding to transitions to the NO-vibration (mainly to $v = 1$ and 2) are found. With 3 SPFs (27 configurations), a nearly quantitative agreement between the MCTDH and the exact result is reached. For 125 configurations (5 SPFs in each degree of freedom), the MCTDH and exact results become identical within graphical resolution. The number of primitive functions is equal[3] to $36 \times 24 \times 60 = 51840$ to be compared with the number of SPFs (125): the subspace built by the SPFs is 414 times smaller than the original primitive space. For NOCl, fast convergence is found when increasing the number of SPFs and few configurations provide the main physical content of the problem.

The $H+H_2$ reaction in Jacobi coordinates displays much stronger correlation effects. The potential energy is depicted in Fig. 8.5. The terms such as $(W_1 - \langle W_1 \rangle)(W_2 - \langle W_2 \rangle)$ become very large when the H-atom reaches the transition state and the TDH approximation cannot even qualitatively describe the fact that one part of the wavepacket can react (from $H^{(1)}+H^{(2)}H^{(3)}$ to $H^{(1)}H^{(2)}+H^{(3)}$) (see Ref. [60]). Few configurations cannot correctly reproduce the reaction probability as a function of the energy. Convergence is obtained [61] for 3024 configurations (18 SPFs in r_d, 14 in r_v, and 12 in θ), i.e. one order of magnitude larger than for the spectrum of

[3] For the primitive basis set of each coordinate, we use a DVR (see Sect. 8.1.3), more precisely HO DVR for R_1 and a sine DVR for R_2 with $R_1 \in [3.8, 5.6]$ a.u. and an unrestricted Legendre DVR for θ. Here, $N_1 = 36$ for r_d, $N_2 = 24$ for r_v, and $N_3 = 60$ for θ.

NOCl. The number of primitive functions is equal[4] to 68 × 48 × 31 = 101,184 (see the caption of Fig. 8.5) to be compared with the number of SPFs (3024): the subspace built by the SPFs is 33.5 times smaller than the original primitive space. For $H+H_2$, the Jacobi coordinates do not provide a single reaction coordinate that describes the chemical process. Both r_d and r_v play the role of a reaction coordinate in one of the two channels resulting in strong correlation around the saddle point and in particular in the reactive exit channel, because here "wrong" Jacobi vectors, namely those of the entrance channel, are used (see the discussion in Sect. 5.2). Of course, the separability of the Hamiltonian operator and hence the correlation *depends upon the set of coordinates*. In particular, it is well-known that a set of so-called hypersherical coordinates leads to a more separable form of the potential energy for a system such as $H+H_2$ (see Refs. [62–64] and Sect. 12.7 of Ref. [2] for the definition of hyperspherical coordinates). However, hypersherical coordinates lead to a complicated, singular KEO, which is difficult to handle. To conclude, the convergence of MCTDH is related to the correlation between coordinates. In addition, a set of inappropriate coordinates can easily introduce some strong artificial correlation and it is this effect that slows down the convergence of MCTDH. This is exemplified by the study of the $H+H_2$ system where the coordinates become inappropriate when the wavepacket reaches the exit channel.

8.3.5 Mode Combination

There is no prescription for SPFs to depend on a single coordinate, and they may depend on several coordinates. Several physical coordinates can be regrouped into one logical coordinate, also called *particle* or *combined mode*

$$Q_\kappa \equiv \left(q_{\kappa,1}, q_{\kappa,2}, \ldots, q_{\kappa,d}\right), \tag{8.218}$$

$$\varphi_j^{(\kappa)}(Q_\kappa, t) = \varphi_j^{(\kappa)}(q_{\kappa,1}, q_{\kappa,2}, \ldots, q_{\kappa,d}, t). \tag{8.219}$$

The MCTDH wavefunction is now expanded as

$$\Psi(q_1, \ldots, q_f, t) \equiv \Psi(Q_1, \ldots, Q_p, t) = \sum_{j_1 \cdots j_p} A_{j_1 \cdots j_p}(t) \prod_{\kappa=1}^{p} \varphi_j^{(\kappa)}(Q_\kappa, t), \tag{8.220}$$

and the SPFs themselves are expanded as:

$$\varphi_j^{(\kappa)}(Q_\kappa, t) = \sum_{i_1 \cdots i_d} C_{i_1 \cdots i_d}^{(\kappa, j)}(t)\, \chi^{(\kappa,1)}(q_{\kappa,1}) \cdots \chi^{(\kappa,d)}(q_{\kappa,d}). \tag{8.221}$$

[4]For the primitive basis set of each coordinate, we use a DVR (see Sect. 8.1.3), more precisely a sine DVR for R_1 and R_2 with $R_1 \in [0.6, 6.24]$ and 48 functions and $R_2 \in [1.0, 9.04]$ a.u. and 68 functions, and a Legendre DVR for θ with 31 functions.

In addition, the number of SPFs per particle required for convergence will increase with mode combination. If $\tilde{n} = n^d$ held, there would be no gain, the A-vector length would not change. Luckily one finds as a rule of thumb:

$$\tilde{n} \approx d \cdot n \,, \tag{8.222}$$

sometimes even less. Here, $\tilde{n}$ denotes the number of SPFs required for convergence when mode-combination is used, while n is the corresponding number of SPFs without mode-combination. For sake of simplicity it is assumed that $n_\kappa = n$ for all κ, and similarly for $\tilde{n}$, N, and $\tilde{N}$. Note that now all correlations between the DOFs within a particle is taken care of at the SPF level. Only the correlation between particles has to be accomplished by the A-vector.

The usefulness of mode-combination originates from the fact that the size of the A-vector is reduced, but the usefulness is limited by the fact that now multi-dimensional SPFs have to be propagated. If one "over-combines", the propagation of the SPFs will take more effort than the propagation of the A-vector and efficiency is lost. However, we know a method that efficiently propagates multi-dimensional wavefunctions: MCTDH! One hence may think of propagating the SPFs of an MCTDH calculation with MCTDH. This idea has led to the development of the multi-layer MCTDH (ML-MCTDH) algorithm.

8.3.6 Summary and Multi-layer MCTDH

Let us summarize the standard and MCTDH methods and introduce the ML-MCTDH method. The latter represents an extension of the standard MCTDH approach that gives access to quantum dynamics simulations for significantly larger systems. The key factor for this powerful extension is the more flexible way of representing the overall wavefunction in the ML-MCTDH method. ML-MCTDH has been explained elsewhere in great detail [65–72]. To explain the basics of ML-MCTDH while avoiding clumsy notations we confine ourselves here to considering a three-dimensional case, q_1, q_2, q_3 denoting the three degrees of freedom. In the standard wavepacket method, the wavefunction is written as:

$$\Psi(q_1, q_2, q_3, t) = \sum_{i_1=1}^{N_1} \sum_{i_2=1}^{N_2} \sum_{i_3=1}^{N_3} C_{i_1,i_2,i_3}(t) \chi_{i_1}^{(1)}(q_1) \chi_{i_2}^{(2)}(q_2) \chi_{i_3}^{(3)}(q_3) \,. \tag{8.223}$$

$\chi_{i_1}^{(1)}(q_1)$, $\chi_{i_2}^{(2)}(q_2)$, and $\chi_{i_3}^{(3)}(q_3)$ are time-independent primitive basis set functions. Thus, only the expansion coefficients vary during the time evolution. In contrast, the MCTDH ansatz (without mode combination) can be written as:

$$\Psi(q_1, q_2, q_3, t) = \sum_{j_1=1}^{n_1} \sum_{j_2=1}^{n_2} \sum_{j_3=1}^{n_3} A_{j_1,j_2,j_3}(t)\varphi_{j_1}^{(1)}(q_1, t)\varphi_{j_2}^{(2)}(q_2, t)\varphi_{j_3}^{(3)}(q_3, t), \quad (8.224)$$

where

$$\varphi_{j_1}^{(1)}(q_1, t) = \sum_{i_1=1}^{N_1} c_{i_1}^{(1,j_1)}(t)\chi_{i_1}^{(1)}(q_1), \quad (8.225)$$

$$\varphi_{j_2}^{(2)}(q_2, t) = \sum_{i_2=1}^{N_2} c_{i_2}^{(2,j_2)}(t)\chi_{i_2}^{(2)}(q_2), \quad (8.226)$$

and

$$\varphi_{j_3}^{(3)}(q_3, t) = \sum_{i_3=1}^{N_3} c_{i_3}^{(3,j_3)}(t)\chi_{i_3}^{(3)}(q_3). \quad (8.227)$$

Now, both the expansion coefficients and the SPFs are time-dependent. They evolve according to the variation principle and the SPFs adapt along the propagation. This greatly reduces the number of parameters compared to standard wavepacket propagation methods since, in general, n_1, n_2, and n_3 are smaller than N_1, N_2, and N_3. However, the scalability to very large systems of the MCTDH method is limited since the SPFs are expanded in terms of time-independent basis functions. It is thus natural to exploit the possibility of applying the basic MCTDH strategy also at the level of the configurations/SPFs. This is the multilayer ML-MCTDH theory. Le us consider an example. A ML-MCTDH ansatz with three layers[5] could be:

$$\Psi(q_1, q_2, q_3, t) = \sum_{j_{12}=1}^{n_{12}} \sum_{j_3=1}^{n_3} A_{j_{12},j_3}(t)\varphi_{j_{12}}^{(12)}(q_1, q_2, t)\varphi_{j_3}^{(3)}(q_3, t), \quad (8.228)$$

where

$$\varphi_{j_{12}}^{(12)}(q_1, q_2, t) = \sum_{k_1=1}^{n_1} \sum_{k_2=1}^{n_2} B_{k_1,k_2}^{(12,j_{12})}(t)\xi_{k_1}^{(1)}(q_1, t)\xi_{k_2}^{(2)}(q_2, t), \quad (8.229)$$

$$\xi_{k_1}^{(1)}(q_1, t) = \sum_{i_1=1}^{N_1} c_{i_1}^{(1,k_1)}(t)\chi_{i_1}^{(1)}(q_1), \quad (8.230)$$

[5]A two-layers ML-MCTDH is identical to standard MCTDH

and

$$\xi_{k_2}^{(2)}(q_2, t) = \sum_{i_2=2}^{N_2} c_{i_2}^{(1,k_2)}(t)\chi_{i_2}^{(2)}(q_2)\,. \tag{8.231}$$

$\varphi_{j_3}^{(3)}(q_3, t)$ is expressed as in Eq. (8.227). In other words, for a combined SPF, one can use a new, additional MCTDH expansion instead of the primitive basis. Thus, using this approach, deeper layer schemes can be constructed [69, 71, 72]. All the time-dependent functions are optimized at the same time according to the variational principle. Finally, as a crude rule of thumb, we note that mode combination usually becomes important if there are 5 or more DOFs and the multi-layer approach usually outperforms standard MCTDH if there are more than 10 DOFs. For large systems ML-MCTDH can be much faster than MCTDH, and it can treat systems of sizes undoable for MCTDH. Systems with several hundreds of DOFs have been successfully treated with ML-MCTDH [71, 73, 74].

8.3.7 Electronic States

If the WF is to be propagated on several electronic states, i.e. when vibronic couplings become important, the WF may be written as follows:

$$|\Psi\rangle = \sum_{j_1}^{n_1} \cdots \sum_{j_f}^{n_f} \sum_{\alpha=1}^{n_s} A_{j_1 \cdots j_f \alpha}\, \varphi_{j_1}^{(1)}(q_1, t) \cdots \varphi_{j_f}^{(f)}(q_f, t)|\alpha\rangle\,. \tag{8.232}$$

This is the so-called single-set formalism. It is called *single-set* because there is one set of SPFs for all electronic states. The single-set formalism closely follows the MCTDH philosophy. In contrast, the *multi-set* formulation uses different sets of SPFs for each state

$$|\Psi\rangle = \sum_{\alpha=1}^{n_s} \psi_\alpha(q_1, \ldots, q_f, t)|\alpha\rangle\,, \tag{8.233}$$

where each component ψ_α is expanded in MCTDH form

$$\psi_\alpha(q_1, \ldots, q_f, t) = \sum_{j_1^\alpha}^{n_1^\alpha} \cdots \sum_{j_f^\alpha}^{n_f^\alpha} A_{j_1^\alpha \cdots j_f^\alpha}^{(\alpha)}(t)\, \varphi_{j_1}^{(1,\alpha)}(q_1, t) \cdots \varphi_{j_f^\alpha}^{(f,\alpha)}(q_f, t)\,. \tag{8.234}$$

The equations of motion must be generalized

$$i\,\dot{A}_J^{(\alpha)} = \sum_{\beta=1}^{n_s}\sum_L \mathcal{K}_{JL}^{(\alpha,\beta)}\,A_L^{(\beta)}\,, \tag{8.235}$$

$$i\,\dot{\varphi}_j^{(\kappa,\alpha)} = \left(1 - P^{(\kappa,\alpha)}\right)\sum_{k=1}^{n_\kappa^\alpha}\left(\rho^{(\kappa,\alpha)}\right)^{-1}_{jk}\sum_{\beta=1}^{n_s}\sum_{l=1}^{n_\kappa^\beta}\mathcal{H}_{kl}^{(\kappa,\alpha,\beta)}\,\varphi_l^{(\kappa,\beta)}\,, \tag{8.236}$$

with the obvious definitions:

$$\mathcal{K}_{JL}^{(\alpha,\beta)} = \left\langle\Phi_J^{(\alpha)}\middle|H^{(\alpha,\beta)}\middle|\Phi_L^{(\beta)}\right\rangle, \tag{8.237}$$

$$\mathcal{H}_{jl}^{(\kappa,\alpha,\beta)} = \left\langle\Psi_j^{(\kappa,\alpha)}\middle|H^{(\alpha,\beta)}\middle|\Psi_l^{(\kappa,\beta)}\right\rangle, \tag{8.238}$$

where $H^{(\alpha,\beta)}$ denotes the electronic $(\alpha,\,\beta)$-block of the full Hamiltonian. The single-set formalism is of advantage if the dynamics in the different electronic states is similar, e.g. when the surfaces are almost parallel. The more complicated multi-set formalism is more efficient when the dynamics on the various diabatic states is rather different. There is then one set of single-particle functions for each electronic state. In most cases multi-set is the preferred scheme. Note that the SPFs of different electronic states are no longer orthogonal to each other, which complicates the algorithm.

8.3.8 Propagation in MCTDH *

As shown above, the MCTDH equations of motion (for $g \equiv 0$ and $H = \sum_\kappa h^{(\kappa)} + H_R$) using

$$\mathcal{K}_{JL} = \left\langle\Phi_J\middle|H\middle|\Phi_L\right\rangle \quad\text{and}\quad \left\langle H_R\right\rangle_{kl}^{(\kappa)} = \sum_{r=1}^{s}\mathcal{H}_{rkl}^{(\kappa)}\,h_r^{(\kappa)}\,,$$

read

$$i\dot{A}_J = \sum_L \mathcal{K}_{JL}\,A_L\,,$$

$$i\dot{\varphi}_j^{(\kappa)} = \left(1 - P^{(\kappa)}\right)\left\{h^{(\kappa)}\varphi_j^{(\kappa)} + \sum_{k,l=1}^{n_\kappa}\left(\rho^{(\kappa)}\right)^{-1}_{jk}\left\langle H_R\right\rangle_{kl}^{(\kappa)}\,\varphi_l^{(\kappa)}\right\}. \tag{8.239}$$

This set of non-linear coupled differential equations can be solved by a standard all-purpose integrator such as the Runge-Kutta or Adams-Bashforth-Moulton integrators. (The latter have not been described in Sect. 8.1.6.) This strategy is called the variable mean-field (VMF) approach. The problem is that the mean-fields matrices $\mathcal{H}^{(\kappa)}_{rkl}$ and the $\mathcal{K}$-matrix $\mathcal{K}_{JL}$ have to be built at every time step. The time steps, however, have to be small as one has to describe an oscillating function.

Formally Ψ is given by

$$\Psi(t) = \sum_n a_n \Psi_n e^{-iE_n t}, \qquad H\Psi_n = E_n \Psi_n\,. \tag{8.240}$$

To integrate $e^{-iE_n t}$ one needs step sizes of the order

$$\Delta t \lesssim \frac{1}{|E_n|}\,. \tag{8.241}$$

Hence the step size is determined by the absolute largest eigenvalue of the matrix representation of the Hamiltonian.

The mean fields, on the other hand, are not that strongly oscillating. It is hence tempting to set the mean-fields constant over a larger update time step τ and to integrate the A-vector and the SPFs with much smaller time steps. This is called the constant mean-field (CMF) approach. Keeping the mean-fields constant yields

$$\begin{aligned}
i\dot{A}_J &= \sum_L \bar{\mathcal{K}}_{JL}\, A_L\,,\\
i\dot{\varphi}^{(1)}_j &= \left(1 - P^{(1)}\right)\left\{h^{(1)}\varphi^{(1)}_j + \sum \bar{\rho}^{(1)^{-1}}_{jk} \sum_r \bar{\mathcal{H}}^{(1)}_{rkl}\, h^{(\kappa)}_r\, \varphi^{(1)}_l\right\},\\
&\;\;\vdots\\
i\dot{\varphi}^{(f)}_j &= \left(1 - P^{(f)}\right)\left\{h^{(f)}\varphi^{(f)}_j + \sum \bar{\rho}^{(f)^{-1}}_{jk} \sum_r \bar{\mathcal{H}}^{(f)}_{rkl}\, h^{(\kappa)}_r\, \varphi^{(f)}_l\right\}.
\end{aligned} \tag{8.242}$$

Note that all the differential equations *decouple*! The bar indicates that the quantities are kept constant over the update time step τ. As the equations decouple, one can use different time steps and in fact different integrators for each set of equations. The EOM for the A-vector is *linear* and now time-independent, thus one may use an adapted integrator such as Short Iterative Lanczos (SIL). The EOM for the SPFs are still non-linear because of the projector $P^{(\kappa)}$. But the main gain is of course that the mean fields must be built less often.

The scheme outlined above is too simple. One needs at least a second-order scheme, i.e. one in which the error for one step scales like $\|\Psi_{ex} - \Psi\| \sim \tau^3$. In the present scheme, the error scales like τ^2.

A higher-order scheme is displayed in Fig. 8.6:

Note that there is an additional propagation of the SPFs from $t = 0$ to $t = \tau/2$ (second line). The additional backward propagation of the coefficients ($A(\tau/2) \rightarrow \tilde{A}(0)$) (last line) is done for error estimation only. This step is virtually cost free, because the Krylov-space, built for the previous forward propagation of A, is used. (The SIL integrator introduced in Sect. 8.1.6 is used for A-propagation). A slightly different CMF-scheme has recently been discussed by Manthe [75].

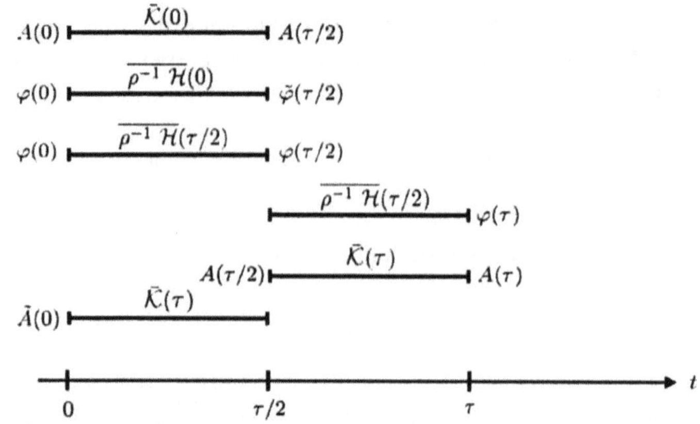

Fig. 8.6 Second-order CMF scheme

The error of the propagation scheme of Fig. 8.6 is given in lowest-order by

$$\|\Psi - \tilde{\Psi}\|^2 = \|\Delta A\|^2 + \sum_{\kappa=1}^{f} tr\big(\Delta \boldsymbol{O} \cdot \boldsymbol{\rho}^{(\kappa)}\big)\,, \tag{8.243}$$

where

$$\begin{aligned} \Delta A &= A(0) - \tilde{A}(0)\,, \\ \Delta\varphi_j &= \varphi_j(\tau/2) - \tilde{\varphi}_j(\tau/2)\,, \\ \Delta O_{jl} &= \big\langle \Delta\varphi_j \big| \Delta\varphi_l \big\rangle\,. \end{aligned} \tag{8.244}$$

This allows for an automatic step-size control. One sets an error limit and the algorithm searches for an appropriate value of τ.

To demonstrate that the scheme on Fig. 8.6 gives an improved scaling of the error, let us consider a one-dimensional differential equation. The Taylor expansion of the solution propagated over one step reads:

$$y(\tau) = y(0) + y'(0) \cdot \tau + \frac{1}{2}\, y''(0) \cdot \tau^2 + \frac{1}{6}\, y'''(0) \cdot \tau^3 + \ldots \tag{8.245}$$

The previous scheme, Eq. (8.242), is equivalent to a Euler integrator

$$y^{\text{app}}(\tau) = y(0) + y'(0) \cdot \tau\,, \tag{8.246}$$

which has an error

$$\text{error} : (y^{\text{app}} - y)(\tau) = -\frac{1}{2}\, y''(0) \cdot \tau^2 + \ldots \tag{8.247}$$

To investigate the error introduced by the scheme shown on Fig. 8.6 for the SPFs, we first note that the time derivative at a half step reads

$$y'(\tau/2) = y'(0) + y''(0) \cdot \tau/2 + \frac{1}{2} y'''(0) \cdot (\tau/2)^2 \,.$$

The one-step propagated solution, using this mid-step derivative, reads

$$y^{\mathrm{app}}(\tau) = y(0) + y'(\tau/2) \cdot \tau = y(0) + y'(0) \cdot \tau + y''(0) \cdot \tau^2/2 + y'''(0) \cdot \tau^3/8 \,, \tag{8.248}$$

and has the error

$$\text{error} : (y^{\mathrm{app}} - y)(\tau) = \Big(\frac{1}{8} - \frac{1}{6}\Big)\, y'''(0) \cdot \tau^3 = -\frac{1}{24}\, y'''(0) \cdot \tau^3 \,. \tag{8.249}$$

Similarly for the propagator of the A-vector, we obtain

$$\begin{aligned} y^{\mathrm{app}}(\tau) &= y(0) + y'(0) \cdot (\tau/2) + y'(\tau) \cdot (\tau/2) \,, \\ &= y(0) + y'(0) \cdot \tau + y''(0) \cdot (\tau^2/2)\, \frac{1}{4} y'''(0) \cdot \tau^3 \,, \end{aligned} \tag{8.250}$$

and the error

$$\text{error} : \Big(\frac{1}{4} - \frac{1}{6}\Big) y'''(0) \cdot \tau^3 = \frac{1}{12}\, y'''(0) \cdot \tau^3 \,. \tag{8.251}$$

Hence the error done in one step scales like τ^3. The total error then scales like τ^2 as the number of steps scales like τ^{-1} (Fig. 8.7).

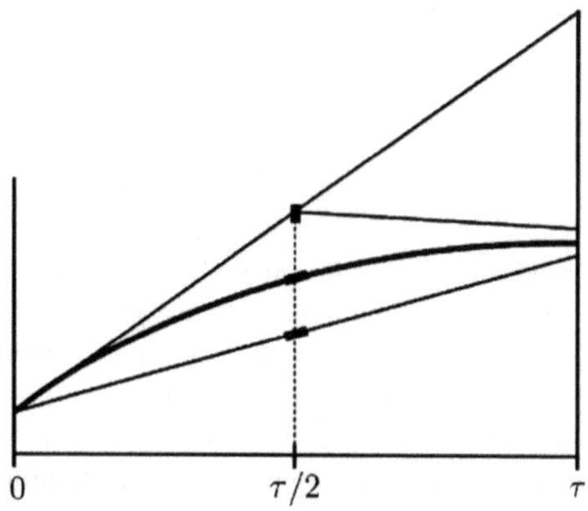

Fig. 8.7 Graphical interpretation of the numerical integration. The *bold line* (*middle*) shows the exact solution of a differential equation. Taking the initial derivative $y'(0)$ throughout the propagation leads to a rather large error at $t = \tau$; see *upper straight line*. Using $y'(\tau/2)$ rather than $y'(0)$ provides a much better solution; see the *lowest straight line*. And using $y'(0)$ for the first half-step and $y'(\tau)$ for the second half-step also provides a good approximate solution

To understand why CMF works, let us consider a separable case

$$H = \sum_{\kappa=1}^{f} h^{(\kappa)}, \tag{8.252}$$

hence $H_R = 0$ and $\mathcal{H}^{(\kappa)} = 0$ and the mean fields $\langle \boldsymbol{H} \rangle^{(\kappa)}$ reduce to $h^{(\kappa)}$. These are obviously constant but what is with $\mathcal{K}_{JL}$? In the following we allow for a general constraint operator $g^{(\kappa)}$, although CMF is usually performed with $g^{(\kappa)} \equiv 0$. The $\mathcal{K}$ matrix is then given (cf. Eq. (8.201)) by

$$\mathcal{K}_{JL} = \langle \Phi_J | H - \sum_{\kappa} g^{(\kappa)} | \Phi_L \rangle = \sum_{\kappa=1}^{f} \langle \varphi_{j_\kappa}^{(\kappa)} | h^{(\kappa)} - g^{(\kappa)} | \varphi_{l_\kappa}^{(\kappa)} \rangle \delta_{J^\kappa L^\kappa}. \tag{8.253}$$

The EOM for the φ's read (we drop κ for simplicity):

$$i\dot{\varphi}_j = g\,\varphi_j + (1 - P) \cdot (h - g)\,\varphi_j\,. \tag{8.254}$$

From this follows:

$$\begin{aligned}
&\frac{d}{dt}\langle \varphi_j | h - g | \varphi_l \rangle = \\
&i\langle g\,\varphi_j + (1-P)(h-g)\varphi_j | h - g | \varphi_l \rangle - i\langle \varphi_j | h - g | g\,\varphi_l + (1-P)(h-g)\,\varphi_l \rangle, \\
&= i\langle \varphi_j | g^\dagger (h-g) + (h^\dagger - g^\dagger)(1-P)(h-g) - (h-g)g - (h-g)(1-P)(h-g) | \varphi_l \rangle, \\
&\overset{\{g^\dagger = g\}}{=} i\langle \varphi_j | (h^\dagger - h)(1-P)(h-g) | \varphi_l \rangle + i\langle \varphi_j | gh - hg | \varphi_l \rangle,
\end{aligned} \tag{8.255}$$

hence

$$\begin{aligned}
\frac{d}{dt}\mathcal{K}_{JL} = 0 \quad &\text{if} \;\; h^{(\kappa)} = h^{(\kappa)\dagger}, \\
&\text{and} \;\; \left[h^{(\kappa)}, g^{(\kappa)}\right] = 0\,.
\end{aligned} \tag{8.256}$$

The commutator vanishes for $g^{(\kappa)} \equiv 0$ and $g^{(\kappa)} = h^{(\kappa)}$, the standard choices! For non-Hermitian $h^{(\kappa)}$ one may be tempted to choose $g^{(\kappa)} = h^{(\kappa)}$ as this will ensure a constant $\mathcal{K}$. However, the choice $g^{(\kappa)} = h^{(\kappa)}$ is only allowed for Hermitian $h^{(\kappa)}$ because the constraint operators have to be Hermitian.

The CMF integrator can take arbitrarily large update steps τ if the Hamiltonian is Hermitian and separable. In a scattering problem, the Hamiltonian often becomes almost separable when the colliding partners are far apart from each other. However, when the scattered particle is finally absorbed by a Complex Absorbing Potential (CAP) the separable Hamiltonian becomes non-Hermitian and the CMF-integrator is forced to take small steps. But our analysis has clearly shown that the assumption of constant mean fields is violated by the non-separable (and non-Hermitian) terms of the Hamiltonian. These terms are usually much smaller than the separable ones, which justifies the assumption that the mean fields can be taken as constant over a small update time τ, which, however, is much larger than the integration steps used to propagate the SPFs.

The CMF integrator scheme violates energy conservation which should hold for constant Hermitian Hamiltonians. Only for $\tau \to 0$ energy conservation is strictly obeyed. If an MCTDH calculation shows an energy deviation which is too high to be acceptable, one must increase the integrator accuracy, in particular the CMF accuracy.

8.4 Relaxation and Improved Relaxation *

The MCTDH algorithm can also be used to solve the time-independent Schrödinger equation. This is a convenient approach to obtain eigenstates or quantum resonances with high accuracy.

For this purpose, we combine iteratively a propagation in imaginary time, a relaxation, of the SPFs with a diagonalisation of the Hamiltonian matrix (using the Davidson algorithm) to determine the A-coefficients. This procedure is essentially a vibrational Multi-Configuration Self-Consistent-Field (MCSCF) algorithm. In other words, the procedure converges to one or several eigenstate(s) that are expanded as linear combinations of products of SPFs. The latter may be viewed as *vibrational orbitals*. Both the orbitals and the coefficients in the expansion are optimized with a time-independent variational procedure as in the MCSCF method in quantum chemistry.

8.4.1 Propagation in Imaginary Time

The ground-state wavefunction can be obtained by a time-dependent method via a relaxation, i.e. a propagation in negative imaginary time. The Schrödinger equation is then turned into

$$\dot{\Psi} = -H\Psi \,. \tag{8.257}$$

To see its effect, we expand the WF in terms of eigenstates and obtain

$$\Psi(t) = \sum_n a_n \, e^{-E_n t} \Psi_n \,. \tag{8.258}$$

We now have decaying exponential functions. The state with the lowest energy (usually E_0) will thus dominate as time increases. Of course, the norm must be restored. To accomplish norm conservation, one may change the Schrödinger equation to

$$\dot{\Psi}(t) = -\big(H - E(t)\big)\Psi(t)\,, \quad \text{where} \quad E(t) = \big\langle \Psi(t)\big|H\big|\Psi(t)\big\rangle\,. \tag{8.259}$$

Then,

$$\big\langle \Psi(t)|\dot{\Psi}(t)\big\rangle = 0 \quad \Rightarrow \quad \frac{d}{dt}\|\Psi\|^2 = 0\,. \tag{8.260}$$

The energy E can be interpreted as a Lagrange parameter introduced to keep the norm of Ψ constant (we assume Ψ to be normalized). Differentiation of $E(t)$ leads to

$$\dot{E} = -2\big\langle \Psi(t)|(H - E(t))^2|\Psi(t)\big\rangle\,. \tag{8.261}$$

Hence the energy decreases with relaxation time and converges if the variance vanishes, i.e. if the wavefunction becomes an eigenstate of H. Usually this will be the ground state. Only if the initial state is orthogonal to the ground state, the algorithm will converge to an excited state.

Relaxation is effective when the initial state Ψ has a reasonable overlap with the ground state and when the ground state is well separated energetically from excited states. However, relaxation may converge slowly if the energy of the first excited state, E_1, is close to the ground state energy E_0. To damp out the contribution of the excited state one needs a propagation time that satisfies $(E_1 - E_0) \cdot t \approx 10 - 30\hbar$. The relaxation can be accelerated, and excited states can be computed as well, if the MCTDH A-vector is not determined by relaxation but by diagonalization. This method is called *improved relaxation*.

8.4.2 Improved Relaxation: An MCSCF Procedure *

The improved relaxation algorithm [76, 77] can be derived via a standard *time-independent* variational principle $\delta\{\langle\Psi|H|\Psi\rangle - \text{constraints}\} = 0$, i.e.

$$\delta\Big\{\big\langle\Psi\big|H\big|\Psi\big\rangle - E\big(\sum_J A_J^* A_J - 1\big) - \sum_{\kappa=1}^{f}\sum_{j,l=1}^{n_\kappa} \epsilon_{jl}^{(\kappa)}\big(\langle\varphi_j^{(\kappa)}|\varphi_l^{(\kappa)}\rangle - \delta_{jl}\big)\Big\} = 0\,. \tag{8.262}$$

This approach is thus similar to the MCSCF method in quantum chemistry.

The first Lagrange parameter, E, ensures that the A-vector is normalized and the $\epsilon_{jl}^{(\kappa)}$ ensure that the SPFs stay orthonormal. We note that

$$\big\langle\Psi\big|H\big|\Psi\big\rangle = \sum_{JK} A_J^*\, H_{JK} A_K\,, \qquad \text{with } H_{JK} = \big\langle\Phi_J\big|H\big|\Phi_K\big\rangle\,. \tag{8.263}$$

Varying A_J^* yields secular equations

$$\sum_K H_{JK} A_K = E A_J \,. \tag{8.264}$$

Hence the coefficient vector is obtained as an eigenvector of the Hamiltonian matrix represented in the basis of the SPFs. To obtain the coefficients we use the Davidson algorithm described in Sect. 8.4.3.

Now, for the SPFs, using

$$\langle\Psi|H|\Psi\rangle = \langle \sum_j \Psi_j^{(\kappa)} \varphi_j^{(\kappa)} | H | \sum_l \Psi_l^{(\kappa)} \varphi_l^{(\kappa)} \rangle = \sum_{j,l} \langle \varphi_j^{(\kappa)} | \langle H \rangle_{jl}^{(\kappa)} | \varphi_l^{(\kappa)} \rangle \,, \tag{8.265}$$

and varying $\langle \varphi_j^{(\kappa)} |$ yields

$$\sum_{l=1}^{n_\kappa} \langle H \rangle_{jl}^{(\kappa)} \varphi_l^{(\kappa)} = \sum_{l=1}^{n_\kappa} \epsilon_{jl}^{(\kappa)} \varphi_l^{(\kappa)} \,. \tag{8.266}$$

Projecting this equation onto $\varphi_k^{(\kappa)}$ leads to

$$\epsilon_{jk}^{(\kappa)} = \sum_l \langle \varphi_k^{(\kappa)} | \langle H \rangle_{jl}^{(\kappa)} | \varphi_l^{(\kappa)} \rangle \,, \tag{8.267}$$

and from that follows

$$\left(1 - P^{(\kappa)}\right) \sum_{l=1}^{n_\kappa} \langle H \rangle_{jl}^{(\kappa)} \varphi_l^{(\kappa)} = 0 \,. \tag{8.268}$$

As this equation holds for any j, it must hold for any linear combination as well. To arrive at a form similar to the MCTDH equations of motion we insert the inverse of the density operator

$$\dot{\varphi}_j^{(\kappa)} := -\left(1 - P^{(\kappa)}\right) \sum_{k,l} \left(\rho^{(\kappa)^{-1}}\right)_{jk} \langle H \rangle_{kl}^{(\kappa)} \varphi_l^{(\kappa)} = 0 \,, \tag{8.269}$$

with

$$\dot{\varphi} = \frac{\partial \varphi}{\partial \tau} \,, \quad \tau = -i\, t \,. \tag{8.270}$$

We thus obtain the updated SPFs simply by relaxation.

In fact, one can show that, when the SPFs are relaxed, the energy changes according to

$$\dot{E} = -2 \sum_{\kappa=1}^{f} \sum_{l=1}^{n_\kappa} \| \sum_{j=1}^{n_\kappa} \left(\rho^{(\kappa)^{1/2}}\right)_{lj} \dot{\varphi}_j^{(\kappa)} \|^2 \leq 0 \,. \tag{8.271}$$

This implies that that the orbital relaxation will always lessen the energy. As the energy cannot decrease indefinitely it follows $\|\dot{\varphi}\| \to 0$ for $\tau \to \infty$, hence Eq. (8.269) will be satisfied for a sufficiently long relaxation.

Proof of Eq. (8.271)**:** The A-vector is kept constant during the SPF-relaxation, i.e. we use a constant mean-field (CMF) approach as described in Sect. 8.3.8. The time derivative of the energy thus reads

$$\begin{aligned}
\dot{E} &= 2\,\mathrm{Re}\langle\dot{\Psi}|H|\Psi\rangle \\
&= 2\,\mathrm{Re}\langle\sum_{\kappa}\sum_{j}\dot{\varphi}_j^{(\kappa)}\,\Psi_j^{(\kappa)}|H|\Psi\rangle \\
&= 2\,\mathrm{Re}\sum_{\kappa}\sum_{j,l}\langle\dot{\varphi}_j^{(\kappa)}\,\Psi_j^{(\kappa)}|H|\varphi_l^{(\kappa)}\,\Psi_l^{(\kappa)}\rangle \\
&= 2\,\mathrm{Re}\sum_{\kappa}\sum_{j,l}\langle\dot{\varphi}_j^{(\kappa)}|\langle H\rangle_{jl}^{(\kappa)}|\varphi_l^{(\kappa)}\rangle \\
&= 2\,\mathrm{Re}\sum_{\kappa}\sum_{j,l}\langle\dot{\varphi}_j^{(\kappa)}|(1-P^{(\kappa)})\langle H\rangle_{jl}^{(\kappa)}|\varphi_l^{(\kappa)}\rangle,
\end{aligned} \tag{8.272}$$

where the projector $P^{(\kappa)}$ could be inserted, because $\langle\dot{\varphi}_j^{(\kappa)}|\varphi_l^{(\kappa)}\rangle = 0$ is the adopted constrained (see Eq. (8.155)). As

$$-\sum_{k}\rho_{jk}^{(\kappa)}\,\dot{\varphi}_k^{(\kappa)} = \left(1-P^{(\kappa)}\right)\sum_{l}\langle H\rangle_{jl}^{(\kappa)}\,\varphi_l^{(\kappa)}, \tag{8.273}$$

we have

$$\begin{aligned}
\dot{E} &= -2\,\mathrm{Re}\sum_{\kappa}\sum_{j,k}\langle\dot{\varphi}_j^{(\kappa)}|\dot{\varphi}_k^{(\kappa)}\rangle\,\rho_{jk}^{(\kappa)} \\
&= -2\,\mathrm{Re}\sum_{\kappa}\sum_{j,k,l}\langle(\rho^{(\kappa)^{1/2}})_{lj}\dot{\varphi}_j^{(\kappa)}|(\rho^{(\kappa)^{1/2}})_{lk}\dot{\varphi}_k^{(\kappa)}\rangle \\
&= -2\sum_{\kappa}\sum_{l}\|\sum_{j}\rho_{lj}^{(\kappa)^{1/2}}\,\dot{\varphi}_j^{(\kappa)}\|^2 .
\end{aligned} \tag{8.274}$$

□

Improved relaxation proceeds as follows: first, an initial state has to be defined. This state should have a reasonable overlap with the target state. Then the matrix representation of the Hamiltonian, H_{JK}, is built and diagonalized by a Davidson routine: see Sect. 8.4.3. Actually H_{JK} is never explicitly built as a full matrix but applied term by term to the A-vector (see Sect. 8.7.1). Then the mean fields are built and the SPFs are relaxed until their time derivatives are sufficiently small. After that, H_{JK} is built in the space of the new SPFs and the whole process is iterated until convergence: this is an MCSCF iterative procedure.

If the ground state is computed, the selection of the eigenvector of the Hamiltonian is simple: one takes the eigenvector of lowest energy. When excited states are to be computed, that eigenvector is taken that corresponds to the wavefunction that has the largest overlap with the initial state.

An MCTDH propagation always provides a result, whatever the number of SPFs. If there are too few configurations, the propagation will be less accurate but usually still describes the overall features correctly. This contrasts with *improved relaxation*, where the diagonalisation/relaxation iteration loop fails to converge when the configuration space is too small. There is never a problem in computing the ground state, but converging to excited states becomes more difficult the higher the excitation energy or, more precisely, the higher the density of states.

The *improved relaxation* algorithm may be used in a block form [78], i.e. one may start with a block of initial vectors, which then converges collectively to a set of eigenstates. Formally the different wavefunctions are treated as electronic states of one "super wavefunction". As the single-set algorithm is used, there is one set of SPFs for all wavefunctions. The mean fields are hence state-averaged mean fields and the Davidson routine is replaced by a block-Davidson one. The block form of *improved relaxation* is more efficient than the single-vector one, when several eigenstates are to be computed. However, the block form requires considerably more memory.

Improved relaxation has been applied quite successfully to a number of problems. For 4-atoms systems (6D) it is in general possible to compute all eigenstates of interest (the lowest 150, say). For a system as large as $H_5O_2^+$ (15D) it was, of course, only possible to converge the first few low-lying states [79, 80].

*8.4.3 The Davidson Algorithm **

The diagonalization of the Hamiltonian matrix to obtain the A-coefficients of Eq. (8.264) is never performed in the full SPF basis set, which is often far too large. It is performed by targeting one or a group of eigenstates and with an efficient algorithm that avoids large matrices, such as the Davidson algorithm. The latter is an iterative diagonalization method. Starting from an initial subspace, the algorithm looks for residue vectors that contain, at least partly, the parts that are missing in the eigenstates we are looking for.

The Davidson algorithm will not be described in full detail here, and we refer to Ref. [81] for the description of the algorithm. Let us first separate the exact eigenstates and the exact eigenenergies into an approximate and an error term:

$$\begin{aligned} \Psi_i &= \varphi_i + \delta_i \,, \\ E_i &= \rho_i + \epsilon_i \,. \end{aligned} \tag{8.275}$$

Ψ_i and E_i denote the target eigenstate and the eigenvalue, φ_i and ρ_i are the corresponding approximate values that are used as initial guesses. $\rho_i = \langle \varphi_i | H | \varphi_i \rangle$. Since φ_i is not an exact eigenstate, we have:

$$H\varphi_i = \rho_i \varphi_i + r_i \,, \tag{8.276}$$

where r_i is called the residual vector. It is given by

$$r_i = (H - \rho_i)\varphi_i \,. \tag{8.277}$$

Since

$$H\Psi_i = E_i \Psi_i \,, \tag{8.278}$$

we have

$$H(\varphi_i + \delta_i) = (\rho_i + \epsilon_i)(\varphi_i + \delta_i) \,, \tag{8.279}$$

and thus

$$\begin{aligned} (H - \rho_i - \epsilon_i)\delta_i &= -(H - \rho_i - \epsilon_i)\varphi_i \\ &= -r_i + \epsilon \varphi_i \,. \end{aligned} \tag{8.280}$$

Finding δ_i, i.e. what is missing in the guess to obtain the eigenstate, and appending it to the subspace is the essence of the Davidson method. Equation (8.280) is formally solved as

$$\delta_i = (H - \rho_i - \epsilon_i)^{-1}(-r_i + \epsilon_i \varphi_i) \,. \tag{8.281}$$

Davidson proposed first to neglect ϵ_i and second to replace H by an approximate operator H^0, called *preconditioner*. Thus we have

$$\delta_i \approx -(H^0 - \rho_i)^{-1}\, r_i \,. \tag{8.282}$$

In general, H^0 is the diagonal part of the Hamiltonian matrix. Better approximate H^0 can be used for instance by taking a block of H with lowest diagonal energies and keeping for the rest the diagonal approximation. Furthermore, the correction vector, δ can be improved with the so-called Olsen correction [82]. Both options are implemented in the Heidelberg MCTDH package.

In concrete terms, the iterative process is as follows. We want to obtain an eigenvector for which we have a first guess Ψ_i^0 that should have a decent overlap with the eigenstate one wants to compute. We then calculate the residue vector $r_i^0 = (H - \rho_i^0)\varphi_i^0$. We calculate δ_0, orthogonalize it to Ψ_i^0 and add the new vector to the one-dimensional subspace. We diagonalize the 2×2 Hamiltonian matrix.

We take the new eigenstate, Ψ_i^1, with the largest overlap with Ψ_i^0 and calculate the residue vector $r_i^1 = (H - \rho_i^1)\varphi_i^1$. We calculate δ_1, orthogonalize it to Ψ_i^1 and add the new vector to the two-dimensional subspace. We then diagonalize the 3×3 Hamiltonian matrix. The process is repeated iteratively. We stop if $||r_i^n||$ is smaller than a tolerance criterion. If this is not the case, we calculate δ_n, orthogonalize it to the previous subspace and append the orthogonalized correction vector to the previous vector and repeat the iteration from the construction of the Hamiltonian matrix until convergence is achieved.

To detect which new eigenstate, Ψ_i^n, is the one we are looking for, there are several possibilities. For instance, the procedure can take the eigenvector that is the closest to the initial wavefunction Ψ_i^0. Another possibility is to take Ψ_i^n as the closest to Ψ_i^{n-1}.

*8.4.4 Conclusion **

In the context of improved relaxation, the fact that the Davidson algorithm has achieved convergence does not mean that we have obtained the eigenstate. It means that we have successfully solved Eq. (8.264) for a given set of SPFs. To achieve full convergence, we need to solve Eqs. (8.264) and (8.269) self-consistently. This is the reason why we then relax, i.e. propagate in imaginary time, the SPFs as explained in Sect. 8.4.2. Relaxation in imaginary time improves the quality of the basis functions by eliminating the components of the space that are higher in energy. Step by step, the procedure converges to the optimal MCSCF basis functions with respect to the eigenstate, Ψ_i, we are looking for. At the end, it becomes possible to obtain the eigenstate with a high accuracy although the size of the matrix that is diagonalized is small compared to the size of the underlying primitive basis. There exist a complex, non-Hermitian form of improved relaxation, which can be used to converge quantum resonances with complex energies (energy position and width). For this, one has to add a complex absorbing potential (CAP) [83] to the Hamiltonian operator following the approach proposed by Jolicard and Austin [84–86]. The CAP parameters must be chosen carefully since energy position and width of the resonance should not depend on them.

8.5 Complex Absorbing Potentials (CAPs)

When dealing with a bound system, there is no problem with grids. Turning to study dissociation or scattering processes one notices that some of the grids may become very long. The minimal propagation time is determined by the time needed for the slow components of the WF to leave the interaction region. Within this time interval the fast components of the WP may have travelled a long distance requiring a long grid.

A solution to this problem is provided by complex absorbing potentials (CAP) [83–92]. A CAP is a negative imaginary potential, usually written as

$$-i\eta W(r) = -i\eta\big(r - r_c\big)^n \,\theta\big(r - r_c\big)\,,$$

where $W(r)$ is a non-negative real function, often of the indicated monomial form, n is 2, 3, or 4, η is a strength parameter, r_c denotes the position where the CAP is switched on, and θ denotes a step function, which ensures that the CAP vanishes for $r \leq r_c$.

Let us investigate how a CAP changes the norm

$$\begin{aligned}\frac{d}{dt}\|\Psi\|^2 &= \frac{d}{dt}\langle\Psi|\Psi\rangle = \langle\dot{\Psi}|\Psi\rangle + \langle\Psi|\dot{\Psi}\rangle \\ &= \langle -iH\Psi|\Psi\rangle + \langle\Psi| -iH\Psi\rangle \\ &= i\langle\Psi|H^\dagger - H|\Psi\rangle\,,\end{aligned} \tag{8.283}$$

with

$$\begin{aligned}H &= H_0 - i\eta W \qquad & H_0 &= H_0^\dagger \\ H^\dagger &= H_0 + i\eta W \qquad & W &= W^\dagger\,,\end{aligned} \tag{8.284}$$

follows

$$\frac{d}{dt}\|\Psi\|^2 = -2\eta\,\langle\Psi|W|\Psi\rangle\,, \tag{8.285}$$

or

$$\frac{d}{dt}\|\Psi\| = -\eta\,\frac{\langle\Psi|W|\Psi\rangle}{\|\Psi\|}\,. \tag{8.286}$$

Hence the norm of the WF decreases when the wavepacket enters the CAP region. We want to inspect in more detail how the CAP annihilates the wavepacket. Using the split operator ideas (see Sect. 8.1.6), we write the formal solution of the Schrödinger equation as

$$\begin{aligned}\Psi(t+\tau) &= e^{(-iH_0-\eta W)\tau}\,\Psi(t) \\ &= e^{-iH_0\frac{\tau}{2}}\,e^{-\eta W\tau}\,e^{-iH_0\frac{\tau}{2}}\,\Psi(t) + O(\tau^3)\,,\end{aligned} \tag{8.287}$$

i.e. in the middle of each time step, the WF is multiplied by $e^{-\eta W\tau}$, a half Gaussian when $W \sim r^2$ (Fig. 8.8).

When is it legitimate to use a CAP? Of course, it is legitimate to annihilate the outgoing parts when they do not enter in the computation of desired observables. For instance, when computing the autocorrelation function

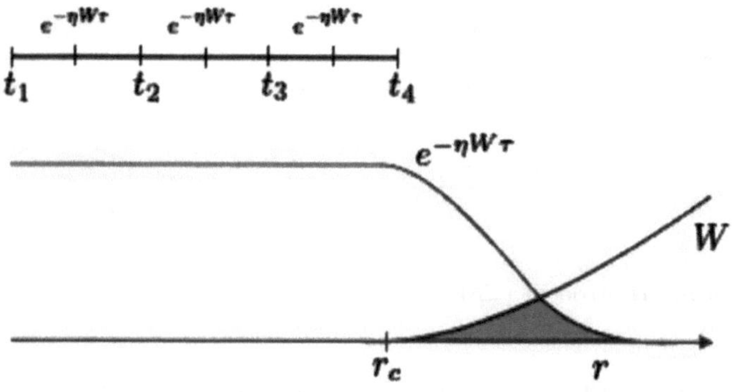

Fig. 8.8 Decrease of the norm of a wavepacket being annihilated by a CAP starting at r_c

Fig. 8.9 Example of the correct location of a CAP

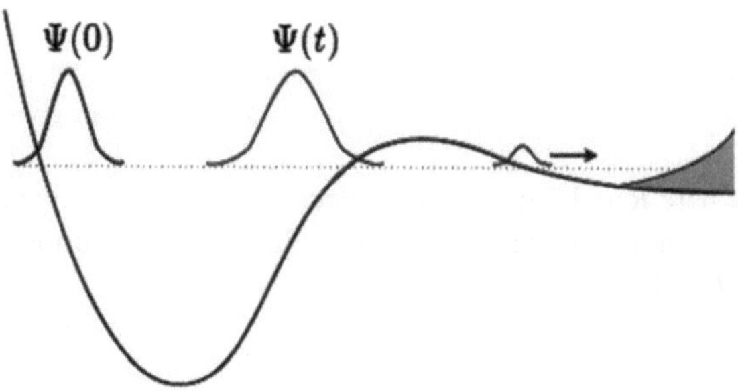

$$a(t) = \langle \Psi(0) | \Psi(t) \rangle , \tag{8.288}$$

then it is clear that those parts of $\Psi(t)$ which do not overlap with $\Psi(0)$ and will never return to overlap with $\Psi(0)$ may be annihilated (Fig. 8.9).

What happens, if we do not introduce a CAP but still work with a small grid? At the end of the grid one automatically introduces a wall, i.e. a grid or a finite basis set has a similar effect as putting the system into a box. Due to the wall, the outgoing part of the WP will be reflected and will overlap again with $\Psi(0)$. This destroys the correctness of the autocorrelation function. However, the CAP does not only annihilate a WF, but also induces some artificial reflections.

The origin of the reflection is easy to understand. It is related to the Heisenberg uncertainty principle. We change the form of the WF, i.e. its coordinate distribution. But this implies that one also changes the momentum distribution which is just the Fourier transform of the coordinate representation and this means reflection. To see this, let us turn to the time-independent picture (Fig. 8.10).

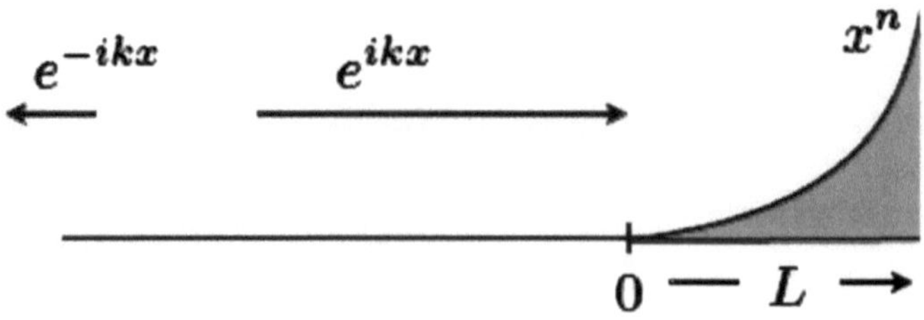

Fig. 8.10 Undesired behavior of a CAP

At energy E the WF must be a linear combination of e^{ikx} and e^{-ikx} where $E = k^2/2m$. Hence

$$\Psi(x) \sim e^{ikx} - Re^{-ikx} \qquad \text{for} \quad x < 0\,, \tag{8.289}$$

and

$$\Psi(x) \sim Te^{ikx} \qquad \text{for} \quad x > L\,, \tag{8.290}$$

where R and T denote the reflection and transmission coefficients, respectively. If we put an infinite wall at $x = 0$, we have total reflection ($R = 1$) and zero transmission ($T = 0$):

$$\Psi(x) \sim e^{ikx} - e^{-ikx} \sim \sin\, kx \qquad \Psi(0) \equiv 0\,. \tag{8.291}$$

The wavefunction, which when undisturbed is proportional to e^{ikx}, has thus completely lost its form by reflection from the end of the grid.

Using scattering theory and semiclassical arguments, approximate but very reliable formulas for the transmission and reflection coefficients of a CAP have been derived in Ref. [90]. (These formulas are evaluated when running the MCTDH script *plcap*.) For the following discussion, we give a simplified form of the lengthy equations of Ref. [90]:

$$\begin{aligned}
|R|^2 &= \Big|\frac{n!}{2^{n+2}}\Big|^2 \cdot \frac{\eta^2}{E^2 \cdot k^{2n}} = \Big|\frac{n!}{2^{n+2}}\Big|^2 \cdot \Big(\frac{1}{2m}\Big)^n \cdot \frac{\eta^2}{E^{n+2}} \\
|T|^2 &= \exp\Big(-\frac{\eta L^{n+1} \cdot 2m}{k(n+1)}\Big) = \exp\Big(-\frac{\eta W(L)}{E} \cdot \frac{k \cdot L}{n+1}\Big)\,.
\end{aligned} \tag{8.292}$$

Of course one wants to achieve $|T|^2 + |R|^2 \ll 1$. To keep the reflection small, one needs a weak CAP (small η), in particular when the energy is low. To achieve a small transmission, for a given CAP-strength η, one needs a long CAP (L large).

Note that $k \cdot L = 2\pi$ is equivalent to saying that L equals one de-Broglie wavelength. A CAP should be at least two de-Broglie wavelengths long. Hence, when a particle enters the CAP with very low energy, e.g. at a threshold for opening a new channel, then a CAP is likely to produce unwanted reflections.

8.6 Flux Analysis

In general there is not much interest in the propagated wavefunction $\Psi(t)$ as such, the quantities of interest are observables like spectra or cross-sections. Spectra are conveniently generated by a Fourier transform of the autocorrelation function, see Sect. 9, and cross-sections can be computed by a flux-analysis of the time evolved wavefunction.

Consider a reactive scattering event, e.g. A + BC → AB + C. There are hence two reaction channels, the educt channel A + BC and the product channel C + AB. We assume that the scattering energy is small enough such that a three-body breakup is not possible, AB must be bound when the distance between AB and C becomes large. For simplicity we also ignore the reaction channel AC + B.

We want to know the probability with which an initial wavepacket, which starts in the educt channel, ends up in the product channel. This reaction probability is defined as

$$P_{\text{react}} = \lim_{t\to\infty} \int_{R_c}^{\infty} dR \int d\boldsymbol{q}\, |\Psi(R, \boldsymbol{q}, t)|^2 , \tag{8.293}$$

where R denotes the dissociative coordinate and $\boldsymbol{q}$ all remaining coordinates, the latter describing bound motion. The separation, reaction coordinate versus bound coordinates, depends on the reaction channel considered.[6] The interpretation of Eq. (8.293) is clear. The reaction probability is given by that part of the probability density $|\Psi|^2$ that for large times is separated from the AB fragment by a distance larger than R_c. The value of R_c must be large enough to ensure that a particle, which has passed the coordinate space dividing surface defined by $R = R_c$, will never again assume distances smaller than R_c. The time limit can be replaced by a time integral over a time derivative[7] and the lower bound of the R-integration by a step function θ

$$\begin{aligned}
P_{\text{react}} &= \int_0^{\infty} dt \int\!\!\int \theta(R - R_c)\, \frac{\partial}{\partial t} \big(\Psi^*(R, \boldsymbol{q}, t)\Psi(R, \boldsymbol{q}, t)\big)\, dR\, d\boldsymbol{q} \\
&= i \int_0^{\infty} dt \int\!\!\int (H\Psi^*)\theta\Psi - \Psi^*\theta H\Psi\, dR\, d\boldsymbol{q} \\
&= i \int_0^{\infty} dt\, \big(\langle H\Psi|\theta|\Psi\rangle - \langle\Psi|\theta H|\Psi\rangle\big) \\
&= i \int_0^{\infty} \langle\Psi|[H, \theta]|\Psi\rangle\, dt \\
&= \int_0^{\infty} \langle\Psi(t)|F|\Psi(t)\rangle\, dt
\end{aligned} \tag{8.294}$$

[6]Actually, we are considering the wavefunction of a particular total angular momentum J (that makes the dissociate coordinate R one-dimensional). Hence Ψ should be replaced with Ψ^J, but for the sake of simplicity we suppress the total angular momentum label J.

[7]The initial state is assumed to have no density beyond R_c, i.e. $\theta(R - R_c)|\Psi(R, \boldsymbol{q}, t = 0)|^2 \equiv 0$.

where we have introduced the flux operator

$$F = i[H, \theta] . \tag{8.295}$$

The reaction probability is hence defined by summing the quantum flux passing through the plane defined by the step function, i.e. by $R = R_c$. As the step function commutes with the potential, the Hamiltonian H in Eq. (8.295) can be replaced with the KEO. When the KEO assumes for the dissociative coordinate the simple form $-1/2\mu \cdot \partial^2/\partial R^2$, the flux operator becomes

$$F = \frac{-i}{2\mu}\left(\frac{\partial}{\partial R}\,\delta(R - R_c) + \delta(R - R_c)\,\frac{\partial}{\partial R}\right) . \tag{8.296}$$

The flux through a surface is independent of the detailed shape of the surface. This is an important point because it allows us to choose the dividing surface (i.e. the surface where θ jumps from zero to one) quite arbitrarily as long as it separates the educt from the product channel. All following equations involving the flux operator remain valid for any such dividing surface.

It is now convenient to switch to a time-independent picture for a short while. For each reaction channel γ there exist a separate set of coordinates $(R_\gamma, \boldsymbol{q}_\gamma)$, where R_γ is the distance between the center of mass of the molecular fragment and the leaving atom (or, for larger systems, the distance between the centers of mass of the two molecular fragments of channel γ). The exact scattering wavefunction with outgoing scattering boundary conditions [93] reads

$$\Psi^{+}_{E\alpha\nu}(R_\gamma, \boldsymbol{q}_\gamma) \overset{R_\gamma \to \infty}{\longrightarrow} \Phi^{-}_{E\alpha\nu}(R_\alpha, \boldsymbol{q}_\alpha)\,\delta_{\alpha\gamma} - \sum_{\nu'} S_{\gamma\nu',\alpha\nu}(E)\,\Phi^{+}_{E\gamma\nu'}(R_\gamma, \boldsymbol{q}_\gamma) \tag{8.297}$$

$$\Phi^{\pm}_{E\gamma\nu}(R_\gamma, \boldsymbol{q}_\gamma) = \chi^{\pm}_{E\gamma\nu}(R_\gamma)\,\xi_{\gamma\nu}(\boldsymbol{q}_\gamma) \tag{8.298}$$

$$\chi^{\pm}_{E\gamma\nu}(R_\gamma) = \sqrt{\frac{\mu_{\gamma R}}{2\pi k_{\gamma\nu}}}\; e^{\pm i k_{\gamma\nu} R_\gamma} , \tag{8.299}$$

where $\xi_{\gamma\nu}$ is a ro-vibrational eigenstate of the fragment molecule of channel γ, and ν denotes collectively all quantum numbers of this state. The channel momentum $k_{\gamma\nu}$, which appears in the definition of the plane waves $\chi^{\pm}_{E\gamma\nu}(R_\gamma)$, is given by $k_{\gamma\nu} = \sqrt{2\mu_{\gamma R}(E - E_{\gamma\nu})}$ with $E_{\gamma\nu}$ being the energy eigenvalue of the fragment wavefunction $\xi_{\gamma\nu}$. $S_{\gamma\nu',\alpha\nu}$ denotes the S-matrix [93] describing a transition from the initial reaction channel α and initial quantum numbers ν to a final reaction channel γ and final quantum numbers ν'. The upper indices $+$ and $-$ refer to outgoing and ingoing scattering boundary conditions, respectively. The fragment states ξ are L^2 normalized eigenfunctions of the internal fragment Hamiltonian, and the scattering states are normalized to δ-functions.

$$
\begin{aligned}
\langle \chi^{\pm}_{E\gamma\nu} | \chi^{\pm}_{E'\gamma\nu} \rangle &= \delta(E - E') \\
\langle \Phi^{\pm}_{E\gamma\nu} | \Phi^{\pm}_{E'\gamma\nu'} \rangle &= \delta(E - E')\, \delta_{\nu\nu'} \\
\langle \Psi^{+}_{E\alpha\nu} | \Phi^{+}_{E'\alpha\nu'} \rangle &= \delta(E - E')\, \delta_{\nu\nu'}
\end{aligned} \tag{8.300}
$$

It is now easy to compute the expectation value of the flux operator with respect to the exact scattering state. For $\gamma \neq \alpha$ follows

$$
\langle \Psi^{+}_{E\alpha\nu} \mid F_\gamma \mid \Psi^{+}_{E\alpha\nu} \rangle = \frac{1}{2\pi} \sum_{\nu'} \left| S_{\gamma\nu',\alpha\nu}(E) \right|^2 , \tag{8.301}
$$

where we have used, Cf. Eq. (8.296), that

$$
\langle \chi^{\pm}_{E\gamma\nu} | F_\gamma | \chi^{\pm}_{E\gamma\nu} \rangle = \pm \frac{1}{2\pi} \tag{8.302}
$$

holds. The sum over the internal states may be removed by employing projectors onto these states,

$$
P_{\gamma\nu} = \mid \xi_{\gamma\nu} \rangle \langle \xi_{\gamma\nu} \mid , \tag{8.303}
$$

yielding[8]

$$
\langle \Psi^{+}_{E\alpha\nu} \mid P_{\gamma\nu'} F_\gamma P_{\gamma\nu'} \mid \Psi^{+}_{E\alpha\nu} \rangle = \frac{1}{2\pi} \mid S_{\gamma\nu',\alpha\nu}(E) \mid^2 . \tag{8.304}
$$

The integral cross-section can be computed from the $|S|^2$ elements. For this we have to remember that, for the sake of simplicity, we have suppressed the total angular momentum index J from wavefunctions and the S-matrix.[9] All equations above are diagonal in J. Re-introducing J we can write the integral cross-section as

$$
\sigma_{\gamma\nu' \leftarrow \alpha\nu} = \frac{\pi}{k^2_{\alpha\nu}} \sum_{J=0}^{\infty} (2J + 1) |S^J_{\gamma\nu',\alpha\nu}(E)|^2 , \tag{8.305}
$$

where $k_{\alpha\nu}$ denotes the initial momentum with which the two fragments collide. Summing over all final quantum numbers ν' one obtains the initial state selected total cross section

$$
\begin{aligned}
\sigma_{\gamma \leftarrow \alpha\nu} &= \sum_{\nu'} \sigma_{\gamma\nu' \leftarrow \alpha\nu} \\
&= \frac{2\pi^2}{k^2_{\gamma\nu}} \sum_{J=0}^{\infty} (2J + 1) \langle \Psi^{J+}_{E\alpha\nu} \mid F_\gamma \mid \Psi^{J+}_{E\alpha\nu} \rangle .
\end{aligned} \tag{8.306}
$$

[8]Actually $P_{\gamma\nu}$ commutes with F_γ. Thus $P_{\gamma\nu}\, F_\gamma = F_\gamma\, P_{\gamma\nu} = P_{\gamma\nu}\, F_\gamma\, P_{\gamma\nu}$. The extra projector is added for symmetry reasons only.

[9]We have also suppressed some J-dependent phase factors in Eqs. (8.297, 8.299). They are irrelevant, because here we consider only the modulus of the S-matrix elements.

Note that the flux formalism automatically sums over all final states if there is no projection.

Turning to a time-dependent approach one starts with an initial wave packet Ψ_0, which is a superposition of infinitely many exact scattering states. Formally Ψ_0 is written as

$$\Psi_0 = \sum_\nu \int c_\nu(E)\, \Psi^+_{E\alpha\nu}\, dE \tag{8.307}$$

This wavepacket may then be propagated. But in the end one is not interested in the time evolved wavepacket $\Psi(t)$, not even in its reaction probability Eq. (8.293), in general one is interested to compute *energy resolved* quantities. For this we define the flux function

$$F_\gamma(E, \Psi_0) = 2\pi\langle\Psi_0|\delta(H-E)\, F_\gamma\, \delta(H-E)|\Psi_0\rangle\ , \tag{8.308}$$

or, state resolved

$$F_\gamma(E, \nu, \Psi_0) = 2\pi\langle\Psi_0|\delta(H-E)\, P_{\gamma\nu} F_\gamma P_{\gamma\nu}\, \delta(H-E)|\Psi_0\rangle\ . \tag{8.309}$$

Before we continue with the flux analysis, let us study the action of the δ-function on the initial wavepacket

$$\delta(H-E)\,\Psi_0 = \sum_\nu \int c_\nu(E)\, \delta(H-E)\, \Psi^+_{E'\alpha\nu}\, dE' = \sum_\nu c_\nu(E)\, \Psi^+_{E\alpha\nu}\ , \tag{8.310}$$

and introduce the energy density function

$$|\Delta(E)|^2 := \langle\Psi_0|\, \delta(H-E)\, |\Psi_0\rangle = \sum_\nu |c_\nu(E)|^2\ , \tag{8.311}$$

where the last equation follows from Eqs. (8.300, 8.310).

Next we want to discuss the flux analysis of a direct dissociation process, e.g photodissociation of NOCl as discussed in Sect. 10.1. Then there is only one channel and we suppress the channel index γ in the following discussion. Using Eqs. (8.310, 8.302), the flux-function takes the form

$$\begin{aligned} F(E, \Psi_0) &= 2\pi \sum_\nu \sum_{\nu'} c^*_\nu(E)\, c_{\nu'}(E)\, \langle\Psi^+_{E\alpha\nu}|\, F\, |\Psi^+_{E\alpha\nu'}\rangle \\ &= 2\pi \sum_\nu |c_\nu(E)|^2\, \langle\chi^+_{E\nu}|F|\chi^+_{E\nu}\rangle \\ &= \sum_\nu |c_\nu(E)|^2\ , \end{aligned} \tag{8.312}$$

or, state resolved

$$F(E, \nu, \Psi_0) = |c_\nu(E)|^2 . \tag{8.313}$$

Hence, in this case the unprojected flux reproduces the energy density function, i.e. the power spectrum. The latter is more easily computed by a Fourier transform of the autocorrelation function. However, the projected flux provides more information, namely the dissociation probability for leaving the fragment molecule (here NO) in a particular ro-vibrational state ν.

Turning back to reactive scattering, it is now easy to show that

$$F_\gamma(E, \Psi_0) = \sum_{\nu,\nu'} |c_\nu(E)|^2 \, |S_{\gamma\nu',\alpha\nu}(E)|^2 , \tag{8.314}$$

and

$$F_\gamma(E, \nu', \Psi_0) = \sum_{\nu} |c_\nu(E)|^2 \, |S_{\gamma\nu',\alpha\nu}(E)|^2 . \tag{8.315}$$

When the initial state, as usual for reactive scattering calculations, is pure in the internal quantum state ν, the sum over ν drops out. In this case the modulus of the S-matrix element is given by

$$|S_{\gamma\nu',\alpha\nu}(E)|^2 = F_\gamma(E, \nu', \Psi_{0\nu})/|\Delta(E)|^2 \tag{8.316}$$

where we have replaced Ψ_0 with $\Psi_{0\nu}$ to indicate that this initial state is an eigenstate with quantum number ν with respect to the internal degrees of freedom. Compare with Sect. 11 for more details.

What is left to be done is to define a stable and efficient method to evaluate the flux-function. After the wavepacket has passed the surface defined by $R = R_c$ it is no longer needed and can be absorbed by a CAP that starts at $R = R_c$. The R-grid then ends at $R_c + L$, where L is the length of the CAP. We know from Eq. (8.285) that the speed of annihilation is proportional to the matrix element of the CAP, hence it should be possible to express the flux by such a matrix element. Indeed, let us introduce the CAP augmented Hamiltonian

$$\tilde{H} = H - iW , \tag{8.317}$$

where H is Hermitian and W a real, non-negative, continuous function[10] of R, which vanishes for $R \leq R_c$. Hence W commutes with the flux defining step function, i.e. $\theta W = W\theta = W$. The flux operator may now be written as (Cf. Eq. (8.295))

$$F = iH\theta - i\theta H = 2W + i\tilde{H}^\dagger\theta - i\theta\tilde{H} . \tag{8.318}$$

[10]Here the strength parameter η is included in the definition of W, in contrast to Sect. 8.5.

Turning to the reaction probability P_{react} of Eqs. (8.293, 8.294) one obtains

$$\begin{aligned} P_{\text{react}} &= \int_0^\infty \langle\Psi(t)|F|\Psi(t)\rangle \, dt \\ &= \int_0^\infty \langle\Psi(t)|2W|\Psi(t)\rangle \, dt \\ &+ \int_0^\infty \frac{d}{dt}\langle\Psi(t)|\theta|\Psi(t)\rangle \, dt \,, \end{aligned} \tag{8.319}$$

where the last term can be understood by noting that $\partial_t\langle\Psi(t)| = i\langle\Psi(t)|\tilde{H}^\dagger$, and $\partial_t|\Psi(t)\rangle = -i\tilde{H}|\Psi(t)\rangle$. This last line vanishes, because $\langle\Psi(t)|\theta|\Psi(t)\rangle = 0$ for both $t = 0$ and $t = \infty$. Initially Ψ has no contributions for $R \geq R_c$ and finally that part of Ψ, which has entered the reaction channel, has vanished because of the CAP. However, in practice one does not propagate to infinite times but stops the propagation at some final time T. In this case it is useful to keep the second term, which is trivial to integrate. The working equation thus reads

$$P_{\text{react}} = 2\int_0^T \langle\Psi(t)|W|\Psi(t)\rangle \, dt \; + \; \langle\Psi(T)|\theta|\Psi(T)\rangle \,. \tag{8.320}$$

The first part accounts for the density which is annihilated by the CAP and the second part stands for the density which still exists on the interval $[R_c, R_c + L]$. If one runs `flux84` of the MCTDH package, then these two terms and their sum are written to the file `iwtt` at every output time step.

We want to use the technique developed here to derive working equations for the flux function $F_\gamma(E, \Psi_0)$. We re-introduce the channel index γ, there will be a CAP, $-iW_\gamma$, in each reaction channel. The δ-functions are expressed in Fourier representation

$$\delta(H - E) = \frac{1}{2\pi}\int_{-\infty}^\infty e^{-i(H-E)t} \, dt \tag{8.321}$$

Inserting this equation into Eq. (8.308), replacing H with $\tilde{H}$, and making use of Eq. (8.318) one obtains

$$\begin{aligned} F_\gamma(E, \Psi_0) &= \frac{1}{2\pi}\int_{-\infty}^\infty dt \int_{-\infty}^\infty dt' \langle\Psi_0|e^{-i(\tilde{H}^\dagger - E)t} F_\gamma e^{-i(\tilde{H}-E)t'}|\Psi_0\rangle \\ &= \frac{1}{2\pi}\int_0^\infty dt \int_0^\infty dt' \langle\Psi(t)|2W_\gamma|\Psi(t)\rangle e^{-iE(t-t')} \\ &+ (\frac{d}{dt} + \frac{d}{dt'})\langle\Psi(t)|\theta_\gamma|\Psi(t')\rangle e^{-iE(t-t')} \,, \end{aligned} \tag{8.322}$$

where $\Psi(t) = \exp(-i\tilde{H}t)\Psi_0$ is used and where again the lower limits of the integrals could be safely replaced with zero, because for negative times the wavepacket cannot reach the CAP region of the γ's reaction channel. We now replace the upper integral limits with the final propagation time T and first evaluate the second term

$$\begin{aligned} F_{\gamma,\theta}(E,\Psi_0) &= \frac{1}{2\pi}\int_0^T dt\,\langle\Psi(t)|\theta_\gamma|\Psi(T)\rangle e^{iE(T-t)} \\ &+\frac{1}{2\pi}\int_0^T dt'\,\langle\Psi(T)|\theta_\gamma|\Psi(t')\rangle e^{-iE(T-t')} \\ &= \frac{1}{\pi}\mathrm{Re}\int_0^T dt\,\langle\Psi(t)|\theta_\gamma|\Psi(T)\rangle e^{iE(T-t)} \\ &= \frac{1}{\pi}\mathrm{Re}\int_0^T d\tau\,\langle\Psi(T-\tau)|\theta_\gamma|\Psi(T)\rangle e^{iE\tau}\,, \end{aligned} \tag{8.323}$$

where we used the substitution $\tau = T - t$.

To evaluate the first term of Eq. (8.322) we substitute $t' = t + \tau$

$$\begin{aligned} F_{\gamma,W}(E,\Psi_0) &= \frac{1}{\pi}\int_0^T d\tau\int_0^{T-\tau} dt\,\langle\Psi(t)|W_\gamma|\Psi(t+\tau)\rangle e^{iE\tau} \\ &+\frac{1}{\pi}\int_{-T}^0 d\tau\int_{-\tau}^T dt\,\langle\Psi(t)|W_\gamma|\Psi(t+\tau)\rangle e^{iE\tau}\,. \end{aligned} \tag{8.324}$$

With the substitutions $\tau \to -\tau$ and then $t \to t + \tau$ the second integral turns into

$$\frac{1}{\pi}\int_0^T d\tau\int_0^{T-\tau} dt\,\langle\Psi(t+\tau)|W_\gamma|\Psi(t)\rangle e^{-iE\tau} \tag{8.325}$$

which is just the complex conjugate of the first integral of Eq. (8.324). Hence

$$F_{\gamma,W}(E,\Psi_0) = \frac{2}{\pi}\mathrm{Re}\int_0^T d\tau\int_0^{T-\tau} dt\,\langle\Psi(t)|W_\gamma|\Psi(t+\tau)\rangle\, e^{iE\tau}\,. \tag{8.326}$$

In practice we define the auxiliary function $g(\tau)$

$$g(\tau) = \int_0^{T-\tau}\langle\Psi(t)|W_\gamma|\Psi(t+\tau)\rangle\,dt\; + \frac{1}{2}\langle\Psi(T-\tau)|\theta_\gamma|\Psi(T)\rangle\,. \tag{8.327}$$

The flux, $F_\gamma = F_{\gamma,W} + F_{\gamma,\theta}$, is then given by the Fourier integral

$$F_\gamma(E,\Psi_0) = \frac{2}{\pi}\mathrm{Re}\int_0^T g(\tau)\, e^{iE\tau} d\tau\,. \tag{8.328}$$

Of course one can make use of projectors here as well. For this one merely replaces W_γ and θ_γ in Eq. (8.327) with $P_{\gamma\nu}W_\gamma P_{\gamma\nu}$ and $P_{\gamma\nu}\theta_\gamma P_{\gamma\nu}$, respectively.

Finally we note that if no projectors are used, the dissociative coordinate may be any coordinate which separates the fragments, it does not have to be Jacobian-like. If projectors are used, one needs a Jacobian-like dissociation coordinate, because otherwise an internal channel Hamiltonian H_γ cannot be defined.

If there is no reaction but inelastic scattering, e.g. ro-vibrational excitation in non-reactive $H_2 + D_2$ scattering [94–98], then there exist an alternative to the flux formalism, the Tannor-Weeks method [99–101]. Note that within this approach there is no summation over non-projected states, only state-to-state probabilities can be computed.

8.7 Representation of the Hamiltonian Operator

8.7.1 The Product Form

We have already mentioned the quadrature problem. At each time step we have to compute the matrix representation of the Hamiltonian

$$H_{JK} = \langle \Phi_J | H | \Phi_K \rangle , \tag{8.329}$$

and the mean fields,

$$\langle H \rangle_{jl}^{(\kappa)} = \langle \Psi_j^{(\kappa)} | H | \Psi_l^{(\kappa)} \rangle . \tag{8.330}$$

If one were to calculate these integrals by straightforward quadrature over the primitive grid, one would have to run over N^f grid points for potential-like operators and N^{2f} points for non-diagonal operators. For example

$$V_{JK} = \langle \Phi_J | V | \Phi_K \rangle = \sum_{i_1=1}^{N_1} \cdots \sum_{i_f=1}^{N_f} \varphi_{j_1}^{(1)*}(q_{i_1}^{(1)}) \cdots \varphi_{j_f}^{(f)*}(q_{i_f}^{(f)}) V(q_{i_1}^{(1)}, \ldots, q_{i_f}^{(f)}) \varphi_{j_1}^{(1)}(q_{i_1}^{(1)}) \cdots \varphi_{j_f}^{(f)}(q_{i_f}^{(f)}) , \tag{8.331}$$

and this integral has to be calculated for each J and K, hence n^{2f} times.

Example:
Let us take $f = 6, n = 6$ and $N = 32$. We obtain for one integral: $N^f = 32^6 = 10^9$ operations and a number of integrals equal to $n^{2f} = 6^{12} = 2 \times 10^9$, hence $\approx 10^{18}$ operations in total. This is impossible!

The trick is to write the Hamiltonian as a sum of products

$$H = \sum_{r=1}^{s} c_r \, h_r^{(1)} \cdots h_r^{(f)} , \tag{8.332}$$

where $h_r^{(\kappa)}$ operates on the κ-th DOF.

If we now calculate the integral we find:

$$H_{JK} = \sum_{r=1}^{s} c_r \langle \varphi_{j_1}^{(1)} | h_r^{(1)} | \varphi_{j_1}^{(1)} \rangle \cdots \langle \varphi_{j_f}^{(f)} | h_r^{(f)} | \varphi_{j_f}^{(f)} \rangle , \tag{8.333}$$

i.e. a sum of products of *one-dimensional* integrals.[11] Calculating all the H_{JK} integrals we can re-use the $\langle h_r^{(\kappa)} \rangle$ integrals. There are

$$s \cdot f \cdot n^2 \tag{8.334}$$

1D integrals to be calculated. Hence, considering potential-like operators, i.e. diagonal operators

$$s \cdot f \cdot N \cdot n^2 \tag{8.335}$$

multiplications, have to be performed.

The final summation is a negligible amount of work. This is to be compared with the work of calculating the integrals directly, i.e. $n^{2f} \cdot N^f$. Going back to our example: $f = 6$, $n = 6$ and $N = 32$, and assuming a rather large number of terms $s = 14{,}000$, we find: $s \cdot f \cdot N \cdot n^2 \approx 10^8$ and $n^{2f} \cdot N^f \approx 10^{18}$. Hence, we gain 10 orders of magnitude!

The question is, how realistic is a product form of the Hamiltonian? Fortunately, KEOs are almost always of product form. For example, in the case of NOCl with the definition of the coordinates of Fig. 5.4, the KEO reads (see Eq. (6.155)):

$$\begin{aligned} T_{J=0}{}' = & -\frac{1}{2\mu_{R_1}} \frac{\partial^2}{\partial R_1^2} - \frac{1}{2\mu_{R_2}} \frac{\partial^2}{\partial R_2^2} \\ & - \left(\frac{1}{2\mu_{R_1} R_1^2} + \frac{1}{2\mu_{R_2} R_2^2} \right) \frac{1}{\sin\theta} \frac{\partial}{\partial\theta} \sin\theta \frac{\partial}{\partial\theta} , \end{aligned} \tag{8.336}$$

Potentials are sometimes given as polynomials or Fourier series. E.g., for NOCl,

$$V(R_1, R_2, \theta) = \sum_{i,j,k} C_{i,j,k} \left(R_2 - R_2^e \right)^i \left(R_1 - R_1^e \right)^j \cos(k\theta) . \tag{8.337}$$

[11] Of course, mode combination can be used. Then $h_r^{(\kappa)}$ operates on the κ-th particle in Eq. (8.333) and the integrals become low-dimensional rather than one-dimensional ones.

Hence the product form is not as unusual as it may look at first glance. In the general case, however, one needs an algorithm which brings a general potential to product form. POTFIT is such an algorithm.

8.7.2 The POTFIT Algorithm *

The most direct way to achieve a product form is an expansion of the potential in a product basis:

$$V^{\text{app}}(q_1, \ldots, q_f) = \sum_{j_1=1}^{m_1} \cdots \sum_{j_f=1}^{m_f} C_{j_1 \ldots j_f}\, v_{j_1}^{(1)}(q_1) \cdots v_{j_f}^{(f)}(q_f)\,. \tag{8.338}$$

(This looks like an MCTDH expansion of a WF!) The expansion orders, m_κ, have to be chosen such that the representation of the *potfitted* potential, V^{app}, is on the one hand as small as possible but on the other hand of sufficient accuracy. As we use DVRs, we must know the potential only at the grid points.

Let $q_i^{(\kappa)}$ denote the position of the i-th grid point of the κ-th grid. Then we define

$$V_{i_1,\ldots,i_f} = V(q_{i_1}^{(1)}, \ldots, q_{i_f}^{(f)})\,, \tag{8.339}$$

i.e. $V_{i_1,\ldots,i_f}$ denotes the value of the potential at the grid points. The approximate potential on the grid is given by

$$V_{i_1,\ldots,i_f}^{\text{app}} = \sum_{j_1=1}^{m_1} \cdots \sum_{j_f=1}^{m_f} C_{j_1 \ldots j_f}\, v_{i_1 j_1}^{(1)} \cdots v_{i_f j_f}^{(f)}\,, \tag{8.340}$$

where

$$v_{i_\kappa j_\kappa}^{(\kappa)} = v_{j_\kappa}^{(\kappa)}(q_{i_\kappa}^{(\kappa)})\,, \tag{8.341}$$

are called the single-particle potentials (SPPs). The SPPs are assumed to be orthogonal on the grid

$$\sum_{i=1}^{N_\kappa} v_{ij}^{(\kappa)} v_{il}^{(\kappa)} = \delta_{jl}\,. \tag{8.342}$$

Throughout this section i and k label grid points and j and l label SPPs. We can, of course, use mode combination. Then the SPPs are defined on multi-dimensional grids and the number of DOFs, f, is to be replaced by the number of MCTDH-particles, p.

To find the optimal coefficients and the optimal SPPs, we minimize

$$\Delta^2 = \sum_{i_1=1}^{N_1} \cdots \sum_{i_f=1}^{N_f} \left(V_{i_1 \ldots i_f} - V_{i_1 \ldots i_f}^{\text{app}} \right)^2 = \sum_I (V_I - V_I^{\text{app}})^2 . \tag{8.343}$$

Minimizing Δ^2 by varying only the coefficients yields:

$$C_{j_1 \cdots j_f} = \sum_{i_1=1}^{N_1} \cdots \sum_{i_f=1}^{N_f} V_{i_1 \ldots i_f} \, v_{i_1 \, j_1}^{(1)} \cdots v_{i_f \, j_f}^{(f)} , \tag{8.344}$$

hence the coefficients are given as overlaps (as expected).

Plugging this into the expression for Δ^2 yields:

$$\Delta^2 = \|\boldsymbol{V}\|^2 - \|\boldsymbol{C}\|^2 = \sum_I V_I^2 - \sum_I C_I^2 . \tag{8.345}$$

Therefore, one has to optimize the (orthonormal) SPPs such that $\|\boldsymbol{C}\|^2$ becomes maximal. The solution of this variational problem is difficult. It is numerically very demanding and likely to converge to a local minimum.

We take a shortcut and define potential density matrices as:

$$\varrho_{nm}^{(\kappa)} = \sum_{i_1=1}^{N_1} \cdots \sum_{i_{\kappa-1}=1}^{N_{\kappa-1}} \sum_{i_{\kappa+1}=1}^{N_{\kappa+1}} \cdots \sum_{i_p=1}^{N_p} V_{i_1 \ldots i_{\kappa-1} n i_{\kappa+1} \ldots i_p} V_{i_1 \ldots i_{\kappa-1} m i_{\kappa+1} \ldots i_p} . \tag{8.346}$$

We then diagonalize the densities $\boldsymbol{\rho}^{(\kappa)}$ and take the eigenvectors with the largest eigenvalues as SPPs. (Note that $\boldsymbol{\rho}^{(\kappa)}$ is positive semi-definite. Hence all eigenvalues are non-negative). The procedure is known to yield the optimal SPPs for a two-dimensional case [102]. For higher dimensions the representation is not optimal but sufficiently close to optimal.

8.7.3 Contraction in POTFIT *

Contraction over one mode is another very useful trick to reduce the numerical effort. We can perform one sum once and for all. Let us, for the sake of simplicity, contract over the first DOF:

$$D_{i_1 j_2 \cdots j_f} := \sum_{j_1=1}^{N_1} C_{j_1 \cdots j_f} \, v_{i_1 j_1}^{(1)} . \tag{8.347}$$

The potential is then given by

$$V^{\text{app}}_{i_1,\ldots,i_f} = \sum_{j_2=1}^{m_1} \cdots \sum_{j_f=1}^{m_f} D_{i_1 j_2 \ldots j_f}\, v^{(2)}_{i_2\, j_2} \cdots v^{(f)}_{i_f\, j_f}\,. \tag{8.348}$$

Hence, rather than $s = m^f$ terms we have only $s = m^{(f-1)}$ terms. Moreover, if we increase m_1 to N_1, which increases the accuracy, one notices that $\boldsymbol{C}$ of that index is a unitary transformation of $\boldsymbol{V}$, which is then transformed back. Hence there is no transformation at all and $\boldsymbol{D}$ is given by

$$D_{i_1 j_2 \ldots j_f} = \sum_{i_2 \cdots i_f} V_{i_1 \cdots i_f}\, v^{(2)}_{i_2\, j_2} \cdots v^{(f)}_{i_f\, j_f}\,. \tag{8.349}$$

Turning to a coordinate representation, we write the contracted potential as

$$V^{\text{app}}(q_1, \ldots, q_f) = \sum_{j_2=1}^{m_2} \cdots \sum_{j_f=1}^{m_f} D_{j_2 \ldots j_f}(q_1) v^{(2)}_{j_2}(q_2) \cdots v^{(f)}_{j_f}(q_f)\,. \tag{8.350}$$

Of course, we can contract over any degree of freedom, not necessarily over the first one. In general one will contract over that mode κ that otherwise has the smallest expansion efficiency N_κ / m_κ. Note that when using contraction, the coefficient vector $\boldsymbol{C}$ and the SPPs of the contracted mode are not computed, see Eq. (8.349).

*8.7.4 Error Estimate **

Letting ν denote the contracted mode, the error can be bounded by [5, 103]

$$\frac{\Lambda}{f-1} \le \Delta^2_{opt} \le \Delta^2 \le \Lambda\,, \tag{8.351}$$

where

$$\Lambda = \sum_{\substack{\kappa=1 \\ \kappa \ne \nu}}^{f} \sum_{j=m_\kappa+1}^{N_\kappa} \lambda^{(\kappa)}_j\,, \tag{8.352}$$

and where Δ^2 denotes the potfit L^2-error and Δ^2_{opt} the L^2-error one would obtain after a full optimization of the SPPs. Note that the error is determined by the eigenvalues of the neglected SPPs while ignoring those of the contracted mode ν. In particular, for $m_\kappa = N_\kappa$ one recovers the exact potential on the grid.

The last inequality of Eq. (8.351) tells us how to choose the expansion orders, m_κ, for a given error to be tolerated.[12] The inequality in the middle is trivial and the last inequality shows that the error bound Λ is at most $(f-1)$ times larger than the optimal error Δ^2_{opt}. For the two-dimensional case, $f=2$, one obtains $\Delta^2_{\text{opt}} = \Delta^2 = \Lambda$, hence, as already noted, POTFIT is optimal in this special case.

8.7.5 Weights *

The inclusion of weights is often important, because one does not need a uniform accuracy. The accuracy may be low when the potential is high, simply because the wavefunction does not go there. On the other hand, we need a high accuracy near the minimum and at transition states (saddle points). Hence, we want to minimise:

$$\Delta^2_w = \sum_I w_I^2 \, (V_I - V_I^{\text{app}})^2 . \tag{8.353}$$

The inclusion of separable weights

$$w_I = w^{(1)}_{i_1} \dots w^{(f)}_{i_f} \tag{8.354}$$

is very simple. One simply potfits $w_I \cdot V_I$ and then divides the SPPs by the weights

$$v_i^{(\kappa)} \;\rightarrow\; v_i^{(\kappa)} / w_i^{(\kappa)} . \tag{8.355}$$

However, separable weights are in general not very helpful. The inclusion of non-separable weights is very difficult. There appear matrices such as

$$\left\langle v^{(1)}_{j_1} \dots v^{(f)}_{j_f} \middle| w \middle| v^{(1)}_{j_1} \dots v^{(f)}_{j_f} \right\rangle , \tag{8.356}$$

which have to be inverted. As their dimension is the full total grid size, this is impossible for most cases.

There is a nice trick to emulate non-separable weights. Assume there is a reference potential V^{ref} such that

$$\left(V_I - V_I^{\text{app}}\right) w_I^2 = V_I^{\text{ref}} - V_I^{\text{app}} \tag{8.357}$$

holds. Then, we simply potfit V^{ref} and hence minimize

$$\sum_I \left(V_I^{\text{ref}} - V_I^{\text{app}}\right)^2 , \tag{8.358}$$

[12] Note that the root-mean-square error is given by $rmse = \sqrt{\Delta^2/N_{\text{tot}}}$, where N_{tot} is the total number of grid points.

which in turn is equal to

$$\sum_I w_I^2 \left(V_I - V_I^{\text{app}}\right)^2 , \tag{8.359}$$

i.e. the weighted sum that we want to minimize! Obviously, V^{ref} is given by

$$V_I^{\text{ref}} = w_I^2 V_I + (1 - w_I^2) V_I^{\text{app}} . \tag{8.360}$$

However, as V_I^{app} is unknown, we have to use an iterative process[13]:

(1) $V_I^{app\ (0)} =$ `potfit(V)`
(2) `for` $k = 1, \ldots, k_{\max}$ `do`

$$V_I^{\text{ref}\ (k)} = w_I^2\ V_I + (1 - w_I^2)\ V_I^{\text{app}\ (k-1)} ,$$
$$V^{\text{app}\ (k)} = \texttt{potfit}(V_I^{\text{ref}\ (k)}) ,$$

(3) `next k`

The question is, of course, does this process converge? In fact, one may multiply w_I by some positive constant. The final converged result must not change. One can show that for sufficiently small w_I the iteration will always converge and for sufficiently large w_I it will always diverge.

For defining the weights, we adopted the concept of a relevant region, i.e.

$$w_I = w(q_I) = \begin{cases} 1 & \text{if } q_I \in \text{relevant region} , \\ 0 & \text{else} . \end{cases}$$

The relevant region is often defined by an energy criterion

$$w_I = w(q_I) = \begin{cases} 1 & \text{if } V_I \leq E_{rel} , \\ 0 & \text{if } V_I > E_{rel} . \end{cases}$$

but it may contain restrictions on the coordinate as well. We also tried to replace w_I with $\alpha \cdot w_I$. The iterative process always converges for $0 < \alpha \leq 1$ and always diverges for $\alpha > 2$. The convergence is slower for smaller α. An improved convergence speed can be obtained for $\alpha \approx 1.5$.

8.8 Complement to Chapter 8: MCTDH-EOM for $g^{(\kappa)} \neq 0$ *

We want to re-derive the EOM but this time for the general case $g^{(\kappa)} \neq 0$. For part 1 of the Eq. (8.185) we obtain (see also (8.190)):

[13]Actually, we loop over the modes and update V^{ref} after each new SPP(k).

$$i \sum_L \langle \Psi_j^{(\kappa)} | \Phi_L \rangle \dot{A}_L =$$

$$\sum_L \langle \Psi_j^{(\kappa)} | \Phi_L \rangle \langle \Phi_L | H | \Psi \rangle - \sum_{\nu=1}^{f} \sum_{k=1}^{n_\nu} \sum_L \langle \Psi_j^{(\kappa)} | \Phi_L \rangle \, g_{l_\nu k}^{(\nu)} \, A_{L_k^\nu} =$$

$$P^{(\kappa)} \langle \Psi_j^{(\kappa)} | H | \Psi \rangle - \sum_{l_\kappa, k} \rho_{jk}^{(\kappa)} \, g_{l_\kappa k}^{(\kappa)} \, \varphi_{l_\kappa}^{(\kappa)} - D \tag{8.361}$$

where

$$D = \sum_{\nu \neq \kappa} \sum_L \sum_k \langle \Psi_j^{(\kappa)} | \Phi_L \rangle g_{l_\kappa k}^{(\nu)} \, A_{L_k^\nu} \tag{8.362}$$

The term $\nu = \kappa$ yields

$$\sum_k \sum_{L^\kappa} \sum_{l_\kappa} \langle \Psi_j^{(\kappa)} | \Phi_{L^\kappa} \, \varphi_{l_\kappa}^{(\kappa)} \rangle g_{l_\kappa k}^{(\kappa)} \, A_{L_k^\kappa} =$$

$$\sum_{k l_\kappa} \langle \Psi_j^{(\kappa)} | \Psi_k^{(\kappa)} \rangle \varphi_{l_\kappa}^{(\kappa)} \, g_{l_\kappa k}^{(\kappa)} = \sum_{k l_\kappa} \rho_{jk}^{(\kappa)} \, g_{l_\kappa k}^{(\kappa)} \, \varphi_{l_\kappa}^{(\kappa)} \tag{8.363}$$

which proves Eq. (8.361). Part 2 of Eq. (8.185) now reads

$$i \langle \Psi_j^{(\kappa)} | \sum_{\nu=1}^{f} \sum_{l=1}^{n_\nu} \dot{\varphi}_l^{(\nu)} \Psi_l^{(\nu)} \rangle = i \sum_l \rho_{jl}^{(\kappa)} \, \dot{\varphi}_l^{(\kappa)} + D' \tag{8.364}$$

where

$$D' = \sum_{\nu \neq \kappa} \sum_l \langle \Psi_j^{(\kappa)} | \dot{\varphi}_l^{(\nu)} \, \Psi_l^{(\nu)} \rangle g_{l_\kappa k}^{(\nu)} \, A_{L_k^\nu} \tag{8.365}$$

We will show later that $D = D'$. Hence adding part 1, Eq. (8.361), and part 2, Eqs. (8.364), (8.185) turns into

$$\sum_l \langle H \rangle_{jl}^{(\kappa)} \, \varphi_l^{(\kappa)} =$$

$$P^{(\kappa)} \sum_l \langle H \rangle_{jl}^{(\kappa)} \, \varphi_l^{(\kappa)} - \sum_{l_\kappa k} \rho_{jk}^{(\kappa)} \, g_{l_\kappa k}^{(\kappa)} \, \varphi_{l_\kappa}^{(\kappa)} - D + i \sum_l \rho_{jl}^{(\kappa)} \, \dot{\varphi}_l^{(\kappa)} + D'$$

or, assuming $D = D'$

$$i \sum_l \rho_{jl}^{(\kappa)} \, \dot{\varphi}_l^{(\kappa)} = \left(1 - P^{(\kappa)}\right) \sum_l \langle H \rangle_{jl}^{(\kappa)} \, \varphi_l^{(\kappa)} + \sum_{l_\kappa k} \rho_{jk}^{(\kappa)} \, g_{l_\kappa k}^{(\kappa)} \, \varphi_{l_\kappa}^{(\kappa)} \tag{8.366}$$

Writing

$$\varphi^{(\kappa)} = \left(\varphi_1^{(\kappa)} \cdots \varphi_{n_\kappa}^{(\kappa)}\right)^T \tag{8.367}$$

and multiplying by ρ^{-1} yields

$$i\dot{\varphi}^{(\kappa)} = \left(\boldsymbol{g}^{(\kappa)^T} + \left(1 - P^{(\kappa)}\right) \boldsymbol{\rho}^{(\kappa)^{-1}} \left\langle \boldsymbol{H} \right\rangle^{(\kappa)}\right) \varphi^{(\kappa)} \tag{8.368}$$

As

$$\left(\boldsymbol{g}^{(\kappa)^T} \varphi^{(\kappa)}\right)_j = \sum_l |\varphi_l\rangle\langle\varphi_l|g^{(\kappa)}|\varphi_j\rangle = P^{(\kappa)}\, g^{(\kappa)}\, \varphi_j^{(\kappa)} \tag{8.369}$$

Hence we also have

$$i\dot{\varphi}^{(\kappa)} = \left(P^{(\kappa)}\, \boldsymbol{g}^{(\kappa)} + \left(1 - P^{(\kappa)}\right) \boldsymbol{\rho}^{(\kappa)^{-1}} \left\langle \boldsymbol{H} \right\rangle^{(\kappa)}\right) \varphi^{(\kappa)} \tag{8.370}$$

and from this all other forms follow.

Finally, we use again the separation

$$H = \sum_\kappa g^{(\kappa)} + H_g$$

yielding

$$\langle \Phi_J | H | \Phi_L \rangle = \langle \Phi_J | H_g | \Phi_L \rangle + \sum_\kappa \sum_{l_\kappa} g^{(\kappa)}_{j_\kappa l_\kappa}\, \delta_{J^\kappa L^\kappa} \tag{8.371}$$

and

$$i\dot{A}_J = \sum_L \langle \Phi_J | H_g | \Phi_L \rangle\, A_L + \sum_\kappa \sum_{l_\kappa} g^{(\kappa)}_{j_\kappa l_\kappa}\, A_{J^\kappa_{l_\kappa}} - \sum_\kappa \sum_l g^{(\kappa)}_{j_\kappa l_\kappa}\, A_{J^\kappa_l} \tag{8.372}$$

Hence

$$\begin{aligned} i\dot{A}_J &= \sum_L \langle \Phi_J | H_g | \Phi_L \rangle\, A_L + \sum_\kappa \sum_l g^{(\kappa)}_{j_\kappa l_\kappa}\, A_{J^\kappa_{l_\kappa}} \\ &= \sum_L \langle \Phi_J | H - \sum_\kappa g^{(\kappa)} | \Phi_L \rangle\, A_L \end{aligned} \tag{8.373}$$

We still have to show that $D = D'$.

$$D = \sum_{\nu \neq \kappa} \sum_L \sum_k \langle \Psi_j^{(\kappa)} | \Phi_L \rangle g^{(\nu)}_{l_\kappa k}\, A_{L^\nu_k}$$

$$D' = i \sum_{\nu \neq \kappa} \sum_l \langle \Psi_j^{(\kappa)} | \dot{\varphi}_l^{(\nu)}\, \Psi_l^{(\nu)} \rangle$$

We insert $P^{(\nu)}$ in the equation for D' and, given that $\langle \Psi_j^{(\kappa)} | P^{(\nu)} = \langle \Psi_j^{(\kappa)} |$ for $\nu \neq \kappa$, we obtain

$$\begin{aligned} D' &= i \sum_{\nu \neq \kappa} \sum_{l} \langle \Psi_j^{(\kappa)} | \sum_{l_\nu} |\varphi_{l_\nu}^{(\nu)}\rangle\langle\varphi_{l_\nu}^{(\nu)}|\dot{\varphi}_l^{(\nu)}\rangle \, \Psi_l^{(\nu)}\rangle \\ &= \sum_{\nu \neq \kappa} \sum_{l} \sum_{l_\nu} g_{l_\nu l}^{(\nu)} \langle \Psi_j^{(\kappa)} | \varphi_{l_\nu}^{(\nu)} \sum_{L^\nu} A_{L_l^\nu} \, \Phi_{L^\nu} \rangle \\ &= \sum_{\nu \neq \kappa} \sum_{l} \sum_{L} \langle \Psi_j^{(\kappa)} | \Phi_L \rangle \, g_{l_\nu l}^{(\nu)} \, A_{L_l^\nu} = D \end{aligned} \tag{8.374}$$

In summary, we again display the EOM in various forms

$$\begin{aligned} i \dot{A}_J &= \sum_L \langle \Phi_J | H | \Phi_L \rangle \, A_L - \sum_{\kappa=1}^{f} \sum_{l=1}^{n_\kappa} g_{j_\kappa l}^{(\kappa)} \, A_{J_l^\kappa} \\ &= \sum_L \langle \Phi_J | H - \sum_{\kappa=1} g^{(\kappa)} | \Phi_L \rangle \, A_L \end{aligned} \tag{8.375}$$

$$\begin{aligned} i\dot{\varphi}^{(\kappa)} &= \left\{ g^{(\kappa)} \, \mathbf{1}_{(\kappa)} + \left(1 - P^{(\kappa)}\right) \left[\boldsymbol{\rho}^{(\kappa)^{-1}} \langle \boldsymbol{H} \rangle^{(\kappa)} - g^{(\kappa)} \, \mathbf{1} \right] \right\} \varphi^{(\kappa)} \\ &= P^{(\kappa)} \, g \, \varphi^{(\kappa)} + \left(1 - P^{(\kappa)}\right) \boldsymbol{\rho}^{(\kappa)^{-1}} \langle \boldsymbol{H} \rangle^{(\kappa)} \, \varphi^{(\kappa)} \\ &= \left[\left(\boldsymbol{g}^{(\kappa)}\right)^T + \left(1 - P^{(\kappa)}\right) \boldsymbol{\rho}^{(\kappa)^{-1}} \langle \boldsymbol{H} \rangle^{(\kappa)} \right] \varphi^{(\kappa)} \\ &= \left\{ g^{(\kappa)} \, \mathbf{1} \right)^T + \left(1 - P^{(\kappa)}\right) \boldsymbol{\rho}^{(\kappa)^{-1}} \langle \boldsymbol{H} - \sum_{\kappa'} g^{(\kappa')} \rangle^{(\kappa)} \right\} \varphi^{(\kappa)} \end{aligned} \tag{8.376}$$

References

1. Cohen-Tannoudji C, Diu B, Laloe F (1992) Quantum mechanics. Wiley-VCH
2. Tannor DJ (2007) Introduction to quantum dynamics: a time-dependent perspective. University Science Books, Sausalito, CA
3. Dirac PAM (1930) Note on exchange phenomena in the Thomas atom. Proc Camb Philos Soc 26:376–385
4. Frenkel J (1934) Wave mechanics. Clarendon Press, Oxford
5. Beck MH, Jäckle A, Worth GA, Meyer H-D (2000) The multi-configuration time-dependent Hartree (MCTDH) method: a highly efficient algorithm for propagating wave packets. Phys Rep 324:1–105
6. Kosloff D, Kosloff R (1983) A Fourier-method solution for the time-dependent Schrödinger equation as a tool in molecular dynamics. J Comput Phys 52:35
7. Kosloff R, Tal-Ezer H (1986) A direct relaxation method for calculating eigenfunctions and eigenvalues of the Schrödinger equation on a grid. Chem Phys Lett 127:223
8. Light JC, Hamilton IP, Lill JV (1985) Generalized discrete variable approximation in quantum mechanics. J Chem Phys 82:1400
9. Lill JV, Parker GA, Light JC (1986) The discrete variable-finite basis approach to quantum scattering. J Chem Phys 85:900
10. Corey GC, Lemoine D (1992) Pseudospectral method for solving the time-dependent Schrödinger equation in spherical coordinates. J Chem Phys 97:4115

11. Bramley MJ, Tromp JW, Carrington T Jr, Corey RC (1994) Efficient calculation of highly excited vibrational energy levels of floppy molecules: the band origins of H_3^+ up to 35 000 cm^{-1}. J Chem Phys 100:6175
12. Lemoine D (1994) The finite basis representation as the primary space in multidimensional pseudospectral schemes. J Chem Phys 101:10526
13. Kosloff R (1996) Quantum molecular dynamics on grids. In: Wyatt RE, Zhang JZH (eds) Dynamics of molecules and chemical reactions. Marcel Dekker, New York, pp 185–230
14. Light JC, Carrington T Jr (2000) Discrete variable representations and their utilization. Adv Chem Phys 114:263
15. Zare RN (1988) Angular momentum. Wiley, New York
16. Worth GA, Beck MH, Jäckle A, Meyer HD (2007) The MCTDH Package, Version 8.2, (2000). Meyer HD (2002), Version 8.3, Version 8.4. Current version: 8.4.12 (2016). http://mctdh.uni-hd.de/
17. Dawes R, Carrington T Jr (2004) A multidimensional discrete variable representation basis obtained by simultaneous diagonalization. J Chem Phys 121:726
18. Li J, Carter S, Bowman JM, Dawes R, Xie D, Guo H (2014) High-level, first-principles, full-dimensional quantum calculation of the ro-vibrational spectrum of the simplest criegee intermediate (CH_2OO). J Phys Chem Lett 5:20364
19. Harris DO, Engerholm GG, Gwinn GW (1965) Calculation of matrix elements for one-dimensional quantum-mechanical problems and the application to anharmonic oscillators. J Chem Phys 43:1515
20. Dickinson AS, Certain PR (1968) Calculation of matrix elements for one-dimensional quantum-mechanical problems. J Chem Phys 49:4209
21. Corey GC, Tromp JW, Lemoine D (1993) Fast pseudospectral algorithm in curvilinear coordinates. In: Cerjan C (ed) Numerical grid methods and their application to Schrödinger's equation. Kluwer Academic Publishers, The Netherlands, pp 1
22. Sukiasyan S, Meyer H-D (2001) On the effect of initial rotation on reactivity. A multi-configuration time-dependent Hartree (MCTDH) wave-packet propagation study on the $H+D_2$ and $D+H_2$ reactive scattering systems. J Phys Chem A 105:2604
23. Echave J, Clary DC (1992) Potential optimized discrete variable representation Chem Phys Lett 190:225
24. Bramley MJ, Handy NC (1993) Efficient calculation of rovibrational eigenstates of sequentially bonded four-atom molecules. J Chem Phys 98:1378
25. Tremblay JC, Carrington T Jr (2006) Calculating vibrational energies and wave functions of vinylidene using a contracted basis with a locally reorthogonalized coupled two-term lanczos eigensolver. J Chem Phys 125:094311
26. Bowman JM, Carrington T Jr, Meyer H-D (2008) Variational quantum approaches for computing vibrational energies of polyatomic molecules. Mol Phys 106:2145
27. Wang X, Carrington T Jr (2008) Using a nondirect product discrete variable representation for angular coordinates to compute vibrational levels of polyatomic molecules. J Chem Phys 128(19):194109
28. Wang X, Carrington T Jr (2008) Vibrational energy levels of CH_5^+. J Chem Phys 129:234102
29. Wang X-G, Carrington T Jr, Dawes R, Jasper AW (2011) The vibration-rotation-tunneling spectrum of the polar and T-shaped-N-in isomers of $(NNO)_2$. J Mol Spec 268:53
30. Wang X-G, Carrington T Jr (2013) Computing rovibrational levels of methane with curvilinear internal vibrational coordinates and an Eckart frame. J Chem Phys 138:104106
31. Wang X-G, Carrington T Jr (2014) Rovibrational levels and wavefunctions of Cl^-H_2O. J Chem Phys 140:204306
32. Bowman J (1978) Self-consistent field energies and wavefunctions for coupled oscillators. J Chem Phys 68:608
33. Bowman J, Christoffel K, Tobin F (1979) Application of SCF-SI theory to vibrational motion in polyatomic molecules. J Phys Chem 83:905
34. Bowman JM, Carter S, Huang X (2003) MULTIMODE: a code to calculate rovibrational energies of polyatomic molecules. Int Rev Phys Chem 22:533

35. Culot F, Laruelle F, Liévin J (1995) A vibrational CASSCF study of stretch-bend interactions and their influence on infrared intensities in the water molecule. Theory Chem Acc 92:211
36. Bégué D, Gohaud N, Pouchan C, Cassam-Chenaï P, Liévin J (2007) A comparison of two methods for selecting vibrational configuration interaction spaces on a heptatomic system: ethylene oxide. J Chem Phys 127:164115
37. Heislbetz S, Rauhut G (2010) Vibrational multiconfiguration self-consistent field theory: implementation and test calculations. J Chem Phys 132:124102
38. Leforestier C, Bisseling RH, Cerjan C, Feit MD, Friesner R, Guldenberg A, Hammerich A, Jolicard G, Karrlein W, Meyer H-D, Lipkin N, Roncero O, Kosloff R (1991) A comparison of different propagation schemes for the time dependent Schrödinger equation. J Comput Phys 94:59
39. Feit MD, Fleck JA Jr, Steiger A (1982) Solution of the Schrödinger equation by a spectral method. J Comput Phys 47:412
40. Feit MD, Fleck JA Jr (1983) Solution of the Schrödinger equation by a spectral method II: vibrational energy levels of triatomic molecules. J Chem Phys 78:301
41. Park TJ, Light JC (1986) Unitary quantum time evolution by iterative Lanczos reduction. J Chem Phys 85:5870
42. Arnoldi WE (1951) The principle of minimized iterations in the solution of the matrix eigenvalue problem. Q Appl Math 9:17
43. Saad Y (1980) Variations on Arnoldi's method for computing eigenelements of large unsymmetric matrices. Linear Algebra Appl 34:269
44. Friesner RA, Tuckerman LS, Dornblaser BC, Russo TV (1989) J Sci Comput 4:327
45. Manthe U, Köppel H, Cederbaum LS (1991) Dissociation and predissociation on coupled electronic potential energy surfaces: a three-dimensional wave packet dynamical study. J Chem Phys 95:1708
46. Gear CW (1971) Numerical initial value problems in ordinary differential equations. Prentice-Hall, Englewood Cliffs
47. Beck MH, Meyer H-D (1997) An efficient and robust integration scheme for the equations of motion of the multiconfiguration time-dependent Hartree (MCTDH) method. Z Phys D 42:113–129
48. Gerber RB, Ratner MA, Buch V (1982) Simplified time-dependent self-consistent field approximation for intramolecular dynamics. Chem Phys Lett 91:173
49. Bisseling RH, Kosloff R, Gerber RB, Ratner MA, Gibson L, Cerjan C (1987) Exact time-dependent quantum mechanical dissociation dynamics of I_2He: comparison of exact time-dependent quantum calculation with the quantum time-dependent self-consistent field (TDSCF) approximation. J Chem Phys 87:2760
50. Meyer H-D, Gatti F, Worth GA (eds) (2009) Multidimensional quantum dynamics: MCTDH theory and applications. Wiley-VCH, Weinheim
51. Kotler Z, Nitzan A, Kosloff R (1988) Multiconfiguration time-dependent self-consistent field approximation for curve crossing in presence of a bath. A fast Fourier transform study. Chem Phys Lett 153:483
52. Makri N, Miller WH (1987) Time-dependent self-consistent (TDSCF) approximation for a reaction coordinate coupled to a harmonic bath: single and multiconfiguration treatments. J Chem Phys 87:5781
53. Meyer H-D, Manthe U, Cederbaum LS (1990) The multi-configurational time-dependent Hartree approach. Chem Phys Lett 165:73
54. Manthe U, Meyer H-D, Cederbaum LS (1992) Wave-packet dynamics within the multiconfiguration Hartree framework: general aspects and application to NOCl. J Chem Phys 97:3199
55. Szabo A, Ostlund NS (1996) Modern quantum chemistry. Dover, Mineola, NY
56. Jensen F (1999) Introduction to computational chemistry. Wiley, Chichester
57. Atkins PW (1983) Molecular quantum mechanics, 2nd edn. OUP, Oxford, UK
58. Zanghellini J, Kitzler M, Fabian C, Brabec T, Scrinzi A (2003) An MCTDHF approach to multi-electron dynamics in laser fields. Laser Phys 13:1064

59. Alon OE, Streltsov AI, Cederbaum LS (2008) Multiconfigurational time-dependent Hartree method for bosons: many-body dynamics of bosonic systems. Phys Rev A 77:033613
60. Hammerich AD, Kosloff R, Ratner MA (1990) Quantum mechanical reactive scattering by a multiconfigurational time-dependent self-consistent field (MCTDSCF) approach. Chem Phys Lett 171:97
61. Jäckle A, Meyer H-D (1998) Calculation of $H+H_2$ and $H+D_2$ reaction probabilities within the multiconfiguration time-dependent Hartree approach employing an adiabatic correction scheme. J Chem Phys 109:2614
62. Launay JM, Dourneuf ML (1989) Hyperspherical close-coupling calculation of integral cross sections for the reaction $H+H_2 \rightarrow H_2+H$. Chem Phys Lett 163:178
63. Pack RT, Parker G (1987) Quantum reactive scattering in three dimensions using hyperspherical (aph) coordinates. theory. J Chem Phys 87:3888
64. Pack RT, Parker G (1989) Quantum reactive scattering in three dimensions using hyperspherical (aph) coordinates. III. Small θ behavior and corrigenda. J Chem Phys 90:3511
65. Wang H, Thoss M (2003) Multilayer formulation of the multiconfiguration time-dependent Hartree theory. J Chem Phys 119:1289
66. Wang H, Thoss M (2006) Quantum-mechanical evaluation of the Boltzmann operator in correlation functions for large molecular systems: a multilayer multiconfiguration time-dependent Hartree approach. J Chem Phys 124:034114
67. Wang H, Skinner DE, Thoss M (2006) Calculation of reactive flux correlation functions for systems in a condensed phase environment: a multilayer multi-configuration time-dependent hartree approach. J Chem Phys 125:174502
68. Wang H, Thoss M (2007) Quantum dynamical simulation of electron-transfer reactions in an anharmonic environment. J Phys Chem A 111:10369
69. Manthe U (2008) A multilayer multiconfigurational time-dependent Hartree approach for quantum dynamics on general potential energy surfaces. J Chem Phys 128:164116
70. Wang H, Thoss M (2009) Numerically exact quantum dynamics for indistinguishable particles: The multilayer multiconfiguration time-dependent Hartree theory in second quantization representation. J Chem Phys 131(2):024114
71. Vendrell O, Meyer H-D (2011) Multilayer multiconfiguration time-dependent Hartree method: Implementation and applications to a Henon-Heiles Hamiltonian and to pyrazine. J Chem Phys 134:044135
72. Wang H (2015) Multilayer multiconfiguration time-dependent Hartree theory. J Phys Chem A 119:7951
73. Westermann T, Brodbeck R, Rozhenko AB, Schoeller W, Manthe U (2011) Photodissociation of methyl iodide embedded in a host-guest complex: a full dimensional (189D) quantum dynamics study of CH_3I@resorc[4]arene. J Chem Phys 135:184102
74. Wang H, Shao J (2012) Dynamics of a two-level system coupled to a bath of spins. J Chem Phys 137:22A504
75. Manthe U (2006) On the integration of the multi-configurational time-dependent Hartree (MCTDH) equations of motion. Chem Phys 329:168
76. Meyer H-D, Worth GA (2003) Quantum molecular dynamics: propagating wavepackets and density operators using the multiconfiguration time-dependent Hartree (MCTDH) method. Theory Chem Acc 109:251
77. Meyer H-D, Le Quéré F, Léonard C, Gatti F (2006) Calculation and selective population of vibrational levels with the multiconfiguration time-dependent Hartree (MCTDH) algorithm. Chem Phys 329:179
78. Joubert Doriol L, Gatti F, Iung C, Meyer HD (2008) Computation of vibrational energy levels and eigenstates of fluoroform using the multiconfiguration time-dependent Hartree method. J Chem Phys 129:224109
79. Vendrell O, Gatti F, Lauvergnat D, Meyer H-D (2007) Full dimensional (15D) quantum-dynamical simulation of the protonated water dimer I: Hamiltonian setup and analysis of the ground vibrational state. J Chem Phys 127:184302

80. Vendrell O, Brill M, Gatti F, Lauvergnat D, Meyer H-D (2009) Full dimensional (15D) quantum-dynamical simulation of the protonated water dimer III: mixed Jacobi-valence parametrization and benchmark results for the zero-point energy, vibrationally excited states and infrared spectrum. J Chem Phys 130:234305
81. Davidson E (1975) The iterative calculation of a few of the lowest eigenvalues and corresponding eigenvectors of large real-symmetric matrices. J Comput Phys 17:87
82. Olsen J, Jørgenson P, Simons J (1990) Passing the one-billion limit in full configuration-interaction (FCI) calculations. Chem Phys Lett 169:493
83. Riss UV, Meyer H-D (1993) Calculation of resonance energies and widths using the complex absorbing potential method. J Phys B 26:4503
84. Jolicard G, Austin E (1985) Optical potential stabilisation method for predicting resonance level. Chem Phys Lett 121:106
85. Jolicard G, Austin E (1986) Optical potential method of caculating resonance energies and widths. Chem Phys 103:295
86. Jolicard G, Leforestier C, Austin E (1988) Resonance states using the optical potential model. Study of Feshbach resonances and broad shape resonances. J Chem Phys 88:1026
87. Kosloff R, Kosloff D (1986) Absorbing boundaries for wave propagation problems. J Comput Phys 63:363
88. Neuhauser D, Baer M (1989) The time-dependent Schrödinger equation: application of absorbing boundary conditions. J Chem Phys 90:4351
89. Riss UV, Meyer H-D (1995) Reflection-free complex absorbing potentials. J Chem Phys 28:1475
90. Riss UV, Meyer H-D (1996) Investigation on the reflection and transmission properties of complex absorbing potentials. J Chem Phys 105:1409
91. Jäckle A, Meyer H-D (1996) Time-dependent calculation of reactive flux employing complex absorbing potentials: general aspects and application within MCTDH. J Chem Phys 105:6778
92. Scheit S, Meyer H-D, Moiseyev N, Cederbaum LS (2006) On the unphysical impact of complex absorbing potentials on the Hamiltonian and its remedy. J Chem Phys 124:034102
93. Taylor JR (1972) Scattering theory: the quantum theory of nonrelativistic collisions. Wiley, New York
94. Gatti F, Otto F, Sukiasyan S, Meyer H-D (2005) Rotational excitation cross sections of para-H_2 + para-H_2 collisions. A full-dimensional wave packet propagation study using an exact form of the kinetic energy. J Chem Phys 123:174311
95. Panda AN, Otto F, Gatti F, Meyer H-D (2007) Rovibrational energy transfer in ortho-H_2 + para-H_2 collisions. J Chem Phys 127:114310
96. Otto F, Gatti F, Meyer H-D (2008) Rotational excitations in para-H_2 + para-H_2 collisions: full- and reduced-dimensional quantum wave packet studies comparing different potential energy surfaces. J Chem Phys 128:064305
97. Otto F, Gatti F, Meyer H-D (2009) Erratum: "Rotational excitations in para-H_2 + para-H_2 collisions: full- and reduced-dimensional quantum wave packet studies comparing different potential energy surfaces". J Chem Phys 131:049901
98. Otto F, Gatti F, Meyer H-D (2012) Rovibrational energy transfer in collisions of H_2 with D_2. A full-dimensional wave packet propagation study. Mol Phys 110:619
99. Tannor DJ, Weeks DE (1993) Wave packet correlation function formulation of scattering theory: the quantum analog of classical S-matrix theory. J Chem Phys 98:3884
100. Weeks DE, Tannor DJ (1993) A time-dependent formulation of the scattering matrix using Møller operators. Chem Phys Lett 207:301
101. Weeks DE, Tannor DJ (1994) A time-dependent formulation of the scattering matrix for the collinear reaction H+H_2 $(\nu) \rightarrow H_2$ (ν')+H. Chem Phys Lett 224:451
102. Schmidt E (1906) Zur Theorie der linearen und nichtlinearen Integralgleichungen. Math Ann 63:433
103. Peláez D, Meyer H-D (2013) The multigrid POTFIT (MGPF) method: grid representations of potentials for quantum dynamics of large systems. J Chem Phys 138:014108

Part II
Applications

Chapter 9
Infrared Spectroscopy

9.1 Introduction

Molecular quantum dynamics has given rise to many *applications of relevance for astrophysics, astrochemistry and atmospheric chemistry*. In the interstellar medium or in the atmosphere of planets, including the Earth, molecules are generally in the gas phase. Many accurate spectroscopic data are collected by astrophysicists: absorption lines, for instance, which must be identified by comparison with theoretical simulations. Such data provide crucial information about the composition of interstellar media, the atmosphere of planets in the solar system, or of exoplanets. They also allow us to quantify processes related to essential environmental issues such as the greenhouse effect or ozone depletion. In this context, much theoretical effort is directed toward computing spectra of small molecules. For instance, in 2012, the High Resolution Transmission (HITRAN) database that focuses on the description of transmission and emission of light in the atmosphere contained 7,400,447 identified spectral lines for 47 different molecules, including 120 isotopomers [1]. The most studied molecules are H_2O, CO_2, O_3, CH_4, CO, or O_2. A quantum-mechanical treatment is unavoidable, because the vibrational and rotational states are quantized and this must be taken into account. Indeed, the observed absorption lines are discrete and correspond to energy differences between specific rovibrational levels. For an Overview of High-resolution Spectroscopy see, for instance [2].

9.2 The Time-Dependent Approach

In Sect. 9.2.1, we introduce *the autocorrelation function* of a wavepacket and show that *its Fourier transform gives the power spectrum*. Such a property will allow us to find the vibrational energies of a molecule and its infrared spectrum after introduction of the dipole moment operator.

F. Gatti et al., *Applications of Quantum Dynamics in Chemistry*,
Lecture Notes in Chemistry 98, DOI 10.1007/978-3-319-53923-2_9

9.2.1 The Autocorrelation Function and Its Properties

Let us return to the time-evolution of the wavepacket given by Eq. (2.37).A quantity that plays a major role in the treatment of spectroscopy and reactive scattering in molecular physics is the *autocorrelation function of a wavepacket*,

$$A(t) = \langle \Psi(0)|\Psi(t)\rangle = \int \Psi^{\star}(\boldsymbol{R}, t=0)\Psi(\boldsymbol{R}, t)d\boldsymbol{R}\,. \tag{9.1}$$

At $t = 0$, the wavepacket reads

$$\Psi(\boldsymbol{R}, t=0) = \sum_n c_n \Psi_n(\boldsymbol{R})\,, \tag{9.2}$$

with

$$H\Psi_n(\boldsymbol{R}) = E_n \Psi_n(\boldsymbol{R})\,, \tag{9.3}$$

Using Eq. (2.37), i. e.[1]

$$\Psi(\boldsymbol{R}, t) = \sum_n c_n e^{-iE_n t} \Psi_n(\boldsymbol{R})\,, \tag{9.4}$$

(with $c_n = \int \Psi_n^{\star}(\boldsymbol{R})\Psi(\boldsymbol{R}, t=0)d\boldsymbol{R}$) leads to

$$\begin{aligned} A(t) &= \sum_n \sum_m c_n^{\star} c_m e^{-iE_m t} \langle \Psi_m|\Psi_n\rangle\,, \\ &= \sum_n |c_n|^2 e^{-iE_n t}\,. \end{aligned} \tag{9.5}$$

If the initial wavefunction is equal to an eigenstate, there is a single term in the sum and *the modulus of the autocorrelation function is constant and equal to one*.

The *Fourier transform of the autocorrelation function*, $\frac{1}{2\pi}\int_{-\infty}^{+\infty} e^{iEt} A(t)\,dt$, has important properties that will be useful for many applications. Let us integrate over a finite time T:

$$\begin{aligned} \frac{1}{2\pi}\int_{-T}^{+T} e^{iEt} A(t)\,dt &= \sum_n |c_n|^2 \frac{1}{2\pi}\int_{-T}^{+T} e^{i(E-E_n)t} dt\,, \\ &= \sum_n |c_n|^2 \frac{1}{2\pi}\left[\frac{e^{(i(E-E_n)T)} - e^{(-i(E-E_n)T)}}{i(E-E_n)}\right], \\ &= \sum_n |c_n|^2 \frac{\sin((E-E_n)T)}{\pi(E-E_n)}\,. \end{aligned} \tag{9.6}$$

[1] Here, we adopt the atomic units system, in particular $\hbar = 1$.

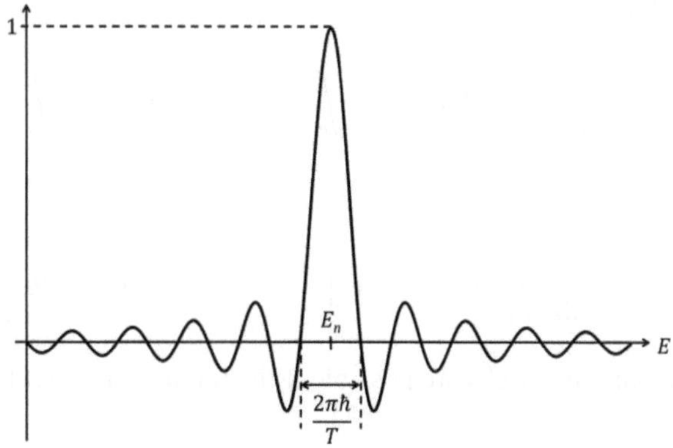

Fig. 9.1 The cardinal sine function

Let us divide the previous expression by T and note that $\frac{\sin(x)}{x}$ is the cardinal sine function. $\frac{\sin((E-E_n)T)}{(E-E_n)T}$ is depicted in Fig. 9.1. We see that $\frac{1}{2\pi T}\int_{-T}^{+T} e^{iEt}A(t)dt$ is a sum of cardinal sine functions centered on the energies of the eigenstates that appear in the wavepacket. The width of the central peak of the cardinal sine function is proportional to $1/T$. Thus, if the time propagation is long enough, $\frac{1}{2\pi}\int_{-T}^{+T} e^{iEt}A(t)dt$ provides the actual values of the eigenenergies with sufficient accuracy. This link between the time propagation and the accuracy of the energy is a direct consequence of the Heisenberg time-energy uncertainty principle. When T tends to infinity, $\frac{\sin((E-E_n)T)}{\pi(E-E_n)}$ tends to the Dirac delta function $\delta(E-E_n)$. The heights of the main peaks are then proportional to the squares of the overlaps between the wavepacket and the eigenstates, $|c_n|^2$. In other words, the propagation of the wavepacket yields information about the eigenstates that constitute it: their eigenvalues and their amplitudes, c_n, in the linear combination.

More precisely, when T tends to infinity we obtain,

$$\begin{aligned}\sigma(E) &= \lim_{T\to\infty}\frac{1}{2\pi}\int_{-T}^{+T} e^{iEt}A(t)dt\,,\\ &= \frac{1}{2\pi}\int_{-\infty}^{+\infty} e^{iEt}A(t)dt\,,\end{aligned} \tag{9.7}$$

and thus

$$\sigma(E) = \sum_n |c_n|^2\delta(E-E_n)\,. \tag{9.8}$$

$\sigma(E)$ is often called the *power spectrum of the wavepacket.*

The integration over negative time is cumbersome, but can be avoided easily. If the Hamiltonian is Hermitian, one shows

$$A(-t) = \langle\Psi|e^{iHt}|\Psi\rangle = \langle e^{-iHt}\,\Psi|\Psi\rangle = \langle\Psi|e^{-iHt}|\Psi\rangle^* = [A(t)]^*\,. \tag{9.9}$$

Thus

$$\int_{-T}^{0} e^{iEt}\, A(t)\, dt = \int_{0}^{T} e^{-iEt}\, A(-t)\, dt = \int_{0}^{T} \left[e^{iEt}\, A(t)\right]^{*}\, dt\,, \tag{9.10}$$

and

$$\frac{1}{2\pi}\int_{0}^{T} \left(\left[e^{iEt}\, A(t)\right]^{*} + e^{iEt}\, A(t)\right) dt = \frac{1}{\pi}\mathrm{Re}\int_{0}^{\mathrm{T}} \mathrm{e}^{\mathrm{iEt}}\, \mathrm{A(t)}\, \mathrm{dt}\,. \tag{9.11}$$

Additional tricks are possible if the initial state is real

$$\begin{aligned} A(t) &= \left\langle \Psi \middle| e^{-iHt} \middle| \Psi \right\rangle, \\ &= \left\langle e^{iH^{\dagger}t/2}\, \Psi \middle| e^{-iHt/2}\, \Psi \right\rangle, \\ &= \left\langle \left(e^{-iH^{\dagger *}t/2}\, \Psi^{*}\right)^{*} \middle| e^{-iHt/2}\, \Psi \right\rangle, \\ &= \left\langle \Psi(t/2)^{*} \middle| \Psi(t/2) \right\rangle, \end{aligned} \tag{9.12}$$

where the last step requires a real initial state ($\Psi^{*} = \Psi$) and a real-symmetric Hamiltonian

$$H = H^{T} = H^{\dagger *}\,. \tag{9.13}$$

This so-called $t/2$-trick [3–6] is very useful because it provides an autocorrelation function over a duration that is twice as long as the propagation time. In general, one wants to use both Eqs. (9.11) and (9.12). This requires a real-symmetric Hamiltonian.[2]

9.2.2 The Window Functions

Now, rather than using the upper integral limit T, we introduce a *window function* $g(t)$. The previous Fourier transform can be written as

$$\sigma_g(E) = \frac{\mathrm{Re}}{\pi}\int_{0}^{\infty} e^{iEt}\, g(t)\, A(t)\, dt = \frac{1}{2\pi}\int_{-\infty}^{\infty} e^{iEt}\, g(t)\, A(t)\, dt\,, \tag{9.14}$$

and we require

$$0 \le g(t) \le 1, \quad g(0) = 1, \quad g(t) = 0 \text{ for } |t| > T, \quad g(t) = g(-t)\,. \tag{9.15}$$

[2]Note, however, that the $t/2$-trick can still be applied to real-symmetric Hamiltonians plus a complex absorbing potential (see Sect. 8.5). This is because the imaginary part of the complex absorbing potential has to change sign when propagating in negative times, in order to absorb the wavefunction when t tends to $-\infty$.

As is well known, the Fourier transform of a product of two functions is equal to the convolution (indicated by a star "*") of the Fourier transforms of the two functions, i.e.

$$\sigma_g(E) = (\sigma * \tilde{g})(E)\,,$$
$$= \int \sigma(\epsilon)\,\tilde{g}(E-\epsilon)\,d\epsilon\,,$$
$$= \int \sigma(E-\epsilon)\,\tilde{g}(\epsilon)\,d\epsilon\,,$$

where

$$\tilde{g}(\epsilon) = \frac{1}{2\pi}\int_{-\infty}^{\infty} e^{i\epsilon t}\,g(t)\,dt\,.$$

The proof is simple: we use

$$\delta(\tau - t) = \frac{1}{2\pi}\int e^{i(\tau-t)\epsilon}\,d\epsilon\,, \tag{9.16}$$

and

$$\begin{aligned}
\sigma_g(E) &= \frac{1}{2\pi}\int_{-\infty}^{\infty} e^{iEt}\,g(t)\,A(t)\,dt\,,\\
&= \frac{1}{2\pi}\int\int e^{iEt}\,g(\tau)\,A(t)\,\delta(\tau-t)\,dt\,d\tau\,,\\
&= \frac{1}{(2\pi)^2}\int\int\int e^{iEt}\,A(t)\,e^{-i\epsilon t}\,e^{i\epsilon\tau}\,g(\tau)\,d\tau\,dt\,d\epsilon\,,\\
&= \frac{1}{2\pi}\int\int e^{i(E-\epsilon)t}\,A(t)\,\tilde{g}(\epsilon)\,dt\,d\epsilon\,,\\
&= \int \sigma(E-\epsilon)\,\tilde{g}(\epsilon)\,d\epsilon\,.
\end{aligned} \tag{9.17}$$

We now have to specify the analytic form of the window function $g(t)$. In particular, we consider four window functions

$$g_k(t) = \cos^k\left(\frac{\pi t}{2T}\right)\theta\big(T - |t|\big)\,, \tag{9.18}$$

with[3] $k = 0, 1, 2, 3$. These are displayed in Fig. 9.2. $k = 0$ corresponds to the case described in Sect. 9.2.1.

Fourier-transforming the (time) window functions g_k yields the energy window functions $\tilde{g}_k$

[3]The symbol $\theta(x)$ denotes the Heaviside step function.

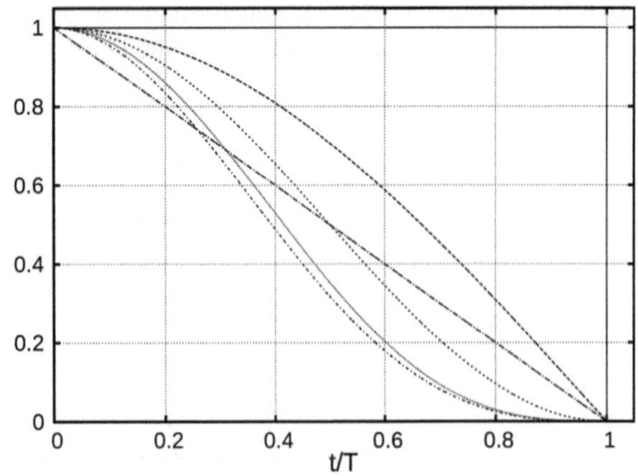

Fig. 9.2 Window functions g_0, g_1, g_2, g_3, and, to be discussed below, g'_0 and g'_1. The straight line is g'_0 whereas g'_1 is the lowest curve in the plot

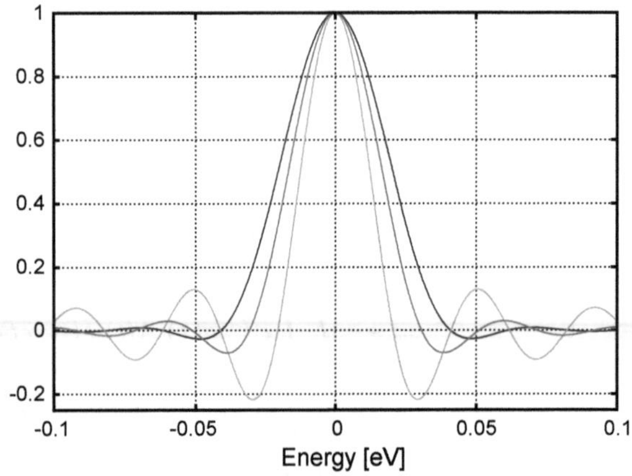

Fig. 9.3 Energy window functions. Reduction of the Gibbs phenomenon by application of time window functions: (i) $\tilde{g}_0$, *green*; (ii) $\tilde{g}_1$, *red line*; (iii) $\tilde{g}_2$, *blue line*. The duration of the autocorrelation function is $T = 100$ fs

$$\begin{aligned}
\tilde{g}_0(\omega) &= \frac{\sin(\omega T)}{\pi\omega},\\
\tilde{g}_1(\omega) &= \frac{2T\cos(\omega T)}{(\pi - 2\omega T)\,(\pi + 2\omega T)},\\
\tilde{g}_2(\omega) &= \frac{\pi\sin(\omega T)}{2\omega\,(\pi - \omega T)(\pi + \omega T)},\\
\tilde{g}_3(\omega) &= \frac{12\pi^2 T\cos(\omega T)}{(\pi - 2\omega T)\,(\pi + 2\omega T)(3\pi - 2\omega T)\,(3\pi + 2\omega T)}.
\end{aligned} \tag{9.19}$$

The oscillations caused by the box-filter ($k = 0$) are known as the *Gibbs phenomenon*. To avoid or at least lessen those we use in general g_1 or g_2. As shown in Fig. 9.3 the better filter leads to broader lines. The energy window functions $\tilde{g}_k$ are normalized[4]

[4] $\int \tilde{g}(\omega)d\omega = \frac{1}{2\pi}\int e^{i\omega t}g(t)dtd\omega = \int \delta(t)g(t)dt = g(0) = 1.$

Table 9.1 FWHM values of the window functions $\tilde{g}_k$ times the duration of the autocorrelation function. Remember that the duration of the autocorrelation function is twice the propagation time, if the $t/2$-trick, Eq. (9.12), is used

$\tilde{g}_0$	$\tilde{g}_1$	$\tilde{g}_2$	$\tilde{g}_3$	$\tilde{g}'_0$	$\tilde{g}'_1$	Unit
2.49	3.38	4.14	4.78	3.66	4.91	eV fs
20.1	27.3	33.2	38.6	29.5	39.6	cm^{-1} ps

as $\int \tilde{g}_k(\omega)\, d\omega = 1$. For better visibility, however, they are re-scaled in Fig. 9.3 at their maxima.

For the convenience of the reader we provide the full widths at half maximum (FWHM) of the window functions $\tilde{g}_k$. (The filter functions $\tilde{g}'$ will be discussed below.) The entries in Table 9.1 are to be divided by the duration of the autocorrelation function to yield the FWHM.

As is well known, the Fourier transform of the convolution of two functions equals the product of the Fourier transforms of the functions (Cf. Eqs. (9.16) and (9.17)). The self-convolution of a filter function thus leads to a filter in ω-space, which is a square and hence non-negative, a very desirable property. We define

$$g'(t) = \frac{a}{2T} \int g(\tau) g(2t - \tau) d\tau \ , \tag{9.20}$$

where a is a normalization factor to be chosen such that $g'(0) = 1$, i.e. $a = 2T / \int [g(\tau)]^2\, d\tau$. One has to use $2t$ rather than t as shift to ensure that the new function, g', has again the support $[-T, T]$. The Fourier transform reads

$$\tilde{g}'(\omega) = \frac{a\pi}{2T} \left[\tilde{g}(\frac{\omega}{2}) \right]^2 \ , \tag{9.21}$$

where the appearance of $\omega/2$ is a consequence of the use of $2t$ in Eq. (9.20). This doubles the width of the filter, but the squaring reduces the widths and the overall increase in width is about $\sqrt{2}$. Applying the procedure just described to g_0 and g_1 yields

$$g'_0(t) = (1 - |t|/T)\, \theta(T - |t|) \, , \tag{9.22}$$

$$g'_1(t) = \left[\left(1 - \frac{|t|}{T}\right) \cos(\frac{\pi t}{T}) + \frac{1}{\pi} \sin(\frac{\pi |t|}{T}) \right] \theta(T - |t|) \ , \tag{9.23}$$

and their Fourier transforms read

$$\tilde{g}'_0(\omega) = \frac{T}{2\pi} \frac{\sin^2(\omega T/2)}{(\omega T/2)^2} \tag{9.24}$$

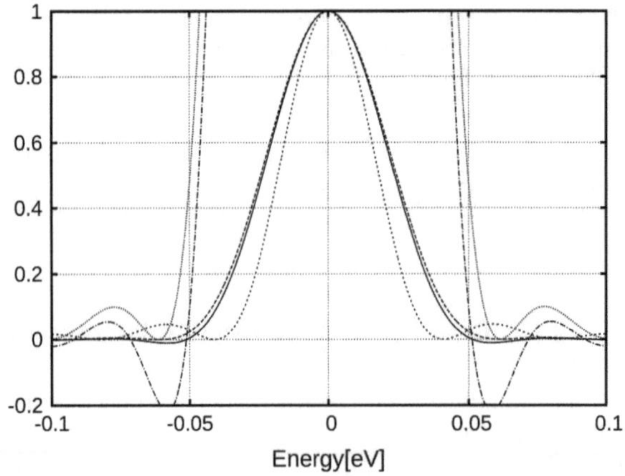

Fig. 9.4 Energy window functions, second set. (i) $\tilde{g}_3$, *solid line*; (ii) $\tilde{g}'_0$, *dotted line*; (iii) $\tilde{g}'_1$, *dashed line*. For $\tilde{g}_3$ and $\tilde{g}'_1$ the wings of the filters are also shown 20 times enlarged. The duration of the autocorrelation function is $T = 100$ fs

$$\tilde{g}'_1(\omega) = \frac{4\pi T\,\cos^2(\omega T/2)}{(\pi - \omega T)^2\,(\pi + \omega T)^2}\,. \tag{9.25}$$

It should be emphasized again, that the $\tilde{g}'$ filters are non-negative. In particular, $\tilde{g}'_1$ is an almost ideal filter with only very small wing-oscillations. See Fig. 9.4 However, it is the broadest of all filters discussed.

In conclusion, let us assume that we have a discrete spectrum with $H\Psi_m(R) = E_m\Psi_m(R)$. The Fourier transform of the autocorrelation function, $A(t)$, associated with the wavepacket $\Psi(R, t = 0) = \sum_m c_m\Psi_m(R)$ multiplied by a window function $g(t)$ is given by

$$\frac{1}{2\pi}\int_{-\infty}^{\infty} e^{iEt}\, g(t)\, A(t)\, dt = \sum_n |c_n|^2 \tilde{g}(E - E_n)\,, \tag{9.26}$$

In particular, Eq. (9.6) corresponds to Eq. (9.26) for the choice $g(t) = g_0(t)$. Several functions g can be employed to analyze the power spectrum depending on the accuracy we want to reach and the propagation time one wants to consider.

9.3 Application to the Vibrations of the Water Molecule

Let us first focus on the electronic ground state within the framework of the Born-Oppenheimer approximation. The Hamiltonian operator for the nuclear problem is given by

$$H = T^{nu} + V_0\,, \tag{9.27}$$

with V_0, the PES of the adiabatic electronic ground state and T^{nu}, the KEO for the nuclei. The total angular momentum of the molecule, J, is a constant of the motion and can thus be fixed in the calculations. We will consider the $J = 0$ case only for the sake of simplicity but all the calculations below could be performed for other values of J. If $J = 0$, the nuclear KEO no longer depends on the Euler angles. The corresponding vibrational eigenstates satisfy (see Sect. 3.2.3)

$$(T^{nu}(\boldsymbol{q}) + V_0(\boldsymbol{q}))\Psi_0^n(\boldsymbol{q}) = E_0^n \Psi_0^n(\boldsymbol{q}), \tag{9.28}$$

where $\boldsymbol{q}$ denotes the $3N$-6 internal degrees of freedom, N being the number of atoms.

We assume here that the system stays below the dissociation energy and thus that the spectrum is discrete. For a given initial vibrational wavepacket:

$$\Psi(\boldsymbol{q}, t = 0) = \sum_n c_n \Psi_0^n(\boldsymbol{q}), \tag{9.29}$$

with

$$c_n = \int \Psi_0^{n\star}(\boldsymbol{q})\Psi(\boldsymbol{q}, t = 0)d\boldsymbol{q}, \tag{9.30}$$

the Fourier transform of the autocorrelation function of the wavepacket for a finite propagation time, T, is given by

$$\sigma(E) = \sum_n |c_n|^2 \tilde{g}(E - E_0^n), \tag{9.31}$$

with $\tilde{g}$, the Fourier transform of the window function, $g(t)$ (Eq. (9.26)).

Let us consider an example, the water molecule. The spectrum of water vapor is of fundamental importance for many reasons [7]. For instance, water plays a major role in the absorption and retention of sunlight in the Earth atmosphere, mainly in the infrared region where water shows strong absorption [8]. Without the greenhouse effect of water that provides some 30 K of heating (the majority of the greenhouse effect!), life on Earth would not be possible, since our planet would be in a permanent ice age. In addition, water is the third most abundant molecule in the universe (after H_2 and CO) and plays an important role in the formation of stars [8]. Understanding the lines and levels of water is crucial to describe comets, planets, exoplanets, interstellar clouds [9].

Here, we consider only the vibrations of the molecule in its electronic ground state. Water is a nonlinear triatomic molecule and its electronic ground state gives an equilibrium structure with C_{2v} point group symmetry: we refer the reader to Sect. 7.2 of Chap. 7 for a description of this point group symmetry.

The three internal coordinates are chosen as the three valence coordinates R_1, R_2 and θ as depicted on Fig. 9.5a. The corresponding KEO is given by Eq. (6.169) of Sect. 6.5.

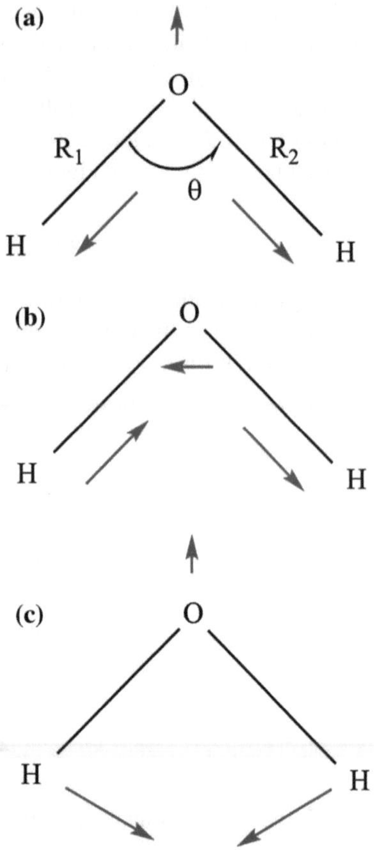

Fig. 9.5 The three normal modes of vibration of water: **a** symmetric stretching, **b** antisymmetric stretching, and **c** bending. The three valence coordinates R_1, R_2, and θ are given in (**a**). R_1 and R_2 are the distances between the oxygen atom and the two hydrogen atoms

The molecule has three normal modes of vibration that are shown on Fig. 9.5: (a) a symmetric stretching mode that changes the lengths of the two chemical bonds symmetrically, (b) an antisymmetric stretching mode that changes the lengths of the two chemical bonds antisymmetrically, and (c) a bending mode that changes the angle between the two bonds. These three normal coordinates have been calculated with the method described in Sect. 5.1 but, again, we do not use normal coordinates in the following calculations. We use valence coordinates for the dynamics.

We used here the PES of Ref. [10] that has been obtained by fitting the parameters in the potential in order to reproduce many experimental rovibrational levels of H_2O. In other words, this is a semi-empirical PES calibrated by comparison with available experimental data.[5] A cut through the potential is shown on Fig. 9.6, where the angle

[5]The PES has been re-expressed in a product form using the POTFIT approach described in Sect. 8.7.2. θ is the contracted degree of freedom (see Sect. 8.7.3) and we used 12 single particle potentials (SPPs) for R_1 and 12 also for R_2. The SPPs are the first eigenvectors of the density matrices of Eq. (8.346). At the end, we have thus 144 terms in the fit (see Eq. (8.350) with three degrees of freedom). We used 20 iterations to improve the relevant region defined as the geometries

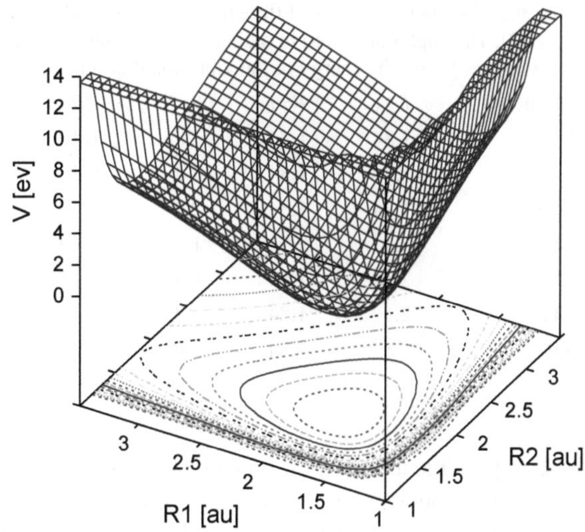

Fig. 9.6 Cut of the three-dimensional PES of H_2O taken from Ref. [10]. The coordinates are the valence coordinates of Fig. 9.5a. Here, θ, the angle between the two chemical bonds is fixed at 1.79 rad

θ is fixed at 1.79 rad. We see that there is a minimum for $R_1 = R_2 \approx 1.8$ a.u. Around this equilibrium geometry the PES can be approximated by harmonic functions. Thus the PES is an elliptic paraboloid (more precisely a paraboloid of revolution since R_1 and R_2 play a symmetric role). For higher values of R_1 and R_2, the chemical bonds tend to break. If we fix R_1 at 1.8 a.u., the corresponding one-dimensional curve along R_1 looks like a Morse function (similar to the blue curve in Fig. 3.12 of Sect. 3.5.6). Note that we have added a energy cutoff at 0.5 hartree ($\approx$ 13.9 eV). This is the reason why the potential is flat above the cutoff. These regions are physically inaccessible and correspond to R_1 and R_2 being very small. It is always very important to cut the potential because very high potential values drastically slow down the numerical integrators (the latter are described in Sect. 8.1.6).

The actual equilibrium values of the coordinates are $R_{1\,eq} = R_{2\,eq} = 1.84$ a.u. and $\theta_{eq} = 1.83$ rad. Let us propagate wavepackets with the following initial form:

$$\Psi(R_1, R_2, \theta, t = 0) = N\, e^{-\frac{(R_1 - R_{1,0})^2}{4\,\Delta R_1{}^2}}\, e^{-\frac{(R_2 - R_{2,0})^2}{4\,\Delta R_2{}^2}} \times e^{-\frac{(\theta-\theta_0)^2}{4\,\Delta\theta^2}}, \tag{9.32}$$

i.e. a product of three Gaussian functions, N being a normalization constant. For the first wavepacket WP1 (see Table 9.2), we take $R_{1,0} = R_{2,0} = R_{1,eq}$, $\theta_0 = \theta_{eq}$. For the widths of the Gaussian functions, we take $\Delta R_1 = \Delta R_2 = 0.13$ a.u. and $\Delta\theta = 0.16$ rad. The parameters of WP1 have been chosen such that this wavepacket is as

(Footnote 5 continued)
corresponding to a potential energy below 2 eV above the minimum (see Sect. 8.7.5). This guarantees a root-mean-square (rms) error of 0.0355 meV on the relevant region and thus a very high accuracy.

Table 9.2 Parameters for the different initial wavepackets. They all refer to the expression of Eq. (9.32). The angles are given in radian and the distances in a.u. The last column shows the mean energy in cm^{-1} with the energy of the vibrational ground state (ZPE) subtracted

Parameters	$R_{1,0}$	$R_{2,0}$	θ_0	ΔR_1	ΔR_2	$\Delta\theta$	$\langle H\rangle$
WP1	1.84	1.84	1.83	0.13	0.13	0.16	41
WP2	1.84	1.84	2.4	0.13	0.13	0.2	3775
WP3	1.84	1.84	2.8	0.13	0.13	0.2	7765
WP4	2.1	2.1	1.83	0.18	0.18	0.2	5399
WP5	2.1	1.65	1.83	0.15	0.10	0.16	5516
WP6	2.1	1.84	2.3	0.16	0.13	0.17	6222
WP7	2.8	1.84	1.82	0.18	0.18	0.18	19,557

close as possible to the vibrational ground state. The previous wavefunction is not the vibrational ground state since the latter cannot be written as a simple product of one-dimensional functions. Even for the vibrational ground state, the three coordinates are correlated due to the presence of cross terms in the Hamiltonian operator, i.e. to the fact that the Hamiltonian operator is not separable. Using normal coordinates would reduce this correlation for the vibrational ground state since the latter corresponds to small vibrations around the equilibrium geometry. However, even with normal coordinates, the vibrational ground state is not exactly a product of one-dimensional functions, since the potential is not purely harmonic. The probability density of the wavepacket starting from the initial condition of Eq. (9.32) will thus depend on time according to Eq. (2.38). We will not use Eq. (2.37) to obtain its time evolution since the latter requires the accurate knowledge of several eigenstates and we do not know them at this stage. The propagation of the wavepacket is performed using the Heidelberg MCTDH package, i.e. by solving Eqs. (8.151) and (8.201). This approach gives the time evolution of the wavepacket without the need of knowing the eigenstates. The wavepacket is here written as

$$\Psi(R_1, R_2, \theta, t) = \sum_{j_1=1}^{n_1}\sum_{j_2=1}^{n_2}\sum_{j_3=1}^{n_3} A_{j_1 j_2 j_3}(t)\,\varphi_{j_1}(R_1, t)\,\varphi_{j_2}(R_2, t)\,\varphi_{j_3}(\theta, t)\,. \tag{9.33}$$

In the following, we use, $n_1 = n_2 = 7$ and $n_3 = 8$. Each single particle function is expressed in a "primitive", time-independent basis set. For the primitive basis set of each coordinate, we use a DVR, more precisely a sine DVR for R_1 and R_2 with R_1 and $R_2 \in [1.0, 3.475]$ a.u. (this DVR is described in detail in Sect. 8.1.3) and a Legendre DVR for θ restricted between 0.5 and 3.2 rad.[6] Here, in Eq. (8.152) $N_1 = N_2 = 34$ and $N_3 = 50$ and we propagate the wavepacket during 150 fs. The average energy value

[6]The Legendre DVR is similar to the DVR in θ introduced for the two-dimensional spherical harmonics DVR in Sect. 8.1.3 with $m = 0$ in Eq. (8.55). It corresponds to the proper DVR associated with Legendre polynomials.

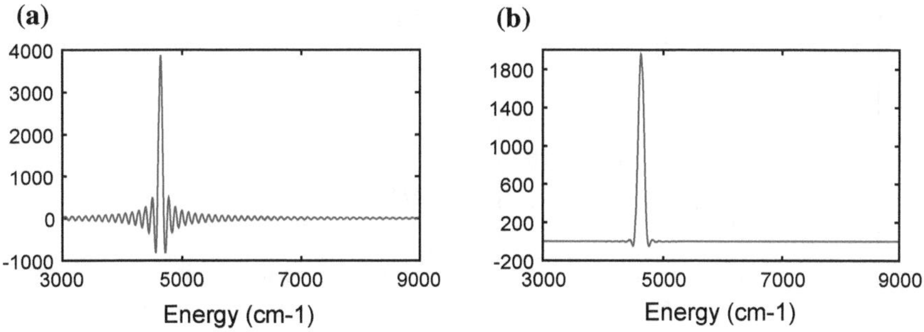

Fig. 9.7 Fourier transform of the autocorrelation function for a propagation time T = 150 fs of the wavepacket WP1. In **a** $\frac{1}{2\pi}\int_{-T}^{+T} e^{iEt}A(t)\,dt$ is used. In **b** $\frac{1}{2\pi}\int_{-T}^{+T}\cos^2\left(\frac{\pi t}{2T}\right)e^{iEt}A(t)\,dt$ is used. The Y-axis is in arbitrary unit

of the wavepacket, $\langle H\rangle$ is equal to 4676 cm^{-1}, a little bit higher than the "zero point energy" (ZPE), i.e. the absolute value of energy of the vibrational ground state that is equal to 4635 cm^{-1} (the zero in energy is the minimum of the PES). The modulus of the autocorrelation function varies little and remains close to one (not shown here). In Fig. 9.7a, we show the Fourier transform of the autocorrelation function over a finite propagation time of $T = 150$ fs: $\frac{1}{2\pi}\int_{-T}^{+T} e^{iEt}A(t)\,dt$ (i.e. $k = 0$ in Eq. (9.18)). One observes a peak located around 4635 cm^{-1} corresponding to the eigenvalue of the vibrational ground state. As discussed in Sect. 9.2.1, the Fourier transform has the form of the cardinal sine function. In other words, $\tilde{g}(E)$ in Eq. (9.31) is a cardinal sine function that is centered on the energy of the ground state. There are other peaks since the wavepacket is not identical to the vibrational ground state but they are very small and hidden within the oscillations of the cardinal sine function.

In Fig. 9.7b the Fourier transform is depicted again over 150 fs, but with a window function $\cos^2\left(\frac{\pi t}{2T}\right)$ (i.e. $k = 2$ in Eq. (9.18)). Here, $\tilde{g}(E)$ in Eq. (9.26) is equal to $\frac{\pi\sin(ET)}{2E\ (\pi-ET)(\pi+ET)}$. The shape of $\tilde{g}(E)$ can be seen on Fig. 9.3. There are lower oscillations due to the presence of the window function but there are still wiggles in the tails of the peak due to the Gibbs phenomenon as explained in Sect. 9.2.2. By comparison of Fig. 9.7a, b, we see that the Gibbs phenomenon is attenuated by the presence of the window function. Note, however, that the price to pay is the broadening of the peak that leads to a loss of accuracy for the determination of the eigenvalue. If we propagate longer, the width of the peak will become smaller as predicted by the time-energy uncertainty principle.

Let us now consider wavepacket WP2 with the expression of Eq. (9.32) and the parameters given in Table 9.2. In other words, the Gaussian function in θ is centered at a larger angle and with a larger width in θ: we create artificially a wavepacket with more energy in the bending mode of vibration. The average energy value of the new wavepacket, $\langle H\rangle$, is equal to 8410 cm^{-1}, which is 3775 cm^{-1} above the ZPE. The propagation time is equal to 300 fs and the numbers of SPFs are slightly increased: $n_1 = n_2 = 9$ and $n_3 = 10$. The modulus of the autocorrelation function

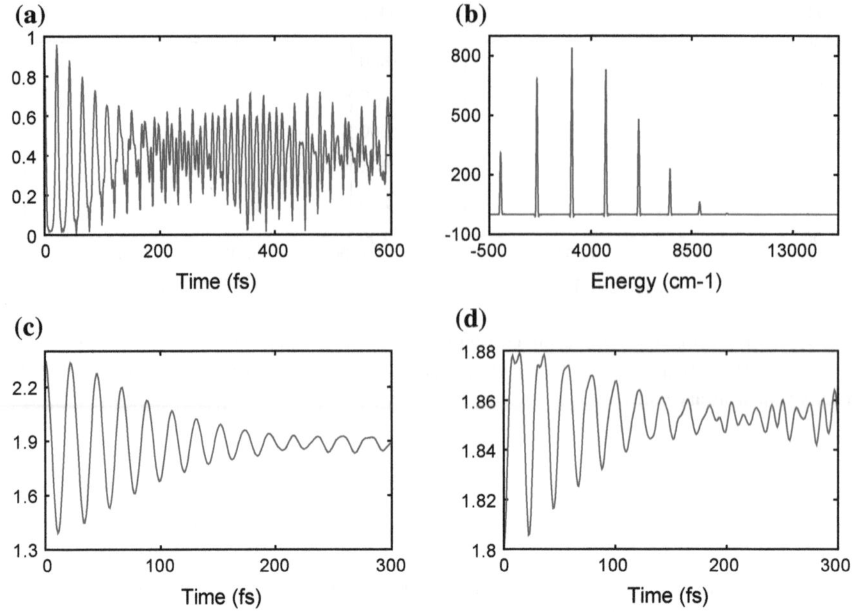

Fig. 9.8 Propagation of a wavepacket WP2 over 300 fs for H_2O, i.e. with some excitation of the bending motion. **a** Modulus of the corresponding autocorrelation function. **b** Fourier transform of the autocorrelation function (arbitrary unit). The peaks correspond to the first eigenvalues of the bending mode of water (the Y-axis is in arbitrary unit). **c** Expectation value $\langle\theta\rangle$ in radian. **d** Expectation value $\langle R_1\rangle$ in a.u.

is given in Fig. 9.8a. Since the initial function is real and the Hamiltonian operator is Hermitian, we can use Eq. (9.12) and obtain the autocorrelation function up to 600 fs, i.e. twice the propagation time. By visual inspection of Fig. 9.8a, we see that the autocorrelation now oscillates, indicating that the wavepacket is moving in the potential well. This motion corresponds mainly (but not only) to a bending motion. The propagation with the MCTDH package presents some oscillations of the wavepacket in θ. In particular, some nodes appear in the one-particle θ-density (i.e. some values in θ for which the wavepacket is equal to zero[7]), indicating that the wavefunction does not remain Gaussian. Figure 9.9 provides the two-dimensional densities (with R_2 being integrated out) of the corresponding wavepacket at different times. The short propagation time in Fig. 9.9a exhibits a simple oscillation from the starting wavefunction at $t = 0$ fs around $\theta = 2.4$ in red and moving to smaller values of θ: around 1.8 rad at 6 fs in purple, around 1.4 rad at 12 fs in black. The wavepacket will then continue oscillating, moving back, first, to larger values of θ. The wavefunction is a Gaussian function in θ at time equal to 0 fs and *a Gaussian wavepacket has the essential property to remain Gaussian in the presence of a harmonic potential* [11,

[7]What we call "nodes" corresponds to geometries for which the one-dimensional density is very small and not necessarily equal to zero.

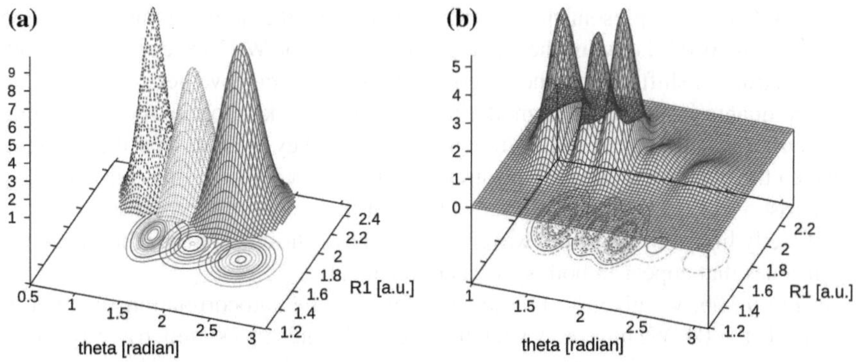

Fig. 9.9 Two-dimensional densities along θ and R_1 starting from WP2. **a** The density at 0 (*red*), 6 (*purple*), and 12 fs (*black*). **b** The density at 80 fs

12]. In other words, the center of the Gaussian wavefunction moves and its width increases and decreases but the wavefunction remains Gaussian. As a consequence, the center of the Gaussian function is the expectation value $\langle\theta\rangle$ that is depicted in Fig. 9.8c. Since the potential is approximately harmonic in θ in the low-energy domain, we see on the short propagation times (Fig. 9.9a) that the density along θ keeps a Gaussian form and simply oscillates. But, due to the anharmonicity of the potential in θ, the density loses its Gaussian shape rapidly and, in particular, several maxima and nodes crop up as shown for $t = 80$ fs in Fig. 9.9b.

By visual inspection of Fig. 9.9b, we can see that the maxima and nodes are rather well localized along θ only and, thus, that the coupling with the stretching modes is small. The propagation also shows some small vibrations of R_1 and R_2 due to the fact that the bending motion is slightly coupled to the stretching modes of vibration. This is confirmed by the expectation values $\langle\theta\rangle$ and $\langle R_1\rangle$ in Fig. 9.8c, d: we see that the average value of θ strongly oscillates, whereas $\langle R_1\rangle$ oscillates very little (we obtain exactly the same picture for $\langle R_2\rangle$, since R_1 and R_2 play a symmetric role).

The Fourier transform of the autocorrelation function with the window function $\cos\left(\frac{\pi t}{2T}\right)$ is given on Fig. 9.8b. As often in spectroscopy, we have taken the ZPE (here, $4635\,\text{cm}^{-1}$) as the reference energy, i.e. we give the energy differences with respect to the vibrational ground state. In the power spectrum, we see seven main peaks. The first one is centered at zero and corresponds to the vibrational ground state, the one that was at the origin of the large peak in Fig. 9.7. The other peaks are rather regularly spaced and thus correspond to vibrational levels with one, two, three, four, five, and six quanta in the bending mode of vibration. The peaks corresponding to modes of stretching of the O-H bonds are too small to be distinguished at this level. Note that the most intense peak lies around $3151\,\text{cm}^{-1}$, not far from the average value of the energy, which is equal to $3775\,\text{cm}^{-1}$. We have thus been able to define an initial wavepacket that is almost entirely a superposition of vibrational bending states with very small contributions from the other (stretching) vibrations.

In Fig. 9.10a, we present the Fourier transform of the autocorrelation function starting from WP3, i.e. with the same parameters as for WP2 except $\theta_0 = 2.8$ rad. The spectrum is shifted to higher energy values, i.e. to eigenvalues corresponding to more quanta in the bending mode. Several small peaks crop up, corresponding to eigenstates with quanta in the stretching modes. They indicate that the coupling between the bending mode and the stretching modes becomes a little bit larger. By comparison of Figs. 9.8b and 9.10a, we see that the height of the peaks is artificial and directly linked to the initial wavepacket we have chosen. On the other hand, the eigenvalues that appear in both spectra are identical.

In Fig. 9.10b, we give the Fourier transform of the autocorrelation function of wavepacket WP4. We have now stretched the O-H bonds in a symmetric way. Logically, a progression of symmetric stretching vibrations is observed. However, there are smaller peaks corresponding to symmetric-stretching/bending combination lines. Note that for this symmetric stretch excitation the eigenvalues corresponding to odd antisymmetric stretching excitations are forbidden by symmetry, but even excitations are possible: the $v = 2$ antisymmetric stretching excitation appears at 7445 cm^{-1} with a small intensity. In Fig. 9.10c, we give the Fourier transform of the autocorrelation function of wavepacket WP5. The symmetry between the two O-H bonds has been broken and a progression of antisymmetric stretching vibrations is observed. Finally, in Fig. 9.10d, we present the Fourier transform of the autocorrelation function of wavepacket WP6 with an excitation of all the modes of vibration.

At this stage, several conclusions can be drawn: (i) with this approach the level of accuracy, limited by the widths of the peaks, is directly linked to the propagation time. Experimental data can be very accurate; this computational approach would require very long propagations to reach such a high accuracy; (ii) this approach provides the eigenvalues but not the eigenstates; (iii) the heights of the peaks depend on the initial wavepacket. Each line corresponds to the energy of a photon that could be absorbed to create a transition between two eigenstates. To obtain the absorption spectrum, an important information is missing: the efficiency of the absorption of the photon, i.e. the efficiency of the interaction between light and the molecule for a given transition. This information is linked to the dipole moment operator. But, before introducing it, let us first calculate the eigenstates of water with a time-independent method.

9.4 The Time-Independent Approach

The first eigenstates and their eigenvalues were calculated with the improved relaxation method of Sect. 8.4. The eigenvalues are given in Tables 9.3 and 9.4 in cm^{-1}. The observed experimental values are also given, taken from Ref. [9]. In Tables 9.3 and 9.4 are also given the labels proposed by the experimentalists to assign the eigenstates. As usual in spectroscopy [13], the conventional picture in terms of normal modes is adopted. In other words, the labels are given in the form $(v_1 v_2 v_3)$ where v_1, v_2, v_3 are the numbers of quanta in the three normal modes. The corresponding symmetric stretching normal coordinate is called q_1 and depicted in Fig. 9.5a; the

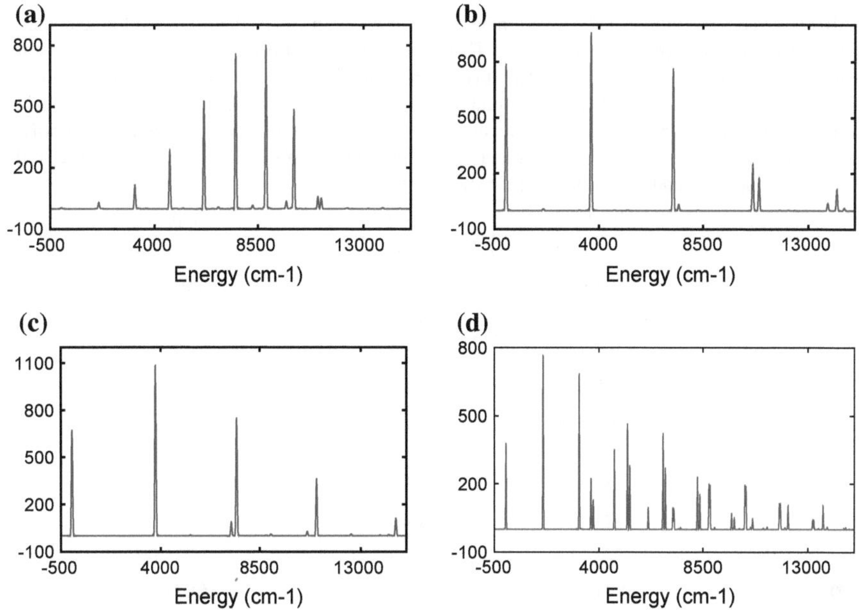

Fig. 9.10 Propagation of wavepackets starting from WP3, WP4, WP5 and WP6 (see Table 9.2 and the text) over 300 fs for H_2O in its electronic ground state. Given are the Fourier transforms of the autoco(rrelation functions of WP3 (**a**), WP4 (**b**), WP5 (**c**), and WP6 (**d**) in arbitrary unit. The window function g_1' (if, Eqs. (9.23, 9.25)) is used here

bending normal coordinate is called q_2 and depicted in Fig. 9.5c; the antisymmetric stretching coordinate is called q_3 and depicted in Fig. 9.5b. We thus use Eqs. (5.9), (5.10) and (5.11), with $3N - 6 = 3$, as a "zero-order" description of the vibrations of water. Such a description is excellent for this molecule in the low-energy domain.

The vibrational ground state bears, of course, the label (000) and its eigenvalue serves as the reference in energy (ZPE). Indeed, only transitions and, thus, energy differences between the eigenvalues are observed experimentally. The two-dimensional reduced density over R_1 and R_2 associated with the eigenfunction (000) is depicted in Fig. 9.11a.[8] The ground state looks like a two-dimensional Gaussian function centered at the equilibrium geometry.

Low in energy, the description in terms of normal modes is perfectly adapted. For instance, the level at $3657.157\,\text{cm}^{-1}$ bears the label 100. The corresponding two-dimensional reduced density is depicted on Fig. 9.11b: we clearly see one node along the direction corresponding to the stretching mode, i.e. the bisector in the (R_1, R_2) plane. For the level at $3755.809\,\text{cm}^{-1}$ with the label 001, we see on Fig. 9.11c one node now along the antisymmetric stretching mode of vibration (the diagonal perpendicular to the previous direction). The level 101 at $7249.792\,\text{cm}^{-1}$ has one node along the stretching mode and another node along the antisymmetric mode as shown

[8]The reduced density are obtained by integrating $|\Psi|^2$ over all the other coordinates, here over θ.

Table 9.3 The first vibrational excitation energies for water in cm^{-1}: theoretical values calculated with the improved relaxation of MCTDH ("Theory") and with the PES of Ref. [10] and experimental values taken from Ref. [9] ("Exp."). The theoretical absolute eigenvalue of the vibrational ground state, or "zero point energy" (ZPE), is equal to 4634.7499 cm^{-1}. The labels correspond to the assignment proposed by the experimentalists based on a normal mode description of the vibrations of the system. v_1, v_2, v_3 denote the numbers of quanta in the symmetric stretching mode (ν_1), the bending mode (ν_2), and the antisymmetric stretching mode (ν_3)

Number	Theory	Label	Exp.
0	0.000	000	0.00
1	1594.671	010	1594.75
2	3151.510	020	3151.63
3	3657.157	100	3657.05
4	3755.809	001	3755.93
5	4666.854	030	4666.79
6	5234.965	110	5234.98
7	5331.344	011	5331.27
8	6133.101	040	6134.01
9	6775.212	120	6775.09
10	6871.536	021	6871.52
11	7202.163	200	7201.54
12	7249.792	101	7249.82
13	7444.710	002	7445.06
14	7536.656	050	7542.37
15	8273.860	130	8273.98
16	8373.924	031	8373.85
17	8761.747	210	8761.58
18	8806.734	111	8807.00
19	8851.331	060	8869.95
20	9000.076	012	9000.14

on Fig. 9.11d. Let us return to the discussion of the spectrum given in Fig. 9.10b. The second strong peak corresponds to the transition to 100. The following strong peak corresponds to the transition to 200 at 7202.163 cm^{-1} and the smaller peak on its right corresponds to the transition to 002. The transition to 101 is hidden by the transition to 200. In the same manner, the second high peak in Fig. 9.10c corresponds to the transition to 001. Again, the transition to 101 is too small to be observed on this spectrum. The peaks between 7100 and 7600 cm^{-1} correspond to the transitions to 200 (the smaller peak) and to 022 (the higher peak on the right). On Fig. 9.10d, the two transitions to 100 and 001 appear just below 4000 cm^{-1}. A zoom around 7250 cm^{-1} would show that there are two lines that are not completely resolved, corresponding to the transitions to 101 and 200.

Table 9.4 First vibrational eigenvalues for water in cm^{-1} (continued)

Number	Theory	Label	Exp.
21	9720.134	140	
22	9832.644	041	9833.58
23	10,044.452	070	
24	10,284.542	220	10,284.36
25	10,328.428	121	10,328.73
26	10,521.781	022	10,521.76
27	10,600.699	300	10,599.69
28	10,613.675	201	10,613.36
29	10,868.470	102	10,868.87
30	11,032.605	003	11,032.40
31	11,067.700	150	
32	11,215.268	080	11,254.00
33	11,237.358	051	11,242.78
34	11,766.707	230	11,767.39
35	11,812.515	131	11,813.21
36	12,007.815	032	12,007.77
37	12,139.554	310	12,139.32
38	12,151.130	211	12,151.25
39	12,324.354	090	
40	12,406.796	112	12,407.66

Similarly to Figs. 9.11, 9.12a–c display the two-dimensional densities along θ and R_1 for the levels 010 (1594.671 cm^{-1}), 020 (3151.510 cm^{-1}) and 050 (7536.656 cm^{-1}), respectively. They present one, two and five nodes along θ, i.e. they correspond to eigenstates with one, two and five quanta in the bending mode of vibration. All the corresponding transitions can be clearly seen on the spectra shown in Figs. 9.8b and 9.10a. Note that the transition to 010 is small on Fig. 9.10a. These transitions appear also on Fig. 9.10d: the first two large peaks after the ground state correspond to the transitions to 010 and 020.

Higher in energy, the situation becomes more complex and the assignment of the vibrational states by counting minima (nodal lines) in the density plots becomes questionable. For instance, the state presented in Fig. 9.12d is labeled 090 by the experimentalists. But one does not clearly see on Fig. 9.12d nine nodes along θ. In other words, the nodes and maxima are not perfectly aligned along θ for the same values of R_1. Some local maxima correspond to higher values of R_1 indicating that there is a coupling between the bending mode and the stretching mode. In the normal mode picture, the bending mode and the stretching modes are supposed to be decoupled but in this energy domain around 12,000 cm^{-1} above the ZPE, it is known, for water, that there is a coupling, a *Fermi resonance* between the bending

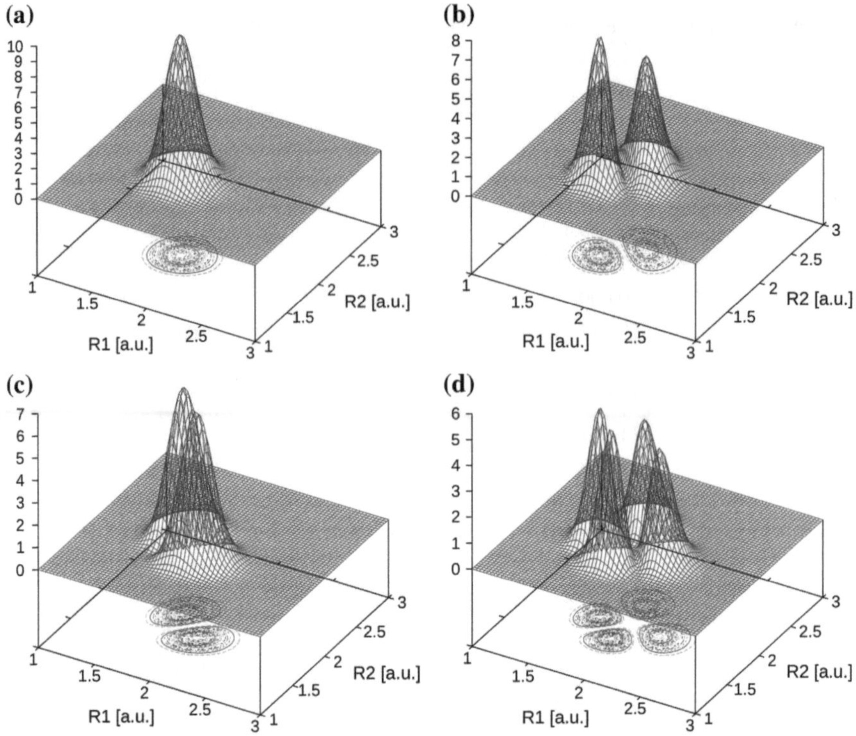

Fig. 9.11 Two-dimensional densities along R_1 and R_2. The reduced densities are obtained by integrating $|\Psi|^2$ over θ. **a** The vibrational ground state 000, **b** the state 100 at 3657.157 cm^{-1}, **c** the state 001 at 3755.809 cm^{-1}, and **d** the state 101 at 7249.792 cm^{-1} (see Table 9.3)

mode and the symmetric stretching mode. They start to vibrate in a correlated way and the corresponding eigenstates mix the vibrations along θ, R_1, and R_2. This Fermi resonance is due to the fact that the harmonic frequency of the stretching mode is roughly twice the frequency of the bending mode inducing an interaction between the corresponding harmonic oscillators.[9] The description in terms of normal modes starts to become inadequate, although such a description is a very good starting point for water as opposed to other molecules where Fermi resonances may complicate the spectrum even for the very first excited states, such as for $H_5O_2^+$ that will be studied in Sect. 9.5.

Another coupling is very interesting in water. As mentioned above, the description in terms of normal modes works very well for the first excited states of the stretching modes: look at the reduced densities for 100, 001 and 101 on Fig. 9.11b–d and

[9]The interaction between the zero-order levels associated with the normal modes can be rationalized by a two-state quantum system: in second-order perturbation theory, it can be shown that the strength of the interaction is proportional to the strength of the coupling divided by the energy difference between the zero-order levels.

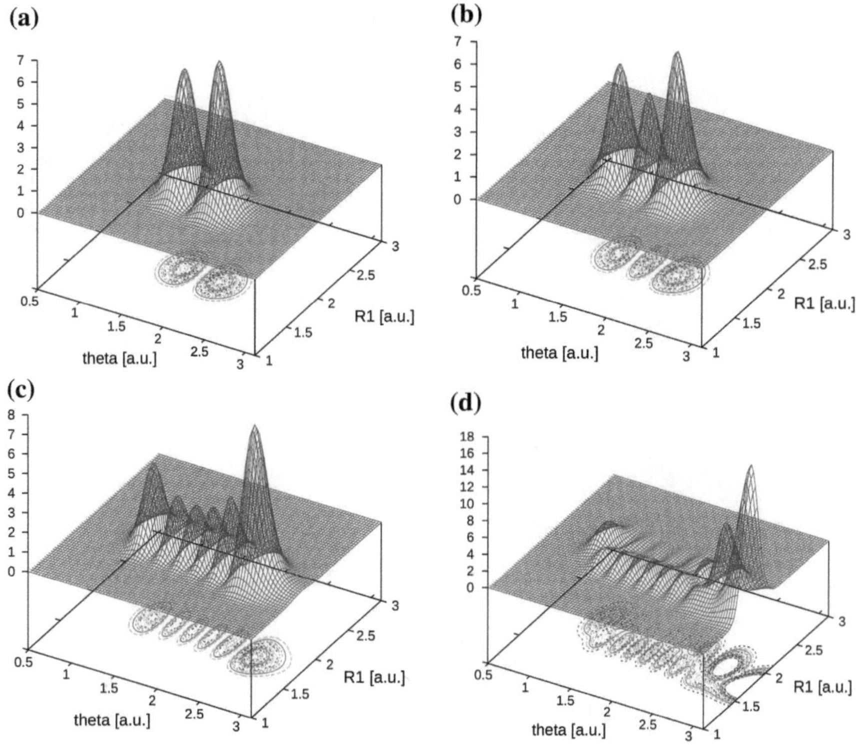

Fig. 9.12 Two-dimensional densities along R_1 and θ. The reduced densities are obtained by integrating $|\Psi|^2$ over R_2. **a** State 010 at 1594.671 cm^{-1} (see Table 9.3), **b** state 020 at 3151.510 cm^{-1}, **c** state 050 at 7536.656 cm^{-1}, and **d** state 090 at 12324.354 cm^{-1}

for 201 on Fig. 9.13a. However, the situation is different higher in energy. For the states labeled 300, 102 given on Fig. 9.13b, c, the distribution of the nodes is no longer regular. For the state 300 on Fig. 9.13b, we would expect three nodes and four maxima along the diagonal corresponding to the symmetric stretching motion, such as in Fig. 9.11b, but with three nodes instead of one. However, this is not exactly what is observed: the presence of two local maxima on the right side of the figure clearly indicates that there is a coupling with the antisymmetric mode of vibration. In the same manner, we would expect for 102 of Fig. 9.13c a distribution of the nodes similar to 201 of Fig. 9.13a, with only a permutation of the role of the symmetric and antisymmetric motions. This is not what is observed. The coupling between the symmetric and antisymmetric stretching modes is not specific to water and occurs in many molecules: this is what is called a *Darling-Dennison resonance*. Such a coupling leads to a profound change of regime in the dynamics: the emergence of *local modes of vibration* [14].

We have already explained in Sect. 5 that, low in energy, the vibrations are often dominated by the normal-mode regime but higher in energy the situation can change:

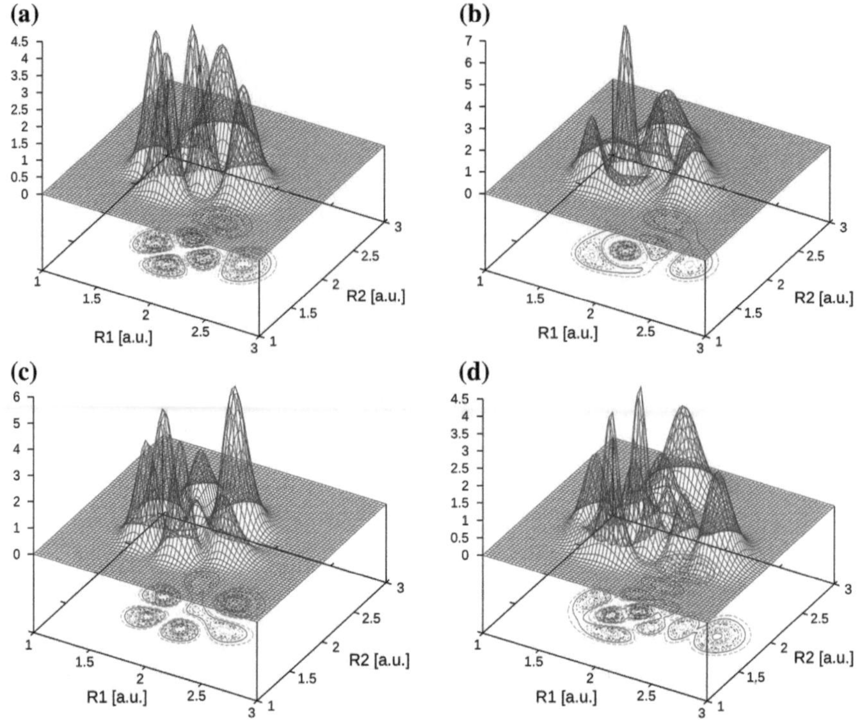

Fig. 9.13 Two-dimensional densities along R_1 and R_2: **a** the state 201 at 10,613.675 cm^{-1}, **b** the state 300 at 10,600.699 cm^{-1}, **c** the state 102 at 10,868.470 cm^{-1} (see Table 9.3), and **d** the state 500 at 16,904.898 cm^{-1} (not given in Tables 9.3 and 9.4)

some chemical bonds may vibrate in a more independent way. Some vibrations are more localized and this situation is referred to as the local-mode regime [15, 16]. In this context, Fig. 9.13d is very interesting. The corresponding state has the label 500 but some of the nodes are clearly localized along the R_1 and R_2 directions and not along the symmetric and antisymmetric motions that are linked to the diagonal directions on the figures. This is quite clear for the nodes for R_1 or $R_2 > 2.0$ a.u. on Fig. 9.13d. The vibrations start to be localized along the bond lengths. This means that the chemical bonds start to vibrate independently, as expected in the local-mode regime. For the vibrations along R_1 and R_2, a zero-order description based on the eigenstates of one-dimensional Morse functions along R_1 and R_2 becomes better-adapted than a zero-order description based on both one-dimensional harmonic oscillators along the normal coordinates (i.e. along the symmetric and antisymmetric stretching motions) [16].

This is illustrated in a more vivid way on Fig. 9.14a, b. Figure 9.14a corresponds to the two-dimensional density of WP4 at 91 fs. We see a clear delocalization of the motion along the symmetric direction. For other times, we see nodes appearing along the antisymmetric direction (not shown). Figure 9.14b corresponds to WP7 at

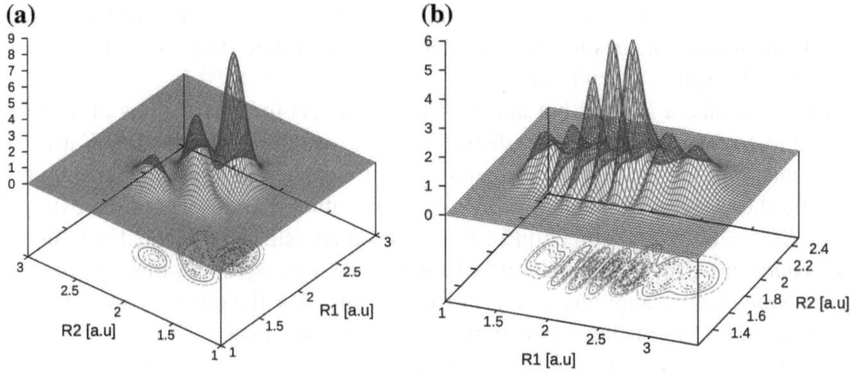

Fig. 9.14 Two-dimensional densities along R_1 and R_2 at $t = 91$ fs for the initial wavepackets WP4 (**a**) and WP7 (**b**)

91 fs. WP7 corresponds to a strong stretching of R_1 alone: $R_{1\,0} = 2.8$ a.u. We see that the wavepacket remains delocalized along R_1 only. This holds during the whole propagation. This shows that in this energy domain (about 20,000 cm^{-1} above the ZPE), the chemical bonds are almost not coupled together. Of course, if we stretch even more R_1, we will break the chemical bond.[10] A symmetric stretching, with $R_{1\,0}$ = 2.8 a.u. and $R_{2\,0}$ = 2.8 a.u., would lead to complex dynamics, with nodes along R_1 and R_2 and not along the two diagonals as for WP4. The dynamics of WP4 in a lower energy domain (around 5400 cm^{-1}) shows that the two chemical bonds always move in a concerted way. If we had stretched R_1 only in this energy domain (by taking $R_{1\,0}$ = 2.1 a.u. and $R_{2\,0}$ = 1.8 a.u. for instance), we would have seen, first, a localization of the nodes along R_1, but after a certain time a delocalization of the nodes along R_2 (not shown here), exemplifying that there is a strong coupling between R_1 and R_2 in the local-mode regime, in sharp contrast with the dynamics of WP7 in the local-mode regime.

Until now, we have only explained how to obtain the eigenstates and the eigenvalues but not how to obtain the infrared spectrum (the intensities). For this purpose, we need to further introduce the dipole moment of the molecule.

9.5 Calculation of the Infrared Spectrum

The interaction of a molecule with an external electric field is given by the term $-\boldsymbol{\mu} \cdot \boldsymbol{E}(t)$, where $\boldsymbol{\mu}$ is the dipole moment operator and $\boldsymbol{E}$ the external electric field (see Sect. 3.4). However, the dipole operator $\boldsymbol{\mu}$ is expressed in the BF-frame, whereas the electric field and its polarisation is known in the LF frame. A correct theoretical description hence requires to rotate the wavefunction from the BF frame to the LF

[10]The dissociation energy above the ZPE is estimated at $D_0 \approx 41{,}128\,\text{cm}^{-1}$ [17].

frame. The resulting theory is rather complicated, but correctly accounts for the fact that the molecule changes its total angular momentum state (from $J = 0$ to $J = 1$, say) after absorption of a photon.

Here, we take a simpler but approximate route. As the orientations of a molecule in free space are statistically distributed, one averages over the three Cartesian components of the dipole operator, and ignores a change in total angular momentum. Furthermore, we assume a sudden approximation with respect to the external field, $\boldsymbol{E}(t) = \boldsymbol{e}\delta(t)$. A δ-function of time is a superposition with equal weights of all frequencies. Due to the linearity of the Schrödinger equation, this approach is hence equivalent to applying a series of fields, one after the other, of well-defined frequency. The latter is the way experiments are usually done.

Assuming a weak field, the IR absorption cross section is given, from first-order perturbation theory, as

$$I(E) = \frac{\pi E}{3\, c\, \epsilon_0} \sum_{\lambda=x,y,z} \sum_n \left| \int \Psi_0^{n\,\star}(\boldsymbol{q}) \mu_{00}^{\lambda}(\boldsymbol{q}) \Psi_0^0(\boldsymbol{q}) d\boldsymbol{q} \right|^2 \delta(E + E_0 - E_n^0), \quad (9.34)$$

where Ψ_n^0 is the n-th exact eigenstate for total angular momentum $J = 0$, μ_{00}^{λ} is the λ-th Cartesian component of the electronic ground state dipole operator, E_0 is the ground state energy, and E the photon energy.

We arrive to the following important result: *applying the three components of the dipole moment vector along the x, y, and z directions on the vibrational ground state gives three wavepackets. The sum of the three corresponding wavepacket spectra multiplied by E is proportional to the infrared absorption spectrum of the molecule.*

It is now clear, how to proceed in the time-dependent picture. One defines the autocorrelation functions:

$$A^{\lambda}(t) = \langle \mu_{00}^{\lambda} \Psi_0^0 | e^{-iHt} | \mu_{00}^{\lambda} \Psi_0^0 \rangle \,, \quad (9.35)$$

i.e. one has to propagate the initial states $\mu_{00}^{\lambda} \Psi_0^0$. For $\lambda = x, y, z$,

$$\sigma^{\lambda}(E) = \frac{1}{\pi} \mathrm{Re} \int_0^{\infty} e^{iEt} A^{\lambda}(t) dt \,, \quad (9.36)$$

where, as discussed before, an appropriate window function $g(t)$ should be added to allow for a finite propagation time with suppressed Gibbs artifacts. Then one obtains

$$I(E) = \frac{\pi E}{3\, c\, \epsilon_0} \sum_{\lambda=x,y,z} \sigma^{\lambda}(E) \,. \quad (9.37)$$

Let us consider an example: the protonated water dimer, $H_5O_2^+$, i.e. two water molecules with a proton in between. The incentive for studying this system exceeds the sole description of this small cluster, since hydrated protons are ubiquitous in nature [18]. For instance, hydrogen cations abound in liquid water but only in the form of complexes and $H_5O_2^+$ is one of the most important of them. The accurate description of this system thus contributes to the understanding of how protons are transported through liquid water. In particular, the protonated dimers are important for environmental reasons since, for instance, they take part in acid-base reactions that cause acid rain; they are important also for biological processes, etc. [18]. Describing correctly the vibrational quantum dynamics of this system has thus implications beyond IR-spectroscopy.

We have seen in the previous section, that for a single molecule of water, the description of the vibrations is very well described by the normal-mode picture at low energy. In the case of $H_5O_2^+$, the situation is completely different. First, the system is more anharmonic for the modes of vibration corresponding to the vibrations of the proton between the two water molecules and for the other intermolecular vibrations. It is natural to think that a description in terms of normal modes that are defined as linear combinations of Cartesian coordinates is not adapted to describe the relative motions of rotations of the molecules of water (this is particularly clear for the internal torsion) and that a description by means of angles for these "motions of large amplitude" is better-adapted. In other words, this system is said to be "floppy", and a description in terms of "curvilinear" coordinates is required, as explained in Sect. 5.2 for smaller systems. The exact definition of the coordinates is not given here. We simply mention the fact that they lead to a complicated KEO with hundreds of terms [19, 20]. We refer here to the 15-dimensional PES and the three dipole moment surfaces of Ref. [21].

Second, and more importantly, the system is impacted by strong couplings between the most important modes of vibration, i.e. strong Fermi resonances play a dominant role even for the first vibrational eigenstates. We have already observed these Fermi resonances for water: the resonance between the bending mode and the symmetric stretching mode and the Darling-Dennison coupling, but their effects appear rather high in energy, at least not in the low-energy domain. Again, for $H_5O_2^+$, these resonances play a role even for low-lying eigenvalues, and $H_5O_2^+$ is a good example of a molecular system for which the conventional picture in terms of normal modes is inadequate.

The system has seven atoms and thus 15 modes of vibration. Let us assume that all the nuclei are in the same plane as shown on Fig. 9.15. The vibrations of the system can be classified in terms of the following 15 vibrational modes[11]:

- Three intramolecular vibrations for each molecule of water. They are similar to the modes of the isolated molecule of water described in Fig. 9.5: the symmetric stretching mode, the antisymmetric stretching mode and the bending mode. Since there are two molecules of water, this yields six intramolecular modes of vibration.
- The proton transfer corresponding to the motion of the proton along the Oxygen-Oxygen axis.

[11]These modes of vibration do not correspond exactly to the normal modes of the system. As already pointed out, the normal mode description is not adapted for this system anyway. However, these 15 modes of vibration can be seen a natural first-order descriptions of the vibrations of the system.

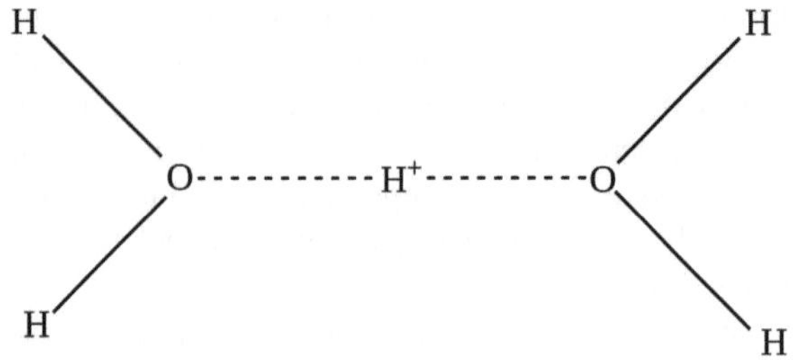

Fig. 9.15 The $H_5O_2^+$ system: geometry corresponding to all the nuclei in the same plane

- The two out-of-axis proton motions corresponding to the motions of the proton perpendicular to the Oxygen-Oxygen axis.
- The Oxygen-Oxygen stretching mode.
- The two wagging motions of the molecules of water: they correspond to the out-of-plane motions of the two molecules of water.
- The two rocking motions of the molecules of water: they correspond to rotations of the molecules of water in the plane.
- The internal rotation corresponding to the rotation of one molecule of water with respect to the other (torsion along O–O).

As aforementioned, it is necessary to use curvilinear coordinates due to the strong anharmonicity of the system, but the modes of vibration listed above fail also to describe qualitatively the nature of the first eigenstates due to the presence of strong "Fermi resonances". In particular, when the proton moves to the molecule on the right, the latter goes out of the plane and a pyramidal H_3O^+ appears whereas the other molecule of water remains in the plane: see Fig. 9.16. Of course, if the proton goes back and moves to the molecule of water on the left, the molecule on the right will go back to the plane. Then, the molecule on the left will go out of the plane and create another pyramid with the proton. In other word, through this process of "pyramidalization", there is a very strong coupling between the mode corresponding to the proton transfer and the wagging modes of vibration of the molecules of water. But that is not all: when H_3O^+ forms a pyramid the HOH angle in the molecule of water changes. There is thus a strong coupling also with the bending mode of

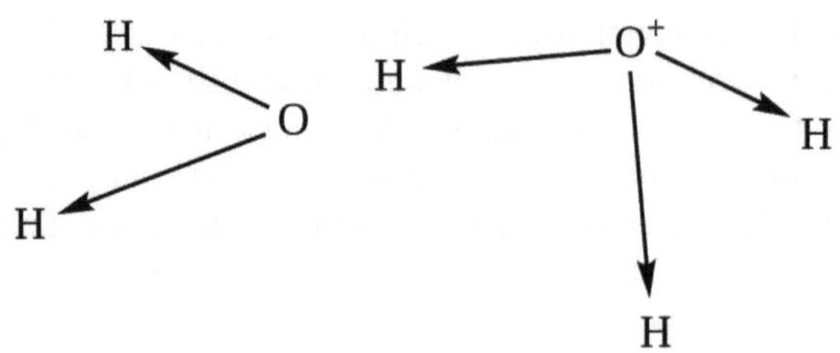

Fig. 9.16 The $H_5O_2^+$ system: starting from Fig. 9.15, the proton moves to the molecule of water on the *right side*. The molecule of water goes out of the plane and the H_3O^+ subsystem forms a pyramid whereas the other molecule of water remains in the original plane

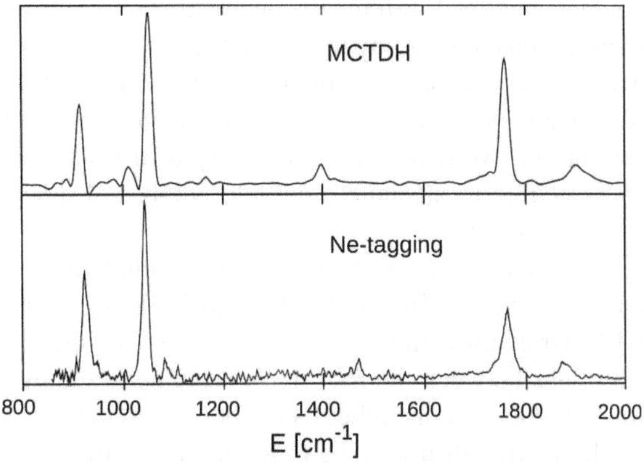

Fig. 9.17 Comparison of the MCTDH spectrum with the available experimental spectrum for $H_5O_2^+$

vibration of the water molecule. In this system, and contrarily to the water dimer [22, 23], there is a strong coupling between the intermolecular modes of vibration (the proton transfer and the wagging modes) and the intramolecular vibrations (the bending modes of vibration of the molecules of water). In addition, the O–O distance changes in the process and thus the O–O stretching mode is also coupled to the modes that are at the origin of the pyramidalization. Another strong coupling exists between the water bending and the proton motion along the O–O axis [24, 25].

The IR predissociation spectrum of the cation has been measured under neon- and argon-solvated [26] conditions. It is expected that the photodissociation spectrum of the $H_5O_2^+{\cdot}Ne_1$ complex is close to the infrared absorption spectrum of the bare cation [26]. Due to experimental conditions, the observed peaks have a natural width and the reproduction of the spectrum does not require a very high accuracy contrarily to many cases in infrared spectroscopy. The problem is thus ideally adapted for a time-dependent approach. Starting from the 15D vibrational ground state, Ψ_0^0, the MCTDH method has been used to propagate the wavepackets $\mu_{00}^x\Psi_0^0$, $\mu_{00}^y\Psi_0^0$ and $\mu_{00}^z\Psi_0^0$. They have been propagated over 500 fs, giving an autocorrelation length of over 1 ps by using the $t/2$-trick of Eq. (9.5). The simulated spectrum is obtained by Fourier transforming the autocorrelation functions of the wavepackets as explained above. This provides a resolution of $30\,\text{cm}^{-1}$ with a full width at half maximum (FWHM) similar to the resolution of the experiments.

The calculated and experimental spectra between 700 and $1900\,\text{cm}^{-1}$ are provided in Fig. 9.17[12]: the excellent agreement reveals the very high quality of the PES and dipole moment surfaces of Bowman and coworkers Ref. [21]. The calculated spectrum covers a larger domain from zero cm^{-1} (corresponding to the vibrational ground state) to $3700\,\text{cm}^{-1}$: see Ref. [27]. However, we present here only a zoom between 700 and $1900\,\text{cm}^{-1}$ because, first, there is available experimental data only in this domain and, second, because this corresponds to the vibrational eigenvalues that combine the modes forming the pyramidalization of H_3O^+. In particular, the

[12]Reprinted with permission from [24]. Copyright 2007, American Institute of Physics.

doublet at about $1000\,cm^{-1}$ is mainly due to the strong coupling between the proton transfer and the wagging motion of the molecules of water. It is also due to a coupling between the two previous modes, the bending modes, and the O–O stretching mode. No assignment of the corresponding eigenstates can be done without a quantitative simulation. We realize here the sharp difference with the water molecule where it was possible to attribute a definite number of quanta in the normal modes to the low-lying peaks.

It is possible to perform the same simulations for the main isotopologues of the system: $D(D_2O)_2^+$, $H(D_2O)_2^+$ and $D(H_2O)_2^+$ just by changing the masses in the KEO. Dramatic differences are found between the spectra [25, 28]. In particular, the couplings between the four modes leading to the pyramidalization is even stronger for $H(D_2O)_2^+$. On the other hand, these couplings are much smaller in $D(H_2O)_2^+$. We see how difficult it may be to predict the nature of the different peaks in a system such as $H_5O_2^+$ and its isotopologues since no qualitative chemical interpretation can be invoked to explain them. The position and the nature of the peaks originate from the values of the different frequencies associated with the modes of vibration and from the possible resonances between them. Only a quantitative quantum-mechanical treatment could disentangle the spectrum.

9.6 Installing MCTDH

Before starting with the lab-session, one has to download and install the Heidelberg `MCTDH` package. Please write an e-mail to Hans-Dieter.Meyer@pci.uni-heidelberg.de and ask for access to the code. You may also wish to visit the `MCTDH` web-site http://mctdh.uni-hd.de. Please download the current 84-version mctdh84.x.tgz (with a suitable integer value for x), as well as the files addsurf.tgz and lab-session.tgz from the page http://mctdh.uni-hd.de/packages/. This page is password protected: you will receive the password together with an extended installation instruction when asking for the MCTDH package.

All the input files for the lab sessions and a version of the code are also provided as supporting material of the present book.

It is most convenient to create a directory called `MCTDH` and store the downloaded files there, then un-tar them. An alternative way, and in fact the recommended one, is to use `SUBVERSION` (svn) to download the package-code, the addsuf, and the lab-session directories. When you ask for the package, you will receive a description of how to use svn. Alternatively, one may download all files needed from the Springer web-server `http://install_mctdh`.

The next step is to run "`./install_mctdh`" while being in the `$MCTDH_DIR/install` directory. (`$MCTDH_DIR` denotes the path of your `MCTDH` directory). The installation will be easy on a modern Linux system. If you are running on Apple OSX, installation should also be painless. However, one first has to install `xcode` and then download and install `gfortran` and `gnuplot`. Please see the HTML documentation under "Installation and Compilation"/"Installation prob-

lems" for more details on OSX installation.
(http://www.pci.uni-heidelberg.de/tc/usr/mctdh/doc.84/mctdh/gen.html)

If you do not want to download the most recent version but rather like to use an already installed one, please make sure that you are using version 8.4.12 or higher (or 8.5.5 and higher). Older versions do not support all options of some analysis routines used here.

In this session, we begin with studying the dynamics of the water molecule. However, no water PES is included by default in the `MCTDH` package. To include it please run "`mklinks h2o`", after having successfully installed the package and the `addsurf` directory. If running `mklinks` does not work, because some environment variables may not be set, please copy the file `addsurf/h2o.f` to `$MCTDH_DIR/source/surfaces`. To include the surface one has to re-compile potfit: `compile -i h2o potfit`. Try `compile -h` for a better understanding of what you are doing. All `MCTDH` programs and scripts know the help-option `-h`, use it!

Now we are ready to run MCTDH. Move to the directory `lab-session`. It contains the file `lab-session.pdf`, which provides a description of the lab-session, and the directory `lab-inputs`. Copy the directory `lab-inputs` to some appropriate location, where you want to perform the calculations. Then follow the instructions below.

9.7 Lab-Session I: H_2O, Computing Vibrational Eigenstates ($J = 0$)

For our studies we will use the PJT2 PES for H_2O by Polyansky, Jensen and Tennyson. See JCP **105**, 6490-6497 (1996) and JCP **101**, 7651 (1994) for details. As `MCTDH` requires a product form for the Hamiltonian, the PES has to be potfitted first. After that we will generate the vibrational ground state by `improved relaxation`, and then vibrational excited states by three sequential `block improved relaxations`. The present example, H_2O, is actually too small to release the full power of `MCTDH`. However, one can learn a lot by studying this system and all calculations will be fast.

To do the calculations please move to `lab-inputs/relax` and type "`./run-rlx`"[13]. Please inspect the script first. One may submit the commands individually, but it is more convenient to run the script. Depending on your computer installation, running the script will take between 1 and 4 min. Before we inspect the results we will discuss the input files. The `POTFIT` input file `h2o.pfit.inp` starts with a `RUN-SECTION` where the directory is specified to which all results are written (we call this the *name-directory*). The next line states that 20 iterations should be performed to improve the fit and the following line ensures that some error measures are written to the file `iteration`. The PES is selected in the

[13]run-rlx is an executable file. Make it executable if necessary.

`OPERATOR-SECTION` where additionally a potential-cut-off is defined. Note that when no unit is specified, atomic units (a.u.) are assumed. It is very important to "cut" a potential, because very high potential values will slow down the integrators. On the other hand, a wavepacket does not go to regions where the potential is very high, hence cutting the potential at some reasonable value is safe. As we are not interested in energies above 3 eV, a cut at 13.6 eV (0.5 a.u.) is reasonable. In the next section, `NATPOT-BASIS-SECTION`, the number of natural potentials is specified. Here 12 natural potentials are used for $r1$ and $r2$, i.e. the `natpot` consists of 144 terms. There is a contraction over the theta-DOF, note that one always must contract over one mode. The `PRIMITIVE-BASIS-SECTION` specifies the DVRs, see the HTML documentation for details. (Access `file://$MCTDH_DIR/doc/index.html` with your preferred browser). Finally, the `CORRELATED-WEIGHT-SECTION` determines the so-called relevant region. During the iterations `POTFIT` tries to improve the representation on the relevant region while the fit may deteriorate elsewhere. All grid points with a potential energy below 2 eV are defined as relevant regions. Since the vibrational energy is distributed over three degrees of freedom, this value is sufficient for calculating vibrational states up to a total energy of 3 eV.

After running `potfit84` we want to inspect the PES. Move to the directory `relax/h2o.pfit` and run `showpot84`. A menu will appear, but first simply accept the default values by typing "1" three times. A blank input will bring back the menu. Next, we may switch to an eV scale (type "90" and then "ev") and to a 3D view (type "160"). Then again accept the defaults. As long as `showpot84` is running one can rotate the figure with the mouse. With menu point 20 one can change the coordinate selection. Type "20" and then "x 1.8 1.9". A cut through the potential along r_1 for $r_2 = 1.8$ a.u. and $\theta = 1.9$ rad is shown. With menu point 150 one may switch on grid lines and with menu point 10 we may select another plot task, e.g. displaying the difference $V_{\text{exact}} - V_{\text{fit}}$. The effect of the iterative improvement of the fit can be seen by running "`plpit -G ur`" and "`plpit -G mr`" which show the root-mean-square and the maximum error, respectively, of the fit within the relevant region.

Next, we inspect the input file for the relaxation to the ground state, `h2o.gs.inp`. In the `RUN-SECTION` we specify the name-directory, give some descriptive text as a title, and request an improved relaxation to the ground-state (`relaxation=0`) while using a preconditioner and the Olsen-correction. The setting of the times for improved relaxation is a bit unusual (as compared to propagation). The meaning of `tfinal` is clear and `tout=all` lets the program produce an output after every update step, rather than at fixed time intervals. The keyword `tpsi` is used here to set an upper limit to the enlargement of the update time. The meaning of the next two lines as well as of the next three sections should be clear. For details see the the HTML documentation, which is part of the package. The `INTEGRATOR-SECTION` needs some explanation, as again it is rather different from one for propagation. The initial update time and the CMF accuracy is set in the first line. Note that the CMF accuracy can be rather low. The accuracies of the integrators for both the SPFs and the A-vector (the latter is not an integrator but a diagonalizer), are rather high. The next statement overwrites the default setting (1.d-8) of the regularization parameter (see, for instance the MCTDH-review [6], Eq. (82) for its definition), and the last line requests a transformation to natural orbitals after each update step. In the `INIT_WF-SECTION` an initial wavefunction is built, which is a Hartree product of three Gaussian functions.

Now move to the name-directory `h2o.gs` and there type "`rdrlx`". This script reads the `rlx_info` file and prints it in a better readable form. The last section of this output consists of several columns. The first column gives the relaxation time (in fs), the second the number of Davidson iterations. In the second row there appears a star (*), indicating that here the Davidson diagonalizer did not converge, because in the input file the maximal number of iterations was limited to 20. It is legitimate if non-convergence happens at low relaxation times, but if it appears later, one cannot trust the results! The column `beta` displays the overlap (times 1000) of an A-vector with the previous one. If this overlap is very close to unity, the difference to 1 is printed. The next column shows the energy. The first row, characterized by a negative time, displays the expectation value of the initial wavefunction. The second row, characterized by time zero, shows the energy after diagonalization but without relaxation of the SPFs. The following rows give the energy after SPF relaxation followed by diagonalization of the Hamiltonian in the basis of the SPFs. (See e.g. Sect. 8.4.1 for a brief description of *improved relaxation*). The final wavefunction is written to the `restart` file and a large amount of information on the computation is written to the ASCII files `log, op.log, speed, steps, timing, update`, and `output`. One may inspect those. Here we discuss only the last block of the `output` file and concentrate on the natural weights. All first three natural weights are very close to unity (≈ 0.998) and all forth three natural weights are negligible. Hence, the ground-state wavefunction of water, represented in valence coordinates, is only weakly correlated. A single Hartree product could be a good approximation. In fact, if one performs a normal relaxation[14] with only one SPF for each DOF, one obtains a ground-state energy of 0.576282 eV, which should be compared with 0.574636 eV, the value obtained with improved relaxation, i.e. the numerically exact value for the given PES. The difference is merely 1.6 meV or 13 cm^{-1}.

We now turn to block improved relaxations and inspect the input file `h2o.1.inp`. It is quite similar to `h2o.gs.inp`, except that there is the new line `packets=35, single-set` in the `SPF-BASIS-SECTION` and the line `autoblock` in the `INIT_WFSECTION`. The packet keyword requests that there are now 35 states which are simultaneously relaxed, and the autoblock keyword enables the creation of a block of orthogonal initial A-vectors. Moreover, the numbers of SPFs and the size of the preconditioner are enlarged. Running `rdrlx` in the directory `h2o.1` prints the lowest 35 eigenenergies of our H_2O Hamiltonian. Note that the ground-state energy is already subtracted,[15] hence shown are the 34 lowest excitation energies. Higher excitations are calculated by running `h2o.2.inp` and `h2o.3.inp`. Here

[14] For normal relaxation replace `relaxation=0 precon=100,olsen` with `relaxation`, set `tfinal=20`, and remove the `INTEGRATOR-SECTION`, a default integrator will then be taken.

[15] This is accomplished by the line "rlxunit=cm-1,4634.749881" which converts the computed energies to the unit cm^{-1} and subtracts 4634.749881 cm^{-1} from all computed energies.

the keyword `rlxemin` guarantees that higher states are computed. It is now of advantage to read the initial SPFs from the previous block improved relaxation (keyword `block-SPF`), rather than to build them.

The computed eigenenergies are compiled in the file `eigenval` together with experimentally measured eigenenergies. For the lowest 13 excitations the agreement is excellent, better than 1 cm^{-1}. This shows the high quality of the PES. A stick spectrum is shown by running "`plgen -z 0 -y 1600 eigenval 2:3 -i`" and the number of states up to a given excitation energy is displayed when typing "`plgen -G eigenval 2:1`" while being in the directory `relax`. However, the excitation energies alone are often not enough: one needs to assign them to fundamental vibrations. For this one requires the wavefunctions. Due to the keyword `split-rst` the block wavefunction is split into individual ones stored in the files `rst000` to `rst034`. To visualize these wavefunctions, more precisely, their 2D reduced densities, move to `h2o.1` and submit the commands:

```
showsys84 -rst -f rst001   ! 20: x 0 y (bend(v=1)
showsys84 -rst -f rst002   ! 20: x 0 y (bend v=2)
showsys84 -rst -f rst003   ! (sym str)
showsys84 -rst -f rst004   ! (asym str)
showsys84 -rst -f rst005   ! 20: x 0 y (bend v=3)
showsys84 -rst -f rst006   ! (sym str + bend)
showsys84 -rst -f rst007   ! (asym str + bend)
```

The exclamation mark (!) and everything after is a comment and should not appear on the command line. The comment "`20: x 0 y`" is to be interpreted as going to menu point 20 and then typing "`x 0 y`", by which one chooses r_1 and θ as coordinates of the plot. The coordinates not assigned an x or y are integrated over to obtain reduced densities. Note that menu point 20 requires as input exactly as many entries as there are DOFs. The DOFs not to be plotted must be assigned a number. The value of this number is irrelevant when reduced densities are plotted, however, these numbers define the location of a cut when cuts are plotted. After one is satisfied with the setting of the menu points, one should type "1" to select 'plot to screen' and then enter again twice a "1" to accept the default plot parameters (or modify them to ones convenience). To get acquainted with `showsys84` please try the menu points 240 and 245 which switch off (or on) the keys and the title, respectively. Try also menu points 150 and 160 which switch on (or off) grid lines and a 3D depiction, respectively. When the GNUPLOT picture appears it can be rotated with the mouse. This is most useful when the surface is toggled on (menu point 160). However, when one inputs something at the `showsys84` command line, the plot will freeze.

The 2D-density plots make it clear that the states are to be assigned to bend(v=1), bend(v=2), symmetric stretch, antisymmetric stretch, and so on. (Simply count the number of nodal lines.) The last two `showsys84` runs from the list above should be done with and without "`20: x 0 y`". One may inspect and assign higher excited states as well. An assignment of the first 43 states is listed in the `eigenval` file. The assignment agrees perfectly with the one provided by the experimentalists, except for state No. 40. The assignment is easy and clear, because the H_2O molecule in valence coordinates is not strongly coupled, at least at lower excitation energies. For highly excited states (or for more strongly correlated systems), the assignment may become ambiguous. (Inspect e.g. states 27 and 29 with `showsys84`).

9.7.1 H_2O : Wavepacket Propagation ($J = 0$)

To study wavepacket propagation move to the directory `propagate` and either run the script `run-prop`, or run the individual commands by hand. The computations should take 10–30 min, depending on your hardware. First a potfit is done. Smaller numbers of natural potentials are used here, because, when propagating, one does not aim at such high accuracy as when computing eigenenergies. This results in a slightly less accurate but a three times more compact potential representation.

The Ground State

The first propagation is done with the input file `h2o.prop.inp`. The main difference to the relaxation inputs is found in the `INTEGRATOR-SECTION`. The setting shown here is rather typical for a propagation run. Next, an initial wavefunction is built as a Hartree product of Gaussian functions, where the parameters of the Gaussian functions are chosen to resemble the ground state most closely. During the propagation the autocorrelation function $A(t) = <\Psi(0)|\Psi(t)>$ is computed[16] from which the power spectrum can be obtained by Fourier transforming it. To do so move to the directory `h2o.prop` and run "`plspec -G -g2 3000 9000 cm-1`". (See file `propagate/commands`). One observes a single peak located at the ground-state energy of 4635 cm^{-1}. The wiggles in the tails of the peak are due to the Gibbs phenomenon, i.e. to the finite length of the autocorrelation function. These "false" structures can be attenuated with the aid of a window function. For a comprehensive discussion on window functions see Sect. 9.2. The different window functions are selected by the `-g` option, `-g0`, `-g1`, ..., `-g5`. (`g4` and `g5` correspond to g_0' and g_1' of Sect. 9.2.2). If none is given, `-g1` is taken as the default. Run `plspec` with different window functions to observe the effect they have on the spectrum. The Gibbs structures can be strongly attenuated by using a window function, but the price to be paid is a broadening of the peaks. If the spectrum consists of individual well resolved lines, the windows `-g2` or `-g5` are recommended, whereas for less structured spectra -g1 is often a better choice.

As the autocorrelation function is known on discrete time-points only, the Fourier transform is actually a discrete Fourier transform that replicates the spectra infinitely. To show this behavior type "`plspec -G -g2 -q 5000 -40000 50000 cm-1`". One observes three peaks the spacing of which is given by $2\pi/\Delta t$,

[16] Actually, the formula $A(t) = <\Psi(t/2)^*|\Psi(t/2)>$ is used because this delivers an autocorrelation function twice as long as the propagation. However, this trick can only be applied if the initial state is real and the Hamiltonian symmetric. See Sect. 9.2.

where the time spacing of the autocorrelation data points, Δt, is in general defined by the `MCTDH` keyword `tout`. If `tout` is chosen too large such that the replicated spectra overlap, the result cannot be trusted. The "true" spectrum out of the series of replicated spectra is the one the central energy of which agrees with the total energy printed in the `output` file.

The initial wavefunction, a simple Hartree product of Gaussian functions, is certainly not an eigenstate. Hence, there should be more than one state appearing in the spectrum. In fact, running "`plspec -y 1.6 -z -0.05 -G -g5 -q 5000 2500 30000 cm-1`" one observes several excited states, the populations of which, however, are about three orders of magnitude smaller than the one of the ground state. This again demonstrates the fact that the vibrational ground state of the H_2O molecule (in valence coordinates) is only weakly correlated.

Bending Vibrations

To compute the IR absorption spectrum one should multiply the ground-state wavefunction with the dipole moment surface and then propagate the dipole-operated ground state. Here we proceed differently and propagate an initial Hartree product where some coordinates are shifted from their ground-state values. The input file `bend.inp` is very similar to `h2o.prop.inp`, except that there are now slightly more SPFs and that the Gaussian function of the θ-DOF is centered at larger angles and has a wider width than in `h2o.prop.inp`. The Gaussian functions of the r-DOFs are centered at slightly smaller distances, because for larger angles the r-equilibrium distances shrink.

To inspect the spectrum move to the directory `bend` and run
"`plspec -z 0 -e -4634.743 cm-1 -G -q 4000 -g5 -500 15000 cm-1`"
where the option "`-e -4634.743 cm-1`" shifts the spectrum by the ground-state energy such that the ground-state peak appears at zero energy. (One may copy the command from the file `propagate/commands`). One clearly observes a progression of bend vibrations. Other vibrational states are excited—if at all—with very low intensity. To see them add the option "`-y 8`" to `plspec`, i.e. run
"`plspec -y 8 -z 0 -e -4634.743 cm-1 -G -q 4000 -g5 -500 15000 cm-1`". Most of the now visible small peaks are noise. The noise appears in the vicinity of large peaks because of the Gibbs phenomenon. For energies above 9000 cm^{-1}, where no large peaks exist, the small peaks seem to be real.

To assign the peaks we compare them with the eigenenergies computed in the previous section. To this end we have to separate Fourier transforming and plotting. The Fourier transformation is performed by running
"`autospec84 -lin -q 5000 -e -4634.743 cm-1 -500 15000 cm-1 0`"
and the result is written to the file `spectrum.pl`. The spectrum and the eigenenergies are then plotted by running "`plgen -n -G -a -500 -x 14000 -y 850 -z 0 spectrum.pl 1:4 ../eigenval 2:3 -i`". The peak locations agree excellently with the computed eigenenergies. Next run "`plgen -n -G -a 8000 -x 14000 -y 8 -z -2 spectrum.pl 1:4 ../eigenval`

`2:3 -i`". The agreement of the peak locations with the computed eigenenergies may indicate that these structures are real. But remember that all these structures are more than three orders of magnitude smaller than the largest peak, which is assigned as (0, 2, 0). Hence we have been able to define an initial wavepacket which is almost entirely a superposition of vibrational bend states with negligible contributions from other vibrations.

Let us remark that the script `plspec` simply calls `autospec84` and then `gnuplot`. The script `plgen` is a `gnuplot` wrapper. For better understanding see "`plspec -h`", "`plgen -h`", and "`autospec84 -h`", as well as the HTML documentation, which comes with the package.

Finally, we want to inspect the wavepacket as it moves in time. To this end run "`showd1d84 -a -y 3.5 -G f3`". Shown is the one-dimensional reduced density for the θ-DOF. Each pressing of "enter" moves the wavepacket one time-step further. Repeated pressing of "enter" can be avoided by setting the option `-M`. For times smaller than 50 fs the wavepacket (more precisely the 1D reduced density) remains roughly Gaussian and oscillates back and forth with roughly the expected time period of $2\pi/\omega_{\text{bend}} = 20.92$ fs. However, for larger times the wavepacket looses its Gaussian shape and its movement looks somewhat erratic. This is the typical behavior of a wave packet moving in an anharmonic well.

The r-motion can be inspected by replacing the argument `f3` of `showd1d84` with `f1`. But in this case the motion is rather uninteresting. Similar as in Sect. 9.7 one can show two-dimensional reduced densities by running `showsys84`, but this time without the option `-rst`, because we want to make use of the `psi` file rather than of the `restart` file. Use the menu point 20 to select the (r_1, θ) pair and repeatedly press "enter" to proceed to the next time-step. In this case the 2D-plots are not much more enlightening than the 1D-plots, but this will change when we discuss stretch vibrations.

Rather than showing the reduced densities, we may investigate the coordinate expectation values. These numbers contain less information as compared to the densities, but can be shown for all times at once. Running "`plqdq -G 1 3`" shows the expectation value of θ. Very clear oscillations are shown,which, however, are damped for larger times. This damping is not an indication of loss of energy (e.g. to other vibrational modes), but a consequence of the already mentioned decoherence caused by the anharmonicities. After pressing "enter", `plqdq` will show the variance or width $<dq>=\sqrt{<q^2> - <q>^2}$. Over the first 100 fs the width roughly doubles.

Next, we may run "`plqdq -G 1 1`" to inspect the expectation value and variance of the r_1-DOF, but more instructive is to run "`plqdq -G 1 0`", which shows the expectation values of all DOFs in one graph. One clearly observes that the r-coordinates oscillate with an amplitude that is about an order of magnitude smaller than that of the θ-oscillation. However, this is not a vibrational r-motion, as this would occur with a more than twice higher frequency, it is an adiabatic following of the r-coordinates with the θ-vibration. If the bend angle is small, the r-coordinates take slightly larger values, whereas for large bend angles the r-distances are slightly compressed.

Symmetric and Antisymmetric Stretch Vibrations

In the input file `sym.inp` we have enlarged both r-distances to excite symmetric stretch vibrations. The width of the Gaussian functions is also slightly enlarged, as—due to anharmonicity—the Gaussian functions should be wider for larger distances and narrower at compressed positions. Move to the directory `sym` and submit similar commands as when studying bend vibrations. (See file `propagate/commands`). A progression of symmetric stretch vibrations is seen, as well as bend and symmetric-stretch/bend combination lines at much lower intensity. To observe the latter, enlarge the ordinate scale by setting the option `-y 100` or even `-y 10` when running `plspec`. Odd antisymmetric stretch excitations are forbidden by symmetry, but even excitations are possible and in fact one observes the $v = 2$ antisymmetric stretch excitation, (0, 0, 2), as a weak line at 7445 cm^{-1}.

Next, we investigate the time-dependent 1D-densities by running `showd1d84`. The r-densities show the expected oscillatory motion, but they quickly loose their simple Gaussian form due to anharmonicities. The θ-density, however, performs a breathing motion. If one runs `plqdq` as before, one observes an amazing similarity between the r-expectation value (first picture of `plqdq -G 1 1`) and the θ-width (second picture of `plqdq -G 1 3`). To investigate this further, run "`rdcheck84 1 1`" in order to create the file `qdq.pl`, which contains the expectation values. Then run "`plgen -G qdq.pl 1:2 - '1:($7*12)'`". The θ-width (here 12 times enlarged) agrees almost perfectly with the r-expectation value. The θ-breathing motion is hence not caused by a bend excitation, but is an adiabatic following of the r-motion, similar to the opposite case discussed previously when studying bend vibrations.

The 1D-density plots cannot distinguish between symmetric and antisymmetric motions. Here 2D-density plots are essential. Run `showsys84` and observe that the wavepacket clearly move along the symmetry axis $r_1 = r_2$. To avoid repeatedly pressing "enter" one may choose the menu point 280.

In the input file `asym.inp` we have enlarged the r_1- and shortened the r_2-distance to excite antisymmetric stretch vibrations. Move to the directory `asym` and submit similar commands as when studying symmetric stretch vibrations. A progression of antisymmetric stretch vibrations is seen, and, at lower intensity, various other lines., e.g. (0, 1, 1) at 5331 cm^{-1}, (2, 0, 0) at 7202 cm^{-1}, (0, 1, 2) at 9000 cm^{-1}, or (2, 0, 1) at 10614 cm^{-1}.

Inspect, as before, the time dependence of the 1D- and 2D-densities with `showd1d84` and `showsys84`, respectively.

In conclusion we note that by defining a suitably chosen initial state one can excite selectively, or at least with strong predominance, vibrations of a particular type. This procedure can be used as an alternative way for assigning states.

Exciting All Vibrations

Finally, we propagate a wavepacket with more general dislocations, such that all modes are excited. The propagation time of this run, `h2o.all.inp`, is doubled to ensure that the now dense lying peaks are resolved. Here we can again discuss window functions. Running `autospec84` (in `h2o.all` directory) with and without the options `-lin -o linspec` but otherwise with the usual options, creates the files `linspec.pl` and `spectrum.pl`. The latter contains the spectra computed with the window functions `g0`, `g1`, and `g2`, and the former the windows `g3`, `g4`, and `g5`. (Remember, `g4` and `g5` stand for g_0' and g_1', respectively). Running "`plgen -G -a 10100 -x 10500 -y 350 -z -50 spectrum.pl 1:3 spectrum.pl 1:4 linspec.pl 1:4`" shows a double peak computed with the windows `g1`, and `g2`, and `g5`. The double peak structure is best resolved using the `g1` window but more washed out when the `g5` window is applied. This comparison becomes even more vivid if one reduces the length of the autocorrelation function from 1200 fs to 800 fs by adding the option `-t 800` to the two `autospec84` commands above. The spectrum obtained with the `g5` window shows a single peak, whereas the spectra of the `g1` and `g2` filters still resolve the double peak. However, `g1` and `g2` introduce false negative intensities.

To demonstrate the importance of being converged with respect to the numbers of SPFs, the same calculation was re-run but with smaller numbers of SPFs. See `all.small.inp`. Move to the directory `all.small`, run `autospec84` there and then "`plgen -n -G -a 8000 -x 12000 -z -40 linspec.pl 1:4 ../h2o.all/linspec.pl 1:4 - 1:4`". The new spectrum is very noisy and this noise does not originate from the Gibbs phenomenon, but is caused by a less accurate autocorrelation function. The new spectrum is not only noisy but is also not able to resolve double peak structures, e.g. near 8800 cm^{-1} and 10300 cm^{-1}, although the width of the peaks is small enough to allow resolution. On the other hand, the inaccurate spectrum shows all main features reasonably well. To see this run "`plgen -n -G -a -300 -x 14000 linspec.pl 1:4 ../h2o.all/linspec.pl '1:(-$4)'` ", which shows the accurate spectrum inverted. This again is a demonstration that MCTDH may be used in a "quick and dirty" modus. The main features of the results are often already obtained while running in low accuracy. Note that the small, inaccurate propagation is more than 7 times faster than the large, accurate one.

One may run additional propagations with various sets of SPFs to observe the convergence of the spectrum. One also should inspect the natural populations by running "`rdcheck84 0`"—or, more vividly, "`plnat`"—in the different name directories.

9.7.2 H_2O : Computing Ro-vibrational Eigenstates with $J > 0$

In the previous sections we have considered the case of vanishing total angular momentum, $J = 0$, exclusively. Now we will compute ro-vibrational eigenstates for $J > 0$. We will do so for three different coordinate systems and as they have different grids, we have to run POTFIT three times. See `h2o.KLeg.pfit.inp`, `h2o.sr.pfit.inp`, and `h2o.jac.pfit.inp`. But similarly as before it is convenient to run the script "`./run-rlxJ`" while being in the directory `rlxJ`.

This time, the computations are more elaborate and the script should run for 20–60 min, depending on your hardware.

Valence Coordinates

We first use, as done before, valence coordinates. However, as we now also consider the overall rotation of the molecule, there appears the Euler angle γ as an additional coordinate. (The other two Euler angles, α and β, can be removed due to the existence of two constants of motion, J^2 and J_z). As γ appears only in the KEO but not in the PES it is convenient to switch to a momentum representation of this DOF and use the discrete variable k instead. Note that k is the projection of the total angular momentum $\boldsymbol{J}$ on the BF z-axis. The angular motion is now represented by the so called KLeg-DVR, which is a DVR representation of a spherical harmonics basis set. See the files `h2o.KLeg.inp` and `h2o.KLeg.op` and compare them with the corresponding input files for the $J = 0$ case.

The present calculation, stored in the name directory `h2o.KLeg`, is performed for the total angular momentum $J = 5$, hence there are $2J + 1 = 11$ rotational states on top of each vibrational state. Running `rdrlx` displays the computed excitation energies. The first eleven energies (0–10) are those of rotational states of the vibrational ground state, the second eleven energies (11–21) belong to rotations of the first bend state, and the last two energies are those of the lowest two rotational excitations on top of the second vibrational bend state. To inspect the wave function submit the command "`showsys84 -rst -f rst0`*xx*" where *xx* is to be replaced with the number of the wavefunction to be plotted. When changing the coordinate selection (menu point 20), one must now give four entries, e.g. "`0 x y 0`" for plotting the (r_2, θ) density, or "`0 0 0 x`" for plotting the k-distribution. In the latter case it is advised to set the menu point 260 in order to switch to a stick representation. Remember that there are $2J+1$ overall rotational states on top of the vibrational states. Hence the states `rst000` - `rst011` differ only marginally in their r_1, r_2, θ distributions, but strongly in their k-distributions.

Valence Coordinates, Semi-rigid

In the previous section the two-dimensional `KLeg`-DVR was used to represent the internal (θ, ϕ) motion. As this special DVR is build on the spherical harmonics functions Y_l^m, it is singularity-free, i.e. the singular terms, $\sin(\theta)^{-1}$ and $\sin(\theta)^{-2}$, which appear in a coordinate representation of a KEO, do not lead to singular DVR matrix elements. However, in the present case of water, the PES hinders the molecule to reach geometries with $\sin(\theta) \approx 0$, at least for the vibrational energies considered. Systems, for which singular points of a KEO are avoided, because of potential barriers, are called *semi-rigid*. For semi-rigid systems there is no need to use the two-dimensional `KLeg`-DVR, one may as well use ordinary one-dimensional DVRs for θ and k separately. Moreover, here we replace the variable θ with $u = \cos\theta$ because this leads to a Wilson normalization of the wavefunction, i.e. the differential element $\sin\theta d\theta$ is replaced with the simple differential element du. All this will speedup the numerical calculations.

Move to the name directory `h2o.sr` and run `rdrlx`. The computed ro-vibrational energies are virtually identical to the ones obtained using the `KLeg`-DVR. The small differences, all below 0.012 cm^{-1}, are due to the different grids and different potfits. Next, we may inspect the wavefunctions by running `showsys84`. But more interesting is to compare the files *.inp, *.op, and `timing` of the two runs h2o.KLeg and h2o.sr. Note that the calculation using the semi-rigid representation is about 3.5 times faster than the one using a `KLeg` representation.

Jacobi Coordinates

As a last example we will study H_2O in Jacobi coordinates, which are depicted in Fig. 9.18. Jacobi coordinates are so-called *orthogonal coordinates*, and, as such, they lead to a much simpler expression for the KEO than the non-orthogonal valence coordinates. This feature makes them very attractive. Compare `h2o.KLeg.op` with `h2o.jac.op`. Moreover, for scattering problems Jacobi coordinates are often the most convenient coordinates. However, for a purely vibrational problem, Jacobi coordinates are rarely a good choice. In fact, we will show that the computation of the ro-vibrational energies of water is much less efficient in Jacobi coordinates compared to valence coordinates.

Move to the name directory `h2o.jac` and run `rdrlx`. Again, the computed ro-vibrational energies are virtually identical with the ones of the h2o.KLeg run. (All differences are smaller than 0.035 cm^{-1}). Next run "`showsys84 -rst -f rst012`" and use menu point 20 to select various cuts. Due to the different coordinate sets, the 2D-density plots have a different appearance than the ones in valence coordinates. The vector $\boldsymbol{r}$ connects the two H-atoms, whereas the vector $\boldsymbol{R}$ connects the H_2 center of mass with the O-atom. The angle θ is the angle between these two vectors. Hence the bend vibration is mainly described by the r-coordinate, the symmetric stretch mainly by R, and the antisymmetric stretch mainly by θ. The (r, R)-cut shows a clear nodal line, indicating that this is the first excited bend vibrational state. On top of this one observes a correlation between the r and R motions. Because the

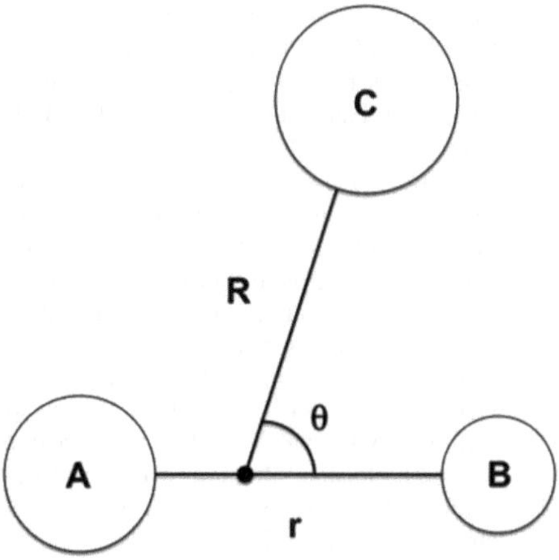

Fig. 9.18 Jacobi coordinates. The vectors **r** and **R** meet at the center of mass of the AB-subsystem. For water: A=H, B=H, C=O, for NOCl: A=N, B=O, C=Cl, and for D + H_2 : A=H, B=H, C=D

HO-distance remains almost constant during the bend vibration, the R-distance has to shrink when the r-distance enlarges. This correlation is the main reason why one needs more SPFs in Jacobi coordinates compared to valence coordinates. Note that the bend excited state shows almost no correlation between r and θ, or R and θ.

Finally, we compare the `timing` file of the `h2o.jac` name directory with those from `h2o.KLeg` and `h2o.sr`. The run using Jacobi coordinates and a KLeg representation takes about 9 times more CPU time than the valence-KLeg run, and about 33 times more CPU time than the valence-semi-rigid run. This again emphasizes the importance of determining a good coordinate system and representation.

References

1. Rothman LS, Gordon IE, Babikov Y, Barbe A, Benner DC, Bernath PF, Birk M, Bizzocchi L, Boudon V, Brown LR, Campargue A, Chance K, Cohen EA, Coudert LH, Devi VM, Drouin BJ, Fayt A, Flaud J-M, Gamache RR, Harrison JJ, Hartmann J-M, Hill C, Hodges JT, Jacquemart D, Jolly A, Lamouroux J, LeRoy RJ, Li G, Long DA, Lyulin O, Mackie C, Massie ST, Mikhailenko S, Müller HS, Naumenko O, Nikitin A, Orphal J, Perevalov VI, Perrin A, Polovtseva ER, Richard C, Smith MAH, Starikova E, Sung K, Tashkun S, Tennyson J, Toon GC, Tyuterev VG, Wagner G (2013) The HITRAN 2012 molecular spectroscopic database. J Quant Spectrosc Radiat Transfer 130:4
2. Quack M, Merkt F (2011) Handbook of High-resolution Spectroscopy. Wiley-VCH
3. Manthe U (1991) Mehrdimensionale Wellenpaketdynamik nach elektronischen Anregungen. PhD thesis, Universität Heidelberg
4. Manthe U, Meyer H-D, Cederbaum LS (1992) Multiconfigurational time-dependent Hartree study of complex dynamics: photodissociation of NO_2. J Chem Phys 97:9062
5. Engel V (1992) The calculation of autocorrelation functions for spectroscopy. Chem Phys Lett 189:76
6. Beck MH, Jäckle A, Worth GA, Meyer H-D (2000) The multi-configuration time-dependent Hartree (MCTDH) method: a highly efficient algorithm for propagating wave packets. Phys Rep 324:1
7. Polyansky OL, Cászár AG, Shririn SV, Zobob NF, Barletta P, Tennyson J, Schenke DW, Knowles PJ (2003) High-accuracy ab initio rotation-vibration transitions for water. Science 299:539
8. Bernath PF (2002) The spectroscopy of water vapour: experiment, theory and applications. Phys Chem Chem Phys 4:1501. Water, spectroscopy
9. Tennyson J, Bernath PF, Brown LR, Campargue A, Császár AG, Daumont L, Gamache RR, Hodges JT, Naumenko OV, Polyansky OL, Rothman LS, Vandaele AC, Zobov NF, Derzi AR, Fábri C, Fazliev AZ, Furtenbacher T, Gordon IE, Lodi L, Mizus II (2013) IUPAC critical evaluation of the rotational-vibrational spectra of water vapor, Part III: energy levels and transition wavenumbers for $H_2{}^{16}O$. J Quant Spectrosc Radiat Transfer 117:29
10. Polyansky OL, Jensen P, Tennyson J (1994) A spectroscopically determined potential energy surface for the ground state of $H_2^{16}O$: a new level of accuracy. J Chem Phys 101:7651
11. Tannor DJ (2007) Introduction to quantum dynamics: a time-dependent perspective. University Science Books, Sausalito, Cal
12. Heller EJ (1975) Time-dependent approach to semiclassical dynamics. J Chem Phys 62:1544
13. Herzberg GH (1945) Infrared and Raman spectra. Van Nostrand Reihnold, Toronto
14. Kellmand ME, Tyng V (2007) The dance of molecules: new dynamical perspectives on highly excited molecular vibrations. Acc Chem Res 40:243
15. Miller WH, Handy NC, Adams JE (1980) Reaction path Hamiltonian for polyatomic molecules. J Chem Phys 72:99

16. Halonen L, Carrington T Jr (1988) Fermi resonances and local modes in water, hydrogen sulfide, and hydrogen selenide. J Chem Phys 88:4171
17. Ruscic B, Wagner AF, Harding LB, Asher RL, Feller D, Dixon DA, Peterson KA, Song Y, Qian X, Ng C-Y, Liu J, Cehn W, Schenke DW (2002) On the enthalpy of formation of hydroxyl radical and gas-phase bond dissociation energies of water and hydroxyl. J Phys Chem A 106:2727
18. Xantheas SS (2009) Computational chemistry dances with hydrogen cations. Nature 457:673
19. Vendrell O, Gatti F, Lauvergnat D, Meyer H-D (2007) Full dimensional (15D) quantum-dynamical simulation of the protonated water dimer I: Hamiltonian setup and analysis of the ground vibrational state. J Chem Phys 127:184302
20. Vendrell O, Brill M, Gatti F, Lauvergnat D, Meyer H-D (2009) Full dimensional (15D) quantum-dynamical simulation of the protonated water dimer III: mixed Jacobi-valence parametrization and benchmark results for the zero-point energy, vibrationally excited states and infrared spectrum. J Chem Phys 130:234305
21. Huang X, Braams BJ, Bowman JM (2005) Ab initio potential energy and dipole moment surfaces for $H_5O_2^+$. J Chem Phys 122:044308
22. Fellers RS, Leforestier C, Braly LB, Brown MG, Saykally RJ (1999) Spectroscopic determination of the water pair potential. Science 284:945
23. Leforestier C, Gatti F, Fellers RS, Saykally RJ (2002) Determination of a flexible (12D) water dimer potential via direct inversion of spectroscopic data. J Chem Phys 117:8710
24. Vendrell O, Gatti F, Meyer H-D (2007) Full dimensional (15D) quantum-dynamical simulation of the protonated water dimer II: infrared spectrum and vibrational dynamics. J Chem Phys 127:184303
25. Vendrell O, Gatti F, Meyer H-D (2009) Full dimensional (15D) quantum-dynamical simulation of the protonated water dimer IV: isotope effects in the infrared spectra of $D(D_2O)_2^+$, $H(D_2O)_2^+$ and $D(H_2O)_2^+$ isotopologues. J Chem Phys 131:034308
26. Hammer NI, Diken EG, Roscioli JR, Johnson MA, Myshakin EM, Jordan KD, McCoy AB, Bowman JM, Carter S (2005) The vibrational predissociation spectra of the $H_5O_2^+ \cdot RG_n$ (RG=Ar, Ne) clusters: correlation of solvent perturbations in the free OH and shared proton transitions of the Zundel ion. J Chem Phys 122:244301
27. Vendrell O, Gatti F, Meyer H-D (2007) Dynamics and infrared spectroscopy of the protonated water dimer. Angew Chem Int Ed 46:6918
28. Vendrell O, Gatti F, Meyer H-D (2009) Strong isotope effects in the infrared spectrum of the zundel cation. Angew Chem Int Ed 48:352

Chapter 10
Photodissociation Spectroscopy

In the previous chapter, we have considered transitions induced by absorption of light between vibrational states in the electronic ground state only. In the present chapter, we stay within the framework of the Born-Oppenheimer approximation but we consider transitions from the electronic ground state to another electronic state that is dissociative. In other words, the absorption of the electromagnetic energy of light, typically in the ultraviolet (UV) domain, induces a fragmentation of a bound molecule. Photodissociations of small polyatomic molecules are the motors for many important chain reactions determining the complex chemistry of the atmosphere and the starting points for many chemical lasers [1]. We consider here two examples: the photodissociations of NOCl and ozone.

10.1 Direct Photodissociation

Let us consider a transition from the electronic ground state to an isolated excited electronic state that is dissociative as shown on Fig. 10.1. The Hamiltonian operator for the nuclear problem in the electronic excited state 1 is given by

$$H^1 = T^{nu} + V_1 \,, \tag{10.1}$$

with V_1, the PES of the electronic excited state and T^{nu} the KEO for the nuclei. We consider the $J = 0$ case only.

If the excited state were bound, the corresponding vibrational eigenstates would satisfy

$$(T^{nu}(\mathbf{q}) + V_1(\mathbf{q}))\Psi_1^n(\mathbf{q}) = E_1^n \Psi_1^n(\mathbf{q}) \,. \tag{10.2}$$

However, for dissociative electronic states, there is no quantization of the energy and the set of eigenstates is no longer discrete:

F. Gatti et al., *Applications of Quantum Dynamics in Chemistry*,
Lecture Notes in Chemistry 98, DOI 10.1007/978-3-319-53923-2_10

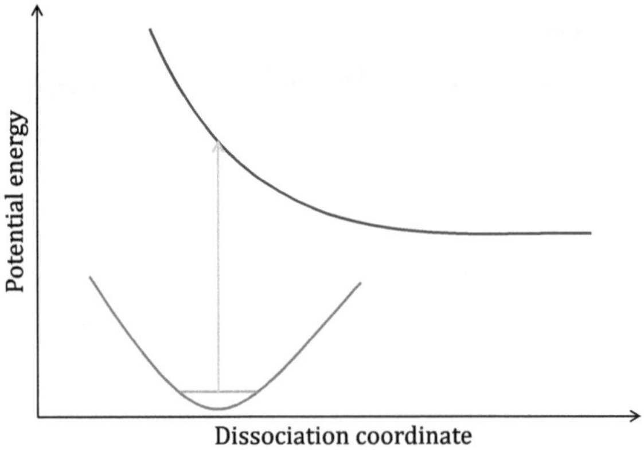

Fig. 10.1 Transition from the electronic ground state to one isolated excited electronic state that is dissociative

$$(T^{nu}(\mathbf{q}) + V_1(\mathbf{q}))\Psi_1^E(\mathbf{q}) = E\Psi_1^E(\mathbf{q})\,, \tag{10.3}$$

where E now varies continuously. The wavefunction Ψ_1^E is no longer square-integrable, but may be normalized to δ-functions $\langle\Psi_1^E|\Psi_1^{E'}\rangle = \delta(E - E')$.

Let us start from the vibrational ground state in the electronic ground state, Ψ_0^0. This wavefunction is a eigenfunction of the Hamiltonian operator for the nuclear problem in the electronic ground state (Eq. 9.28). On the other hand, Ψ_0^0 is a wavepacket and not an eigenstate with respect to H^1. In particular, we can write[1]:

$$\Psi_0^0(\mathbf{q}) = \int_0^{+\infty} c(E)\Psi_1^E(\mathbf{q})dE\,, \tag{10.4}$$

with

$$c(E) = \int \Psi_1^{E\star}(\mathbf{q})\Psi_0^0(\mathbf{q})d\mathbf{q}\,, \tag{10.5}$$

Thus, *a wavefunction can be either a wavepacket or an eigenfunction according to a Hamiltonian operator and not per se*. In particular, *the status of the wavefunction, i.e. the fact that it is a eigenfunction or a wavepacket, can change depending on the physical situation*. The Hamiltonian reflects the physical situation and can change if the molecule, which is first isolated, interacts with another molecule or an external field for instance.

The absorption spectrum is given by an equation similar to Eq. (9.35) except that we now have to sum over the eigenstates of the electronic excited state with coefficients with the following expression

$$|\int \Psi_1^{E\star}(\mathbf{q})\mu_{10}^{\lambda}(\mathbf{q})\Psi_0^0(\mathbf{q})d\mathbf{q}|^2\,, \tag{10.6}$$

[1] $E = 0$ corresponds to the dissociation limit.

with $\mu_{10}^{\lambda}(\mathbf{q})$ the matrix element of the dipole moment corresponding to the transition from the electronic ground state to the electronic state 1 (Eq. 3.158). The discrete sum in Eq. (9.5) is replaced by an integral if the excited state is dissociative since there is no longer any quantization.

We have mentioned in Sect. 3.4 that the dependence on $\mathbf{q}$ of $\boldsymbol{\mu}_{nl}(\mathbf{q}) = \boldsymbol{\mu}_{nl}^{el}(\mathbf{q})$, if $n \neq l$, is, in general, weak. Thus, it is possible to make the following approximation:

$$\mu_{10}^{\lambda}(\mathbf{q}) \approx \mu_{10}^{\lambda}(\mathbf{q}_{eq}), \tag{10.7}$$

with $\mathbf{q}_{eq}$ corresponding to the equilibrium geometry in the electronic ground state, also called the Franck-Condon geometry. Then, we have

$$|\int \Psi_1^{E\star}(\mathbf{q})\mu_{10}^{\lambda}(\mathbf{q})\Psi_0^0(\mathbf{q})d\mathbf{q}|^2 \approx (\mu_{10}^{\lambda}(\mathbf{q}_{eq}))^2 \, |\int \Psi_1^{E\star}(\mathbf{q})\Psi_0^0(\mathbf{q})d\mathbf{q}|^2, \tag{10.8}$$

also called *the Condon approximation*. This approximation would not be valid for the transitions between the vibrational states within a single electronic state. The quantity

$$|\int \Psi_1^{E\star}(\mathbf{q})\Psi_0^0(\mathbf{q})d\mathbf{q}|^2, \tag{10.9}$$

is called *Franck-Condon factor*

Now, the Franck-Condon factors are the squares of the overlaps of Eq. (10.5). Thus, if $\Psi_0^0(\mathbf{q})$ is propagated on the electronic excited state and the resulting auto-correlation function is Fourier-transformed, a power spectrum (see Eq. 9.32) will be obtained with the overlaps of Eq. (10.5). *The propagation of the vibrational ground state of the electronic ground state on the electronic excited PES yields a power spectrum proportional to the absorption spectrum from the electronic ground state to the electronic excited state in the limit of the Franck-Condon approximation. If the latter approximation is no longer valid, $\mu_{10}^{\lambda}(\mathbf{q})\Psi_0^0(\mathbf{q})$ instead of $\Psi_0^0(\mathbf{q})$ must be propagated to obtain the spectrum.*

Let us consider the dissociation of NOCl to NO + Cl induced by the $S_0 \rightarrow S_1$ electronic transition as an example. S_1 is the first singlet excited electronic state of the molecule. For the nuclear degrees of freedom of NOCl, we use the three Jacobi coordinates of Fig. 5.4: R_1 is the distance between the N and the O nuclei, R_2 is the length of the vector joining the center of mass of NO to the Cl nucleus and θ is the angle between the two Jacobi vectors $\mathbf{R}_1$ and $\mathbf{R}_2$. The KEO for $J = 0$ is given by Eq. (6.155) of Sect. 6.5. We use the PESs of Ref. [2]. The PES of the electronic ground state, S_0, is qualitatively similar to the PES of the electronic ground state of H_2O: see Fig. 9.6. The vibrational ground state, $\Psi_0^0(\mathbf{q})$, looks similar to the one for H_2O in Fig. 9.11a.

The PES of the electronic excited state, S_1, is dissociative and leads to NO + Cl. A two-dimensional cut of this PES is shown on Fig. 10.2. The PES looks like a

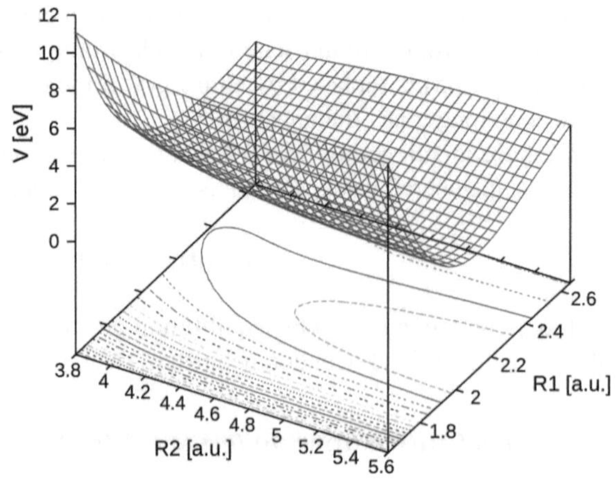

Fig. 10.2 Cut of the three-dimensional PES of the electronic excited state S_1 of NOCl taken from Ref. [2, 3]. The coordinates are the Jacobi coordinates of Fig. 5.4. Here, R_1 is the distance between the N and the O atoms, R_2 is the length of the vector joining the center of mass of NO to the Cl atom and θ, the angle between the two Jacobi vectors, is fixed at 2.1 rad

"gutter" since the potential decreases along R_2. The dissociation of NOCl is a *direct photodissociation*,[2] i.e. *the molecule dissociates immediately after the photon has promoted it to the excited electronic state.*

Let us start from the vibrational ground state, $\Psi_0^0(\mathbf{q})$, and place this wavepacket on the PES of the excited state, S_1. The propagation has been performed with MCTDH.[3] For the primitive basis set, HO DVR for R_1, a sine DVR for R_2 with $R_1 \in [3.8, 5.6]$ a.u. and a Legendre DVR for θ are used.[4] For the SPF basis set, $n_1 = n_2 = n_3 = 5$ have been used. As explained in Sect. 8.3.4, the convergence of MCTDH with respect to the number of SPFs is very fast for this system.[5] A complex absorbing potential (CAP) has been added to the Hamiltonian operator to avoid a long grid.[6]

The modulus of the autocorrelation function decreases very fast as shown on Fig. 10.3a. There is nevertheless a slight recurrence due to the vibrational motion in the NO stretching coordinate around 30 fs almost invisible in Fig. 10.3a. An approximate treatment can show that the width of the autocorrelation function is inversely proportional to the steepness of the potential: the steeper the potential, the faster the autocorrelation function decreases [1].

The evolution of the two-dimensional density in R_1 and R_2 is shown at several instants on Fig. 8.4 in Sect. 8.2: the wavepacket "glides" along R_2 but some oscillations along R_1, the NO distance, are also observed. This explains why the average value, $\langle R_1 \rangle$, oscillates as depicted on Fig. 10.3b.

[2]See Chap. 6 in Ref. [1].

[3]See Ref. [4] for the first calculation with MCTDH.

[4]For HO DVR, we use an equilibrium geometry at 2.134 a.u., a frequency of 0.272 eV and a mass of 7.4667 atomic mass unit.

[5]The PES has been re-expressed in a product form without using POTFIT: see Ref. [4].

[6]See Sect. 8.5 for a discussions about CAPs.

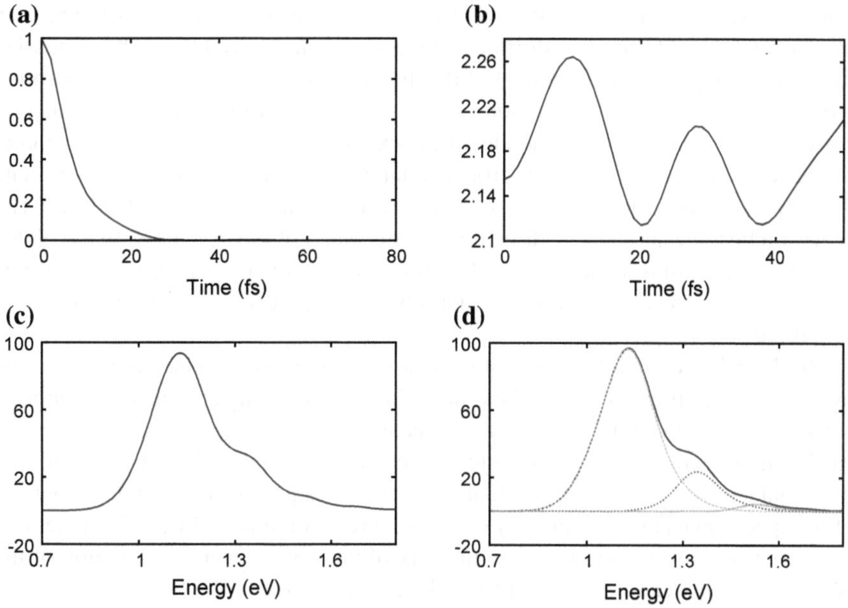

Fig. 10.3 **a** The modulus of the autocorrelation function for the NOCl dissociation, **b** average value of $\langle R_1 \rangle$, the NO distance, in a.u. **c** Fourier transform of the autocorrelation function over 100 fs in arbitrary unit, **d** projection of the exit "flux" onto the vibrational states of the molecule of NO: vibrational ground state, $v = 0$ (*green*), first vibrational excited state, $v = 1$ (*blue*), second vibrational excited state, $v = 2$ (*purple*), third vibrational excited state, $v = 3$ (*turquoise*). In red is the total flux that is identical to the power spectrum of (**c**). Here, the zero point of the energy scale is the minimum of the S_0 PES. The photon energy is obtained by subjecting the S_0 ground state energy of 0.1688 eV from the total energy displayed in Fig. 10.3c and d

The Fourier transform of the autocorrelation function,[7] is given in Fig. 10.3c. It is proportional to the absorption spectrum, provided that we use the approximation of Eq. (10.7). The power spectrum is continuous due to the dissociation of the molecule contrarily to the cases introduced in Chap. 9. More precisely, it shows very diffuse vibrational structures consisting of a main peak with two shoulders at its high-energy side.

For a direct dissociation, as in our case, the spectrum can be obtained approximately by the so-called "reflection principle".[8] Let us first consider a one-dimensional dissociative system (as if we were considering only the coordinate R_1 for NOCl). For the one-dimensional problem, the reflection principle states that the absorption spectrum reflects the coordinate distribution of the molecule in the electronic ground state, with the reflection being mediated by the upper-state potential [1].

[7] Since the autocorrelation function decreases very fast, the choice of the window function does not have a strong impact on the spectrum. This is one of the rare cases where the window function g_0 is a good choice.

[8] See Chap. 6 of Ref. [1].

Since the vibrational ground state $\Psi_0^0(\mathbf{q})$ is approximately a Gaussian function in R_1 and since the Fourier transform of a Gaussian function yields another Gaussian function, an approximate treatment of the dynamics would show that the absorption cross section is a Gaussian function in energy. The center of the Gaussian function corresponds to the energy of the electronic excited state at the equilibrium geometry.[9] The width of the spectrum is proportional to the width of the original coordinate distribution mapped to energies by reflection from the dissociating PES [1]. The spectrum becomes broader with increasing steepness of the S_1 – PES. In the NOCl case, we expect thus a spectrum with approximately a Gaussian structure and this is what we see for the "main band" that also corresponds to the curve in green in Fig. 10.3d.

But NOCl is a three-dimensional nuclear system. However, since the PES of the excited state is rather flat around the Franck-Condon geometry and since the coupling between the three vibrational degrees of freedom is rather weak, the spectrum can be easily interpreted using an "adiabatic" picture with respect to the vibration of NO, the shoulders corresponding to vibrational structures [3]. A more detailed analysis of the spectrum can be obtained by calculating reaction probabilities. The latter can be extracted by evaluating the quantum flux of the wavepacket as a function of the energy. This flux can be obtained from the average value of the "flux operator" [5, 6]: see Sect. 8.6 of Chap. 8. The quantum flux through the dissociative channel of the wavepacket $\Psi_0^0(\mathbf{q})$ will give the power spectrum of Fig. 10.3c. Now, the quantum flux can be projected onto the different vibrational eigenstates of the isolated NO molecule providing the "parts" of the spectrum corresponding to the different vibrational eigenstates of NO also called the vibrational cross sections. Dividing the cross section for a given vibrational state by the power spectrum gives the probability to obtain the vibrational state as a function of the energy that has been absorbed by the molecule. More details about the flux operator and how to use it in conjunction with MCTDH can be found in Sect. 8.6 of Ref. [7] or in Sect. 6.5 of Ref. [8].

The result is presented in Fig. 10.3d, which shows fluxes projected onto different vibrational states: the curve in green corresponds to the quantum flux leading to the vibrational ground state of NO ($v = 0$, v being the quantum number associated with the NO oscillator). The projection onto the first vibrational excited state, $v = 1$, the second vibrational excited state, $v = 2$, and the third vibrational excited state, $v = 3$ are in blue, purple and turquoise, respectively. The projection onto $v = 3$ is almost invisible and corresponds to the very small shoulder around 1.7 eV. In other words, exciting by light in the first maximum of the absorption spectrum yields preferentially NO ($v = 0$), excitation in the first shoulder produces mainly NO ($v = 1$), etc. The fact that the different contributions are rather well separated is linked to the fact that the potential is rather flat around the Franck-Condon geometries leading to several contributions with approximately Gaussian shapes and rather small widths. If the potential had been steeper, the widths of the cross sections would have been larger and the spectrum would have been structure-less, i.e. without shoulders. In addition,

[9] More precisely, the energy of the electronic excited state at the equilibrium geometry minus the ZPE of the ground state.

the fact that we can distinguish the different contributions clearly is linked to the very weak coupling between the dissociation mode and the vibrational mode of NO: the system dissociates in an adiabatic way with respect to the vibrations of NO [3]. The spectrum can be viewed as the different contributions of the NO vibrational states as in Fig. 9.8b for the bending states of H_2O. But instead of having resolved peaks as in Fig. 9.8b, the peaks are "convoluted" by approximately Gaussian functions due to the dissociative channel. We have, thus, to first order a sum of Gaussian functions, the heights of the Gaussian functions being linked to the overlap between the wavepacket, $\Psi_0^0(\mathbf{q})$, and the vibrational eigenfunctions of NO. The analysis could be pursued further by projecting the quantum flux on the rotational states of NO. We could then distinguish for each vibrational cross section the different contributions of j the angular momentum of NO (see Sect. IV in Ref. [3]).

Here, two approximations were made in addition to Eq. (10.7): the Born-Oppenheimer approximation was used and only the $J = 0$ case was considered. In these two approximations, the first one is perhaps the most questionable. Indeed, the experimental spectrum consists of several overlapping electronic bands [9]. However, the peaks labeled D and C by the experimentalists can be assigned to the $S_0 \to S_1$ electronic transitions and to the band NO ($v = 0$) and NO ($v = 1$), i.e. to the green and blue curves on Fig. 10.3d. They appear in the 2.4–3.1 eV domain, thus in the visible spectrum. The agreement between the spacing between the two peaks and the ratio of the two peak intensities are rather well reproduced on Fig. 10.3d [2, 3]. The ratio is overestimated, i.e. the shoulder corresponding to $v = 1$ is higher experimentally. But this discrepancy is mainly due to a small error in the theoretical equilibrium value of R_1 and can be easily corrected [2].

10.2 Indirect Photodynamics

Let us now consider the more complex situation shown in Fig. 10.4.[10] The latter corresponds to a cut along one of the O–O bond lengths through the three 3-dimensional diabatic PESs of ozone (O_3), V_0, V_1, and V_2, corresponding to X, the ground state, B, a weakly-bound excited electronic state, and R, a purely dissociative excited state, respectively. The X, B, and R states are labelled 0, 1, and 2, respectively, in the equations below. The ground state is well isolated and we can work within the framework of the Born-Oppenheimer approximation for this state, i.e. V_0 is also an adiabatic PES. The states B and R are coupled together and the Born-Oppenheimer approximation is no longer valid for these two states: the R-state intersects the B-state close to its minimum. The transition from the ground state to the R state is not dipole-allowed. On the other hand, the matrix element of the dipole moment vector $\boldsymbol{\mu}_{01}$

[10]This section is based mainly on Steve Ndengué's PhD work [10]. The authors gratefully thank him for his contribution and his help. Very helpful discussions with R. Jost are also gratefully acknowledged.

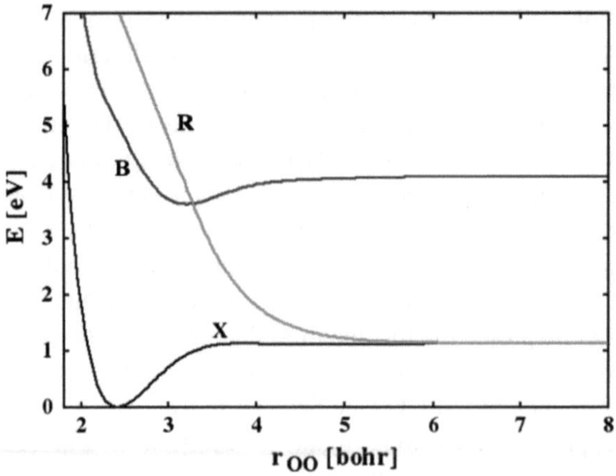

Fig. 10.4 One-dimensional cut through the X (*red*), B (*green*), and R (*blue*) PESs of ozone along one of the O–O bond lengths. In the text, the X, B and R states are denoted 0, 1 and 2, respectively. The other bond length is fixed to its equilibrium value in the electronic ground state: 1.278 Å. The angle between the two chemical bonds is also fixed to its equilibrium value: 116.8°. The X and R PESs lead to $O+O_2$ with the oxygen atom and the dioxygen molecule both in a triplet state (spin equal to one for both species). On the other hand the B state leads to an oxygen atom and the molecule both in a singlet state (spin equal to zero)

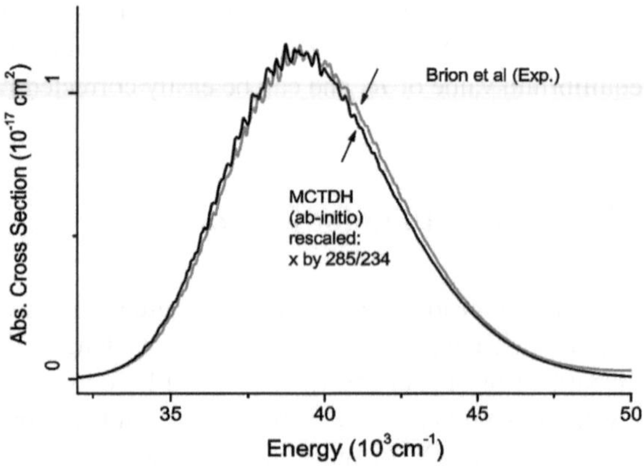

Fig. 10.5 The Hartley band of ozone corresponding to the transition from the ground state X to the B state (see Fig. 10.4). Both the theoretical (*black*) and experimental (*red*) spectra are shown

corresponding to the electronic transition to the B state is very large resulting in a very intense absorption (210–310 nm), called the Hartley band shown on Fig. 10.5.[11]

The presence of this intense absorption band explains why ozone is the main filter of the UV radiation in the atmosphere.[12] The presence of ozone in the stratosphere, i.e. between 10 and 50 km, acts as a shield to protect the Earth surface from the

[11]Reprinted with permission from [11]. Copyright 2010, American Chemical Society.

[12]Together with O_2 that absorbs very strongly between 100 and 210 nm. Ozone also absorbs also above 310 nm but less strongly.

harmful UV radiation of the sun. Ozone in this region is commonly known as the "ozone layer". The experimental and theoretical absorption spectra corresponding to the Hartley band are shown on Fig. 10.5: the theoretical spectrum has been calculated with MCTDH. Ozone, is a bent molecule, with C_{2v} symmetry (similar to the water molecule, see Sect. 7.2 of Chap. 7). We choose "valence" coordinates for ozone as for water (See Fig. 9.5a): R_1 and R_2 are the two O–O bond lengths and θ is the angle between the two chemical bonds. We use the diabatic PESs from Ref. [12] and the KEO is the same as for H_2O except that the masses are different. We have considered the $J = 0$ case only: several calculations have proved that the effect of rotation[13] smoothens the cross sections without significantly altering the envelope of the cross section or the superimposed vibronic structure. This is true not only for the Hartley band [13] but also for the Huggins band that will be discussed later [14]. To obtain the theoretical spectrum of Fig. 10.5, we start from the vibrational ground state in the electronic ground state, Ψ_0^0. For ozone, the approximation of Eq. (10.7) is too crude. The matrix element of the dipole moment vector μ_{01} must thus be applied before propagating the wavepacket on the B state. It can be shown that the modulus of the dipole moment vector $||\mu_{01}||$ can be used to reproduce correctly the spectrum.[14]

In other words, $||\mu_{01}||\Psi_0^0$ is placed onto the B state and propagated. The eigenstates are given by the Schrödinger equation involving two coupled electronic states[15]:

$$\begin{bmatrix} T^{nu}(\mathbf{q}) + V_1(\mathbf{q}) & V_{12}(\mathbf{q}) \\ V_{12}(\mathbf{q}) & T^{nu}(\mathbf{q}) + V_2(\mathbf{q}) \end{bmatrix} \begin{bmatrix} \Psi_1^E(\mathbf{q}) \\ \Psi_1^E(\mathbf{q}) \end{bmatrix} = E \begin{bmatrix} \Psi_1^E(\mathbf{q}) \\ \Psi_2^E(\mathbf{q}) \end{bmatrix}, \tag{10.10}$$

where V_{12} is the potential coupling between the two diabatic states. For propagating the wavepacket, we have to solve

$$\begin{bmatrix} T^{nu}(\mathbf{q}) + V_1(\mathbf{q}) & V_{12}(\mathbf{q}) \\ V_{12}(\mathbf{q}) & T^{nu}(\mathbf{q}) + V_2(\mathbf{q}) \end{bmatrix} \begin{bmatrix} \Psi_1(\mathbf{q}, t) \\ \Psi_2(\mathbf{q}, t) \end{bmatrix} = i\frac{\partial}{\partial t} \begin{bmatrix} \Psi_1(\mathbf{q}, t) \\ \Psi_2(\mathbf{q}, t) \end{bmatrix}, \tag{10.11}$$

starting from the initial condition:

$$\begin{aligned} \Psi_1(\mathbf{q}, t = 0) &= ||\mu_{01}(\mathbf{q})||\Psi_0^0(\mathbf{q}) \\ \Psi_2(\mathbf{q}, t = 0) &= 0\,. \end{aligned} \tag{10.12}$$

---

[13]Including (nuclear) spin statistics.

[14]More rigorously, it would be necessary to define a Body-Fixed frame and to apply each component of the dipole moment vector onto Ψ_0^0 and to propagate the corresponding wavepackets on the B state. We would obtain three spectra and their sum would provide the absorption spectrum.

[15]For the B (or 1) state, the spectrum is discrete below the dissociation limit.

To rationalize the discussion of the results, let us first assume that the two electronic states are not coupled. V_1 displays two minima.[16] The dissociation energy is not very high but there are nevertheless several bound states in the potential wells. If there were no coupling, the absorption spectrum would have a discrete part for low energies that would be followed by a broad continuous band corresponding to a direct photodissociation. The continuous band is only a little bit affected by the presence of the R state. However, the presence of the R state has a stronger effect on the part of the band associated with the potential well of V_1 since the corresponding bound states become metastable states as explained below.

The combined presence of the potential well and the R state leads to an indirect photodissociation. In the low energy domain, V_1 is binding with bound states $\Psi_1^n(\mathbf{q})$ and the corresponding eigenvalues E_1^n, while V_2 is repulsive with outgoing continuum states $\Psi_2^E(\mathbf{q})$. In this kind of situation, it is convenient to simplify the general picture by assuming that the photodissociation separates into two consecutive steps (see Chap. 7 in Ref. [1]). First, the light promotes the molecule from the vibrational ground state, $\Psi_0^0(\mathbf{q})$, to the vibrational states $\Psi_1^n(\mathbf{q})$. Second, the $\Psi_1^n(\mathbf{q})$ states are coupled to the continuum states, $\Psi_2^E(\mathbf{q})$, and the system eventually dissociates. An approximate treatment of the problem shows that the coupling to the continuum states broadens the discrete absorption lines [1]. The discrete spectrum becomes a continuous spectrum with peaks. The centers of the peaks are not exactly the eigenvalues E_1^n: the peaks are slightly shifted due to the coupling with the continuum. Each absorption peak has the shape of a Lorentzian with a width that is linked to the coupling of the bound state and the the continuum states [1].

If the coupling were zero, the bound states would live forever but since the coupling is not equal to zero, the populations of the initially excited bound states decay exponentially. The corresponding quantum states are continuum states but possess some bound state characteristics such as locality.[17] They have a lifetime associated with the exponential decay and do not have a specific energy since the corresponding peaks have a width. These states are called "resonances" and are inherently quantum effects [15]. Physically, they can be seen as metastable states embedded in a continuum. These quantum resonances can be measured experimentally as explained below.

For ozone, the presence of theses resonances results in a highly structured part of the absorption spectrum in Fig. 10.6[18] called the Huggins band between 310 and 370 nm.[19] The Huggins band appears below 31,000 cm^{-1} in Fig. 10.5 but the structure of the band cannot be distinguished on it. A zoom in this energy domain is presented in Fig. 10.6. The experimental cross section was obtained at 218 K. At this temperature, all the peaks corresponding to the resonance overlap due to the presence of many rotational states but also to the fact that there are many resonances with widths that are

[16] A single minimum appears in Fig. 10.4 since only a one-dimensional cut is displayed.

[17] In the potential well of V_1 for ozone.

[18] Reprinted with permission from [16]. Copyright 2012, American Chemical Society.

[19] The distinction between the Hartley and Huggins bands is somehow artificial, the Huggins band being actually the "tail" of the Hartley band in the low energy domain.

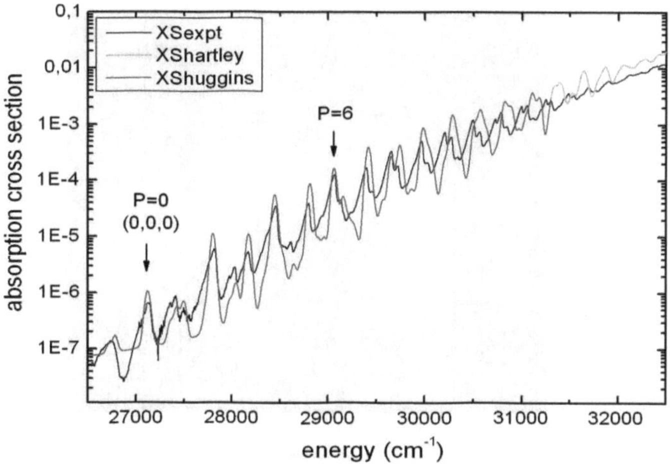

Fig. 10.6 The theoretical Huggins band of ozone in *blue* corresponding to the tail of the Hartley band (in *red* here) for low energies. The present figure is a zoom of Fig. 10.5 for low energies. The experimental band is in *black*

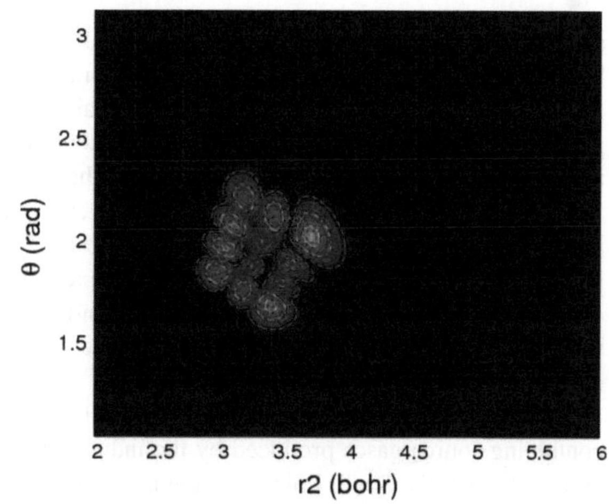

Fig. 10.7 Quantum resonant state of ozone corresponding to 2 quanta in the "long-bond" stretching and three quanta in the bending and thus to the label (2, 3, 0). r_2 is one of the two O–O bond lengths and θ is the angle between the chemical bonds

rather large. However, we clearly see a structure that is the signature of the quantum resonances.

Figures 10.7 and 10.8[20] present two bound states, $\psi_1^n(\mathbf{q})$, of the diabatic state B. The first eigenstates are generally labeled using a local-mode terminology, i.e. the long-bond stretching mode that describes vibration along the dissociation coordinate, i.e., O_2-O, the short-bond stretching mode that describes the vibration of the O_2 entity and the bending mode [14]. The first states are well-labeled using this terminology, for example the state depicted in Fig. 10.7 corresponds to two quanta in the long-bond

[20]Reprinted with permission from [16]. Copyright 2012, American Chemical Society.

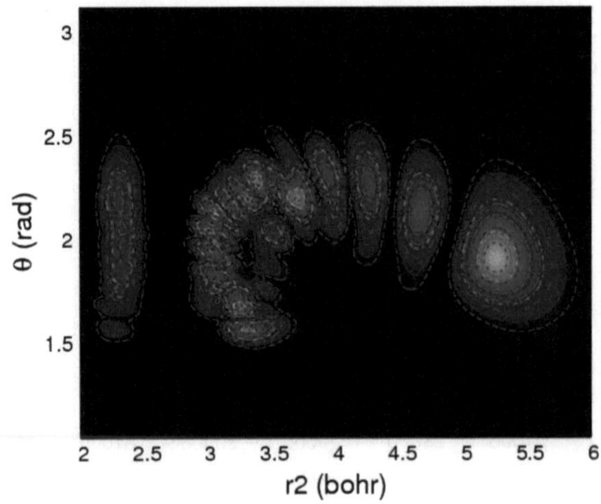

Fig. 10.8 Quantum resonant state of ozone corresponding to a "horseshoe" state. r_2 is one of the two O–O bond lengths and θ is the angle between the chemical bonds

stretching and three quanta in the bending and thus to the label (2, 3, 0). We saw for water in Sect. 9.2 that the labeling of the vibrational states can change depending on the energy regime: starting from the normal-mode description, the local-mode regime appears due to the Darling-Dennison coupling between the symmetric and antisymmetric stretching modes. For ozone, the coupling between the long-bond stretching mode and the bending mode makes a new family of states emerge higher in energy, several states corresponding to "horseshoe" states such as the one shown in Fig. 10.8. This new type of eigenstates has no excitation in the short-bond mode but cannot be interpreted in terms of long-bond and bending vibrations. Instead, their wavefunctions are elongated along a new type of motion toward the dissociation channel.

Due to the facts that the ozone layer is vital for life on Earth and that chlorine-containing source gases produced by the industry tend to create holes in this layer especially over the South Pole (the famous "ozone hole"), its composition has been studied in great detail. In this context, it has been noticed that the ozone layer presents anomalous isotopologue ratios. At natural abundance, besides the main isotopologue $^{16}O_3$ (noted 666), ozone has four relevant isotopologues, $^{16}O^{17}O^{16}O$ (676), $^{16}O^{18}O^{16}O$ (686), $^{16}O_2{}^{17}O$ (667), $^{16}O_2{}^{18}O$ (668). The others are very rare. In the stratosphere, the corresponding isotopologue ratios do not correspond to the natural ones and this anomaly has not been fully explained yet. It reveals that there must be a very specific equilibrium between the formation process and the photolysis of the isotopologues in the stratosphere.

Let us focus on the photolysis process. Figures 10.5 and 10.6 correspond to the absorption spectra of ozone 666 only. The ratios between the absorption bands of

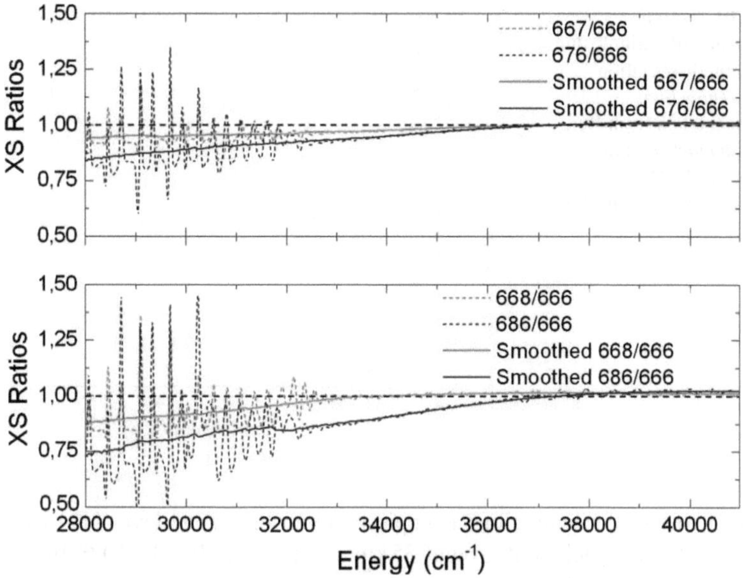

Fig. 10.9 Ratios of the absorption bands for different isotopologues

676, 686, 667 and 668 with the absorption band of 666 are given on Fig. 10.9[21] between 28,000 and 42,000 cm^{-1}.

By visual inspection of Fig. 10.9, we see that the Hartley band varies very little from one isotopologue to another: the ratios are very close to one above 37,000 cm^{-1} and the same feature is observed higher in energy (not shown here) [17]. On the other hand, as regards the Huggins band, the eigenvalues, E_m^1, and thus the positions of the peaks of the resonances and the structure of the Huggins band strongly depend on the isotopologues [16]. Thus, the efficiency of the photolysis process that destroys the molecules is different for each species leading to a change in the isotopic ratios. We, indeed, see that the ratios are smaller than one below 37,000 cm^{-1} on Fig. 10.9. This means that, for a given photon in this energy domain, the isotopologues 676, 686, 667 and 668 dissociate less than 666. A lower efficiency of the dissociation process leads to an isotopic enrichment in the atmosphere.

It can be argued that the absorption in the Huggins band is much less intense than in the Hartley band. However, the concentration of UV photons ("actinic flux") with different energies varies strongly with the altitude in the atmosphere. In particular, almost all the photons that correspond to the Hartley band are absorbed by ozone in the highest part of the stratosphere. Below 40 km, the photons that correspond to the Hartley band have almost completely disappeared.[22] Thus, the role of the Huggins band becomes dominant for altitudes between 20 and 40 km and this band plays an

[21] Reprinted with permission from [16]. Copyright 2012, American Chemical Society.

[22] See Fig. 5 in Ref. [17] that shows the concentration of photons for a given energy as a function of the altitude.

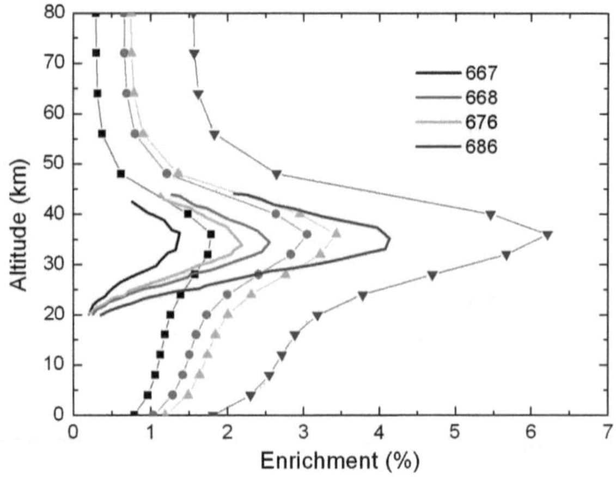

Fig. 10.10 Isotopic enrichment with altitude due to the photodissociation process. *Thick lines* enrichment taken from Ref. [18]. *With markers* more accurate MCTDH calculations taken from Ref. [17]

important role in the anomalous isotopic ratios in the ozone layer [17] leading to a strong isotopic enrichment around 35 km as shown on Fig. 10.10.[23] In the latter figure, the evolution of concentration of photons with the altitude has been taken into account together with the relative absorption spectra of Fig. 10.9. To explain completely the anomalous isotopologue ratios of ozone in the atmosphere, it would be necessary to include also the influence of the formation process, which is not shown in Fig. 10.10.

To conclude, the role of the photolysis process in the anomalous isotopologue ratios in the ozone layer is an illustration of the presence and role of quantum resonances in an indirect photodissociation process.

10.3 Lab-Session II: Photodissociation of NOCl

The photodissociation of NOCl is a simple photochemical reaction. After excitation from the ground to the first excited state, $S_0 \rightarrow S_1$, the chlorine atom dissociates on a femtosecond time scale. This results in a broad band for the absorption spectrum. We will generate the spectrum by wavepacket propagation and Fourier transformation of the autocorrelation function, similarly to what was done for water (Sect. 9.7.1). The spectrum will be further analyzed with a flux method.

[23]Reprinted with permission from [17]. Copyright 2014, American Geophysical Union.

10.3.1 Absorption Spectrum

The calculation consists in two stages. First, the ground-state wavefunction is generated from energy relaxation of an initial guess wavefunction on the ground-state surface (S_0 state). Assuming the Condon approximation, the second stage then places this wavepacket vertically on the excited-state surface (S_1 state), leading to photodissociation. The system is described by three Jacobi coordinates as shown in Fig. 9.18.

To perform the calculations please move to `lab-inputs/NOCl` and type "`mctdh84 -mnd nocl0`".[24] This will generate the vibrational ground state S_0 surface by standard relaxation, i.e. propagation in negative imaginary time. Here a potfit of the PES was not needed, because the PES is given in a sum-of-product form. The PES is defined through the file `nocl0um.srf` which is simply included in the operator file `nocl0.op` via the `srffile` command. Please inspect the files `nocl0.inp`, `nocl0.op`, and `nocl0um.srf` for the details of the calculation. Next, we move (`cd`) to the `nocl0` directory, open the `output` file and inspect the natural weights at final time. One notices that the vibrational ground state is very little correlated. The wavepacket itself, more precisely its reduced density, can be inspected with `showsys84 -rst`. See Sect. 9.7 for details on the use of this program. Try in particular the menu points 20, 150, 160, and 240.

To visualize the PES we cannot run `showpot84` as in the water study, because there is no `natpot` file. However, one can create a so-called `pes` file and visualize it with `showsys84`. To this end return to the parent directory `NOCl` and run "`mctdh84 -pes nocl0`". The WARNING message that appears can be ignored, it merely tells that the option `-pes` has overwritten some keywords of the input file. Then move again to the `nocl0` directory and run "`showsys84 -pes`". The handling of the program is similar to before, but the numbers entered for menu point 20 now have a meaning, they define the location of the cut. Try different cuts. One also may plot 1D cuts by entering "x" and two numbers.

More interesting is, of course, the propagation on the excited-state surface (S_1 state). Return to the `NOCl` directory and run first "`mctdh84 -mnd nocl1`" and then "`mctdh84 -pes nocl1`". Let us begin with inspecting the dissociative motion along `rd` (In Fig. 9.18 this coordinate is called `R`), i.e. run "`showd1d84 -a -G -y 5.0 f1`" in the `nocl1` directory. Each pressing of "Enter" advances the reduced density by one fs. The electronically-excited molecule is unbound, hence the wavepacket will finally reach the end of the grid and there it is artificially reflected. This will destroy the quality of the propagated wavefunction. To avoid this, a complex absorbing potential (CAP) is artificially added to Hamiltonian. The CAP annihilates the wavepacket before it reaches the end of the grid and thus avoids reflection. The CAP used here is defined as $W(q) = -i\eta\,\theta(q - q_c)(q - q_c)^n$, where the symbol θ denotes the Heaviside step function and q_c is the point where the CAP is switched on, i.e. q_c and the end of the grid determine the length of the CAP. In our case the CAP parameters are chosen as CAP-strength $\eta = 0.03$ a.u., starting point of the CAP

[24]Hint: one may copy most of the commands from the file `commands` and paste them to the command line. One may even copy and paste whole blocks of commands.

$q_c = 5.0$ a.u., exponent $n = 3$, and $q = rd$ (compare with `nocl1.inp`). The length of the CAP is 1 a.u. because the grid ends at 6 a.u. The script `plcap` is quite helpful for determining reasonable CAP parameters. Because the CAP starts very slowly, as a third power in distance, a markedly absorption begins only at 5.2 a.u. This is clearly seen in the `showd1d` plots. Up to $t = 20$ fs there is virtually no absorption. But at $t = 30$ fs the wavepacket appears clearly distorted, indicating that some part at the right hand side of the wavepacket is annihilated. The loss of norm is shown by running "`plstate -G -z 0.9`". In general, `plstate` plots the electronic state populations of a *multi-set* run. As here there is only one electronic state, `plstate` merely plots the squared norm of the wavefunction.

Actually, for computing the spectrum the inclusion of a CAP is not really necessary here. The reflected part of the wavepacket does not reach the initial wavepacket, and hence does not contribute to the autocorrelation function, within the rather short time interval of propagation. This short propagation time is sufficient, as can be shown by inspecting the modulus of the autocorrelation function. Run "`plauto`" or, for more details, "`plauto -l -G -z 1.e-4`". The behavior of the autocorrelation function is rather uninteresting. There is a fast decrease to almost zero within the first 30 fs, and a small recovery (due to `rd` motion) near 50 fs. Remember, that due to the t/2-trick (see footnote of Sect. 9.7.1) the autocorrelation function is twice as long as the propagation time. It may be tempting to assume that an autocorrelation duration of 30 fs (i.e. 15 fs propagation time) would be sufficient. In fact, such a short autocorrelation function will reproduce the spectrum rather closely, but some fine details, namely the steps at the high energy side, are a bit washed out. After 60 fs, however, the autocorrelation drops to insignificantly small values and a longer propagation does not influence the spectrum.

To demonstrate that the inclusion of a CAP indeed becomes essential when propagating for a longer period, we may edit the input file `nocl1.inp` by removing the CAP, setting `tfinal` to 100 fs, and changing the name of the name-directory to e.g. `nocl1.nocap`. After running `mctdh84` with this input and moving to the new name-directory, run "`plauto -G -y 0.1`" and observe that there appears a non-negligible contribution at large times. This false contribution introduces some spurious oscillations in the spectrum. The false contribution is caused by the reflected part of the wavepacket. This can be seen by running "`showd1d84 -a f1`". Re-introducing a CAP, the false part will not appear.

Turning to the spectrum please move back to the `nocl1` directory and run "`plspec -g0 -G 0.7 2.2 ev`". The spectrum consists of a broad peak with some shoulders at the high energy side. At higher energies these shoulders turn into peaks, which can be observed by running "`plspec -g0 -G 1.6 2.4 ev`". Note that here we have used the box-filter `g0`, i.e. no filter function at all. Because the autocorrelation function virtually has vanished at final time, a more sophisticated filter is not needed. In fact, using e.g. the `g2` filter would smear out the spectrum a bit, but not improve it otherwise.

The shoulders in the photoabsorption spectrum appear at energies where a new channel opens, namely the dissociation into Cl and vibrationally excited NO. We will investigate this more carefully when discussing flux analysis. But before doing so, we shall visualize the motion of the wavepacket in 2D. As already discussed, `showsys` can plot both 2D potential cuts and 2D reduced wavefunction densities. But it can do more: it can plot the density on top of a PES cut, which provides a vivid picture of the dynamics. To produce such a plot type "`showsys84`" and then go to menu point 10, and change the plot task to `2=plot pes` (type 2). The default coordinate selection is in this case `rd=x rv=y theta=1.545` (see menu point 20). Hence a 2D plot of the PES will be produced for the coordinates (rd, rv) while keeping θ fixed at the value $\theta = 1.545$ rad. One may view this PES cut by entering "1" three times, try different menu points (e.g. 160, 240), or different coordinate selections. When you are done, please make sure that the surface and the keys are both *off* (menu points 160, 240). Then go to menu point 20 and change the coordinate selection to `x y 2.1`, because the wavepacket is located near $\theta = 2.1$ rad. (One may run "`showd1d84 -a f3`" or "`plqdq 1 3`" to check this.) Now use menu point 5. You will be asked for a file name. Chose any convenient name, e. g. `xyz`. The plot data is then written to the file `xyz` for later use.

Next, we want to inspect the wavefunction. Go to menu point 10 and chose `5=plot reduced density`. The density, i.e. $|\Psi|^2$ integrated over all coordinates, except those specified by `x` and `y` (that is integrated over the angle in the present case), will be shown. Because we want to show the wavepacket density on top of the potential, choose menu point 400 (overlay plots) and then 410 and then enter the name of the file which contains the PES data, here `xyz`. Then input "1" five times and you will see the initial density. Pressing RETURN will display the density propagated by one time step, and so on.

10.3.2 Flux Analysis

To further analyze the photoabsorption spectrum and to better understand its structure, we will perform a flux analysis. The program `flux84` measures the energy resolved quantum flux going through a plane defined by `rd=constant`. By default this plane is located at the starting point of the CAP, but it can be moved to another position with the `-s` option. `flux84` is a rather complicated program, for a comprehensive description of its use please see the HTML documentation and the guide, (both come with the package). For a description of the flux algorithm see the MCTDH-review, the MCTDH-book, or Sect. 8.6.

Before running `flux84` one has to propagate the wavepacket for a longer time to ensure that the wavepacket has passed the dividing plane. Rather than starting a propagation anew, we may perform a continuation run. Hence type "`mctdh84 -c -tfinal 70 nocl1`" while being in the directory `NOCl`, or alternatively type "`mctdh84 -c -tfinal 70 -I.`" while being in the name-directory `nocl1`. Read the output of "`mctdh84 -h`" to understand what you are doing. The wavepacket is now propagated up to the final time of 70 fs. Run "`plstate -l -G -z 1.e-3`" to observe that the wavepacket is now almost completely absorbed, only 0.6% of the population has remained. One also may run `showd1d84` to view the absorption process.

We will perform several flux calculations, and in order not to overwrite their outputs, we will store those in separate directories. Type "`mkdir fxall fx0 fx1 fx2 fx3 fx4`" and then run "`flux84 -o fxall 0.7 2.0 ev rd`". To visualize the output run "`plflux -f fxall/flux`". The plot shows the spectrum that we already know. To prove that we have indeed reproduced the old spectrum run "`autospec84 0.7 2.0 ev 0`" and then "`plgen fxall/flux spectrum.pl`". The two lines agree. We have hence found another way of computing the spectrum, which, however, is more elaborate than simply Fourier transforming the autocorrelation function. The flux approach reproduces the power spectrum because the molecule decays completely into a single channel only. However, `flux84` can do more than measuring the total flux, it can operate or project the wavepacket before measuring the flux. We will project onto the first five vibrational states of the NO fragment. For this, run
"`flux84 -P 2 eigenf rvib 1 % -o fx0 0.7 2.0 ev rd`",
and
"`flux84 -P 2 eigenf rvib 2 % -o fx1 0.7 2.0 ev rd`",
and so on till
"`flux84 -P 2 eigenf rvib 5 % -o fx4 0.7 2.0 ev rd`".
Here we have added a projector for the second degree of freedom, rv, which projects onto the eigenstates of the operator "rvib", defined in `nocl1.op`. Note that the counting of the eigenfunctions generated via the keyword `eigenf` starts with 1. Hence the second eigenfunction represents the first excited state. The resulting spectra are best shown in comparison with the unprojected one. Type
"`plgen -G -a 0.8 -x 1.8 -z 0 -u 1:2 fxall/flux fx0/flux fx1/flux fx2/flux fx3/flux`",
and to better visualize the high energy part
"`plgen -G -a 1.4 -x 2.0 -z 0 -u 1:2 fxall/flux fx0/flux fx1/flux fx2/flux fx3/flux fx4/flux`",
or even
"`plgen -G -a 1.7 -x 2.0 -z 0 -u 1:2 fxall/flux fx1/flux fx2/flux fx3/flux fx4/flux`".
It is now clear that the main peak near 1.1 eV is due to the vibrational $v = 0$ channel, the shoulder near 1.3 eV is caused by the opening of the $v = 1$ channel, and so on.

The flux algorithm sums over all internal states, but by projecting one can select an individual state or a group of states. When projecting on a vibrational state one still sums over all rotational states. However, because vibration cannot be separated from rotation, there are no "universal" vibrational states, they depend on the rotational quantum number through a centrifugal potential. Fortunately, the vibrational eigenfunction depends only weakly on the rotational quantum number j. (This is not true for the eigenvalue, but the latter is unimportant for the projection). We simply have set $j = 30$ in the Hamiltonian rvib (see `nocl1.op`), because this choice resembles roughly the maximum of the final j-distribution. (One may replace `rvib` with `vib` in the flux commands above, and observe that the dependence on the rotational state is indeed very weak. See `nocl1.op` for the definitions of the Hamiltonians `rvib` and `vib`.)

To project onto final rotational states, but sum over vibrational ones, run "`mkdir fxr20 fxr30 fxr40 fxr30v0 fxr30v1`",
"`flux84 -P 3 leg 20 % -o fxr20 0.7 2.0 ev rd`",
"`flux84 -P 3 leg 30 % -o fxr30 0.7 2.0 ev rd`",
"`flux84 -P 3 leg 40 % -o fxr40 0.7 2.0 ev rd`",
"`plgen -G -a 0.8 -x 1.8 -z 0 -u 1:2 fxr20/flux fxr30/flux fxr40/flux`". Maxima now appear: it is thus the rotational motion that has smeared out the peaks and turned them into shoulders.

Finally, we may project on an internal quantum state of NO.
"`flux84 -P 2 eigenf rvib 1 % -P 3 leg 30 % -o fxr30v0 0.7 2.0 ev rd`",
"`flux84 -P 2 eigenf rvib 2 % -P 3 leg 30 % -o fxr30v1 0.7 2.0 ev rd`",
"`plgen -G -a 0.8 -x 1.8 -z 0 -u 1:2 fxr30/flux fxr30v0/flux fxr30v1/flux`.
The computed rotational distributions are not very accurate. This is because the angular dependence of the PES at $rd = 5$ a.u.—the starting point of the CAP—is still quite strong and continues to cause rotational transitions when the wavepacket is propagated to larger rd-distances. A longer rd-grid (e.g. 10.0 a.u.) is required to fully converge the rotational transitions. The vibrational distributions, however, are converged with the current grid, and to compute the spectrum by Fourier transforming the autocorrelation function, even a shorter grid suffices. (See e.g. $MCTDH_DIR/inputs/nocl1.inp).

References

1. Schinke R (1993) Photodissociation dynamics. Cambridge University Press, Cambridge
2. Schinke R, Nonella M, Suter HU, Huber JR (1990) Photodisociation of ClNO in the S_1-state: a quantum mechanical ab initio study. J Chem Phys 93:1098
3. Untch A, Weide K, Schinke R (1991) The direct photodissociation of ClNO (S_1): an exact three-dimensional wave packet analysis. J Chem Phys 95:6496
4. Manthe U, Meyer H-D, Cederbaum LS (1992) Wave-packet dynamics within the multiconfiguration Hartree framework: general aspects and application to NOCl. J Chem Phys 97:3199–3213
5. Miller WH, Schwartz SD, Tromp JW (1983) Quantum mechanical rate constants for bimolecular reactions. J Chem Phys 79:4889
6. Thompson WH, Miller WH (1993) Initial state-selected reaction probabilities for OH+H_2 → H+H_2O and photodetachment intensities for HOH_2^-. Chem Phys Lett 206:123
7. Beck MH, Jäckle A, Worth GA, Meyer H-D (2000) The multi-configuration time-dependent Hartree (MCTDH) method: a highly efficient algorithm for propagating wave packets. Phys Rep 324:1
8. Meyer H-D, Gatti F, Worth GA (eds) (2009) Multidimensional quantum dynamics: MCTDH theory and applications. Wiley-VCH, Weinheim

9. Bai YY, Ogai A, Qiand CXW, Iwata L, Segal GA, Reisler H (1989) The electronic spectrum of NOCl: photofragment spectroscopy, vector correlations, and ab initio calculations. J Chem Phys 90:3903
10. Ndengué S (2011) Photodissociation de l'ozone: sélectivité isotopique. PhD thesis, Université de Gernoble et Université de Douala
11. Ndengué SA, Gatti F, Schinke R, Meyer H-D, Jost R (2010) Absorption cross section of ozone Isotopologues calculated with the multiconfiguration time-dependent Hartree (MCTDH) method: I. The Hartley and Huggins bands. J Phys Chem A 114:9855
12. Schinke R, McBane GC (2010) Photodissociation of ozone in the Hartley band: potential energy surfaces, nonadiabatic couplings, and singlet/triplet branching ratio. J Chem Phys 132:044305
13. Alacid M, Leforestier C (2001) The role of rotation in the calculated ultraviolet photodissociation spectrum of ozone. J Chem Phys 114:1685
14. Grebenshchikov SY, Qu Z-W, Zhu H, Schinke R (2007) New theoretical investigations of the photodissociation of ozone in the Hartley, Huggins, Chappuis, and Wulf bands. Phys Chem Chem Phys 9:2044
15. Zhang JZH (1999) Theory and application of quantum molecular dynamics. World Scientific, Singapore
16. Ndengué SA, Schinke R, Gatti F, Meyer H-D, Jost R (2012) Comparison of the Huggins band for six ozone isotopologues: vibrational levels and absorption cross section. J Phys Chem A 116:12260
17. Ndengue S, Madronich S, Gatti F, Meyer H-D, Motapon O, Jost R (2014) Ozone photolysis: strong isotopologue/isotopomer selectivity in the stratosphere. J Geophys Res Atmos 119:4286
18. Liang MC, Irion FW, Weibel JD, Blake GA, Miller CE, Young YL (2006) Isotopic constitution of stratospheric ozone. J Geophys Res Atmos 111:D02302

Chapter 11
Bimolecular Reactions

11.1 Introduction

In Chaps. 9 and 10, we have given several examples of photoabsorption spectra of molecules. In other words, we have addressed the question of how molecules interact with light. The absorption of light can lead to unimolecular reactions when the molecule dissociates for instance as shown in Chap. 10. Another class of important elementary processes in chemistry are bimolecular processes, i.e. processes where two molecules collide and exchange energy, atoms or groups of atoms. Understanding these elementary processes at their most fundamental level is a challenging task of tremendous practical importance for industrial reasons, if the elementary process is the rate determining step of an important industrial chemical reaction. For instance, the endothermic reaction $H + O_2 \rightarrow OH + O$ is considered as the "most important combustion reaction" since it is a dominant molecular-dioxygen-consuming step in hydrogen-oxygen and methane-oxygen combustion mechanisms [1].

11.2 The $H + H_2 \rightarrow H_2 + H$ Reaction

Let us focus on the prototypical $H + H_2 \rightarrow H_2 + H$ reaction in the electronic ground state. We stay within the Born-Oppenheimer approximation. Even though this is the simplest neutral bimolecular reaction, its physics is very rich due to the importance of tunneling and the presence of quantum resonances. We use Jacobi coordinates to describe the process as those depicted in Fig. 5.6a with R_1 the distance between the two hydrogen atoms in the initial diatomic molecule, R_2 the distance between the center of mass of H_2 and the colliding hydrogen atom, and θ the angle between the two Jacobi vectors. T^{nu}, the KEO for $J = 0$, is given by Eq. (6.155) with the masses of Eq. (6.156). The reaction is collinearly-dominated, i.e. with the minimum-energy

F. Gatti et al., *Applications of Quantum Dynamics in Chemistry*,
Lecture Notes in Chemistry 98, DOI 10.1007/978-3-319-53923-2_11

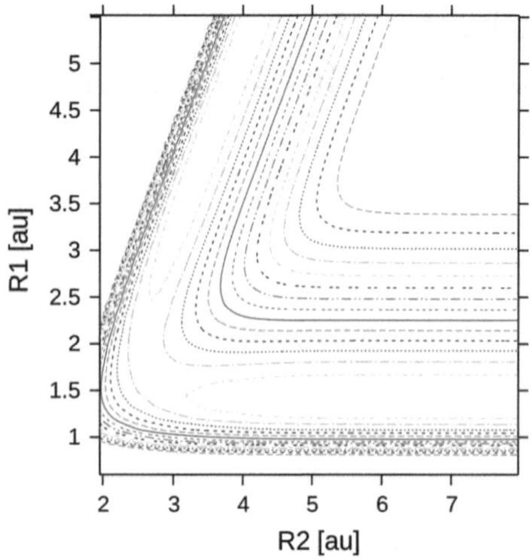

Fig. 11.1 Cut through the three-dimensional PES of H + H_2 taken from Refs. [2–4]. R_1 is the distance between the two hydrogen atoms in the initial diatomic molecule and R_2 is the distance between the center of mass of H_2 and the colliding hydrogen atom. θ, the angle between the two corresponding Jacobi vectors, is fixed at 0.0 rad (collinear geometries)

path along the collinear geometry. For the primitive basis set of each coordinate, a sine DVR for R_1 an FFT DVR for R_2 and a Legendre DVR for θ are used.[1]

Here, V_0, is the "lsth" PES [2–4] and the potential has been re-expressed in a direct product form using the POTFIT algorithm described in Sect. 8.7.[2] A cut through the PES is shown on Fig. 11.1, where the angle θ is fixed at 0.0 (collinear geometry). The PES is purely repulsive and has a transition barrier of 0.36 eV above the asymptote.

Let us propagate wavefunctions with the following initial form:

$$\Psi(R_1, R_2, \theta, t = 0) = N\, e^{-\frac{(R_2 - R_{2,0})^2}{4\Delta R_2{}^2}}\, e^{ip_0(R_2 - R_{20})} \times \varphi_{j_0\, v_0}(R_1) P_{j_0}(\cos\theta)\,, \tag{11.1}$$

where $P_{j_0}(\cos\theta)$ is a Legendre polynomial, $\varphi_{j_0\, v_0}(R_1)$ a vibrational eigenfunction of the isolated H_2 molecule. j_0 is the angular momentum and v_0 the vibrational quantum number of H_2. p_0 is the initial momentum given to the H atom: if $p_0 < 0$, the hydrogen atom will collide with the molecule.

The Hamiltonian operator is $H = T^{nu} + V_0$. The wavefunctions of Eq. (11.1) are wavepackets with respect to this operator. Note that the functions $\varphi_{j_0\, v_0}(R_1)\, P_{j_0}(\cos\theta)$ are eigenfunctions of the diatomic part of the operator when R_2 is very large.

Two complex absorbing potentials have been added to absorb the wavepackets for large values of R_1 and R_2. As for ozone and NOCl, we have calculated the

[1] $R_1 \in [0.6, 6.60]$, $R_2 \in [1.0, 9.80]$ a.u., 51, 81, and 31 functions are used for R_1, R_2, and θ, respectively.

[2] R_2 is the contracted degree of freedom and we used 12 single particle potentials (SPPs) for R_2 and 7 also for θ. At the end, we have thus 84 terms in the fit (see Eq. (8.350z)) with three degrees of freedom. We used 16 iterations to improve the relevant region defined as the geometries corresponding to a potential energy below 4 eV above the minimum. This guarantees a root-mean-square (rms) error of 3.84 meV on the relevant region.

"flux" of the wavepacket. More precisely, we have calculated two fluxes: one for large values of R_2 after collision corresponding to the "part" of the wavepacket that returns without reacting and another flux for the large values of R_1 that correspond to the "part" of the wavepacket that has reacted to form a new molecule of H_2. Let us consider $j_0 = 0$, $v_0 = 0$, $R_{2,0} = 4.8$ a.u., $p_0 = -7.5$ a.u., and $\Delta R_2 = 0.25$ a.u. In Fig. 11.1, the initial wavefunction is thus located around $R_{20} = 4.5$ a.u. and $R_1 = 1.448$ a.u. The wavepacket has been propagated using MCTDH. As explained in Sect. 8.3.4, the convergence of the SPF basis set is more difficult than for NOCl due to the strong coupling that occurs after reaction especially because of the fact that the Jacobi coordinates are not well adapted to describe the reactive channel (see also the discussion of Sect. 5.2). The MCTDH convergence is obtained with 3024 configurations (18 SPFs in R_2, 14 in R_1, and 12 in θ). The reduced density is given on Fig. 8.5 for $t = 20$ fs: we clearly see that the wavepacket on the left, corresponding to large values of R_1, reacts.[3]

A Gaussian function $\varphi(x) \propto \exp(((x - x_0)/2\sigma)^2 + ip_0(x - x_0))$ has the energy distribution $|\Delta(E)|^2 \propto \exp(-4\mu\sigma^2[\sqrt{E} - \sqrt{E_0}]^2)/\sqrt{E}$. The energy distribution for the Gaussian function in R_2 is given on Fig. 11.2 in green.[4] It corresponds to the relative kinetic energy distribution of the colliding H atom. The parameters have been chosen so that the energy distribution covers the energies where the cross sections have to be calculated.

Figure 11.2a gives also the flux corresponding to the reactive part of the wavepacket (in red). For each energy, the flux is smaller than the initial distribution. Dividing the flux by the initial distribution gives the *reaction probability* for a given initial kinetic energy and for the initial conditions $j_0 = 0$ and $v_0 = 0$. The probability does not depend on the initial kinetic energy distribution, provided that the propagation has been correctly converged. The total reaction probability is given in red on Fig. 8.5b.

This reaction probability has several properties that cannot be explained classically. The potential has a barrier of 0.36 eV. If there were no tunneling, we should observe an increase of the probability for initial energies larger than 0.36 eV corresponding to the energy that is necessary to surmount the barrier. On the contrary, *the probability curve as a function of the energy rises smoothly from zero as the collision energy increases and the reaction starts below the transition barrier of 0.36 eV*. This is a clear indication of *quantum tunneling effects*.[5]

In addition, the reaction probability presents *oscillatory structures* due to the presence of *quantum resonances*, i.e. to quantum metastable states as those already encountered for ozone [1]. In other words, for specific domains of collisional energies, the system remains trapped during a certain time around the saddle point although the potential is purely repulsive. For these energies, there is a resonant transfer to the vibrational mode of the H_2 molecule and the hydrogen atom has no sufficient energy

[3] In Fig. 8.5 r_d and r_v correspond to R_2 and R_1, respectively.

[4] The formula given for $|\Delta(E)|^2$ is valid only for a vanishing potential. In practice, $|\Delta(E)|^2$ is computed numerically by incorporating the tail of the potential.

[5] The presence of the zero point energy of H_2, which changes along the reaction coordinate, can also impact the process in a way that has no classical counterpart.

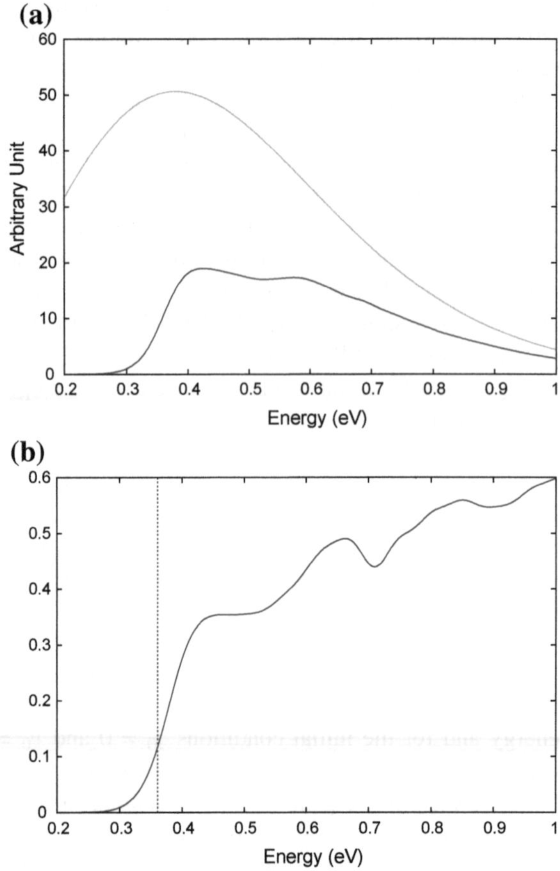

Fig. 11.2 **a** Kinetic energy distribution of the initial wavepacket (*green*) for H + H_2 and $j_0 = 0$, $v_0 = 0$ and the flux corresponding to the reactive part of the wavepacket (*red*). b The corresponding reaction probability obtained by the ratio of both quantities. The transition state lies at 0.36 eV above the dissociation limit (*dashed line*)

to leave the H_3 complex. However, after a certain time, the energy in the vibrational mode of H_2 is transferred back to the hydrogen atom that can either go back or snatch one of the two atoms of hydrogen in the molecule. The resonances that are present in H + H_2 are particular cases of what is called *Feshbach resonances*. Generally speaking, for a Feshbach resonance, the energy associated with the scattering coordinate (here R_2) is temporarily depleted through the coupling with other degrees of freedom (here R_1 and θ).

The reaction probability cannot be measured experimentally. The quantities that can be measured by the experimentalists are cross sections. They involve a summation over all the values of the total angular momentum J and of its projection onto a BF axis K. The initial-state-selected cross sections for H + H_2 are given by

$$\sigma_{v_0\, j_0} = \frac{\pi}{k^2_{v_0\, j_0}} \sum_{|K|}^{j_0} \sum_{J \geq K}^{J_{max}} (2J+1) P^{J\,K}_{v_0\, j_0}(E)\,, \tag{11.2}$$

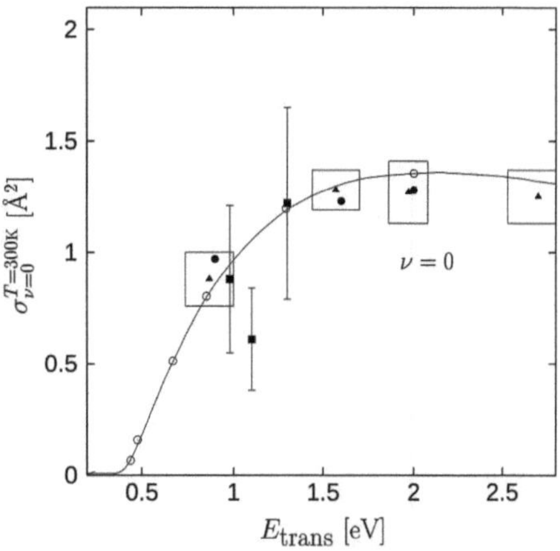

Fig. 11.3 Comparison of the theoretical (*solid line*) and experimental cross sections for H + $D_2(v = 0)$ cross sections. The experimental values are given by *filled circles*, *squares* and *triangles* (see Ref. [8]). The *boxes* indicate error bars in both energy and cross-section. The *open circles* represent results of a time-independent calculation [11]

where E is the energy and $k_{v\,j_0} = \sqrt{2m_R(E - E_{v_0\,j_0})}$ the "local momentum", $E_{v_0\,j_0}$ is the initial energy of H_2 and $m_R = 2/3m_H$. J_{max} is the maximum value of J necessary to converge the cross section for a given energy domain and $P^{J\,K}_{v_0\,j_0}(E)$ is the reaction probability for given values of J and K. The line of Fig. 11.2b corresponds to $P^{0\,0}_{0\,0}(E)$. Even if there are quantum resonances, the summation in Eq. (11.2) tends to average out their contributions and masks their effect. The corresponding cross sections increase monotonically with the energy.

Several cross sections have been measured for the H + D_2 collision [5–7]. The use of isotopes allows experimentalists to easily distinguish the products from the reactants. The cross section calculated with MCTDH (see Ref. [8]) for $v = 0$ is given on Fig. 11.3[6] for an initial Boltzmann distribution of the rotational levels of D_2 at temperature $T = 300$ K. The boxes and the bars give the estimated experimental errors. We clearly see that the oscillations due to the quantum resonances are quenched by the different summations in the calculation of the cross sections. This quenching of the effects of quantum resonances is often observed in cross sections and in reaction rate constants [9, 10].

11.3 Quantum Resonances and Cold Chemistry

In the previous example, the effect of the quantum resonances was quenched in the experimental data. It is now possible to detect experimentally these effects of quantum resonances in molecular processes. Exploiting these effects could lead to a completely different chemistry.

In traditional chemistry, the usual way to explain reaction rates is mainly based on transition-state theory [12]. Let us consider the energy profile of Fig. 11.4 for the

[6]Reprinted with permission from [8]. Copyright 1999, American Institute of Physics.

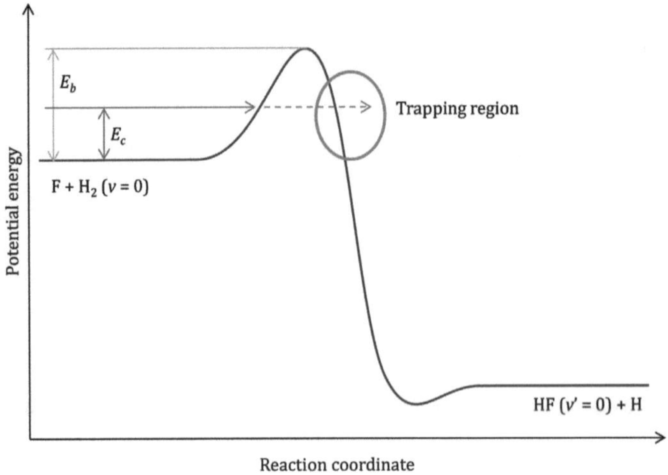

Fig. 11.4 Energy profile for the $F + H_2 \rightarrow HF + H$ reaction as a function of a "reaction coordinate", inspired from Refs. [13, 14]. The physical process indicated by the *red arrow* is very different from what is expected in transition state theory

$F + H_2 \rightarrow HF + H$ reaction as a function of a reaction coordinate. As for the H + H_2 reaction, the potential is purely repulsive in the interaction region. In transition-state theory, the important criterion is that colliding molecules must have sufficient energy to surmount the potential-energy barrier to react. The transition state is the unstable transitory complex that occurs at a potential-energy maximum along the reaction coordinate at the energy E_b in Fig. 11.4. Within transition state theory, it is assumed that if the system reaches the activation energy and goes in the direction of the products, it will not reach the transition state again.

However, the message conveyed by recent measurements of differential cross sections for this system is another one: the system can tunnel below the transition state energy, for the energy E_c in Fig. 11.4. For certain energies, the system remains temporarily trapped due to the presence of quantum resonances [13, 14]. The "trapping region" is shown in yellow on Fig. 11.4. As explained by Z. Sun, D.-H. Zhang and coworkers [13, 14], for the F + HD system, the cross sections display several peaks that have been clearly assigned to very specific Feshbach resonances as for $H + H_2$ in the previous section.

At normal temperatures, the effect of these metastable states is often masked as in Fig. 11.3 at $T = 300$ K for $H + H_2$ but, at lower temperatures, the picture can be very different. In this context, it should be stressed that, on the experimental front, decisive progress has been achieved to reach low temperatures. In particular, "ultracold" chemistry is emerging as a very important field of research [15–20]. Measuring and controlling the reactivity of "cold" (1 mK $< T <$ 1 K) and "ultracold" ($T <$ 1 mK) molecules is now possible [21, 22]. At these temperatures, quantum effects are strongly amplified. For instance, even two atoms with a large mass such as Cesium can be combined through tunneling to form Cs_2 at around 300 μK [23–25].

As explained for $H + H_2$ and $F + H_2$, chemical reactions at low energies can occur along pathways that deviate significantly from the minimum-energy path. The resulting chemistry could give rise to many applications that are not possible otherwise. In particular, it is conceivable to target the resonant states [26] and to guide the chemical reactions in a way that is completely different from chemistry at higher temperature [27].

11.4 Lab-Session III: Reactive Scattering: $D + H_2 \rightarrow DH + H$

The computation of cross sections of a reactive scattering event is at the center of quantum molecular dynamics. However, the description of reactive scattering is a difficult task numerically when state-to-state cross sections are to be computed. For the definition of a state-specific initial state one has to use reactant coordinates, whereas for the projection onto state-specific final states, product coordinates are required. Hence a coordinate transformation must be performed. The wavefunction is expanded on the full product grid, interpolated and evaluated on the product grid of the other coordinate system. Employing the full product grid is somehow against the "philosophy" of MCTDH, but recently very important advantages have been brought forward for computing state-to-state cross sections with MCTDH [28–30]. Here, however, we will avoid a coordinate transformation, use reactant coordinates throughout, and compute initial-state-selected total reaction cross sections. Hence, we are specific with respect to the initial state, but sum over all final states and integrate over all scattering angles of the product. We use Jacobi coordinates, shown in Fig. 9.18, with A = H, B = H, and C = D. A product Jacobi coordinate system would correspond to A = H, B = D, and C = H. In this figure R is the dissociative coordinate, in the following called rd, and r is the internal vibrational coordinate, in the following called rv.

After the reaction, the wavefunction is described with an unsuitable coordinate system, which leads to artificial correlation. To avoid this problem, the reactive part of the wavefunction is absorbed as early as possible by a CAP and the reaction probability is determined from flux analysis. Complex absorbing potentials (CAPs) and the `flux` program are already discussed in Sect. 10.3 when studying the photodissocation of NOCl.

For our investigation we will use the lsth potential energy surface and the first step is, as usual, to potfit the analytic potential. Run "`potfit84 -mnd lsth`" to generate a `natpot` file. As usual, all commands are listed in a commands file, see `lab-inputs/reaction/commands`. Inspect the input file `lsth.inp`. Here we use not only correlated weights but also separable weights and subtract the H_2 curve (`oned_rv` keyword), because we want to arrive at a reasonable fit with only few potential terms. As several propagations have to be performed to cover a range of total angular momenta, the propagations should be fast. See the HTML documentation of Potfit and Chap. 12 of the guide to understand the input in detail.

To visualize the PES run "`showpot84`" and choose in coordinate section (menu point 20) `x y 0` to plot the PES in collinear configuration, $\theta = 0$. Then input "1" three times to accept the defaults. One observes a tilted L-shaped form (ignore the upper left corner, it describes an nonphysical situation where the D-atom has passed the first H-atom). There is an entrance channel near rv = 1.4 a.u. and rd>4.0 a.u., a transition point near rd = 2.7 a.u. and rv = 1.7 a.u., and an exit channel for large rv. Clearly, there is little correlation between rd and rv in the entrance channel, but considerable correlation between rd and rv in the exit channel, because the contour lines near the exit channel are parallel to neither the rd nor the rv axis. To visualize the perpendicular configuration, $\theta = \pi/2$, choose again menu point 20 and type `x y 1.57`. Then, again, input "1" three times to accept the defaults. From the contour lines shown we may be led to the conclusion that the lower rd boundary, 1.0 a.u., is chosen too large. However, the transition point near rd = 1.75, rv = 2.0 a.u. lies more than 2 eV above the energy of the entrance channel. Hence, a perpendicular approach of the particle towards the molecule does not lead to reaction, at least not for the low energy region studied here. One may study different potential cuts and use menu point 160 to switch to a 3D visualization.

Now, we can start the propagation. Run "`mctdh84 -mnd dh2-0 &`", and while it is running, inspect the input file `dh2-0.inp`. The keyword `normstop= 0.02` let the propagation stop when the norm (not its square) is below 0.02. Then there are two CAPs, one for rd, one for rv. The former is an automatic CAP, called ACAP. It is switched on when the intensity at the beginning of the CAP increases, but here not before 6.5 fs propagation time, as given by the last entry of the ACAP bracket. As a primitive basis for rd we use an FFT-DVR. Often we prefer to use a sin-DVR, because an FFT is subject to periodic boundary conditions. Here we do not use an FFT because of the (small) increase in computational speed, but because an FFT allows to plot the momentum distribution of the 1D reduced density. (Remember, the number of grid points on an FFT grid must be a product of powers of 2, 3, and 5, otherwise the performance is slow). For rv we use a sin-DVR. Usually a HO-DVR is more appropriate for vibrational motion, but here the rv coordinate visits regions of the potential which are far from harmonic. The angular degree of freedom, θ, is represented by a Legendre DVR with even symmetry. This means that the DVR is built from Legendre functions with $l = 0, 2, 4, \ldots$. Due to a symmetry operation that exchanges the two H atoms, even angular momentum states couple only with even ones and odd states only with odd ones. Which one of the two situations is realized depends on the nuclear spin. As well know, the two realizations are called *para-* and *ortho*-hydrogen. In MCTDH it is in general difficult to make use of symmetry, and symmetry is often ignored. Here it is easy to account for symmetry, because it can be expressed in terms of symmetry-adapted primitive basis functions.

As an initial SPF for the rd-DOF we choose a Gaussian function augmented with an in-going momentum. The initial wavepacket should be well localized in both coordinate and momentum spaces, hence a Gaussian function is a natural choice. Often one uses an even more localized initial state, such that its momentum distribution, and thus the initial kinetic energy range covered, becomes wider. However, a wider energy distribution requires more SPFs for convergence. For specific cases, we have used two different but slightly overlapping initial energy distributions and run two separate sets of calculations to cover a wider energy range [31, 32]. For the present calculations we have set the input parameters such that initial kinetic energies up to 1 eV are covered. Note that the initial wavepacket has to be carefully designed, as the weight of positive (out-going) momenta should be negligible. We will return to this point later.

The initial SPF of the rv-DOF is chosen as the vibrational ground state of the H_2 potential curve, the latter is given by the lsth PES for $rd \rightarrow \infty$. Finally, the initial SPF of the θ-DOF is an angular momentum state with $j_0 = 0$ and $m_0 = 0$. The MCTDH program will evaluate the energy distribution of the initial wavepacket. To this end the keyword `correction=dia` is given. A Gaussian function $\varphi(x) \propto \exp(((x - x_0)/2\sigma)^2 + ip_0(x - x_0))$ has the energy distribution $|\Delta(E)|^2 \propto \exp(-4\mu\sigma^2[\sqrt{E} - \sqrt{E_0}]^2)/\sqrt{E}$. (This distribution is plotted by running the script "`pledstr`"). Here μ is the reduced mass, and σ and p_0 are the width and momentum of the Gaussian function as given in input. However, we do not use this formula as this would require to place the initial wavepacket far outside where the potential has essentially vanished. To be able to place the initial wavepacket much closer to the scattering center, which obviously reduces the propagation time, we compute the energy distribution numerically with a correction scheme. Now, the condition on the initial location of the wavepacket is that on the interval (x_0, ∞) the interaction potential does not change the ro-vibrational state of the wavepacket when propagated from infinity to the initial location x_0.

Also inspect the operator file `dh2-0.op`. It is a typical one for a three-atom molecule in Jacobian coordinates with vanishing total angular momentum. There is a second Hamiltonian-Section for computing the eigenstates of H_2.

After having discussed the input files we move to the `dh2-0` name directory, open the `log` file for reading, and inspect the

`*** (A)Diabatic Corrections ***`

section. The zero point of the lsth PES is reached when all three particles are infinitely apart from each other. But in our case H_2 is in a bound state and when rd goes to infinity there remains the binding energy of H_2 in its particular initial state. As one usually wants to present the data versus initial kinetic energy rather than total energy, one has to shift the energy scale. The shift is printed to the `log` file. The line of interest reads:

`Energy shift (E_total - E_kinetic) : -4.4777478 eV`

This energy is to be subtracted and in the following we will simply use 4.4777 which is of sufficient accuracy. First let us view the translational mean field, which is the potential averaged over the internal coordinates where the average is weighted by the density of the initial state. Type "`plenerd -G -t -E 4.4777`". The averaged 1D potential shows a boring appearance, it seems to be simply repulsive. (The decrease in slope for $rd < 1.5$ a.u. originates from cutting the potential at 0.0 eV before the shift, i.e. at 4.4777 eV after the shift.) But a closer look, "`plenerd -G`

`-a 5.5 -t -E 4.4777`", shows that there is a weak van-der-Waals minimum at $rd > 6.0$ a.u.

To view the energy distribution type "`plenerd -G -x 1.2 -E 4.4777`". One sees that energies up to 1 eV are covered. Finally we inspect the momentum distribution of the initial state. Type "`showd1d84 -a -G -sm -pop2 f1`". The momentum distribution seems to look perfect as at $t = 0$ the distribution does not cover positive momenta.[7] However, a closer look, "`showd1d84 -a -G -sm -l -pop2 f1`" shows that there are positive momenta in the distribution, albeit at negligible intensities.

To study the reaction dynamics, run `showd1d84` for all three DOFs. Type "`showd1d84 -a -G -y 1.2 f1`" to observe that the rd-density moves towards the scattering center and is reflected back. The point of closest approach appears near 18 fs. Likewise type "`showd1d84 -a -G -y 0.25 -pop2 f1`" to see that the rd momentum distribution moves from negative values to positive ones. Near $t = 18$ fs the average momentum is zero. Next type "`showd1d84 -a -G -y 0.3 f2`" to observe that part of the vibrational density dissociates. Or run "`showd1d84 -a -G f3`". Initially the angular distribution is constant, as for a $j = 0$ rotational state all directions are equally probable. However, when the D-projectile approaches the H_2 fragment, a collinear arrangement is preferred, as indicated by the strong peak at small θ-values, which is formed in particular between 18 and 40 fs. One also may inspect the expectation values by running `plqdq`.

We now run the flux analysis. Type "`mkdir frv frd`" to create two directories that will contain the output data of the flux calculations. There are two CAPs, one on the rd-DOF, the other on the rv-DOF. Hence we shall discuss two fluxes. Run "`flux84 -e 4.4777 eV -lo 9 -o frv 0. 1. ev rv`" and "`flux84 -e 4.4777 eV -lo 9 -o frd 0. 1. ev rd`". Then move to `frv` and type "`plflux -G`". The lower line shows the energy resolved flux that has been absorbed by the CAP. The upper line is the energy distribution that we have already seen before. The quotient of these two lines provides the reaction probability. To show it, type "`plflux -G -r`". Below 0.2 eV there is no reaction and between 0.3 and 0.4 eV one observes a sharp increase in reaction probability, which is followed by a slower increase at higher scattering energies. At 1 eV 63.5% of the initial flux is reactive.

One may move to the `frd` directory and investigate the rd-flux in a similar manner. Then move back to the name-directory `dh2-0` and submit the command "`sumspec84 frv/flux 1.0 frd/flux 1.0`". By this the two fluxes are summed. With "`plgen -z 0.96 -G sumspec.pl 1:4`" the sum of reactive and non-reactive probability is shown, which, of course, should add up to 1. In fact, the plot shows a line that deviates by less than 1% from 1 in the interval 0.1–1.0 eV and even by less than 0.5% in the interval 0.2–0.94 eV. The larger deviations at low energy are unimportant, because there is no reaction below 0.2 eV. In fact, a time-dependent method will always have difficulties to converge reaction probabilities at

[7] Here and for the following plot we are only interested in the $t = 0$ figure. Hence do not press 'enter' to advance the wavepacket. Type 'Ctrl C' to exit.

very low energies, because the low energy parts of a wavepacket move very slowly and thus require very long propagation times for convergence.

The nice agreement of the summed probabilities with unity demonstrates the accuracy of the calculations. In practice, however, one will not compute the rd-flux, the rv-flux alone provides the desired reaction probability. But this is still not a measurable quantity, the quantity of interest is the reaction cross section, which, for the present initial conditions, $j_0 = 0$ and $v_0 = 0$, reads

$$\sigma(E_{kin}) = \frac{\pi}{2\mu_{rd} E_{kin}} \sum_{J=0}^{\infty} (2J+1) P_J(E_{kin})$$

where μ_{rd} is the reduced mass of the rd-DOF, E_{kin} the collisional energy, and P_J the reaction probability. For formulas that are valid for general initial conditions, see e.g. Refs. [33, 34]. The sum converges, since P_J vanishes for large J, as the centrifugal potential hinders a reaction. To evaluate the cross section, σ, one has to perform several propagations with different total angular momenta, J.

To converge the cross section for scattering energies up to 1 eV, one has to include reaction probabilities for total J up to 28. To avoid running 28 propagations, we will compute reaction probabilities on a rather coarse J-grid and obtain the missing ones by interpolation. To this end we generate a directory `FLUX` which shall contain all the computed `flux` files; "`mkdir ../FLUX`". Before we copy the `flux` file of the $J = 0$ calculation to `FLUX`, we will re-do the flux calculation for the energy interval 0.2–1.0 eV, because we know that there is no reaction for scattering energies below 0.2 eV. Hence type "`flux84 -e 4.4777 eV -lo 9 -o frv 0.2 1.0 ev rv`". The flux calculation is now very fast because the `flux84` program notices that there already exists a so-called `gtau` file, which contains the matrix elements of the CAP. Thus `flux84` performs only the Fourier transform anew. Next copy the `flux` file to `FLUX`, "`cp frv/flux ../FLUX/flux000`" and move back, "`cd ..`". The first two zeros of `flux000` stand for the initial conditions $j_0 = 0$ and $k_0 = 0$, and the last zero indicates $J = 0$.

Next, we will submit propagations for J = 5, 10, 15, 20, 25, and 28. To avoid a considerable amount of typing, we will make use of environment variables.
Define: "`cmd1='(mctdh84 -mnd dh2-'`",
"`cmd2='&& cd dh2-'`",
"`cmd3='&& mkdir frv&& flux84-e 4.4777 eV -lo 9 -o frv'`",
"`cmd4='ev rv >/dev/null && cp frv/flux ../FLUX/flux00'`"
and "`J=5`". Then type
"`echo"$cmd1$J $cmd2$J $cmd3 0.2 1.1 $cmd4$J )&"`"
and you will read:

```
(mctdh84 -mnd dh2-5 && cd dh2-5 && mkdir frv &&
flux84-e 4.4777 eV-lo 9-o frv 0.2 1.1 ev rv >/dev/null
&& cp frv/flux ../FLUX/flux005 )&.
```

Hence, when this command is submitted, `mctdh84` will run with the `dh2-5.inp` input file. After the run is completed, one moves to the `dh2-5` name-directory,

creates there a `frv` directory, and runs `flux84`. The screen-output of `flux84` is directed to `/dev/null` to avoid that one is disturbed by this output when inspecting the input files. (No information is lost as the redirected data can also be found on `flux.log`). Finally the `flux` file is copied to the `FLUX` directory, while renaming it to `flux005`. The final energy of the flux evaluation, 1.1 eV, is chosen slightly larger than the energy window we are interested in, because the interpolation process works better on a slightly extended energy interval. The set of commands is bracketed to let it run in a separate shell, and the final and puts the command to background. To actually execute the command submit
"`eval"$cmd1$J $cmd2$J $cmd3 0.2 1.1 $cmd4$J )&"`".
And to run the computations for the other J values, type
"`J=10; eval"$cmd1$J $cmd2$J $cmd3 0.2 1.15 $cmd4$J )&"`".
"`J=15; eval"$cmd1$J $cmd2$J $cmd3 0.3 1.2 $cmd4$J )&"`".
"`J=20; eval"$cmd1$J $cmd2$J $cmd3 0.4 1.2 $cmd4$J )&"`".
"`J=25; eval"$cmd1$J $cmd2$J $cmd3 0.5 1.3 $cmd4$J )&"`".
"`J=28; eval"$cmd1$J $cmd2$J $cmd3 0.6 1.3 $cmd4$J )&"`".
Note that the energy interval for flux evaluation shifts to higher energies with increasing J because the centrifugal barrier increases the reaction threshold with growing J.

While the propagations are running (a single propagation takes between 10 and 20 min, depending on your hardware and on the particular J value), one should study the `*.inp` and `*.op` files. For $J > 0$ there appears an additional dynamical variable, k, which is the projection of the total angular momentum J onto the body-fixed z-axis. It is also the Fourier transform of the third Euler angle γ. To treat the dynamics correctly we combine θ and k into one mode and employ the extended Legendre DVR, KLeg. See the appendix of Ref. [33] for a description of KLeg. The input files, e.g. `dh2-5.inp` for $J = 5$, are similar to the $J = 0$ case, `dh2-0.inp`, except, of course, for the additional definition of the k-DOF. The range of k values is restricted to $|k| \leq J$, but for larger J values a smaller k range suffices. As $|k| \leq j$ likewise holds, the initial k value, k_0, must be zero, because the initial angular momentum of the H_2 fragment, j_0, was chosen to be zero.

For larger J, the centrifugal potential shifts the reaction threshold to higher values, namely to about 0.5, 0.7, and 0.8 eV for $J = 20$, 25, and 28, respectively. To put more emphasis on the high energy region, we slightly increased the initial momentum from $p_0 = 7.5$ a.u. to 7.8 a.u. when $J \geq 25$.

After all propagations are finished,[8] we generate the missing reaction probabilities by interpolation using `jinpol84`. Try "`jinpol84 -h`". A description of the algorithm is given in the appendix of Ref. [34]. To reduce the amount of typing we again make use of environment variables, define
"`jpl='jinpol84 -f flux -o fx -ef flux000'`",
move to the directory `FLUX`, and run

```
"eval $jpl -j1  0 -j2  5 -e1 0.2 -e2 1.0"
"eval $jpl -j1  5 -j2 10 -e1 0.2 -e2 1.0"
"eval $jpl -j1 10 -j2 15 -e1 0.2 -e2 1.0"
```

[8] Submit the command "`ps -r`" to check which jobs are still running.

"`eval $jpl -j1 15 -j2 20 -e1 0.3 -e2 1.1`"
"`eval $jpl -j1 20 -j2 25 -e1 0.4 -e2 1.1`"
"`eval $jpl -j1 25 -j2 28 -e1 0.6 -e2 1.2`"
The first option, `-f flux`, tells the program to search for files `flux` *extension* as input, `-o fx` let the program write the output to `fx` *extension*, and through the `-ef flux000` option, the energies are taken from the file `flux000`. The options `-j1` and `-j2` define the range of J values to be interpolated, and `-e1` and `-e2` define an energy range used to fit some parameters of the interpolating scheme. For larger J we avoided energies for which the reaction probability vanishes, but slightly increased the upper range. To visualize the result, type "`plgen -G -n -x 1.0 fx00*`". All 29 reaction probabilities appear on one plot. To check how well the interpolation works, we may run propagations while setting J = 3, 7, 12, or 27 (the input files for these J values are provided). Then run e.g. "`plgen -G -a 0.2 -x 1.0 flux0012 1:4 fx0012 1:2`" to observe the differences.

Next, we have to sum the reaction probabilities weighted with $2J+1$. For this we use `sumspec84` and to avoid much typing we generate the input of `sumspec84` with a bash script. Type
"`J=0, unset list`"
"`while ["$J" -lt 29 ]; do nn=$((2*$J + 1)); list="$list fx00$J $nn"; J=$(($J + 1)); done`"
To see what we got, run "`echo $list`". Next run "`sumspec84 -o fxall $list`". The weighted sum of reaction probabilities looks rater boring, "`plgen -G fxall`". But this is not the required reaction cross section, to obtain this one has to multiply with some constant and divide by E_{kin} as explained in the equation above. The arithmetics can be performed with `plgen`. Type
"`plgen -G fxall '1:($2*0.0065177/$1)'`",
where the factor 0.0065177 stands for $\pi/2\mu_{rd}$ and for the conversion factors 'eV to a.u.' and 'a.u. to angstroem-squared'. A nicer plot is generated by the command
"`plgen -t 'D + H2 total reactive cross section for initial state j=0,v=0'-X 'E-kin[eV]'-Y'Sigma [Angstroem^2]'-n-G fxall '1:($2*0.0065177/$1)'`"

One may run similar calculations for initial conditions with $j_0 > 0$ or $v_0 > 0$. Note, if $j_0 > 0$, one must average the cross section over all m_0 states with $|m_0| \leq j_0$. The initial value for k is to be set to m_0. In fact $k = m$ holds throughout, because in the body-fixed frame $\ell_z = 0$ holds, and thus $j_z = J_z$. See for example Refs. [33, 34] for further explanations.

References

1. Zhang JZH (1999) Theory and application of quantum molecular dynamics. World Scientific, Singapore
2. Siegbahn P, Liu B (1978) An accurate three-dimensional potential energy surface for H_3. J Chem Phys 68:2457
3. Truhlar DG, Horowitz CJ (1978) Functional representation of Liu and Siegbahn's accurate ab initio potential energy calculations for $H+H_2$. J Chem Phys 68:2466
4. Truhlar DG, Horowitz CJ (1979) Erratum: Functional representation of Liu and Siegbahn's accurate ab initio potential energy calculations for $H+H_2$. J Chem Phys 71:1514
5. Johnston GW, Katz B, Tsukiyama K, Bersohn R (1987) Isotopic variants of the $H+H_2$ reaction. 1. Total reaction cross sections of the $H+D_2$ and H+HD reactions as a function of relative energy. J Chem Phys 91:5445
6. Levene HB, Phillips DL, Nieh J-C, Gerrity DP, Valentini JJ (1988) Measurement of absolute partial reaction cross sections for the hydrogen exchange reaction. Chem Phys Lett 143:317
7. Brownsword RA, Hillenkamp M, Laurent T, Volpp H-R, Wolfrum J, Vatsa RK, Yoo H-S (1998) Excitation function and reaction threshold studies of isotope exchange reactions: $H+D_2 \rightarrow D+HD$ and $H+D_2O \rightarrow D+HOD$. J Phys Chem A 101:6448
8. Jäckle A, Heitz M-C, Meyer H-D (1999) Reaction cross sections for the $H+D_2(\nu = 0, 1)$ system for collision up to 2.5 eV: A multiconfiguration time-dependent Hartree wave-packet propagation study. J Chem Phys 110:241
9. Miller W, Zhang J (1991) How to abserve the elusive resonances in H or $D+H_2 \rightarrow H_2$ or HD+H reactive scattering. J Phys Chem 95:12
10. Miller WH (2006) Including quantum effects in the dynamics of complex (i.e., large) molecular systems. J Chem Phys 125:132305
11. Charutz DM, Last I, Baer M (1997) The Toeplitz approach to treating three-dimensional reactive exchange processes: Quantum mechanical cross sections and rate constants for the $D+H_2$ and $H+D_2$ reactions. J Chem Phys 106:7654
12. Truhlar DG, Garrett BC, Klippenstein SJ (1996) Current status of transition-state theory. J Phys Chem 100:12771
13. Qiu M, Ren Z, Che L, Dai D, Harich SA, Wang X, Yang X, Xu C, Xie D, Gustafsson M, Skoedje RT, Sun Z, Zhang D-H (2006) Observation of Feshbach resonances in the $F+H_2 \rightarrow$ HF+H reaction. Science 311:1440
14. Sun Z, Zhao B, Liu S, Zhang D-H (2014) Reactive scattering and resonance. In: Molecular quantum dynamics, Gatti F (ed). Springer, Heidelberg
15. Dulieu O, Krems R, Weidemüller M, Willitsch S (2011) Physics and chemistry of cold molecules. Phys Chem Chem Phys 13:18703
16. Quéméner G, Julienne PS (2012) Ultracold molecules under control!. Chem Rev 112:4949
17. Jin DS, Ye J (2012) Introduction to ultracold molecules: new frontiers in quantum and chemical physics. Chem Rev 112:4801
18. Baranov MA, Dalmonte M, Pupillo G, Zoller P (2012) Condensed matter theory of dipolar quantum gases. Chem Rev 112:5012
19. Narevicius E, Raizen MG (2012) Toward cold chemistry with magnetically decelerated supersonic beams. Chem Rev 112:4879
20. Stuhl BK, Hummon MT, Yeo M, Quéméner G, Bohn JL, Ye J (2012) Evaporative cooling of the dipolar hydroxyl radical. Nature 492:396
21. Hutson JM (2000) Ultracold chemsitry. Science 327:788
22. Lang F, Winkler K, Strauss C, Grimm R, Hescher J (2008) Denschler. Ultracold triplet molecules in the rovibrational ground state. Phys Rev Lett 101:133005
23. Fioretti A, Comparat D, Crubellier A, Dulieu O, Masnou-Seuuws F, Pillet P (1998) Formation of cold Cs_2 molecules through photoassociation. Phys Rev Lett 80:4402
24. Vatasescu M, Dulieu O, Amiot C, Comparat D, Drag C, Kokooline V, Masnou-Seeuws F, Pillet P (2000) Multichannel tunneling in the Cs_2 $O_{(g)}^{(-)}$ photoassociation spectrum. Phys Rev A 61:044701

25. Dion CM, Drag C, Dulieu O, Torla BL, Masnou-Seuuws F, Pillet P (2001) Resonant coupling in the formation of ultracold ground state molecules via photoassociation. Phys Rev Lett 86:2253
26. Althorpe SC (2010) Setting the trap for reactive resonances. Science 327:1460
27. Yang X, Minton TK, Zhang D-H (2012) Rethinking chemical reactions at hyperthermal energies. Science 336:1650
28. Welsch R, Huarte-Larrañaga F, Manthe U (2012) State-to-state reaction probabilities within the quantum transition state framework. J Chem Phys 136:064177
29. Zhao B, Zhang DH, Lee S-Y, Sun Z (2014) Calculation of state-to-state cross sections for triatomic reaction by the multi-configuration time-dependent Hartree method. J Chem Phys 140:164108
30. Manthe U, Welsch R (2014) Correlation functions for fully or partially state-resolved reactive scattering calculations. J Chem Phys 140:244113
31. Gatti F, Otto F, Sukiasyan S, Meyer H-D (2005) Rotational excitation cross sections of para-H_2 + para-H_2 collisions. A full-dimensional wave packet propagation study using an exact form of the kinetic energy. J Chem Phys 123:174311
32. Otto F, Gatti F, Meyer H-D (2012) Rovibrational energy transfer in collisions of H_2 with D_2. A full-dimensional wave packet propagation study. Mol Phys 110:619
33. Sukiasyan S, Meyer H-D (2001) On the effect of initial rotation on reactivity. A multi-configuration time-dependent Hartree (MCTDH) wave-packet propagation study on the H+D_2 and D+H_2 reactive scattering systems. J Phys Chem A 105:2604
34. Sukiasyan S, Meyer H-D (2002) Reaction cross section for the H+$D_2(\nu_0 = 1) \rightarrow$ HD+D and D+$H_2(\nu_0 = 1) \rightarrow$ DH+H systems. A multi-configuration time-dependent Hartree (MCTDH) wave-packet propagation study. J Chem Phys 116:10641

Chapter 12
Vibronic Coupling

12.1 Introduction

Photons in the ultraviolet-visible region can promote molecules to electronic excited states. If no photodissociation occurs, an excited molecule can release its excess energy and return to the electronic ground state through spontaneous light emission: fluorescence from a singlet state or phosphorescence from a triplet state if the ground state is a singlet state. However, in many cases, the light emission is quenched by a much faster radiationless process that is due to the presence of *a non-adiabatic couplings between electronic states* (see Fig. 12.1). Typical phenomena associated with such situations are the radiationless relaxation of electronic excited states to the ground state, indirect photodissociations as introduced in Sect. 10.2, and isomerization processes of polyatomic molecules [1].

In polyatomic molecules, non-Born-Oppenheimer processes occur mainly around topographical features joining different PESs, known as conical intersections [1–3] and introduced in Chap. 4. A *conical intersection* induces strong couplings between electrons and nuclei. *It acts as a funnel through which a new chemical reaction can occur and enables rapid conversion of the excess electronic energy into nuclear motion* [4]. Thus, conical intersections are a central paradigm for understanding reaction mechanisms in photochemistry and photobiology, as important as transitions states in thermal chemistry [5].

The presence of conical intersections gives rise to new effects that cannot be described classically. First, the fact that a system can be transferred from one quantized state to another is per se a quantum effect. Second, the electronic states can remain coherent with respect to each other during a non-negligible period of time, leading to new quantum interference effects.

F. Gatti et al., *Applications of Quantum Dynamics in Chemistry*,
Lecture Notes in Chemistry 98, DOI 10.1007/978-3-319-53923-2_12

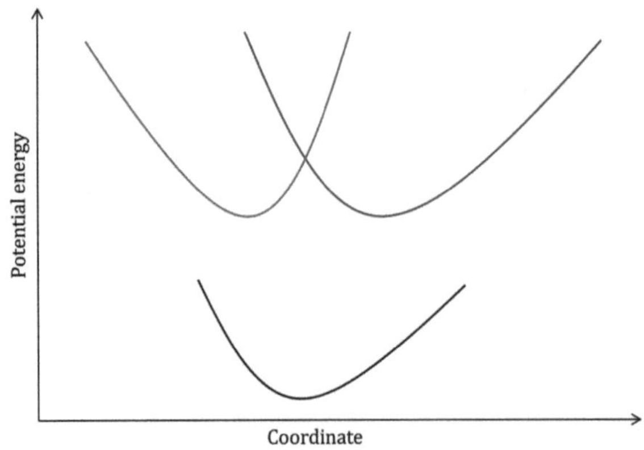

Fig. 12.1 Schematic illustration of the physical situation where the molecular system is first in its electronic ground state (in *blue*) and after absorption of light is promoted to an bound electronic excited that is in turn coupled through a conical intersection to another electronic bound state. The *curves* in *green* and *red* correspond to the potential energies of the two diabatic excited states

Fig. 12.2 The experimental spectrum of butatriene (see Ref. [7])

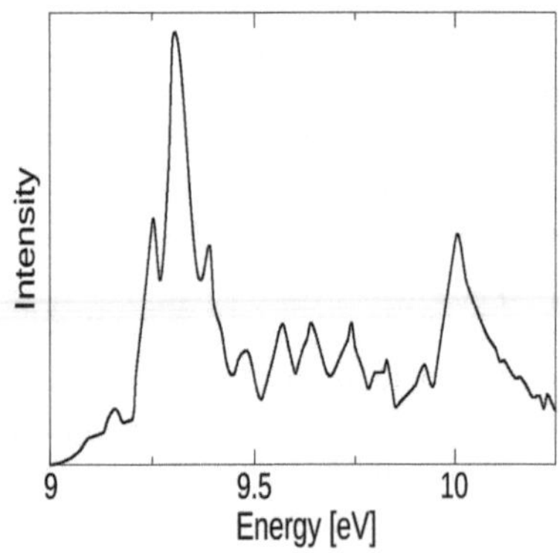

12.2 Photoelectron Spectrum of the Butatriene Molecule

A well-known striking evidence of the presence of a conical intersection is the photoelectron spectrum of the butatriene molecule shown on Fig. 7.5 in Sect. 7.3 of Chap. 7. The calculation of the ionization potentials indicates that only *two* electronic states, called B_{2g} and B_{2u}, due to their symmetry, are present up to 11 eV [6]. The symmetry of butatriene is explained in detail in Sect. 7.3 of Chap. 7. In 1974, the spectrum was measured for the first time [7] and *three* electronic bands were observed in the ionization energy range from 9 to 10 eV. The experimental spectrum is given in Fig. 12.2[1]: the two bands predicted by the calculation of the ionization potentials are centered around 9.3 and 10.1 eV. The band between 9.4 and 9.9 eV does not correspond to any identified electronic state and was called for that reason the *mystery band* of

[1]Reprinted with permission from [8]. Copyright 2001, American Institute of Physics.

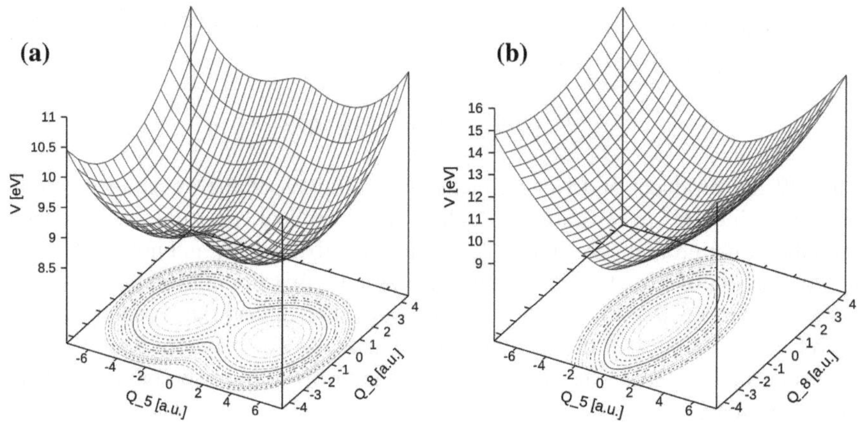

Fig. 12.3 The adiabatic potential energy surfaces of the two excited electronic states of butatriene obtained with the vibronic model of Ref. [8]

butatriene. It was later established by Cederbaum et al. that *the mystery band represents a vibronic effect, i.e. a strong non-adiabatic coupling between the two electronic states* [9].

A one-dimensional schematic illustration of the physical situation is depicted on Fig. 12.1. The system is first in the vibrational ground state of the electronic ground state. There is then absorption of light that promotes the system to an excited bound electronic state that is coupled through a conical intersection to another bound excited electronic state. In the case of the photoelectron spectrum of the butatriene molecule, the systems is first in the ground state of the molecule and the two "excited" states are electronic states of the cation. In fact, the first "excited" state is the electronic ground state of the cation but we will continue to call it the first excited state for the photoionization process. The chemical interpretation of the two excited states of butatriene in terms of molecular orbitals can be seen on Fig. 7.9 in Sect. 7.3 of Chap. 7. A 2D plot of the adiabatic potential energy surface is shown in Fig. 12.3. In the following, we assume that the ejection of the electron during the photoionization is instantaneous (sudden approximation).

The Hamiltonian operator for the nuclear problem in the electronic ground state 0 is given by

$$H^0 = T^{nu} + V_0 \,. \tag{12.1}$$

We use the normal modes of the butatriene molecule in the electronic ground state and a harmonic approximation for the potential energy operator. The system has 8 atoms and thus 18 modes of vibration. Using mass- and frequency-scaled normal modes (see Eq. 5.8 in Chap. 5 for the definition of the normal modes), H^0 resorts to

$$H^0(\mathbf{q}) = \sum_{l=1}^{18} \frac{w_l}{2} \left(q_l^2 - \frac{\partial^2}{\partial q_l^2} \right) \,. \tag{12.2}$$

The normal vibrations are given in Tables 7.6–7.8 in Sect. 7.3. The corresponding vibrational eigenstates satisfy

$$H^0(\mathbf{q})\Psi_0^m(\mathbf{q}) = E_0^m \Psi_0^m(\mathbf{q})\,. \tag{12.3}$$

The Ψ_0^m are products of 18 one-dimensional solutions of the quantum harmonic oscillator since H^0 is separable. To obtain the theoretical spectrum, we start from the vibrational ground state in the electronic ground state, Ψ_0^0 and use the Condon approximation of Eq. (10.7). Since we also assume that the ionization is instantaneous, we simply have to place Ψ_0^0 onto one of the two diabatic states and propagate it. We do the same with the second diabatic state. As for ozone in Sect. 10.2, the eigenstates are given by the Schrödinger equation involving two coupled diabatic electronic states:

$$\begin{bmatrix} T^{nu}(\mathbf{q}) + V_1(\mathbf{q}) & V_{12}(\mathbf{q}) \\ V_{12}(\mathbf{q}) & T^{nu}(\mathbf{q}) + V_2(\mathbf{q}) \end{bmatrix} \begin{bmatrix} \Psi_1^n(\mathbf{q}) \\ \Psi_2^n(\mathbf{q}) \end{bmatrix} = E_n \begin{bmatrix} \Psi_1^n(\mathbf{q}) \\ \Psi_2^n(\mathbf{q}) \end{bmatrix}, \tag{12.4}$$

where, V_{12} is the potential coupling between the two diabatic states. However, contrarily to ozone, the set of eigenstates is now discrete since the two electronic states do not lead to dissociation in the energy domain of interest. The expression of the potential terms are given in Sect. 7.3 of Chap. 7: see Eq. (7.20). The molecular eigenstates can no longer be written as a direct product form, they correspond to a sum of two products:

$$\Psi_1^n(\mathbf{q})\Phi_1^{el/diab}(\mathbf{r};\mathbf{q}) + \Psi_2^1(\mathbf{q})\Phi_2^{el/diab}(\mathbf{r};\mathbf{q})\,, \tag{12.5}$$

where $\Phi_1^{el/diab}$ and $\Phi_2^{el/diab}$ denote the diabatic electronic functions of states one and two, respectively.

In Eq. (12.4), T^{nu} is identical to the kinetic energy operator in Eq. (12.2). For the diabatic potential matrix, we use a vibronic coupling model as described in Sect. 4.4: we use a second-order Taylor expansion around the equilibrium geometry of V_1, V_2 and V_{12}.

Starting from the initial condition

$$\Psi^{mol}(\mathbf{r};\mathbf{q},t=0) = a_1\Psi_1(\mathbf{q},t=0)\Phi_1^{el/diab}(\mathbf{r};\mathbf{q}) + a_2\Psi_2(\mathbf{q},t=0)\Phi_2^{el/diab}(\mathbf{r};\mathbf{q})\,, \tag{12.6}$$

where a_1 and a_2 are complex numbers, the values of which depends on details of the ionization process. Assuming that the phases of a_1 and a_2 randomize over a sample of ionization processes, cross-terms are averaged out and one can study the ionization to the B_{2g} and B_{2u} of $C_4H_4^+$ separately. Thus, one can perform individual propagations of Ψ_1 and Ψ_2 and adds the resulting spectra while weighting them with $|a_1|^2$ and $|a_2|^2$, respectively. In the following, we assume that populating the two ionic states is equally probable and set $a_1 = a_2 = 1$. We hence set the initial wavefunctions to

$$\Psi_1(\mathbf{q}, t=0) = \Psi_0^0(\mathbf{q}),$$
$$\Psi_2(\mathbf{q}, t=0) = 0, \tag{12.7}$$

for B_{2g}: it corresponds to the transition with the yellow arrow Fig. 7.9 in Sect. 7.3 of Chap. 7; or

$$\Psi_1(\mathbf{q}, t=0) = 0,$$
$$\Psi_2(\mathbf{q}, t=0) = \Psi_0^0(\mathbf{q}), \tag{12.8}$$

for B_{2u}: it corresponds to the transition with the purple arrow in Fig. 7.9 in Sect. 7.3 of Chap. 7.

The propagation has been performed with MCTDH (see Ref. [8]). For the primitive basis set of each coordinate, we use a Harmonic oscillator DVR for each degree of freedom (see Sect. 8.1.3).

The wavefunction has two nuclear components $\Psi_1(\mathbf{q}, t)$ and $\Psi_2(\mathbf{q}, t)$. They are obtained by solving the following coupled equations of motion with MCTDH:

$$\begin{bmatrix} T^{nu}(\mathbf{q}) + V_1(\mathbf{q}) & V_{12}(\mathbf{q}) \\ V_{12}(\mathbf{q}) & T^{nu}(\mathbf{q}) + V_2(\mathbf{q}) \end{bmatrix} \begin{bmatrix} \Psi_1(\mathbf{q}, t) \\ \Psi_2(\mathbf{q}, t) \end{bmatrix} = i\frac{\partial}{\partial t} \begin{bmatrix} \Psi_1(\mathbf{q}, t) \\ \Psi_2(\mathbf{q}, t) \end{bmatrix}. \tag{12.9}$$

The MCTDH expressions $\Psi_1(\mathbf{q}, t)$ and $\Psi_2(\mathbf{q}, t)$ are given by Eq. (8.233) of Sect. 8.3.7. Here, we use five combined modes (see Sect. 8.3.5 for the definition of the combined modes):

$$\Psi_\alpha(q_1, \ldots, q_{18}, t) = \sum_{j_1^\alpha}^{n_1^\alpha} \ldots \sum_{j_5^\alpha}^{n_5^\alpha} A^{(\alpha)}_{j_1^\alpha \ldots j_5^\alpha}(t)\, \varphi^{(1,\alpha)}_{j_1^\alpha}(Q_1, t) \ldots \varphi^{(5,\alpha)}_{j_5^\alpha}(Q_5, t), \tag{12.10}$$

with $Q_1, \ldots, Q_5$ corresponding to $(q_{15}, q_{16}, q_{17}, q_{18}, q_{11}, q_{13})$, (q_1, q_4, q_7), (q_2, q_3, q_6), (q_5, q_{14}, q_8), and (q_9, q_{10}, q_{12}), respectively. Here, $n_1^1 = 7, n_2^1 = 9, n_3^1 = 9, n_4^1 = 11, n_5^1 = 9, n_1^2 = 6, n_2^2 = 9, n_3^2 = 9, n_4^2 = 11$, and $n_5^2 = 8$. We use almost 6×10^{16} DVR functions[2] to converge the calculation. The number of SPFs functions is much smaller: 56,133 for $\Psi_1(\mathbf{q}, t)$ and 42,768 for $\Psi_2(\mathbf{q}, t)$. We see that a calculation in the primitive basis set (as in Eq. 8.7), i.e. the DVR, would be impossible since the number of DVR functions is far too large. It is the MCTDH algorithm, i.e. the use of the variational principle to build the optimized SPFs functions, that makes the calculation feasible.

[2]For an HO DVR, since we use mass- and frequency-weighted normal coordinates, the equilibrium geometries are at 0, the frequencies and the masses are fixed to 1. The number of DVR functions in the MCTDH expansion is given by $N_1 = 19, N_2 = 16, N_3 = 18, N_4 = 9, N_5 = 34, N_6 = 9, N_7 = 8, N_8 = 9, N_9 = 7, N_{10} = 7, N_{11} = 5, N_{12} = 6, N_{13} = 5, N_{14} = 10, N_{15} = 5, N_{16} = 5, N_{17} = 5$, and $N_{18} = 6$. Note that most of the figures below are generated with a 5D model of butatriene as in lab-session IV.

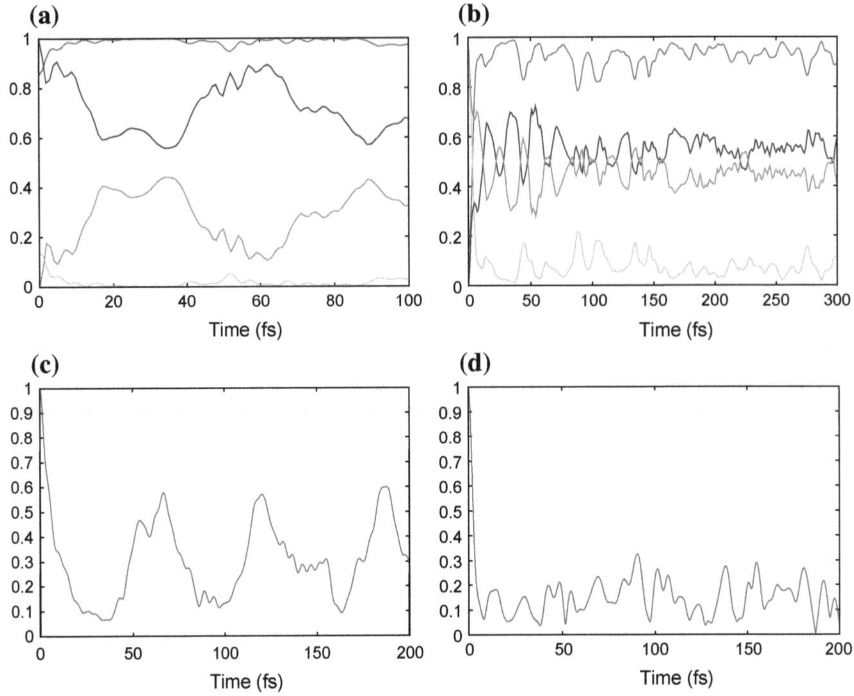

Fig. 12.4 Adiabatic and diabatic electronic populations and autocorrelation functions for the calculation of the photoelectron spectrum of butatriene. **a** and **c** starting from diabatic state B_{2g} (Eq. 12.7), **b** and **d** starting from diabatic state B_{2u} (Eq. 12.8). For (**a**) and (**b**), the populations of the B_{2g} and B_{2u} diabatic states are in *blue* and *indigo*, respectively. The populations of the lowest and highest adiabatic states are in *red* and *green*, respectively

The electronic populations, i.e. $\int |\Psi^{(\alpha)}(q_1, \ldots, q_{18}, t)|^2$, are given in Fig. 12.4a with the wavepacket starting from electronic (diabatic) state B_{2g} (condition of Eq. 12.7) and in Fig. 12.4b with the wavepacket starting from electronic (diabatic) state B_{2u} (condition of Eq. 12.8). The populations of the B_{2g} and B_{2u} diabatic states are in blue and indigo, respectively. The populations of the lowest and highest adiabatic states are in red and green, respectively.

For the B_{2g} state, the adiabatic population is close to one for the lowest electronic state. It means that the non-adiabatic coupling is small is this case. The spectrum will be almost purely vibrational (and not vibronic) for B_{2g}. On the other hand, for B_{2u}, we see that the two electronic states are strongly coupled since *the adiabatic population of the highest adiabatic states deviates from one very rapidly and is close to 0.1 before 10 fs! This is due to the presence of the conical intersection.*

The corresponding autocorrelation functions are shown on Fig. 12.4c and d. The autocorrelation function of Fig. 12.4c is rather regular and is rather similar to the autocorrelation functions corresponding to the vibrations of the H_2O molecule as in Fig. 9.8a. On the other hand, the autocorrelation function of Fig. 12.4d is characteristic of the autocorrelation function of a wavepacket that explores two electronic states: it decreases very fast because of the rapid transfer from one electronic state to the other. If there is a transfer from the initially populated state to the other, the transferred part of the wavepacket does not overlap with the initial state and does not contribute to the autocorrelation function, because the two electronic states are orthonormal. Again, the transfer is due to the presence of the conical intersection close to the initial geometry and occurs in less than 10 fs. Contrarily to the autocorrelation function for the photodissociation of NOCl shown on Fig. 10.3a, the autocorrelation function of Fig. 12.4d does not tend to zero since the system is bound. Instead, it oscillates but less regularly and with smaller maxima than in (c). The comparison between the B_{2g} and B_{2u} spectra indicates that the B_{2u} initial wavepacket decays into many more states as compared to the B_{2g} one.

The Fourier transform of the autocorrelation functions using

$$\frac{Re}{\pi} \int_0^{\infty} e^{iEt}\, g(t)\, A(t)\, dt\,, \tag{12.11}$$

with

$$g(t) = e^{-(t/\tau)} \cos^2\left(\frac{\pi t}{2T}\right) \theta\left(1 - \frac{|t|}{T}\right), \tag{12.12}$$

is given in Fig. 12.5 for $T = 100$ fs. The damping function, $e^{-(t/\tau)}$ with $\tau = 55$ fs has been added to the window function to mimic the experimental line broadening. The latter is due to the presence of many rotational states but also to the experimental

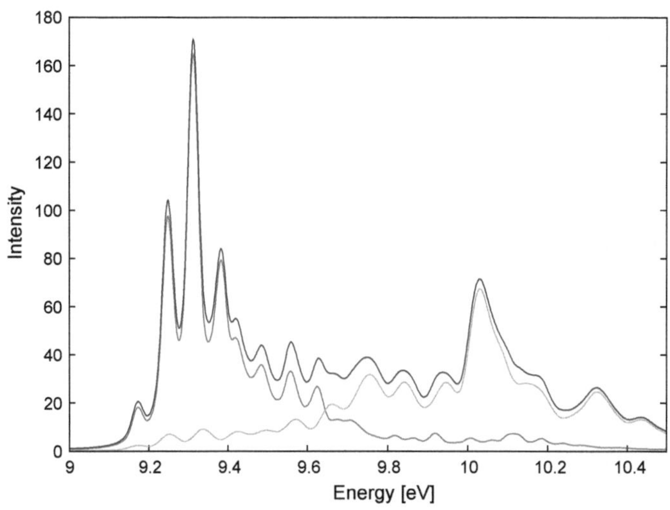

Fig. 12.5 The calculated photoionisation spectra of butatriene: for state B_{2g} in *red*, for state B_{2u} in *green*. The sum is shown in *blue*. The spectra have been shifted to lower energy values of the energy to compare with by the ZPE of the eletronic ground state, $\sum_{l=1}^{18} \frac{w_l}{2}$, experiment

resolution. In Fig. 12.5, the spectrum starting from state B_{2g} is in red and the spectrum starting from state B_{2u} is in green. The two spectra have been shifted by the ZPE of the eletronic ground state quantity to lower energy values to compare with the experiment. The spectrum in red is mainly purely vibrational since the adiabatic population of the lowest electronic state is close to one. On the other hand, the effect of the non-adiabatic coupling is non-negligible and explains the small intensities above 9.8 eV. The spectrum in green is strongly impacted by the non-adiabatic coupling. The peaks in the mystery band correspond to absorption to states that have high components on the two electronic states at the same time.

It is assumed that both electronic states are ionized by photon impact with the same probability. As discussed above, the total spectrum is thus the sum of the two individual spectra. This sum is shown in blue on Fig. 12.5. The comparison with the experimental spectrum (see Fig. 12.2) is excellent.

For the sake of completeness, we have also performed the dynamics on the two *adiabatic* PESs without any non-adiabatic coupling, i.e. *using the Born-Oppenheimer approximation*. The result is shown on Fig. 12.6. The spectrum for the lowest adiabatic state is in red, the one for the highest state is in green and the sum in blue. We see now two different spectra. As expected *the mystery band has almost completely disappeared*. For instance, we see nothing around 9.8 eV and the small peaks between 9.4 and 9.8 eV clearly correspond to highly excited vibrational states of the lowest electronic adiabatic state. The spectra in red in Figs. 12.5 and 12.6 are not very different confirming that the B_{2g} spectrum is almost purely vibrational. On the other hand, the spectra in green are very different since there is now nothing below

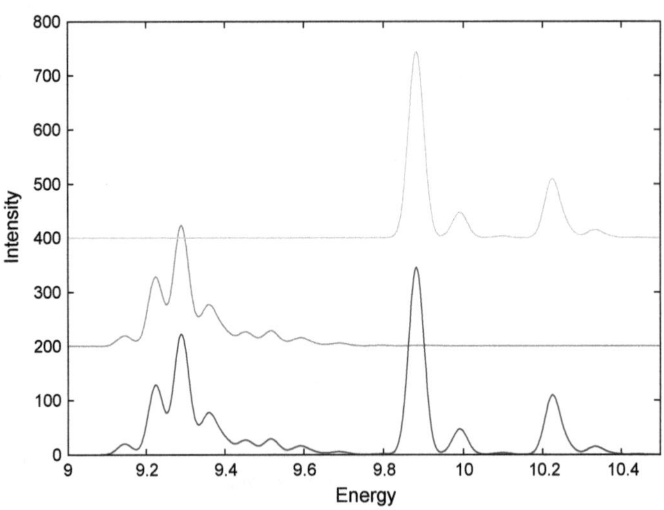

Fig. 12.6 The calculated photoionisation spectra of butatriene using the Born-Oppenheimer approximation: for the lower adiabatic state in *red*, for the higher adiabatic state in *green*. The sum is shown in *blue*. The spectra have been shifted by the ZPE of the electronic ground state to lower energy values to compare with the experiment

the first peak and the spectrum of the higher adiabatic uncoupled state does no longer contribute to the mystery band. This definitively proves that this spectrum reflects a strong vibronic effect.

12.3 Importance of Interferences Around Conical Intersections

Conical intersections lead not only to non-adiabatic transitions but also to strong interferences in the molecular wavefunction. This is very well illustrated by the appearance of the so-called geometric phase or *Berry phase effect* in both electronic and nuclear wave-functions [10, 11]. The Berry phase leads to interference effects.[3]

The conical nature of the PES topography leads to a sign change of the electronic wavefunction whenever the point of degeneracy is encircled in a closed loop. This sign change in the electronic wavefunction must be compensated by a sign change in the corresponding nuclear wavefunction in order to preserve the single-valued character of the total wavefunction [3, 12, 13]. In other words, if the adiabatic nuclear function encircles a conical intersection, it changes sign and this is expected even when the encirclement is very far from the CI. The latter can thus have a strong impact on the dynamics even if it is not reached during the nuclear motion. For instance, in the case of the $H+H_2$ collision discussed in Sect. 11.2, the cross sections between 3.5 and 4.4 eV[4] are strongly impacted by the geometric phase although the non-adiabatic transitions are very small [14]. In this energy domain, the reaction probabilities can be calculated employing just the the ground-state adiabatic surface but if and only if the geometric phase is included in the dynamics [14]. This is particularly true for the differential cross sections.

Let us return to the butatriene cation. Several snapshots of the two-dimensional reduced density along q_5 and q_{14} of the wavepacket corresponding to the B_{2u} spectrum on the upper adiabatic surface are shown on Fig. 12.7. In other words, Fig. 12.7 shows several snapshots of the reduced density with the initial condition of Eq. (12.8) projected onto the upper adiabatic electronic state. Since the position of the conical intersection in the (q_5, q_{14}) plane depends on the values of the other degrees of freedom and since the latter are not fixed for the reduced density (the reduced density is obtained by integrating over the other degrees of freedom), the conical intersection does not correspond to one single well-defined geometry in Fig. 12.7. However, we can say that the conical intersection is located approximately on the middle of the figures. At 6 fs, the wavepacket displays a maximum at roughly $q_5 = 0$ and $q_{14} = 2.8$. At 10 fs, we see that it starts to encircle the conical intersection. At 18 fs, the processes goes on and we see a important result: the wavepacket vanishes for the geometries corresponding to $q_5 = 0$. There are two maxima for approximately

[3]It must be clear that other strong interference effects may appear around conical intersections that are not linked to the Berry phase.

[4]This corresponds to energies higher than in Fig. 11.2.

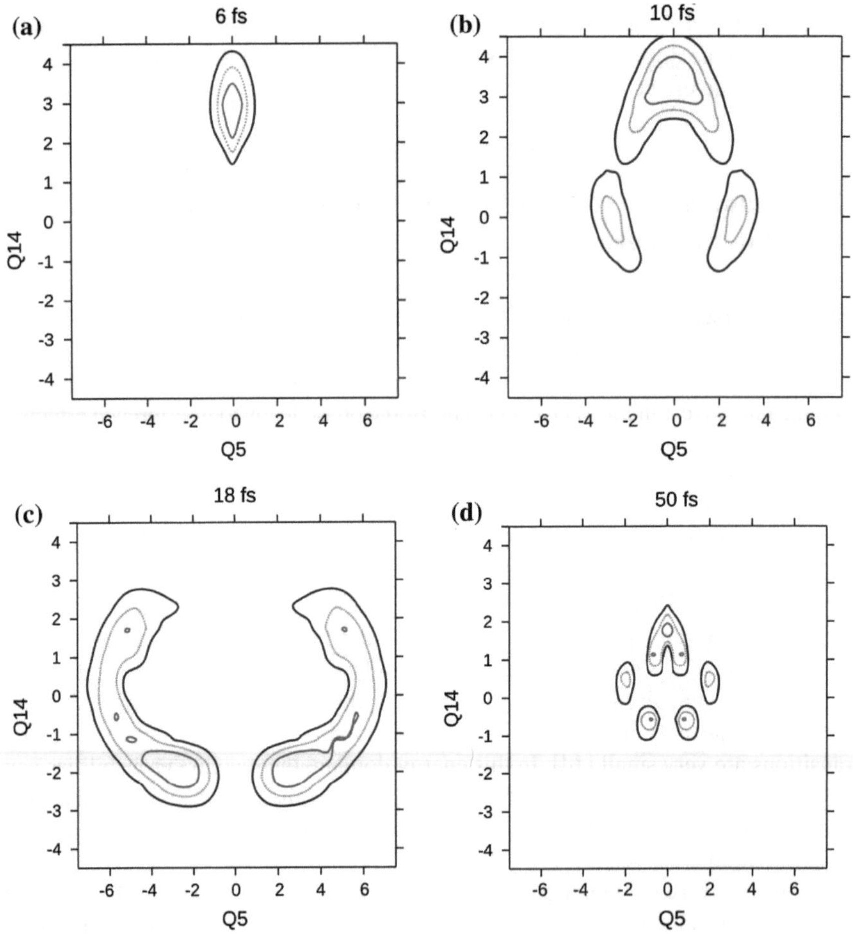

Fig. 12.7 Snapshots for 6, 10, 18 and 50 fs of the time-dependent wavepacket evolving on the upper adiabatic surface of butatriene. The conical intersection is located approximately on the *middle* of the figures

$q_{14} = -1.8$ and $q_5 = -3.5$ and 3.5. Thus, from 6 to 18 fs, there is an destructive interference that explains why starting from a maximum for positive values of q_{14} and $q_5 = 0$, we observe a density that is zero for negative values of q_{14} and $q_5 = 0$ after encirclement of the conical intersection. This is due to the geometric phase that leads to a sign change in the nuclear wavefunction. At 50 fs, we see again that for $q_5 = 0$, there is a maximum for positive values of q_{14} and nothing on the other side of the conical intersection for negative values of q_{14}. Here, the interference effect is clear: the wavefunction on the left has a different sign than the wavefunction on the right. When the two wavefunctions meet the corresponding destructive interference prevents the system to reach the geometries at the bottom of Fig. 12.7.

The destructive interference generated by the geometric phase can have a strong impact on the dynamics: it may make a reaction possible or prevent it [13, 15, 16]. It is suspected that this effect is preserved even if the system is embedded in a complex environment [13, 15, 16]. Thus, neglecting this phenomenon even for low energy nuclear dynamics on the electronic ground state around a conical intersection even if the latter is much higher in energy can result in qualitatively wrong predictions [17]. Note that the geometric phase effect is automatically included when, as done here, working in a diabatic representation.

The example of the Berry phase illustrates the importance of quantum interferences that may occur due to the presence of a non-adiabatic effect. Only quantum mechanics can correctly describe these effects.

12.4 Photoabsorption Spectrum of the Pyrazine Molecule

The previous vibronic coupling mechanism that is at the origin of the mystery band in butatriene can be qualitatively described by a model involving two normal modes only. In other words, the other (16) modes of vibration do not significantly contribute to the mechanism. In addition, the experimental spectrum can be interpreted rather easily with two bands corresponding to the B_{2g} and B_{2u} states and a third band, the mystery band, linked to the vibronic coupling. The situation is more complex for the pyrazine molecule [18, 19]. The molecule is shown on Fig. 12.8. The system has 10 atoms and thus 24 modes of vibration. The experimental spectrum corresponding to the absorption from the ground state S_0 to the $S_1(\pi, \pi^\star)$ yields a set of discrete lines as expected for a an excitation to a single bound electronic state. The chemical interpretation of the two excited states in terms of molecular orbitals can be seen on Fig. 4.11 of Sect. 4.4 in Chap. 4. On the other hand, the excitation to the $S_2(n, \pi^\star)$ state results in a broad band with little structure [20, 21] indicating a fast relaxation due to the presence of a conical intersection between the S_2 and S_1 states [22]. Contrarily to the butatriene case, here all the 24 modes of vibration are strongly coupled [18]

The wavefunction has two nuclear components

$$\Psi_\alpha(q_1, \ldots, q_{24}, t)\,, \tag{12.13}$$

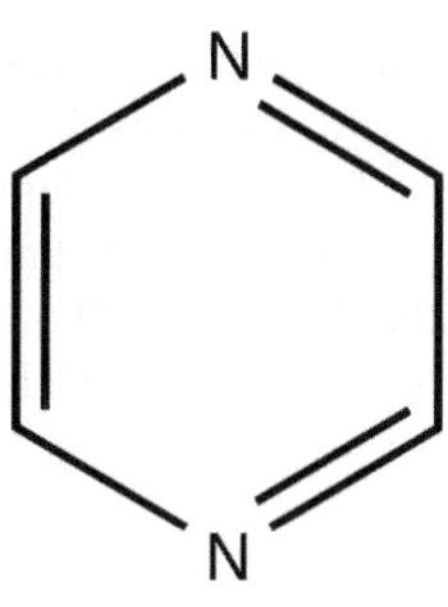

Fig. 12.8 The pyrazine molecule

with $\alpha = 1, 2$. The system is described by the 24 normal coordinates of the electronic ground state. Here, eight combined modes were used to propagate the wavepacket [18]

$$\Psi_\alpha(q_1, \ldots, q_{24}, t) = = \sum_{j_1^\alpha}^{n_1^\alpha} \cdots \sum_{j_5^\alpha}^{n_8^\alpha} A^{(\alpha)}_{j_1^\alpha \ldots j_8^\alpha}(t)\, \varphi^{(1,\alpha)}_{j_1^\alpha}(Q_1, t) \ldots \varphi^{(5,\alpha)}_{j_8^\alpha}(Q_5, t)\,, \tag{12.14}$$

with $Q_1, \ldots, Q_8$ corresponding to (using the notations of Ref. [18]) (q_{10a}, q_{6a}), (q_1, q_{9a}, q_{8a}), (q_2, q_{6b}, q_{8b}), (q_4, q_5, q_3), $(q_{16a}, q_{12}, q_{13})$, (q_{19b}, q_{18b}), $(q_{18a}, q_{14}, q_{19a}, q_{17a})$, and $(q_{20b}, q_{16b}, q_{11}, q_{7b})$, respectively. Here, $n_1^1 = 14, n_2^1 = 8, n_3^1 = 6, n_4^1 = 6,\ n_5^1 = 4,\ n_6^1 = 7,\ n_7^1 = 5,\ n_8^1 = 3,\ n_1^2 = 11, n_2^2 = 7, n_3^2 = 5, n_4^2 = 4, n_5^2 = 5n_6^2 = 7, n_7^2 = 5$, and $n_8^2 = 4$. For the primitive basis set, the following values have been used [18]: $N_{10a} = 40$, $N_{6a} = 32$, $N_1 = 20$, $N_{9a} = 12$, $N_2 = 4$, $N_{6b} = 8$, $N_{8b} = 24$, $N_4 = 24$, $N_5 = 8$, $N_3 = 8$, $N_{16a} = 24$, $N_{12} = 20$, $N_{13} = 4$, $N_{19b} = 72$, $N_{18b} = 80$, $N_{18a} = 6$, $N_{14} = 20$, $N_{19a} = 6$, $N_{17a} = 6$, $N_{20b} = 6$, $N_{11} = 6$, and $N_{7b} = 4$. This results in more than 1.27×10^{27} primitive functions. Thanks to the variational principle, only 2,771,440 SPFs are used in Eq. (12.14) to converge the theoretical spectrum depicted in Fig. (12.9) along with the experimental one [18]. The dense structure of the spectrum is due to the strong coupling between all the modes of vibration and due to the fact that the whole S_2 absorption spectrum is essentially influenced by nonadiabatic effects.

To demonstrate the power of the ML-MCTDH approach (see Sect. 8.3.6) we discuss a very small ML-MCTDH calculation on pyrazine, as outlined in Ref. [23]. Only 2,3 or 4 SPFs were used, depending on the position in the ML-tree. This leads to a total of only 22,444 coefficients describing the wave function. Remember, the underlying product grid consists of 1.27×10^{27} points, the standard method would thus describe the wave function by 1.27×10^{27} parameters.

The resulting ML-MCTDH propagation is extremely fast, it took only 7 min on a standard 2.7 GHz PC. This is to be compared with 70 hours, which are needed for a typical MCTDH propagation on pyrazine. The ML-MCTDH wave function is not of very high quality, but it reproduces the spectrum amazingly well, except for some wiggles on the highenergy tail. Larger deviations can be observed when comparing state populations. But keeping in mind the enormous speed of the 24D ML-MCTDH calculation, these deviations are fully acceptable. Of course, as discussed in Ref. [23], with more computational effort highly accurate ML-MCTDH propagations can be performed as well.

This example demonstrates the high efficiency of the ML-MCTDH approach for propagating high dimensional wave functions. As a crude rule of thumb one may say that ML-MCTDH outperforms MCTDH if there are more than 9 degrees of freedom. Moreever, ML-MCTDH becomes very fast, if a high accuracy is not needed.

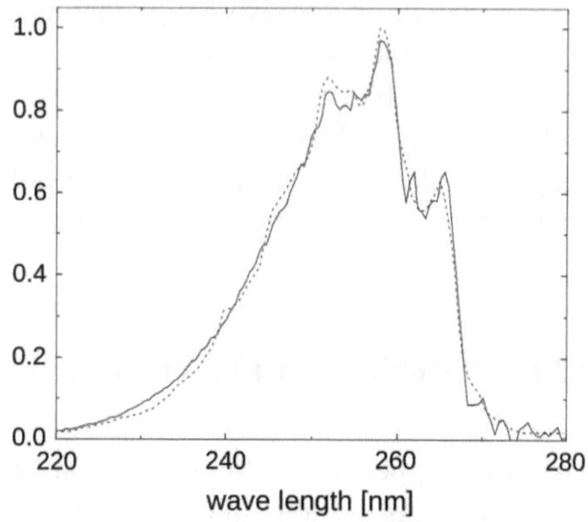

Fig. 12.9 Theoretical spectrum of the pyrazine molecule calculated with MCTDH in *full line* versus experimental spectrum in *dotted line*. The experimental spectrum is taken from Ref. [20]. The wavepacket was propagated during 150 fs [18]

12.5 Importance of Non-adiabatic Effects

After absorption in the UV domain corresponding to the spectrum of Fig. 12.9, the molecule of pyrazine will simply return to the electronic ground state through fluorescence [24] or through an internal conversion via a conical intersection with the electronic ground state [25]. For many other cases, a photochemical reaction such as a photo-isomerization or a photodissociation may occur after deactivation through a conical intersection. Since conical intersections lead to very fast and efficient transfers, it is suspected that they have been selected by evolution in important biological processes such as vision [26, 27] and photosynthesis [28].

The processes induced by conical intersections can be classified in photoreactive and photostable processes. Generally, photoreactive processes involve conical intersections with a peaked topology as shown in Fig. 12.10 (see also Fig. 4.6 in Sect. 4.1). The first step of vision is taken as an example: after a $\pi\pi^\star$ excitation induced by absorption in the visible, the molecule of retinal encounters a conical intersection with a *peaked topology*. The molecule can then isomerize initiating the process of vision. Peaked conical intersections can lead to photoproducts as one of the two gradients leads to a new reactive channel [4].

On the other hand, Fig. 12.11 shows a photostable process that may explain the photostability of DNA with respect to damage by the harmful UV components of sunlight. Here, the 2-aminopyridine dimer is used to model a Watson-Crick base pairs in DNA. A $\pi\pi^\star$ excitation through absorption of UV light leads the system from the ground state (GS) to a locally excited $\pi\pi^\star$ (LE) state. There is then an electron-proton transfer through a conical intersection with a peaked topology from LE to the charge transfer $\pi\pi^\star$ (CT) state. However, the system encounters a second conical intersection with a sloped topology forcing the system to go back to GS (see also Fig. 4.6 in Sect. 4.1 for a sloped topology). As we see, sloped conical intersections tend to favor regeneration of the starting material because both excited-state and ground-state gradients point toward the reactant side of the potential energy surfaces [4].

To conclude, it is worth noting that the presence of a solvent can have a drastic influence on the topography around a conical intersection. In particular, a polar solvent can stabilize the electronic states with a polar nature. On the other hand, it will have little influence on non-polar electronic states. If the conical intersection couples a polar state with a non-polar one, the presence of a polar solvent can change the position of the CI and strongly modify the non-adiabatic transfer.

12.6 Lab-Session IV: Vibronic Coupling

The Born-Oppenheimer approximation breaks down in the vicinity of a conical intersection. As the non-adiabatic coupling elements diverge at the conical point, the adiabatic picture is numerically inadequate and a diabatic description is in general used. The vibronic coupling Hamiltonian (VCH) model will be used here. It is ideally suited for MCTDH because it is expressed in the requested sum-of-products from the onset. Despite its simplicity, the VCH model often describes non-adiabatic dynamics with amazing accuracy.

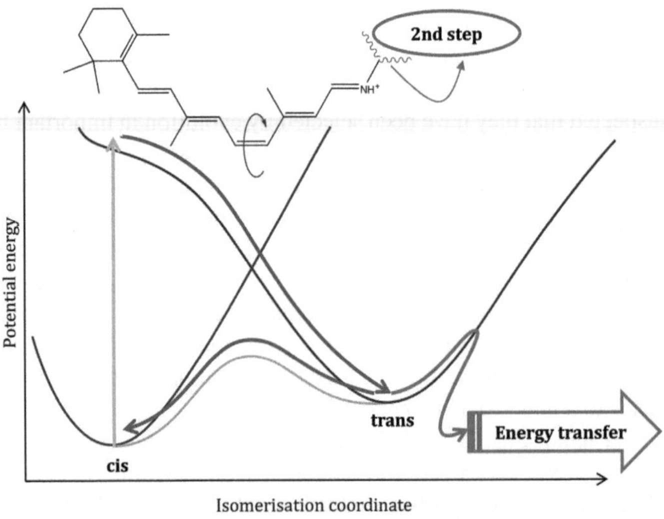

Fig. 12.10 Schematic illustration of a photoreactive process through internal conversion via a conical intersection. The first step of vision is used as a paradigmatic example: after absorption of light in the visible (*arrow* in *yellow*), the molecule of retinal can isomerize (*arrow* in *grey*). The cis-trans isomerization perturbs the active site of the protein and initiates the process of vision (in *green*). The conical intersection is said to be peaked

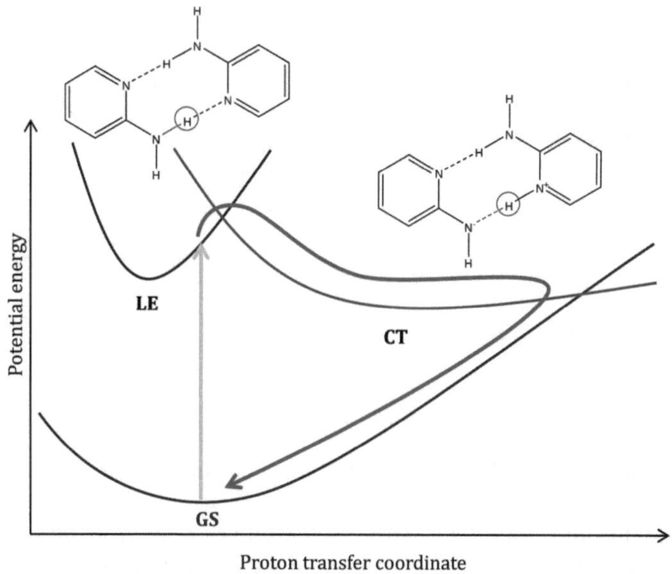

Fig. 12.11 Schematic illustration of a photostable process through a conical intersection. The absorption of Watson-Crick base pairs in DNA is used as a paradigmatic example. The process can be studied in the model compound 2-aminopyridine dimer. The system is first in its electronic ground state (GS). After a $\pi\pi^\star$ excitation due to absorption of UV light (*arrow* in *yellow*), there is an electron-proton transfer through a conical intersection with a peaked topology from the locally excited $\pi\pi^\star$ (LE) state to the charge transfer $\pi\pi^\star$ (CT) state. However, the system encounters a second conical intersection with a sloped topology forcing the system to go back to GS (*arrow* in *grey*). The presence of the sloped topology prevents the DNA bases from damage by the harmful UV components of sunlight

12.6.1 Photoionisation Spectrum of Butatriene

The outstanding importance that vibronic coupling effects may have on photoionisation and photoabsorption spectra was first recognized when theoretically investigating the photoionisation spectrum of butatriene. It was well known that in the considered energy range, 9–11 eV, there are only two ionic potential energy surfaces. But the measured spectrum seemed to show three bands, the one in the middle was called "mystery band"; see Fig. 12.12.[5] It was later shown [9] that the mystery band is created by vibronic coupling effects. Here we will thus investigate the vibronic coupling effects in butatriene. We assume a linear vibronic coupling (LVC) model where, due to symmetry, only five modes are active. In the quadratic vibronic coupling (QVC) model this restriction is lifted and there are non-vanishing coupling constants for all 18 modes. Those who are interested in the QVC model of butatriene may run the corresponding input files from the MCTDH `inputs` directory. Here

[5]Reprinted with permission from [8]. Copyright 2001, American Institute of Physics.

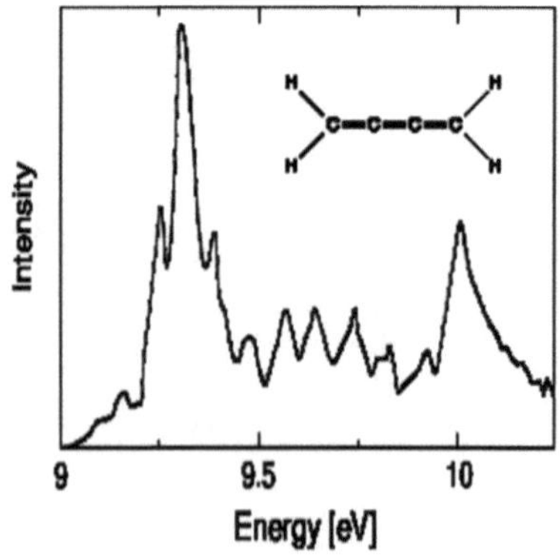

Fig. 12.12 Measured photoelectron spectrum of butatriene (see Ref. [7]). The band between 9.5 and 9.8 eV is called the mystery band

we concentrate on the LVC model because these calculations are much faster than the QVC ones.

Move to the directory `lab-inputs/butatriene`. You may source the `commands` file, "`.commands`", and most of the calculations will be executed by this script. But it may be more instructive to submit the commands "by hand". Type "`mctdh84 -mnd C4H4.B2g`" and "`mctdh84 -mnd C4H4.B2u`". While these calculations are running, you should inspect the input file `C4H4.B2g.inp`. The `B2u` input file is virtually identical, except that there are slightly more SPFs and, of course, that the second electronic state rather than the first one is now populated initially. As mass and frequency scaled normal mode coordinates are used, the mass and frequency parameters are set to 1.0 in both primitive-basis-section and init_WF-section. The LVC Hamiltonian is defined in the operator file `butatriene.op`. Note that q_5 is the coupling mode, which is responsible of the interstate coupling. All the other modes are tuning modes of which q_{14} has the largest coupling constant. There are six additional Hamiltonians for measuring the vibrational energy and twice the kinetic energy, respectively. For those, all constants are multiplied with the conversion factor from a.u. to eV. The output data will thus be in eV for easier interpretation.

When the runs are finished submit "`mctdh84 -pes C4H4.B2g`" and "`mctdh84 -pes C4H4.B2u`". The two runs produce (identical) `pes`-files, which are used when visualizing the PES and when computing adiabatic quantities. Next, move to the name-directory `C4H4.B2g` and submit there the command "`adpop84 &`". This produces the file `adp`, which contains the adiabatic and diabatic state-populations. (Remember, to visualize the diabatic populations alone we may simply run `plstate`, running `adpop84` is then not needed). While `adpop84` is running, one should inspect the PES. Type "`showsys84 -pes`" and use menu point 20 to select q_5 and q_{14} as coordinates to be plotted while setting values of the coordinates not to be shown to zero. Use menu point 160 to toggle between 2D and 3D visualizations, and use menu point 10 to select an adiabatic (default) or diabatic representation of the PES. Use menu point 60 to select plotting of both PESs simultaneously. Remember that one can do 1D plots as well; these are often more informative. Compare the adiabatic and diabatic 1D graphs along q_5 and q_{14}, respectively.

Before leaving this directory please run, for later use,
"`adproj84 -noproj &`",
"`autospec84 -q 3000 -lin -o ../fine-B2g 9 11 ev 0`", and
"`autospec84 -o ../coarse-B2g 9 11 ev 50 2`".
The first command computes the adiabatic PESs from the provided diabatic ones, and the latter two commands compute the spectrum in high and low resolution, respectively.

Next, we move to the `C4H4.B2u` name-directory, "`cd ../C4H4.B2u`", and submit there the command "`adpop84 Q5,Q14 &`". This will again compute the adiabatic populations, but this time, in addition, the adiabatic 2D reduced densities for coordinates q_5 and q_{14}. While the program is running, we can inspect the auto-correlation functions for comparison. Type
"`plgen -G -u 1:4 auto ../C4H4.B2g/auto`".
Putting the initial Gaussian function on the first electronic state, B2g, yields an auto-correlation function, which on average is about a factor of 2.5 larger than the one of the second state, B2u. This indicates that the B2u initial wavefunction decays into many more states as compared to the B2g one. (For a more precise analysis run `probsq84` and see the HTML documentation of `probsq84`). This is understandable, because the wavepacket, which initially is placed on the B2u surface, is more energetic, by 0.544 eV, compared to the one starting on the B2g surface. Compare the total energies of the two runs, which are printed in the `output` files. Finally, we compute, as before, high- and low-resolution spectra:
"`autospec84 -q 3000 -lin -o ../fine-B2u 9 11 ev 0`", and
"`autospec84 -o ../coarse-B2u 9 11 ev 50 2`".
To proceed we move back to the `butatriene` directory, "`cd ..`", and, for later use, submit "`mctdh84 -mnd rlx &`" and "`mctdh84 -mnd uncoup.B2g &`" and "`mctdh84 -mnd uncoup.B2u &`". Now we turn to discuss the spectra. It is assumed that both electronic states, B2g and B2u, are ionized by photon impact with the same probability. The total photoelectron spectrum is thus the sum of the two individual spectra. To perform the sum, type
"`sumspec84 -o fine.pl fine-B2g.pl 1.0 fine-B2u.pl 1.0`"
and "`sumspec84 -o coarse.pl coarse-B2g.pl 1.0 coarse-B2u.pl 1.0`". This produces the files `fine.pl` and `coarse.pl` while the files `fine-B2g.pl` etc. have been created previously by running `autospec84`. To plot the high-resolution spectra, type "`plgen -z 0 -G fine-B2g.pl'1:($4+400)' fine-B2u.pl '1:($4+800)' fine.pl 1:4`". The summed spectrum is on the bottom whereas the B2g and B2u spectra are vertically shifted by 400 and 800 units, respectively. Please note that the B2u spectrum exhibits intensities at energies much below the minimum of the B2u surface, in fact at energies as low as exhibited by the B2g spectrum. This, of course, is due to vibronic coupling. The summed spectrum, however, does not look similar to the experimental one, Fig. 12.12, which is much less structured. To reproduce the experimental spectrum we thus convolute the computed one with a Gaussian

function. To do so, we multiply the autocorrelation function with $\exp(-(t/\tau)^2)$ before Fourier transforming it. Here we have chosen $\tau = 50$ fs (the number before last of the `autospec84` command), which leads to a resolution of 0.044 eV FWHM. (The conversion factor from damping time to energy width is, in atomic units, $4\ln(2)$, which amounts to 2.2 eV fs. The high resolution spectrum has a resolution of about 0.0082 eV FWHM, see Table 9.1). To plot the low resolution spectrum type "`plgen -z 0 -G coarse-B2g.pl'1:($4+150)' coarse-B2u.pl '1:($4+300)' coarse.pl 1:4`". Again the B2u, B2g, and summed spectra are shown in comparison. The computed summed spectrum is now indeed very close to the experimental one, Fig. 12.12. The broadening of the computed spectrum can be justified by the finite experimental resolution, as well as by the neglect of 13 of the in total 18 DOFs and by the neglect of the overall rotation (the computation is for $J = 0$). However, experimental resolution seems to be the most prominent contribution to the broadening.

The time-dependent method is very suitable for simulating photoelectron spectra, because the experimental resolution is in general not very high, when electrons rather than photons are detected. Note that for generating low resolution spectra a typical propagation time of 100 fs or even only 50 fs (rather than the previously used 300 fs) is sufficient. This makes the propagation calculations faster. However, it is instructive to try the time-independent approach as well. We have already submitted an *improved relaxation* run for computing the lowest 25 vibronic states. To inspect the eigenenergies type "`rdrlx rlx/ | tail -n 32`". The energy spacing is rather dense and the first 25 states go up in energy to only 9.446 eV and thus span an energy interval of only 0.3 eV. If one wants to compute all eigenenergies up to 11 eV one would need to converge very many states as the density of states increases strongly with the energy. The line intensities are given by $| < \Psi_n|\Psi(0) > |^2$, where Ψ_n denotes the nth eigenstate and $\Psi(0)$ the initial wavefunction. The intensities may be computed with the command "`crosscorr84 -o no -R -f rlx/rst000 C4H4.B2g/psi`". Here `rst000` stands for Ψ_0 and we may use other restart files, `rst000...rst024`, to overlap with other eigenstates. Also `C4H4.B2g` may be exchanged with `C4H4.B2u` to compute the intensities that originate from putting the initial Gaussian function on the B2u diabatic state. For the convenience of the reader we have compiled the energies and intensities in the file `stickspec`. Just type "`less stickspec`" to read them. The data show that each eigenstate overlaps with only one initial state, either the one of B2g or the one of B2u. (An intensity below 10^{-10} is numerical noise and should be read as zero). It is instructive to compare the energies and intensities of the time-independent approach with the spectra of the time-dependent one. The spectra of the propagation calculations give the intensities as areas under the peaks when the energy unit is a.u. To compare both types of spectra we scale down the continuous one by a factor 3000, which is roughly the inverse peak width in a.u. Please type

"`plgen -z 0 -a 9.1 -x 9.45 -T fine-B2g.pl'1:($4/3000)' stickspec 1:2 -i`"

and

"`plgen -z 0 -a 9.1 -x 9.45 -T fine-B2u.pl'1:($4/3000)'`

stickspec 1:3 -i".
The two types of spectra nicely agree. If one turns to the summed spectrum, "plgen -z 0 -a 9.1 -x 9.45 -T fine.pl'1:($4/3000)' stickspec 1:2 -i - 1:3 -i",
one notices that one peak often covers two states, one originating from the B2g, the other from the B2u initial state.

In passing we note that the density of states increases strongly when turning from the LVC to the QVC model, because the latter includes all 18 degrees of freedom. This makes the computation of the spectrum by improved relaxation rather cumbersome. The time-dependent approach then shows its power.

To investigate the vibronic coupling effect further, we study the vibrational energies on the two diabatic states. It is difficult to unambiguously split the total energy in an electronic and vibrational part. Here we make use of the kinetic energy, which clearly is purely vibrational. According to the virial theorem, the average of twice the kinetic energy equals the total vibrational energy. This can be nicely seen by turning to the uncoupled propagations, where the interstate coupling constant lambda is set to zero and the electronic energy becomes constant and can easily be subtracted. See the supplementary Hamiltonians vib and kin on butatriene.op. Type "plgen -G uncoup.B2g/expectation 1:4 - 1:7" and "plgen -G uncoup.B2u/expectation 1:5 - 1:8".

The numbers correspond to the columns of the expectation file, run "head uncoup.B2g/expectation" to see the labels of the columns. Hence we are plotting the expectation values of the operators vib1 and kin1 for B2g and vib2 and kin2 for B2u. In the uncoupled case the vibrational energy on each electronic surface stays constant, and we use these plots to demonstrate that the expectation value of twice the kinetic energy oscillates around its average value, which is the total vibrational energy. These average energy values are 0.57016 and 0.74824 eV when placing the ground state wavefunction on the B2g and B2u state, respectively. Note that these are also the initial vibrational energies for the coupled case. To study the latter, type
"plgen -G C4H4.B2g/expectation 1:6 - 1:7 - 1:8".
As the B2u state is not populated initially, its vibrational energy is zero initially (see the lowest of the three lines). In the following, the B2u kinetic energy follows roughly the B2u population, investigated earlier. The middle line shows the kinetic energy of the B2g state and the top line the total kinetic energy. The average energy displayed by this line is about 0.74 eV which should be compared with the initial vibrational energy of 0.57 eV. (The kinetic energy expectation values discussed here are always doubled, as noted above). Hence, about 0.17 eV is transferred from electronic to vibrational energy when populating the B2g state initially. One may wonder that electronic energy is transferred to vibrations as we started on the lower B2g surface. However, this is the diabatic one, the first adiabatic surface is still lower.

More interesting is to study the B2u excitation. Type
"plgen -G C4H4.B2u/expectation 1:6 - 1:7 - 1:8".
Now the B2g state is initially unpopulated and hence has vanishing kinetic energy

initially. However, its kinetic energy raises quickly and already after 12 fs it surpasses the `B2u` kinetic energy. The total kinetic energy has an average of about 1.3 eV and is thus 0.55 eV higher than its initial value of 0.75 eV. Hence, in this case a quite substantial amount of electronic energy, 0.55 eV, is converted to vibrational energy.

In addition we may study the expectation values of the `vib` supplementary Hamiltonians, which appear on columns 3, 4, and 5 of the `expectation` file. However, these values are more difficult to interpret.

Next, we investigate the spectra of the uncoupled propagation, where the interstate coupling parameter `lambda` is set to zero. Type
"`plspec -G -z 0 -g5 -q 3000 -f uncoup.B2g/auto 9. 11. ev`" and
"`plspec -G -z 0 -g5 -q 3000 -f uncoup.B2u/auto 9. 11. ev`".
These spectra are much less structured as compared to the ones of the full, coupled calculations. In particular, the spectrum of the uncoupled B2u calculation has no intensity below 9.7 eV, whereas the coupled one reaches down to 9.15 eV. The mystery band has disappeared in the spectrum of the uncoupled calculation.

Next, we investigate adiabatic and diabatic state populations. Type
"`plgen -G C4H4.B2g/adp 1:2 - 1:3 - 1:4 - 1:5`" and
"`plgen -G C4H4.B2u/adp 1:2 - 1:3 - 1:4 - 1:5`".
The first plot shows the populations when initially populating the B2g state. The diabatic B2g and B2u populations start at 1.0 and 0.0 and then oscillate around 0.7 and 0.3, respectively. The adiabatic populations, however, are amazingly close to 1.0 and 0.0, respectively. The lower state adiabatic population starts initially at 0.86, grows within the first 10 fs to 0.98 and then continues to vary between 0.95 an 1.0. As the lower adiabatic state population is always large, we may wonder if an adiabatic treatment would give reasonable results. In fact, as will be shown below, this is the case.

Turning to the second plot where the B2u state is initially populated, one notices that the diabatic populations are near 0.5. The population of the lower adiabatic state now starts at 0.14 and increases rapidly, within the first 10 fs, to 0.93 and then varies between 0.80 and 0.98. The decay from the upper to the lower adiabatic state is thus very fast. Here an adiabatic treatment cannot work because one does not stay on either of the adiabatic states.

To verify our assumption on the validity of the adiabatic picture, we will perform propagations on uncoupled adiabatic surfaces. The adiabatic surfaces have already been generated on the total product grid when running `adproj84`. To bring them into a suitable form one has to potfit these data files. Run "`potfit84 -mnd ad1fit`" and "`potfit84 -mnd ad2fit`" and then the MCTDH propagations "`mctdh84 -mnd C4H4.adiab1`" and "`mctdh84 -mnd C4H4.adiab2`". Next, we need to Fourier transform the autocorrelation functions
"`autospec84 -f C4H4.adiab1/auto -o adiab1 9 11 ev 50 2`"
"`autospec84 -f C4H4.adiab2/auto -o adiab2 9 11 ev 50 2`"
and sum the spectra "`sumspec84 -o adiab.pl adiab1.pl 1.0 adiab2.pl 1.0`" The spectra are plotted by typing "`plgen -z 0-G adiab1.pl'1:($3+200)' adiab2.pl'1:($3+400)' adiab.pl 1:4`".
The upper state spectrum is on top, the lower state one in the middle, and the

summed spectrum at the bottom. Note that the upper state spectrum shows no intensity below 9.85 eV and thus does not contribute to the so-called mystery band. The summed spectrum looks quite different from the experimental one. To compare the spectra of the uncoupled adiabatic approach with the ones of the coupled diabatic calculations, run "`plgen -G -z 0 -u 1:4 -x 10.2 adiab1.pl coarse-B2g.pl`". As expected, both spectra are similar and the adiabatic approximation is not totally off when initially populating the lower surface. On the contrary, for the upper state, "`plgen -G -z 0 -u 1:4 adiab2.pl coarse-B2u.pl`", the adiabatic approach is unreliable.

Finally we want to discuss a geometric phase effect. In the adiabatic picture there appears a so-called geometric phase that let the wavefunction changes sign when a conical intersection is circumcircled. A wavepacket that encircles a conical point will have a nodal line which is crossed when going along a loop. To show this, run "`pladpop -G -c C4H4.B2u/adp_Q5_Q14`", which produces a gnuplot file from the already computed (by `adpop84`) reduced 2D adiabatic densities (The option `-c` stores the gnuplot file. If you want to re-run the plot, type "`gnuplot C4H4.B2u/adp_Q5_Q14.pl`"). As usual, pressing "Enter" increases the time by one step. The clearest figure is probably the one at 9 fs, for times above 30 fs the wavepacket shows a strongly fractionated pattern.

12.6.2 Photoabsorption Spectrum of Pyrazine

The computation of the photoabsorption spectrum of pyrazine including all 24 vibrational modes established the first breakthrough of the MCTDH approach [18]. We thus want to discuss this molecule here briefly too. If one wants to study the original 24D QVC model, one may run the input file `p24+.inp` from the MCTDH `inputs` directory. This, however, will take about 70 h on a single core. If you have access to a multi-core machine, you may try parallelization. (See the HTML documentation and the guide). Here we will discuss reduced dimensionality LVC models of pyrazine. However, some 24D data are provided for comparison.

In contrast to butatriene, pyrazine is very strongly coupled vibronically. The conical intersection of pyrazine lies close to the Frank-Condon region and is thus easily accessible. This strong coupling leads to a spectrum that consists of a broad peak with only little structure. In contrast to butatriene, the pyrazine molecule is photo-excited to an S_1/S_2 manifold, but not ionized. Selection rules make the S_1 state dark, and only the S_2 state can be photo-excited. As before, this is accomplished by taking an initial state where the S_0 ground-state wavepacket is vertically placed on the diabatic S_2 state.

Move to the directory `lab-inputs/pyrazine`. As before, all the commands to be executed below can be taken (copied/pasted) from the `commands` file. Type "`(mctdh84 -mnd pyr4 && mctdh84 -pes pyr4 && cd pyr4 && adpop84)&`"

The first command starts a propagation run. Due to the && construct, the second command, generation of a `pes` file, waits till the propagation is finished successfully. After the `pes` file is generated, a move (`cd`) to the `pyr4` name-directory is performed and `adpop84` is started. The brackets "()" let the commands run in a separate shell & the final and puts the execution of all commands to the background. Next submit a similar command for the second model to be investigated. Type "`(mctdh84 -mnd pyr6 && mctdh84 -pes pyr6 && cd pyr6 && adpop84)&`".

To discuss the 4D model move (`cd`) to the `pyr4` name-directory. Inspect the `input` file. It describes a 4D model of pyrazine. Check if the `stop` file still exist. If it does, the propagation run has not yet finished and one has to wait for a few seconds. Run "`showsys84 -pes`" to inspect the surfaces. Use menu point 60 and input "2" and then "1 2" to enable to simultaneous plotting of both electronic PESs. Use menu point 160 to enable 3D plotting (surface on) and type "1" three times to accept the default settings. A 2D cut of the *adiabatic* surfaces is shown along the coupling mode (ν_{10a}) and the most important tuning mode (ν_{6a}). One may rotate the figure with the mouse. But more instructive is to look at a 1D cut of the diabatic surfaces. Use again menu point 160 to toggle surface off, use menu point 10 to switch to a diabatic PES-representation (type "2"), use menu point 150 to toggle grid lines on, and use menu point 20 to select a cut along ν_{6a} (type "0 × 0 0"). Then type "1" to plot to screen. A wavepacket located at the Frank-Condon point ($\nu_{6a} = 0$) will slide towards the conical point, which is easily reachable, because it lies only slightly higher than the Frank-Condon point. The population transfer will hence be fast. To see this, quit `showsys84` (type "q"), and type "`plstate -G`". After 40 fs, 80% of the population is transferred from state 2 to state 1. Between 80 and 90 fs there is a recurrence and the population of the S_2 state reaches 0.48. Another, but weaker, recurrence appears near 145 fs.

To visit the spectrum type "`plspec -G -0.5 1.5 ev`". One sees traces of individual lines. Because the model is of so low dimensionality, it cannot reproduce a quasi-continuum as observed experimentally. The structures below 0.4 eV are of no interest here, they are created by vibrational states that live (almost) entirely on the lower state and are not subject to vibronic coupling. Hence to concentrate on the main part of the spectrum type "`plspec -G 0.4 1.4 ev`". To mimic the influence of the 20 ignored vibrational modes we artificially broaden the spectrum by convoluting it with a Gaussian function. Type "`plspec -G 0.4 1.4 ev 40 2`". As we will see later, there is now some similarity with the correct spectrum. (Remember that a Gaussian damping of the autocorrelation function with $\tau = 40$ fs is equivalent to a convolution with a Gaussian function of 55 meV FWHM). For later use we generate a spectrum file by running "`autospec84 -o spec40.pl -e -0.687485 eV -0.4 0.8 ev 40 2`" Here we have shifted the spectrum by its mean energy (taken from the `output` file) such that the spectrum is centered at zero. This is done to make spectra of different model Hamiltonians comparable. Also for later use please run
"`rdcheck84 -on no`". This produces the `chk.pl` file, which contains the diabatic populations.

Next move to the `pyr6` directory, "`cd ../py6`", and run "`autospec84-o spec80.pl-e-1.030635 eV-0.4 0.8ev 80 2`",and "`rdcheck84 -on no`". This is for comparison with the `pyr4` data. But, first, let us inspect the spectrum: "`plspec -G 0.6 2.0 ev`". Due to the addition of two vibrational modes, the spectrum is much denser but still far from a quasi-continuum. With `autospec` we have just generated a spectrum which is convoluted with a Gaussian function with 27.5 meV FWHM due to a damping time of $\tau = 80$ fs. Due to the higher density of states of the 6D model, as compared to the 4D one, the broadening width could be reduced by a factor of two. The energy shift 1.030635 eV is the mean energy taken from the output file. To compare both spectra type "`plgen -G -u 1:4 spec80.pl ../pyr4/spec40.pl`". Although the original 4D and 6D spectra are very different, the broadened or smeared out ones are quite similar.

Because a propagation with the full 24D QVC Hamiltonian takes too long for a lab session, we have compiled the most important output files from such a calculation in the directory `p24+`. To compare all three spectra type
"`plgen -G -u 1:4 spec80.pl ../pyr4/spec40.pl ../p24+/spec120.pl`",
and, to see the differences more clearly, you may compare only the 4D spectrum with the full 24D one and similarly for the 6D spectrum. The 24D spectrum is very dense and individual lines cannot be identified. The additional broadening via a damping of $\tau = 120$ fs, resulting in 18.33 meV FWHM, changes the spectrum only little. Run "`plgen -G -u 1:4 ../p24+/spec.pl ../p24+/spec120.pl`"
to observe the differences.

Next, we discuss the autocorrelation functions of the different models. Type "`plgen -G -u 1:4 auto ../pyr4/auto ../p24+/auto`", or, to ignore the first 20 fs, "`plgen -a 20 -G -u 1:4 auto ../pyr4/auto ../p24+/auto`". The autocorrelation function (actually its absolute value) becomes the smaller the more vibrational modes are involved. If the electronic energy, released by vibronic coupling, is distributed over more modes, then the wavefunction becomes less similar to the initial one, $\Psi(0)$, and the autocorrelation function, $<\Psi(0)|\Psi(t)>$ becomes smaller. Note that the 24D autocorrelation function shows almost no recurrences, after 40 fs its absolute value always stays below 0.05. The 4D and 6D autocorrelation functions, however, show pronounced recurrences even at larger times, *e.g.* near 175 and 235 fs.

Let us define the weight that an exact vibronic eigenstate Ψ_n has in the initial wavefunction $\Psi(0)$ as $p_n = |<\Psi(0)|\Psi_n>|^2$. By completeness of the eigenstates we have $\sum_n p_n = 1$. In addition, the sum of p_n^2 can be computed from the autocorrelation function. One can show $S := \sum_n p_n^2 = \lim_{T\to\infty} \frac{1}{T} \int_0^T |c(t)|^2 dt$, where $c(t)$ denotes the autocorrelation function. As we know the autocorrelation function only for finite times, we can evaluate S only approximately. To improve the accuracy one can apply filter functions and start the evaluation of the integral at some offset to avoid the large values of $|c(t)|$ near $t = 0$. If one assumes $p_n = p$ for $n = 1 \ldots N$ and $p_n = 0$ for $n > N$ then it follows $N = 1/S$. Hence *1/S* is a rough measure of the number of states with non-negligible weight in which the initial state decays. Run "`probsq84 -a 20`". The first 20 fs of the autocorrelation function are ignored due to the option `-a 20`.

Printed to screen is the result obtained with four different filter functions. The SIN filter is probably the most reliable one (compare with HTML documentation). Next run "`probsq84 -a 20 -i ../pyr4/auto`" and "`probsq84 -a 20 -i ../p24+/auto`". The so obtained numbers of relevant decay states are 217, 696, and 3037 for the 4D, 6D, and 24D model, respectively. As expected, the number of states increases strongly with dimensionality. As the spectrum is about 1 eV wide, these numbers give also the density of states (per eV). Note, however, that – in particular for the 24D case—the real density of states is much higher, because there are many states with very small intensities.

We close this section by discussing the diabatic and adiabatic populations. To visit the diabatic population of the S_2 state type
"`plgen -G -x 150 -u 1:3 chk.pl ../pyr4/chk.pl ../p24+/chk.pl`"
The diabatic state populations of the 4D and 6D models are very similar, except that the 6D one consistently lies below the 4D one, as expected. Both populations exhibit recurrences near 85 and 145 fs. Surprisingly the 24D model provides a diabatic state population that is most of the time larger then the ones of the other models. However, the initial decay is faster and the recurrences are less pronounced. We attribute this different behavior to the use of different kinds of models. Remember that the 4D and 6D models are of LVC type, whereas the 24D model is a QVC one. This example shows that discussing diabatic state populations is less meaningful, and the adiabatic state populations are more relevant. The calculation of the latter, however, is expensive and we could not do that for the 24D model. To view the adiabatic state populations type
"`plgen -G -x 150 -u 1:3 adp ../pyr4/adp`". For times above 20 fs the adiabatic populations are considerably smaller than the corresponding diabatic ones, and the adiabatic population of the 6D model is considerably smaller than the one of the 4D model. The recurrences near 85 and 145 fs are much weaker in the adiabatic representation. To compare diabatic with adiabatic populations while suppressing the less interesting initial part, type
"`plgen -a 20 -x 150 -y 0.5 -G adp 1:3 - 1:5 ../pyr4/adp 1:3 - 1:5`".
For times above 40 fs the adiabatic population of the 6D model is already very small, smaller than 0.05. For the 24D model one expects even smaller values. The non-adiabatic transition occurs mainly during the first 20 fs and is completed after 40 fs.

Note: If the plotting of the adiabatic curves stops before $t = 150$ fs then the calculation of the adiabatic populations is still running. For the 6D model `adpop84` takes about 3 h to finish. A much faster way, less than 3 min CPU time, is provided by the `adproj84` program. This way, however, requires to write several input files and is hence more cumbersome to use. If you want to try `adproj84`, please copy the directory `ADPROJ` from `lab-inputs/pyrazine` to `pyr6`. Read the `ADPROJ/README` file and execute the commands listed there.

References

1. Domcke W, Yarkony DR, Köppel H (eds) (2004) Conical intersections: electronic structure, dynamics and spectroscopy. World Scientific, New Jersey
2. Domcke W, Yarkony DR, Köppel H (eds) (2004) Conical intersections: theory., Computation and experimentWorld Scientific, New Jersey
3. Köppel H (2014) Vibronic coupling effects in spectroscopy and non-adiabatic transitions in molecular photodynamics. In: Gatti F (ed) Molecular quantum dynamics. Springer, Heidelberg
4. Lasorne B, Worth GA, Robb MA (2011) Excited-state dynamics. WIREs Comput Mol Sci 1:460–475
5. Lasorne B, Worth GA (2014) Non-adiabatic photochemistry: Ultrafast electronic state transitions and nuclear wavepacket coherence. In: Gatti F (ed) Molecular quantum dynamics. Springer, Heidelberg
6. von Niessen W, Diercksen GHF, Cederbaum LS, Domcke W (1976) Ionization potentials of ethylene, allene and butatriene by a Green function method. Chem Phys 18:469
7. Brogli F, Heilbronner E, Kloster-Jensen E, Schmelzer A, Manocha AS, Pople JA, Radom L (1974) The photoelectron spectrum of butatriene. Chem Phys 4:107
8. Cattarius C, Worth GA, Meyer H-D, Cederbaum LS (2001) All mode dynamics at the conical intersection of an octa-atomic molecule: multi-configuration time-dependent Hartree (MCTDH) investigation on the butatriene cation. J Chem Phys 115:2088
9. Cederbaum LS, Domcke W, Köppel H, von Niessen W (1977) Strong vibronic coupling effects in ionization spectra: The "mystery band" of butatriene. Chem Phys 26:169
10. Berry MV (1984) Quantal phase factors accompanying adiabatic changes. Proc R Soc A 392:45
11. Mead CA, Truhlar DG (1979) On the determination of Born-Oppenheimer nuclear motion wavefunctions including complications due to conical intersections and identical nuclei. J Chem Phys 70:2284
12. Althorpe SC (2006) General explanation of geometric phase effects in reactive systems: unwinding the nuclear wave function using simple topology. J Chem Phys 124:084105
13. Ryabinkin IG, Joubert-Doriol L, Izmaylov AF (2014) When do we need to account for the geometric phase in excited state dynamics? J Chem Phys 140:214116
14. Bouakline F, Althorpe SC, Peláez D (2008) Strong geometric-phase effects in the hydrogen-exchange reaction at high collision energies. J Chem Phys 12:124322
15. Ryabinkin IG, Izmaylov AF (2013) Geometric phase effects in dynamics near conical intersections: symmetry breaking and spatial localization. Phys Rev Lett 111:220406
16. Joubert-Doriol L, Ryabinkin IG, Izmaylov AF (2013) Geometric phase effects in low-energy dynamics near conical intersections: a study of the multidimensional linear vibronic coupling model. J Chem Phys 139:234103
17. Babikov D, Kendrick BK, Zhang P, Morokuma K (2005) Cyclic-N_3: II. significant geometric phase effects in the vibrational spectra. J Chem Phys 122:044315
18. Raab A, Worth G, Meyer H-D, Cederbaum LS (1999) Molecular dynamics of pyrazine after excitation to the S_2 electronic state using a realistic 24-mode model Hamiltonian. J Chem Phys 110:936
19. Worth GA, Cederbaum LS (2004) Beyond Born-Oppenheimer: conical intersections and their impact on molecular dynamics. Ann Rev Phys Chem 55:127
20. Yamazaki I, Murao T, Yamanaka T, Yoshihara K (1983) Intramolecular electronic relaxation and photoisomerisation processes in the isolated Azabenzene molecules Pyridine, Pyrazine and Pyrimidine. Faraday Discuss Chem Soc 75:395
21. Innes KK, Ross IG, Moonaw WR (1988) Electronic sates of Azabenzenes and Azanaphthalenes: a revised and extended critical review. J Mol Spec 132:492
22. Köppel H, Domcke W, Cederbaum LS (1984) Multimode molecular dynamics beyond the Born-Oppenheimer approximation. Adv Chem Phys 57:59
23. Vendrell O, Meyer H-D (2011) Multilayer multiconfiguration time-dependent Hartree method: Implementation and applications to a Henon-Heiles Hamiltonian and to pyrazine. J Chem Phys 134:044135

24. Seidner L, Stock G, Sobolewski AL, Domcke W (1992) Ab initio characterization of the s1–s2 conical intersection in pyrazine and calculation of spectra. J Chem Phys 96:5298
25. Sala M, Guérin S, Gatti F (2015) Quantum dynamics of the photostability of pyrazine. Phys Chem Chem Phys 17:29518
26. González-Luque M, Garavelli M, Bernardi F, Mechán M, Robb MA, Olivucci M (2010) 97:9379
27. Polli D, Altoè P, Weingart O, Spillane KM, Manzoni C, Brida D, Tomasello G, Orlandi G, Kukura P, Mathies RA, Garavelli M, Cerullo G (2010) Conical intersection dynamics of the primary photoisomerization event in vision. Nature 467:440
28. Worth GA, Cederbaum LS (2001) Mediation of ultrafast electron transfer in biological systems by conical intersections. Chem Phys Lett 338:219

Chapter 13
Control of Molecular Processes

13.1 Introduction

In the previous applications, we have considered only "artificial" wavepackets, i.e. wavepackets that generally do not correspond to any realistic experimental situation. The mathematical properties of the wavepackets allowed us to obtain absorption spectra and cross sections including all the quantum effect that can impact a molecular process. All the systems were assumed to be isolated and no quantum decoherence occurred during the propagations of the wavepackets. However, *wavepackets are not only mathematical tools to obtain some measured physical quantities, they can be created experimentally*. Since the advent of *lasers*, coherent sources of light can be produced that can in turn create coherent superpositions of molecular states and thus molecular wavepackets. Quantum coherence will finally be dissipated by interaction with the environment but, before this, quantum coherence may be preserved during a time that is sufficient to trigger a new type of chemical process.

It is now well-established that quantum phenomena can yield states of matter with remarkable properties such as superfluidity and superconductivity. It is also thought that quantum computers that exploit quantum superposition, and thus quantum coherence, will be able to solve certain problems much faster than any classical computers. In the same manner, *a systematic use of the quantum coherence created by laser pulses could lead to a much higher level of control and efficiency of chemical processes* [1–4]. In this context, decisive progress has been achieved with the ability to use time-resolved pump-probe laser methods to study chemical processes on the femtosecond time scale (10^{-15} s), i.e., the typical period of molecular vibrations [5–7]. The promise of this *coherent control* is to create the right quantum interferences to guide the molecule to the desired product[1] [3, 4, 14].

---

[1]One method to create the right interferences is to optimize the laser pulse in a systematic way in order to guide the molecule to the desired product. The theory to optimize the laser pulses is called optimal control theory [8–10]. Note that iterative algorithms for optimal control have been implemented in the Heidelberg MCTDH package [11–13].

F. Gatti et al., *Applications of Quantum Dynamics in Chemistry*,
Lecture Notes in Chemistry 98, DOI 10.1007/978-3-319-53923-2_13

One of the main problems of coherent control of chemical reactions is, as for quantum computers, the problem of decoherence. Even if the molecule is isolated the quantum coherence can lose its efficiency because of the rapid intramolecular energy redistribution (IVR) into the numerous vibrational degrees of freedom. In addition to that, since molecules rotate freely in the gas phase there is a blurring of the control due to random orientations.

However, more recently, possibilities have emerged to generate sub-femtosecond or attosecond laser pulses [15–17] for observing and guiding electrons on their natural time scale. These techniques open the door to guiding the reactivity much before IVR can dissipate quantum coherence. One can speak about the advent of "attochemistry" by extension of the research area known nowadays as femtochemistry [18]. Another major advance has been the possibility to orientate or align molecules, which allows one to excite molecular systems with laser pulses in a much more efficient way [19–22].

The theoretical description of a molecular system in interaction with an external field is a difficult task. As explained in Chap. 3, in our approach, the external fields are treated classically, i.e. they are not quantized. But the external field is not only a "driving force" that allows the molecule to go from one state to another. If the external field is intense, it also changes the state of the molecule during the interaction. It is thus necessary to develop a theoretical framework based not only on the quantum states of the isolated molecular system but on the "light dressed" quantum states of a molecular system in interaction with an intense field.[2]

In the present chapter, we use atomic units. In particular, $\hbar = 1$.

13.2 The Light Dressed States *

*13.2.1 Periodic Fields **

We briefly present here the Floquet theory that provides a rigorous framework for describing the dynamics of a quantum system in interaction with a periodic radiation [25–27]. This formalism can be generalized to more realistic situations where the quantum system is in interaction with laser pulses with slowly varying envelopes, frequencies or polarization directions [23, 28–31]. In the latter case, the formalism is called the adiabatic Floquet theory [29]. In this section, we do not give all the demonstrations that can be found in several publication by S. Guérin et al. (see Ref. [29] for a review) and we follow the presentation given by M. Sala in his Ph.D. thesis [23, 24].

The Floquet theory makes a clear link to the purely quantum description of a situation where both the system and the field are quantized [32] (the "theory of

[2]This chapter is based mainly on Matthieu Sala's PhD work [23, 24]. The authors gratefully thank him for his contribution and his help. The authors also thank S. Guérin for having read and corrected this chapter.

dressed states in a cavity"). A classical field can be understood as a coherent state with a very large number of photons. In the Floquet theory, the processes are interpreted in terms of a precise number of photons that are absorbed or emitted by the system and thus taken from or given to the field [23].

An isolated molecular system is described by a Hamiltonian operator $H_0(\boldsymbol{q})$. H_0 could be one of the operators used in the previous chapters. Now, the dynamics of the system in interaction with a periodic field $\boldsymbol{E}(t) = \boldsymbol{A}_0 \cos(\omega t + \theta)$ is given by the time-dependent Schrödinger equation:

$$H(t)\Psi(\boldsymbol{q}, t) = i\frac{\partial}{\partial t}\Psi(\boldsymbol{q}, t)\,, \tag{13.1}$$

and

$$H(t) = H_0(\boldsymbol{q}) - \boldsymbol{\mu}(\boldsymbol{q}) \cdot \boldsymbol{A}_0 \cos(\theta + \omega t) \tag{13.2}$$

is the Hamiltonian operator describing the system in interaction with the field. Here, θ is 2π-periodic and corresponds physically to the phase of the field at $t = 0$ but must be also understood as a *dynamical variable* that will allow us to transform the time-dependent problem to a time-independent problem [23]. We now introduce the square integrable functions $e^{ik\theta}$, k being an integer and $\theta E[0, 2\pi[$. These functions build a new Hilbert space $\mathcal{L}$.

The eigenfunctions of H_0, are given by (a discrete spectrum is assumed),

$$H_0(\boldsymbol{q})\Psi_l^0(\boldsymbol{q}) = E_l^0\Psi_l^0(\boldsymbol{q})\,. \tag{13.3}$$

The products $\Psi_l^0(\boldsymbol{q}) \otimes e^{ik\theta}$ create an enlarged Hilbert space $\mathcal{K} = \mathcal{H} \otimes \mathcal{L}$, with $\mathcal{H}$ the usual Hilbert space, in which acts the *Floquet Hamiltonian* (or quasi-energy Hamiltonian):

$$K(\boldsymbol{q}, \theta) = -i\omega\frac{\partial}{\partial\theta} + H_0(\boldsymbol{q}) - \boldsymbol{\mu}(\boldsymbol{q}) \cdot \boldsymbol{A}_0\, \cos(\theta)\,. \tag{13.4}$$

At this level, a link with the theory of dressed states in a cavity [32], where both the system and the light are quantized, can be made. Starting from the theory of dressed states in a cavity, it can be proven [33] that K represents the Hamiltonian of the molecule interacting in free space (limit of an infinite cavity volume) with a field containing a large number of photons (limit of a large photon number average).[3] In other words, in this limit, the Hamiltonian operator including the quantization of the fields resorts[4] to K. In the same manner [33], in the limit of a large number of photons, the operator $N_r = -i\omega\frac{\partial}{\partial\theta}$ can be interpreted as *the relative photon number operator*. The variation of the average of N_r in the Floquet formalism gives the number of photons absorbed from (or emitted to) the laser field. This number is linked to the

---

[3]The photon density must be taken constant.

[4]More precisely, the Hamiltonian operator including the quantization of the fields gives K up to an additive constant.

quantum number k in the functions $e^{ik\theta}$. For instance, $k = -1$ *corresponds to a photon absorbed by the molecule at the end of the process if one starts with* $k = 0$.

It can be proven [23, 29] that the dynamics given by Eq. (13.1) is equivalent to the dynamics given by

$$K(\boldsymbol{q}, \theta)\Psi_K(\boldsymbol{q}, \theta; t) = i\frac{\partial}{\partial t}\Psi_K(\boldsymbol{q}, \theta; t). \tag{13.5}$$

The advantage is that the Floquet Hamiltonian is now time-independent, leading to the simple propagator

$$U_K(t, t_0; \theta) = e^{-iK(t-t_0)}, \tag{13.6}$$

verifying

$$\Psi_K(\boldsymbol{q}, \theta; t) = U_K(t, t_0, \theta)\Psi_K(\boldsymbol{q}, \theta; t_0). \tag{13.7}$$

The link between the "Floquet wavefunction", $\Psi_K(\boldsymbol{q}, \theta; t)$, and the wavefunction of the system,[5] $\psi(\boldsymbol{q}, \theta, t)$, is obtained by a phase translation [23]:

$$\psi(\boldsymbol{q}; \theta, t) = \Psi_K(\boldsymbol{q}, \theta + \omega t; t). \tag{13.8}$$

The eigenstates of the Floquet Hamiltonian are given by

$$K(\boldsymbol{q}, \theta)\Phi_\nu(\boldsymbol{q}, \theta) = \lambda_\nu \Phi_\nu(\boldsymbol{q}, \theta), \tag{13.9}$$

and are called the *Floquet states*. By analogy with the theory of the dressed states in a cavity, the *Floquet states* are sometimes referred to as the *(light) dressed states*.

Let us assume that the system at $t = 0$ is described by the wavefunction $\Psi(\boldsymbol{q}; t = 0)$. Expanding $\Psi(\boldsymbol{q}; t = 0) \otimes 1_{\mathcal{L}}$ in terms of the eigenfunctions of K, we arrive at

$$\psi(\boldsymbol{q}, \theta; t) = \sum_\nu c_\nu e^{-i\lambda_\nu t}\Phi_\nu(\boldsymbol{q}, \theta + \omega t), \tag{13.10}$$

where the coefficients c_ν can be expressed as

$$\begin{aligned} c_\nu &= \int d\boldsymbol{q} \int \frac{d\theta}{2\pi}\Phi_\nu^*(\boldsymbol{q}, \theta)\psi(\boldsymbol{q}, t = 0), \\ &= \int d\boldsymbol{q}\overline{\Phi_\nu^*}(\boldsymbol{q})\psi(\boldsymbol{q}, t = 0), \end{aligned} \tag{13.11}$$

where $\overline{\Phi_\nu}(\boldsymbol{q}) = \int \frac{d\theta}{2\pi}\Phi_\nu(\boldsymbol{q}, \theta)$ is the average of $\Phi_\nu(\boldsymbol{q}, \theta)$ over the phase.

The eigenfunctions of the Floquet Hamiltonian have a periodic structure, i.e. $\Phi_\nu(\boldsymbol{q}, \theta) \equiv \Phi_{n,k}(\boldsymbol{q}, \theta) = \Phi_{n,0}(\boldsymbol{q}, \theta)e^{ik\theta}$ and $\lambda_\nu \equiv \lambda_{n,k} = \lambda_{n,0} + k\omega$, where the

[5] $\psi(\boldsymbol{q}; \theta, t)$ depends parametrically on θ. Here, θ is not a variable but the phase in the field at t = 0, since we use the expression $\boldsymbol{E}(t) = \boldsymbol{A}_0 \cos(\omega t + \theta)$. In practice, this phase is often set to zero.

index n refers to the Hilbert space $\mathcal{H}$. It must be clear that, in general, there is no direct link between the sets of the $\Phi_{n,0}(\boldsymbol{q},\theta)$ and the sets of the eigenfunctions of $H_0(\boldsymbol{q})$, $\Psi_l^0(\boldsymbol{q})$. However, if there is no coupling between the system and the external field, we have $\Phi_{l,0}(\boldsymbol{q},\theta) = \Psi_l^0(\boldsymbol{q}) \otimes 1_{\mathcal{L}}$ and $\lambda_{l,0} = E_l^0$.

Thus, the Floquet eigenstates can be classified in families labeled by n with individual elements of a given family distinguished by the index k. This property allows one to simplify the expansion in terms of the eigenstates of the Floquet Hamiltonian by considering only a single member of each family, e.g. the $k = 0$ member, as shown below. Eq. (13.10) can be recast as

$$\begin{aligned}\psi(\boldsymbol{q},\theta;t) &= \sum_{n,k} c_{n,k} e^{-i(\lambda_{n,0}+k\omega t)} \Phi_{n,0}(\boldsymbol{q},\theta+\omega t) e^{ik(\theta+\omega t)}\,, \\ &= \sum_{n,k} c_{n,k} e^{-i\lambda_{n,0}t} \Phi_{n,0}(\boldsymbol{q},\theta+\omega t) e^{ik\theta}\,, \\ &= \sum_{n} \tilde{c}_n(\theta) e^{-i\lambda_{n,0}t} \Phi_{n,0}(\boldsymbol{q},\theta+\omega t)\,, \end{aligned} \tag{13.12}$$

with $\tilde{c}_n(\theta) = \sum_k c_{n,k} e^{ik\theta}$. In addition, it can be proven that [23, 29]

$$\tilde{c}_n(\theta) = \langle \Phi_{n,0}(\boldsymbol{q},\theta) | \psi(\boldsymbol{q}; t=0) \rangle_{\mathcal{H}}\,. \tag{13.13}$$

By comparison with Sect. 2.1.4, we see that there is a complete analogy between the eigenfunctions for the isolated molecule and the Floquet states for the molecule in interaction with a field.

We arrive at the very important following property: *the role of the Floquet states in systems driven by a periodic electric field is analogous to that of energy eigenstates in time-independent systems* [27].

*13.2.2 Adiabatic Picture **

In practice, the field is never periodic since is has finite start and end times. However, as aforementioned, the Floquet formalism can be extended to the study of the interaction of quantum systems with *laser pulses with slowly varying parameters* [23, 28, 29].

Now $\boldsymbol{A}_0$ can depend on time, i.e. $\boldsymbol{E}(t) = \boldsymbol{A}_0(t)\cos(\omega t + \theta) = \boldsymbol{e} A_0(t)\cos(\omega t + \theta)$. The unit vector $\boldsymbol{e}$ describes the polarization direction of the linearly-polarized laser field. The total Hamiltonian operator of the system reads

$$H(\theta, \omega(t), A_0(t)) = H_0 - \boldsymbol{\mu}\cdot\boldsymbol{e}\, A_0(t)\cos(\theta + g(t)), \tag{13.14}$$

where. $A_0(t)$ and $g(t) = \omega t$ are the time-dependent envelope and phase frequency, respectively.

The Floquet Hamiltonian reads then, similarly to Eq. (13.4),

$$K^{\eta}(\boldsymbol{q},\theta) = -i\dot{g}(t)\frac{\partial}{\partial\theta} + H_0(\boldsymbol{q}) - \boldsymbol{\mu}(\boldsymbol{q})\cdot\boldsymbol{A}_0(t)\,\cos(\theta)\,, \tag{13.15}$$

where the vector $\boldsymbol{\eta}$ gathers the time-dependent parameters $\boldsymbol{\eta} = [\boldsymbol{A}_0(t), \dot{g}(t)]$. One remarks that the relevant frequency appears as the derivative of the phase.

At this level, it is convenient to introduce a characteristic time τ for the slow parameters [23, 28, 29]. τ can be interpreted as the pulse duration. The adiabatic limit corresponds to $\tau \to +\infty$. Gathering the slow parameters in a vector $\boldsymbol{\eta}(s)$, with $s = t/\tau$, the above Floquet theory can be recast adding $\boldsymbol{\eta}(s)$ as a parameter. For the sake of simplicity, we consider here that only the envelope of the pulse, A_0, can change in time, i.e. $\boldsymbol{\eta}(s) = A_0(s\tau) = \tilde{A}_0(s)$. The polarization of the field and the carrier frequency, $\omega = \dot{g}$, are taken constant. The corresponding adiabatic Floquet states can be written as $\Phi_{n,0}(\boldsymbol{q},\theta;s)e^{ik\theta}$. The system can be treated using an adiabatic approximation, i.e. by studying the instantaneous eigenfunctions of the Floquet Hamiltonian as a function of the slow parameters. In other words, one defines *quasienergy surfaces* as the instantaneous eigenvalues (eigenvalues at fixed time of the slow parameters of the field) of a Floquet Hamiltonian, $\lambda_m(\tilde{A}_0(s)) = \lambda_{n,k}(\tilde{A}_0(s)) = \lambda_{n,0}(\tilde{A}_0(s)) + k\omega$. $\lambda_{n,k}(\tilde{A}_0(s))$ is called a quasienergy surface that depends on parameters $\tilde{A}_0(s)$. At $t = 0$, when the field is off, $\Phi_{n,0}(\boldsymbol{q},\theta;t/\tau = 0) = \Psi_n^0(\boldsymbol{q}) \otimes 1_{\mathcal{L}}$. In the adiabatic picture, the Floquet states will thus connect to the eigenstates of H_0 at $t = 0$ and then change adiabatically due to the action of the field. When the field is off, let us say at t_{final}, the Floquet states again connect to the eigenstates of H_0. But starting from one Floquet state, let say $\Phi_{0,0}(\boldsymbol{q},\theta;t/\tau = 0) = \Psi_0^0(\boldsymbol{q}) \otimes 1_{\mathcal{L}}$, the Floquet state can continuously connect to another eigenstate of H_0, for instance $\Phi_{0,0}(\boldsymbol{q},\theta;t/\tau = t_{final}/\tau) = \Psi_1^0(\boldsymbol{q})e^{-i\theta}$. The process from $\Phi_{0,0}(\boldsymbol{q},\theta;t/\tau = 0)$ to $\Phi_{0,0}(\boldsymbol{q},\theta;t/\tau = t_{final}/\tau) = \Psi_1^0(\boldsymbol{q})e^{-i\theta}$ corresponds to a population transfer from $\Psi_0^0(\boldsymbol{q})$ to $\Psi_1^0(\boldsymbol{q})$ after absorption of one photon ($k = -1$).

It is worth noting that there is an analogy between the Floquet theory and the Born-Oppenheimer approximation introduced in Sect. 3.2.3. The role of the electronic potential energy surfaces is played by the quasienergy surfaces in the adiabatic Floquet theory. If the parameters of the laser field vary slowly enough in time, the dynamics of the system can be interpreted as a trajectory on a single quasienergy surface, which depends on the (time-dependent) parameters of the laser pulse. The slowly varying quantities, s, defining the pulsed laser field become parameters in the Floquet states like the nuclear coordinates in the adiabatic electronic states. The non-adiabatic transitions between one Floquet state to another can occur if the corresponding quasienergies become close to one another during the interaction with the field. This is analogous to the non-adiabatic transitions between the adiabatic electronic states when the corresponding potential energy surfaces are close to one another for certain nuclear configurations.

Let us be more explicit. The Floquet Schrödinger equation (see Eq. (13.5)) can be recast as

$$K(\boldsymbol{q}, \theta; \tilde{A}_0(s))\Psi_K(\boldsymbol{q}, \theta; \tau s) = \frac{i}{\tau}\frac{\partial}{\partial s}\Psi_K(\boldsymbol{q}, \theta; \tau s). \tag{13.16}$$

Let $\left\{\Phi_m(\boldsymbol{q}, \theta; \tilde{A}_0(s))\right\}$ be an orthonormal basis of instantaneous eigenvectors of the Floquet Hamiltonian with the associated eigenvalues $\lambda_m(\tilde{A}_0(s))$. The time-dependent wavefunction can be expanded in the basis of the instantaneous eigenfunctions of the Floquet operator

$$\Psi_K(\boldsymbol{q}, \theta, \tau s) = \sum_m c_m(t)\Phi_m(\boldsymbol{q}, \theta; \tilde{A}_0(s)). \tag{13.17}$$

Introducing Eq. (13.17) into Eq. (13.16) leads to a system of coupled differential equations

$$c_n(t)\lambda_n(\tilde{A}_0(s)) - \frac{i}{\tau}\sum_m c_m(t)\langle\Phi_n(\boldsymbol{q}, \theta; \tilde{A}_0(s))|\frac{\partial}{\partial s}|\Phi_m(\boldsymbol{q}, \theta; \tilde{A}_0(s))\rangle = \frac{i}{\tau}\frac{\partial c_n(\tau s)}{\partial s}, \tag{13.18}$$

with non-adiabatic coupling terms $-\frac{i}{\tau}\langle\Phi_n(\boldsymbol{q}, \theta; \tilde{A}_0(s))|\frac{\partial}{\partial s}|\Phi_m(\boldsymbol{q}, \theta; \tilde{A}_0(s))\rangle$ with $n \neq m$. In the adiabatic limit $\tau \to \infty$, these terms can be neglected and one obtains

$$\left[\lambda_n(\tilde{A}_0(s)) - \frac{i}{\tau}\langle\Phi_n(\boldsymbol{q}, \theta; \tilde{A}_0(s))|\frac{\partial}{\partial s}|\Phi_n(\boldsymbol{q}, \theta; \tilde{A}_0(s))\rangle\right] c_n(\tau s) = \frac{i}{\tau}\frac{\partial c_n(\tau s)}{\partial s}, \tag{13.19}$$

or equivalently

$$\left[\lambda_n(A_0(t)) - i\langle\Phi_n(\boldsymbol{q}, \theta; A_0(t))|\frac{\partial}{\partial t}|\Phi_n(\boldsymbol{q}, \theta; A_0(t))\rangle\right] c_n(t) = i\frac{\partial c_n(t)}{\partial t}. \tag{13.20}$$

In the adiabatic limit, when the non-adiabatic coupling can be neglected, we obtain: $c_n(t) = c_n(0)e^{-i\int_0^t \lambda_n(A_0(s))ds}$. For physical reasons, it may be interesting to create an adiabatic transition between one quantum state, for instance $\Psi_0^0(\boldsymbol{q})$ to another quantum state $\Psi_1^0(\boldsymbol{q})$. Then, it is necessary to choose the parameters of the pulse such that we will avoid the regions where the quasienergy $\lambda_{0,0}(\tilde{A}_0(s))$ is close to the other quasienergies. Using the vocabulary of non-adiabatic electronic transitions, we need to avoid "light-induced conical intersections" between the quasienergies. In principle, this is always possible by using parameters that vary slowly, for instance a long envelope and thus a long pulse [34]. One way to test the robustness of the parameters is to calculate the non-adiabatic couplings to check whether there are close to zero or not. It is even possible to minimize them by appropriately choosing the parameters of the pulse [35].

On the other hand, a "diabatic" transition may be preferred. In this case, we want to populate another Floquet state $\Phi_{n,k}(\boldsymbol{q}, \theta; \tilde{A}_0(s))$ with n and k different from 0. Then, it may be necessary to choose parameters of the field to induce a fast transition through a light-induced conical intersection. Even in this case, the adiabatic Floquet picture can serve as a framework to rationalize the process. Again there is complete analogy with the Born-Oppenheimer picture that serves as a starting point even to describe the non-adiabatic transitions. Note that if non-adiabatic transitions are induced, the wavefuntion at the end is not necessarily an eigenstate $\Psi_l^0(\boldsymbol{q})$ but can be a linear combination of eigenstates and thus a wavepacket. To conclude, *the adiabatic Floquet picture can serve as a reference provided that the parameters of the pulse are not changed abruptly, in particular provided that the laser is not switched on and off abruptly.*

13.2.3 The Rotating Wave Approximation (RWA) *

Let us consider two eigenstates only of H_0: Ψ_0^0 and Ψ_1^0 with eigenvalues E_0^0 and E_1^0, respectively. The Hamiltonian operator of Eq. (13.2) reads in matrix form

$$H(t) = \begin{bmatrix} E_0 & \Omega(t)\cos(\omega t + \theta) \\ \Omega(t)\cos(\omega t + \theta) & E_1 \end{bmatrix}, \tag{13.21}$$

where

$$\Omega(t) = -A_0(t)\langle \Psi_1^0|\boldsymbol{\mu}|\Psi_0^0\rangle \cdot \boldsymbol{e}, \tag{13.22}$$

is called the Rabi frequency. Note that we allow the envelope, $A_0(t)$, to be time-dependent. We now introduce the rotating or resonant wave approximation (RWA) that is often used to rationalize processes involving near-resonant laser pulses (in our case $\omega \approx E_1 - E_0$). More precisely, this approximation is valid when the pulsation of the electric field is close to the resonance $|E_2 - E_1 - \omega| \ll |\Omega(t)|$. Using the Floquet picture, the RWA corresponds to considering only *one photon* processes involving only the resonant transition, i.e. the transition between $\Psi_0^0(\boldsymbol{q})$ and $\Psi_1^0(\boldsymbol{q})$. Formally, RWA consists in neglecting the anti-resonant terms, corresponding to applying the approximation in the upper right term of Eq. (13.21):

$$\cos(\omega t + \theta) = \frac{1}{2}(e^{i(\omega t+\theta)} + e^{-i(\omega t+\theta)}) \approx \frac{1}{2}e^{i(\omega t+\theta)}. \tag{13.23}$$

If we now consider the Floquet operator of Eq. (13.4), applying the RWA amounts to keeping only the two functions[6] $\Psi_0^0(\boldsymbol{q}) \times 1_{\mathcal{L}}$ and $\Psi_1^0(\boldsymbol{q}) \times e^{-i\theta}$.

[6]Considering $\Psi_0^0(\boldsymbol{q}) \times e^{ik\theta}$ and $\Psi_1^0(\boldsymbol{q}) \times e^{i(k-1)\theta}$ with any value of k would give the same results.

Thus, the Floquet operator reads using the RWA and in matrix form

$$K(t) = \begin{bmatrix} E_0 & \Omega(t)/2 \\ \Omega(t)/2 & E_1 - \omega \end{bmatrix} - i\omega\frac{\partial}{\partial\theta}, \tag{13.24}$$

since $\int \frac{d\theta}{2\pi} \cos\theta e^{-i\theta} = 1/2$ and $\int \frac{d\theta}{2\pi} e^{i\theta}(-i\omega\frac{\partial}{\partial\theta})e^{-i\theta} = -\omega$.

Ω constant corresponds to the well-known Rabi problem [36]. $\Delta = E_1 - E_0 - \omega$ is called the detuning from the resonance.

Within the RWA, we assume $\Delta \approx 0$. If we take $E_0 = 0$, Eq. (13.24) resorts to

$$K(t) = \begin{bmatrix} 0 & \Omega(t)/2 \\ \Omega(t)/2 & 0 \end{bmatrix} - i\omega\frac{\partial}{\partial\theta}. \tag{13.25}$$

To obtain the Floquet states, we must diagonalize $K(t)$

$$T_1^\dagger K(t) T_1 = D(t) - i\omega\frac{\partial}{\partial\theta} \tag{13.26}$$

with

$$D(t) = \begin{pmatrix} -\frac{1}{2}\Omega(t) & 0 \\ 0 & \frac{1}{2}\Omega(t) \end{pmatrix} \tag{13.27}$$

using the time-independent transformation

$$T_1 = \begin{pmatrix} -\frac{1}{\sqrt{2}} & \frac{1}{\sqrt{2}} \\ \frac{1}{\sqrt{2}} & \frac{1}{\sqrt{2}} \end{pmatrix}. \tag{13.28}$$

The two states in the initial basis set are $\Psi_0^0(\boldsymbol{q})$ and $\Psi_1^0(\boldsymbol{q})e^{-i\theta}$. Therefore, the two Floquet states are (within the RWA)

$$\begin{aligned} \Phi_1(\boldsymbol{q}, \theta) &= \frac{1}{\sqrt{2}}(-\Psi_0^0(\boldsymbol{q}) + \Psi_1^0(\boldsymbol{q})e^{-i\theta}), \\ \Phi_2(\boldsymbol{q}, \theta) &= \frac{1}{\sqrt{2}}(\Psi_0^0(\boldsymbol{q}) + \Psi_1^0(\boldsymbol{q})e^{-i\theta}). \end{aligned} \tag{13.29}$$

Generally speaking, the evolution of the wavefunction of the system, $\psi(\boldsymbol{q}, \theta, t)$ is given by Eq. (13.10) if the envelope, Ω, is constant. Since Ω depends on time, the time evolution of the "Floquet" wavefunction is given by

$$i\frac{\partial}{\partial t}\Psi_K(\boldsymbol{q}, \theta, t) = D(t)\Psi_K(\boldsymbol{q}, \theta, t), \tag{13.30}$$

and thus

$$\Psi_K(\boldsymbol{q},\theta,t) = ie^{\left(-i\int_0^t D(t')dt'\right)}\Psi_K(\boldsymbol{q},\theta,t=0)\,. \tag{13.31}$$

Eq. (13.8) leads to

$$\psi(\boldsymbol{q},\theta,t) = \sum_\nu c_\nu e^{-i\int_0^t \lambda_\nu(t')dt'}\Phi_\nu(\boldsymbol{q},\theta+\omega t)\,. \tag{13.32}$$

In our particular case, the initial wavefunction is $\Psi_0^0(\boldsymbol{q})$ with $c_1 = -\frac{1}{\sqrt{2}}$, $c_2 = \frac{1}{\sqrt{2}}$, which corresponds to a superposition of the two Floquet states in which we apply the adiabatic principle:

$$\begin{aligned}\psi(\boldsymbol{q},\theta,t) &= c_1 e^{i/2\int_0^t \Omega(t')dt'}\Phi_1(\boldsymbol{q},\theta+\omega t) + c_2 e^{-i/2\int_0^t \Omega(t')dt'}\Phi_2(\boldsymbol{q},\theta+\omega t)\\ &= \frac{1}{2}(e^{i/2\int_0^t \Omega(t')dt'}(\Psi_0^0(\boldsymbol{q}) - \Psi_1^0(\boldsymbol{q})e^{-i(\theta+\omega t)}\\ &+ e^{-i/2\int_0^t \Omega(t')dt'}(\Psi_0^0(\mathbf{r},\mathbf{R}) + \Psi_1^0(\boldsymbol{q})e^{-i(\theta+\omega t)}))\\ &= \Psi_0^0(\boldsymbol{q})\cos\left(\int_0^t \frac{1}{2}\Omega(t')dt'\right) - i\Psi_1^0(\boldsymbol{q})\sin\left(\int_0^t \frac{1}{2}\Omega(t')dt'\right)e^{-i(\theta+\omega t)}\,.\end{aligned} \tag{13.33}$$

We can remark that for the resonant two-level problem, the non-adiabatic coupling is exactly 0. The above result is thus exact. If the initial phase of the field is equal to zero, i.e. $\theta = 0$, we obtain

$$\Psi(\boldsymbol{q},t) = \Psi_0^0(\boldsymbol{q})\cos\left(\int_0^t \frac{1}{2}\Omega(t')dt'\right) - i\Psi_1^0(\boldsymbol{q})\sin\left(\int_0^t \frac{1}{2}\Omega(t')dt'\right)e^{-i\omega t}\,, \tag{13.34}$$

and the probability to find the system in state $\Psi_1^0(\boldsymbol{q})$ at time t is given by

$$|\langle\Phi_1^0|\Psi(t)\rangle|^2 = \sin^2\left(\int_0^t \frac{\Omega(t')}{2}dt'\right)\,. \tag{13.35}$$

If $\Omega(t)$ is a constant Ω_0, the result is a particular case of the Rabi formula [36]

$$P_2(t) = \frac{\Omega_0^2}{\Omega_0^2+\Delta^2}\sin^2\left(\sqrt{\Omega_0^2+\Delta^2}\frac{t}{2}\right)\,. \tag{13.36}$$

A general, two-level model, with an arbitrary expression for the Rabi frequency $\Omega(t)$ and the (time-dependent) detuning $\Delta(t)$, is not analytically solvable. However, there exist models with specific expressions for $\Omega(t)$ and $\Delta(t)$, for with analytical solutions of the time-dependent Schrödinger equation are known [37].

13.3 Application of the RWA to the Enhancement of Tunneling in NHD_2 *

NHD_2, as its isotopologue ammonia, is a prototype for systems with a double-well type potential energy leading to a non-negligible tunneling splitting even for the vibrational ground state. NHD_2 has six internal degrees of freedom. A schematic illustration of the cut along the "umbrella" or inversion mode is shown on Fig. 13.1. The vibrational symmetric ground state is shown schematically in red and the first vibrational excited is the antisymmetric function shown in green. There are both delocalized on the two wells. The fact that the wavefunctions are not equal to zero in regions that are classically forbidden (when the potential is higher than the eigenvalue) is a clear sign of the presence of tunneling, see for instance the function in red at the geometry corresponding to the transition state. The assignment is explained in Sect. 7.4 of Chap. 7. As explained in Sect. 7.4 of Chap. 7, due to the presence of the double well, the relevant molecular-symmetry group for NHD_2 is not the point group at the equilibrium geometry. The relevant molecular-symmetry group connects continuously the two equivalent but distinct $\mathcal{C}_s$ minima through the $\mathcal{C}_{2v}$ transition structure is the nuclear-permutation inversion group $\mathcal{G}_4$, also known as $\mathcal{MS}_4$.

The Heidelberg MCTDH program and the PES of Ref. [38, 39] are used to calculate the 6-dimensional vibrational states of the molecule [40]. The system is described by six valence-type coordinates[7] $\boldsymbol{q}$ (similar to the coordinates for water of Fig. 5.5 but with one more vector).

Several theoretical eigenvalues for $J = 0$ are given in Table 13.1[8] along with the experimental ones. The PES gives a inversion barrier at 1781 cm^{-1}. The first antisymmetric state, Ψ_0^a (with the label $(0)^a$ in Table 13.1), that corresponds to the function in green in Fig. 13.1, lies 0.159 cm^{-1} above the ground state, Ψ_0^s (with the label $(0)^s$ in Table 13.1), that corresponds to the function in red in Fig. 13.1. This energy difference is due to tunneling (the corresponding experimental value is 0.171 cm^{-1}).

In our model, we assume the molecule being "ideally oriented" [44], i.e. the molecule experiences a three-dimensional alignment during the excitation process. Molecular orientation is a current active research field [20, 45–51]. This could be achieved,

[7]More precisely, "Radau" coordinates [40] are used. The Radau coordinates will not be defined in the present book. We briefly mention that they are close to the valence coordinates but have the advantage of being orthogonal coordinates and thus lead to a simpler KEO (see Sect. 5.4 for a discussion about orthogonal and non-orthogonal coordinates).

[8]Reprinted with permission from [40]. Copyright 2012, American Institute of Physics.

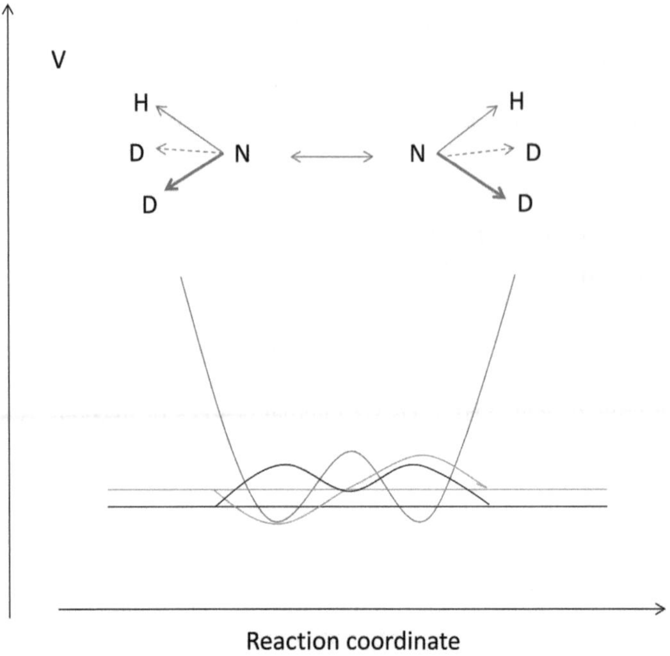

Fig. 13.1 Schematic illustration of the double-well type potential in NHD_2. In *red* is the vibrational symmetric ground state, $\Psi_0^s(\boldsymbol{q})$, in *green* the first (antisymmetric) vibrational excited state, $\Psi_0^a(\boldsymbol{q})$. The splitting in energy between the two levels is due to tunneling

for instance, using simultaneously an adiabatic pulse, that has been switched on before the excitation and is switched off after the excitation [52]. This model allows us to assume that a linearly polarized laser pulse is parallel to one of the Body-Fixed axes. The definition of the Body-Fixed plane is given on Fig. 13.2[9] and we choose a polarization of the laser parallel to the y-axis.

Let us assume that the molecule is initially localized on the left side of Fig. 13.1 and is a linear combination of the symmetric and antisymmetric ground states. Mathematically, the initial wavefunction is thus given by

$$\Psi(\boldsymbol{q}, t = 0) = \frac{1}{\sqrt{2}}(\Psi_0^s(\boldsymbol{q}) + \Psi_0^a(\boldsymbol{q})) . \tag{13.37}$$

It is well known [36] that the system with such an initial condition can tunnel to the right side with the linear combination:

$$\Psi(\boldsymbol{q}) = \frac{1}{\sqrt{2}}(\Psi_0^s(\boldsymbol{q}) - \Psi_0^a(\boldsymbol{q})) . \tag{13.38}$$

[9] Reprinted with permission from [40]. Copyright 2012, American Institute of Physics.

Table 13.1 Several theoretical low-lying 6D vibrational eigenvalues compared with available experimental values (in cm^{-1}). Assignment in terms of $\mathcal{C}_{2v}$(M) normal modes and symmetry labels in the $\mathcal{G}_4$ permutation-inversion symmetry group are given. The assignment is explained in Sect. 7.4 of Chap. 7. E_{MCTDH} correspond to the calculated values of Ref. [40] using the *improved relaxation* method of Sect. 8.4, "Exp." to the experimental values

level	sym. ($\mathcal{C}_{2v}$(M)/$\mathcal{G}_4$)	E_{MCTDH}	Exp.
$(0)^s$	A_1/A^+	0.00	0.00
$(0)^a$	B_1/B^-	0.159	0.171 [41]
$(2^1)^s$	A_1/A^+	808.81	810.23 [41]
$(2^1)^a$	B_1/B^-	817.68	819.56 [41]
$(4_a^1)^s$	A_1/A^+	1235.67	1233.27 [42, 43]
$(4_a^1)^a$	B_1/B^-	1238.11	1235.89 [42, 43]
$(2^2)^s$	A_1/A^+	1450.38	–
$(4_b^1)^s$	B_2/B^+	1461.55	1461.79 [42, 43]
$(4_b^1)^a$	A_2/A^-	1461.72	1461.99 [42, 43]
$(2^2)^a$	B_1/B^-	1575.56	–
$(2^3)^s$	A_1/A^+	1957.55	–
$(2^3)^a$	B_1/B^-	2360.99	–
$(3_a^1)^s$	A_1/A^+	2435.87	2430.80 [42, 43]
$(3_a^1)^a$	B_1/B^-	2437.49	2434.62 [42, 43]
$(3_b^1)^s$	B_2/B^+	2564.05	2559.81 [42, 43]
$(3_b^1)^a$	A_2/A^-	2564.16	2559.96 [42, 43]
$(1^1)^s$	A_1/A^+	3406.08	3404.24 [42, 43]
$(1^1)^a$	B_1/B^-	3406.15	3404.32 [42, 43]

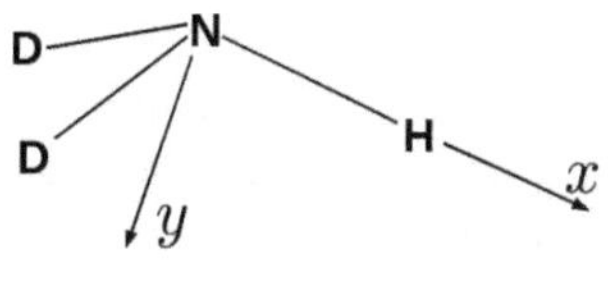

Fig. 13.2 Definition of the BF frame. The x-axis is parallel to the NH bond and the y-axis lies in one of the two NHD planes

with a tunneling time given by (using $c = 2.9979\ 10^{10}$ cm s^{-1} and $\Delta E = 0.159$ cm^{-1})

$$\tau_{\text{tunnel}} = 1/(2\,(\Delta E)\,c) \approx 105 \text{ ps}\,. \tag{13.39}$$

Our goal is here to induce a full population inversion between the two localized superpositions by a laser pulse in a time much smaller than the field-free tunneling time of 105 ps. Mathematically, this corresponds to replacing the "+" sign by a "−" sign in the linear combination. One possibility is to create a phase of π in the linear combination by achieving a complete Rabi oscillation between the Ψ_0^a state with an intermediate state. The latter must be well isolated (if not, other states can be populated and interfere in the process), and not coupled to the Ψ_0^s state since we want to keep Ψ_0^s unchanged.

Table 13.2 Calculated transition dipole moments (in ea_0 unit where e is the elementary charge and a_0 is the Bohr radius) of the selected low-lying vibrational eigenstates with the two tunneling components of the fundamental eigenstate. Values smaller than 10^{-4} ea_0 are not given. $\mu_{x1} = \langle(0)^s|\mu_x|\chi_i\rangle$, $\mu_{x2} = \langle(0)^a|\mu_x|\chi_i\rangle$, $\mu_{y1} = \langle(0)^s|\mu_y|\chi_i\rangle$, $\mu_{y2} = \langle(0)^a|\mu_y|\chi_i\rangle$, $\mu_{z1} = \langle(0)^s|\mu_z|\chi_i\rangle$, $\mu_{z2} = \langle(0)^a|\mu_z|\chi_i\rangle$

level	μ_{x1}	μ_{x2}	μ_{y1}	μ_{y2}	μ_{z1}	μ_{z2}
$(0)^s$	0.3278	–	0.5562	–	–	0.3028
$(0)^a$	–	0.3279	–	0.5561	0.3028	–
$(2^1)^s$	−0.0151	–	0.0448	–	–	−0.0381
$(2^1)^a$	–	−0.0150	–	0.0448	−0.0375	–
$(4_a^1)^s$	−0.0270	–	0.0086	–	–	0.0188
$(4_a^1)^a$	–	0.0275	–	−0.0094	−0.0188	–
$(2^2)^s$	−0.0065	–	0.0079	–	–	0.0046
$(4_b^1)^s$	–	–	−0.0114	–	–	0.0199
$(4_b^1)^a$	–	–	–	0.0114	−0.0199	–
$(2^2)^a$	–	0.0045	–	−0.0058	−0.0045	—
$(2^3)^s$	−0.0017	–	0.0019	–	–	0.0001
$(2^3)^a$	–	–	–	0.0005	−0.0021	–
$(3_a^1)^s$	0.0058	–	−0.0247	–	–	−0.0150
$(3_a^1)^a$	–	0.0061	–	−0.0253	−0.0153	–
$(3_b^1)^s$	–	–	0.0084	–	–	−0.0147
$(3_b^1)^a$	–	–	–	0.0085	−0.0147	–
$(1^1)^s$	–	0.0382	–	−0.0030	−0.0014	–
$(1^1)^a$	0.0381	–	−0.0031	–	–	−0.0014

Table 13.2[10] gives the transition moments between the first two vibrational states, Ψ_0^s and Ψ_0^a, and several excited states. They are given for the three axes of the body-fixed frame.

Table 13.2 shows that a field polarized along the z-axis couples states with different symmetries with respect to the inversion and this is what we want to avoid since we do not want to alter the Ψ_0^s state. On the other hand, a field polarized along the other two axes couples only states of the same parity. We choose here the state with the label $(2^2)^a$ corresponding to the energy $E_1 = 1575.56$ cm^{-1} since $(2^2)^a$ is well isolated. In addition, we choose a polarization along the y-axis since the transition moment between Ψ_0^s and $\Psi_{2^2}^a$ is larger along y than along x. Let us now consider a resonant transition between Ψ_0^s and $\Psi_{2^2}^a$ the state with the label $(2^2)^a$: the transition is shown with a blue arrow on Fig. 13.3.[11]

Let us consider the three eigenstates Ψ_s^0, Ψ_a^0, and $\Psi_{2^2}^a$ only. The Hamiltonian operator reads in matrix form in this basis set:

[10]Reprinted with permission from [40]. Copyright 2012, American Institute of Physics.

[11]Reprinted with permission from [40]. Copyright 2012, American Institute of Physics.

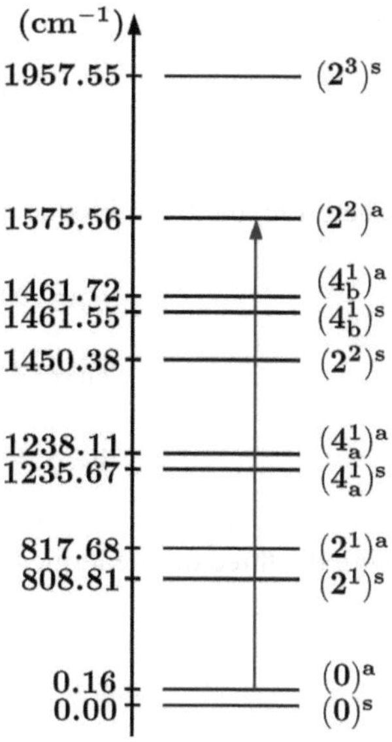

Fig. 13.3 The *blue arrow* features the resonant 2π-pulse

$$H(t) = \begin{bmatrix} E_0^s & 0 & 0 \\ 0 & E_0^a & \Omega(t)\cos(\omega t + \theta) \\ 0 & \Omega(t)\cos(\omega t + \theta) & E_{2^2}^a \end{bmatrix}, \tag{13.40}$$

with

$$\Omega(t) = -A_0(t)\langle \Phi_0^a | \mu_y | \Phi_{2^2}^a \rangle \,. \tag{13.41}$$

The quantity $A_0(t)$ is defined as in Eq. (13.2). The transitions between Ψ_0^s and Ψ_0^a and between Ψ_0^s and $\Psi_{2^2}^a$ are zero by symmetry. If we now consider the Floquet operator of Eq. (13.5), applying the RWA amounts to keeping only the three functions $\Psi_0^s \times 1_{\mathcal{L}}$, $\Psi_0^a \times 1_{\mathcal{L}}$ and $\Psi_{2^2}^a \times e^{-i\theta}$. Thus, the Floquet operator reads within the RWA and in matrix form

$$K(t) = \begin{bmatrix} E_0^s & 0 & 0 \\ 0 & E_0^a & \Omega(t)/2 \\ 0 & \Omega(t)/2 & E_{2^2}^a - \omega \end{bmatrix}, \tag{13.42}$$

Now, within the RWA, we assume $E_{2^2}^a - E_0^a \approx \omega$ and if we put $E_0^s = 0$ and $E_0^s - E_0^a = \delta$, we obtain

$$K(t) = \begin{bmatrix} \delta & 0 & 0 \\ 0 & 0 & \Omega(t)/2 \\ 0 & \Omega(t)/2 & 0 \end{bmatrix}. \tag{13.43}$$

To obtain the Floquet states, we must diagonalize $K(t)$,

$$T_1^\dagger K(t) T_1 = D(t)\,, \tag{13.44}$$

with

$$D(t) = \begin{pmatrix} \delta & 0 & 0 \\ 0 & -\frac{1}{2}\Omega(t) & 0 \\ 0 & 0 & \frac{1}{2}\Omega(t) \end{pmatrix} . \tag{13.45}$$

Therefore, the three Floquet states are (within the RWA)

$$\begin{aligned} \Phi_0^s(\boldsymbol{q}, \theta) &= \Psi_0^s(\boldsymbol{q}) , \\ \Phi_0^a(\boldsymbol{q}, \theta) &= \frac{1}{\sqrt{2}}(-\Psi_0^a(\boldsymbol{q}) + \Psi_{2^2}^a(\boldsymbol{q})e^{-i\theta}) , \\ \Phi_3(\boldsymbol{q}, \theta) &= \frac{1}{\sqrt{2}}(\Psi_0^a(\boldsymbol{q}) + \Psi_{2^2}^a(\boldsymbol{q})e^{-i\theta}) , \end{aligned} \tag{13.46}$$

with the three quasienergies

$$\begin{aligned} \lambda_1 &= \delta , \\ \lambda_2 &= -\frac{1}{2}\Omega(t) , \\ \lambda_3 &= \frac{1}{2}\Omega(t) . \end{aligned} \tag{13.47}$$

In our particular case, the initial "Floquet" wavefunction is

$$\frac{1}{\sqrt{2}}(\Psi_0^s(\boldsymbol{q}) + \Psi_0^a(\boldsymbol{q})) . \tag{13.48}$$

Thus, the evolution of the wavepacket is given by

$$\begin{aligned} \psi(\boldsymbol{q}, \theta, t) &= c_1 e^{-i\int_0^{t_p} \delta dt'} \Phi_0^s(\boldsymbol{q}, \theta + \omega t) \\ &+ c_2 e^{(i/2)\int_0^{t_p} \Omega(t')dt'} \Phi_0^a(\boldsymbol{q}, \theta + \omega t) + c_3 e^{-(i/2)\int_0^{t_p} \Omega(t')dt'} \Phi_{2^2}^a(\boldsymbol{q}, \theta + \omega t) , \\ &= \frac{1}{\sqrt{2}} e^{-i\delta t_p} \Psi_0^s(\boldsymbol{q}) - \frac{1}{2} e^{iA(t_p)/2} (\frac{1}{\sqrt{2}}(-\Psi_0^a(\boldsymbol{q}) + \Psi_{2^2}^a(\boldsymbol{q})e^{-i\theta}) \\ &+ \frac{1}{2} e^{-iA(t_p)/2} (\frac{1}{\sqrt{2}}(\Psi_0^a(\boldsymbol{q}) + \Psi_{2^2}^a(\boldsymbol{q})e^{-i\theta}) , \end{aligned} \tag{13.49}$$

with $A(t_p) = \int_0^{t_p} \Omega(t')dt'$ and t_p the pulse duration. Thus, we obtain

$$\begin{aligned} \psi(\boldsymbol{q}, \theta, t_p) &= \frac{1}{\sqrt{2}} e^{-i\delta t_p} \Psi_0^s(\boldsymbol{q}) + \frac{1}{\sqrt{2}} \cos(A(t_p)/2) \Psi_0^a(\boldsymbol{q}) \\ &+ \frac{i}{\sqrt{2}} \sin(A(t_p)/2) e^{-i\theta} \Psi_{2^2}^a(\boldsymbol{q}) . \end{aligned} \tag{13.50}$$

A complete tunneling inversion corresponds to

$$\psi(\boldsymbol{q}, \theta, t_p) = \frac{1}{\sqrt{2}}(\Psi_0^s(\boldsymbol{q}) - \Psi_0^a(\boldsymbol{q})), \tag{13.51}$$

From these equations, one can see that the tunneling inversion probability is close to unity if the conditions

$$\int_0^{t_p} |\Omega(t)|\, dt = A(t_p) = 2\pi, \quad \delta t_p \ll \pi, \tag{13.52}$$

are fulfilled.

The first condition defines a relation between the duration t_p and the peak amplitude E_0 of the pulse. If we assume a $\sin^2$ pulse envelope:

$$E_0 \left|\mu_{(0)^a/(2^2)^a}\right| \int_0^{t_p} \sin^2\left(\frac{\pi t}{t_p}\right) dt = 2\pi, \tag{13.53}$$

i.e.

$$t_p = \frac{4\pi}{E_0 \left|\mu_{(0)^a/(2^2)^a}\right|}. \tag{13.54}$$

We then set the peak amplitude at $E_0 = 10^{-2}\ E_h/ea_0$. This amplitude is high enough to lead to a relatively short duration of tunneling, and low enough to avoid detrimental effects such as ionization and to guarantee a negligible influence of the electronic polarizability (not included in our model). With this constraint, the parameters found within our model are

$$\begin{aligned} E_0 &= 10^{-2}\ E_h/ea_0, \\ \omega &= 1575.56 \times 2\pi c\ \text{cm}^{-1}, \\ t_p &= 5280\ \text{fs}. \end{aligned} \tag{13.55}$$

The simulations show that the population transfer is improved by slightly detuning ω from the transition frequency. In particular, a six-dimensional simulation with $\omega = 1574 \times 2\pi c\ \text{cm}^{-1}$ yields a final population of the target superposition of approximately 0.995 [40].

The fact that the frequency must be detuned is called the Stark effect. We cannot predict this value in our model since we have calculated the Floquet states and the quasienergies using three states only. The Stark effect is due to the presence of the other vibrational states. An exact calculation of the quasienergies would show that $\lambda_{2^2}^a$ would slightly decrease with respect to λ_0^a when the intensity increases.[12] To obtain those quasienergies it would be necessary to diagonalize the Floquet operator numerically since no analytical solution could be found any more.

[12] I.e., $\lambda_{2^2}^a - \lambda_0^a$ decreases with A_0, all the other parameters being fixed.

13.4 Application of the RWA to the Destruction of Tunneling in NHD_2 *

In the previous section, we have induced a population inversion of 0.995 after 5.28 ps. Here, our goal is the opposite, i.e. to achieve a coherent destruction of tunneling (CDT).

The tunneling splitting for the ground state is equal to

$$E_0^a - E_0^s = 0.159\ \mathrm{cm}^{-1}\,.$$

Since our goal is to prevent the system from tunneling, we must find the parameters of a laser pulse such that the first two quasienergies have the same value:

$$\lambda_0^a - \lambda_0^s \approx 0\ \mathrm{cm}^{-1}\,.$$

The laser field is linearly polarized along the z-axis of the BF frame (see Fig. 13.2 for the definition of the BF axes). The CDT is achieved in the situation where the system remains in the state defined by Eq. (13.37). We first consider a quasi-resonant regime, the frequency of the laser field (featured by the red arrow in Fig. 13.4[13]) is close to the transition between Ψ_0^a and $\Psi_{2^1}^s$. Since the field is parallel to the z BF axis, for symmetry reasons, Ψ_0^a is coupled with $\Psi_{2^1}^s$ and not with $\Psi_{2^1}^a$ and Ψ_0^s is coupled with $\Psi_{2^1}^a$ and not with $\Psi_{2^1}^s$. In addition, because the frequency of the laser field is much higher than the energy difference between the eigenvalues of Ψ_0^a and Ψ_0^s, a laser driven transition between these two states can be neglected.

The Floquet operator in the basis $\left\{\Psi_0^s \times 1_{\mathcal{L}}, \Psi_0^a \times 1_{\mathcal{L}}, \Psi_{2^1}^s \times e^{-i\theta}, \Psi_{2^1}^a \times e^{-i\theta}\right\}$ reads in the framework of the RWA

$$K^{RWA} = \begin{pmatrix} E_0^s & 0 & 0 & \Omega_{(0)^s,(2^1)^a}(t)/2 \\ 0 & E_0^a & \Omega_{(0)^a,(2^1)^s}(t)/2 & 0 \\ 0 & \Omega_{(0)^a,(2^1)^s}(t)/2 & E_{2^1}^s - \omega & 0 \\ \Omega_{(0)^s,(2^1)^a}(t)/2 & 0 & 0 & E_{2^1}^a - \omega \end{pmatrix}. \tag{13.56}$$

For the electric field, we turned on the laser field adiabatically and then an envelope that is constant:

$$E(t) = \begin{cases} A_0 \sin^2\left(\frac{\pi t}{2t_{\mathrm{ramp}}}\right)\cos(\omega t) & \text{for } t \le t_{\mathrm{ramp}} \\ A_0 \cos(\omega t) & \text{for } t > t_{\mathrm{ramp}} \end{cases}, \tag{13.57}$$

where t_{ramp} the duration of the ramp of the electric field. A ramp of fifty optical cycles is considered, *i.e* $t_{\mathrm{ramp}} = 100\pi/\omega$.

For $t > t_{\mathrm{ramp}}$, $\Omega_{(0)^s,(2^1)^a}(t)/2$ and $\Omega_{(0)^a,(2^1)^s}(t)/2$ can be considered as fixed, and diagonalising the matrix of (13.56) gives for the first two quasienergies

[13] Reprinted with permission from [53]. Copyright 2014, American Institute of Physics.

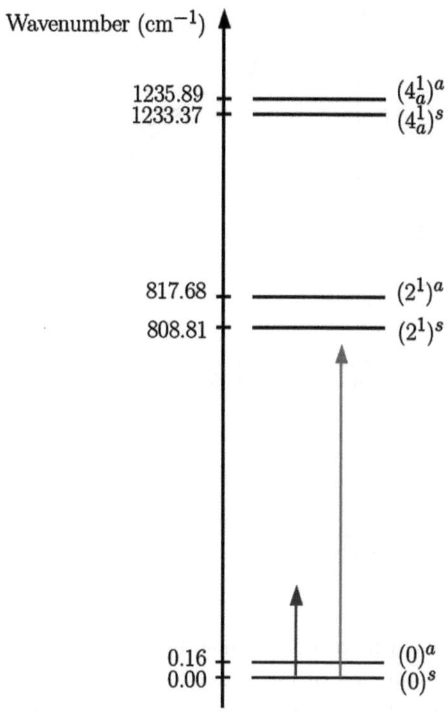

Fig. 13.4 Energy diagram illustrating the two laser field frequency regimes considered for the coherent destruction of tunneling in NHD_2. The *blue* (*red*) *arrow* features the non-resonant (quasi-resonant) regime

$$\lambda_0^s = \frac{1}{2}(E_0^a + E_{2^1}^s - \omega) - \frac{1}{2}\sqrt{(E_{2^1}^s - \omega - E_0^a)^2 + \Omega^2_{(0)^a,(2^1)^s}}\,, \tag{13.58}$$

and

$$\lambda_0^a = \frac{1}{2}(E_0^s + E_{2^1}^a - \omega) - \frac{1}{2}\sqrt{(E_{2^1}^a - \omega - E_0^s)^2 + \Omega^2_{(0)^s,(2^1)^a}}\,. \tag{13.59}$$

The quasienergies λ_0^s and λ_0^a are continuously connected to E_0^a and E_0^s. They are shown as functions of the amplitude of the electric field and for three different field frequencies of 795, 800, and 805 cm^{-1} in Fig. 13.5.[14] These frequencies are close to resonance between the $(0)^a$ and the $(2^1)^s$ states at 808.65 cm^{-1}. The results obtained from the RWA model using Eqs. (13.58) and (13.59) are shown in blue dashed lines and the results obtained from the "exact" numerical diagonalization of the Floquet Hamiltonian in black full lines (taken from Ref. [53]). The comparison between the numerical and analytical values proves the validity of the RWA for the frequencies considered here. The differences increase with intensity and are mainly due to the Stark effect as explained in the previous section.

In Fig. 13.5, we see that the quasienergies show a single crossing. The parameters of the pulses must correspond to these crossing points since the levels are degenerate

[14]Reprinted with permission from [53]. Copyright 2014, American Institute of Physics.

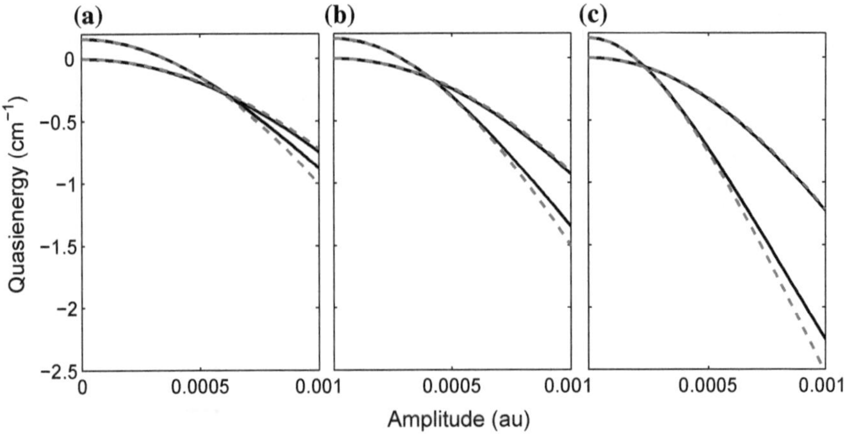

Fig. 13.5 Quasienergies λ_0^s and λ_0^a plotted as a function of the laser field amplitude for three different frequencies **a** ω =795 cm^{-1}, **b** 800 cm^{-1}, and **c** 805 cm^{-1}. The results obtained from the model using Eqs. (13.58) and (13.59) are shown in *red dotted lines* and the results obtained from the numerical diagonalization of the Floquet Hamiltonian are shown in *black full lines*

when there is no tunneling: for instance $\omega = 800$ and $A_0 = 4.17 \times 10^{-4}$. The population of the initial state, $\frac{1}{\sqrt{2}}(\Psi_0^s(\boldsymbol{q}) + \Psi_0^a(\boldsymbol{q}))$, located on the left side of the double-well shows oscillations (not shown here). The minimum of the oscillations remains above 0.95 even after 10 ps [53].

13.5 Destruction of Tunneling in NHD_2: Non-resonant Regime *

We now consider a non-resonant regime where the frequency of the laser field (featured by the blue arrow in Fig. 13.4) is much larger than the ground-state splitting of 0.159 cm^{-1} but much smaller than the transition frequencies with the other vibrational states. As in the previous section, the laser field is linearly polarized along the z-axis of the BF frame (see Fig. 13.2 for the definition of the BF axes).

The Floquet operator in the basis $\left\{\Psi_0^s \times e^{ik\theta}, \Psi_0^a \times e^{ik\theta}\right\}$, reads

$$K(t) = -i\omega\frac{\partial}{\partial\theta} + \begin{pmatrix} E_0^s & \Omega(t)\cos\theta \\ \Omega(t)\cos\theta & E_0^a \end{pmatrix}. \tag{13.60}$$

If we assume that the frequency of the laser field is much higher than the tunneling splitting, $\omega \gg E_0^a - E_0^s$, the diagonalization of the matrix of Eq. (13.60) has analytic solutions (see Sect. III A 1 of Ref. [53]). In addition, the Stark effect, i.e. the influence of the other states on the first two quasienergies, λ_0^s and λ_0^a, can be calculated by

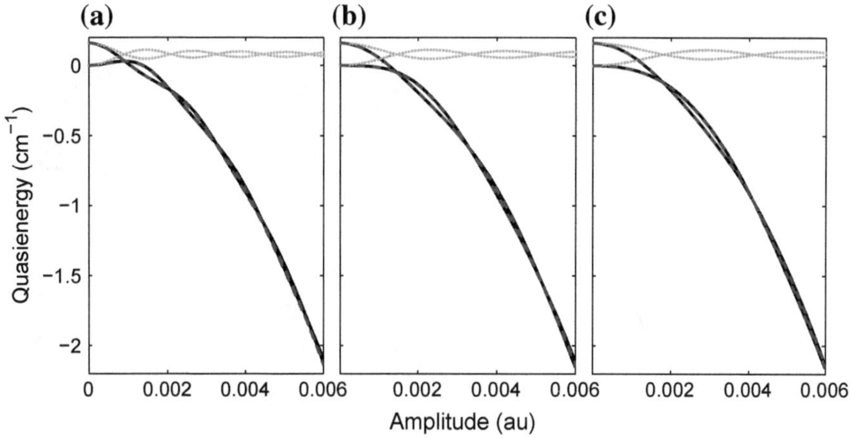

Fig. 13.6 Quasienergies λ_0^s and λ_0^a ($k = 0$) plotted as functions of the laser field amplitude for three different frequencies **a** ω =50 cm^{-1}, **b** 80 cm^{-1}, and **c** 100 cm^{-1}. The results obtained from the two-level model are shown in *red dotted lines*, the results obtained from the perturbative model including the Stark shifts are shown in *blue dashed lines* and the results obtained from the numerical diagonalization of the Floquet operator are shown in *black full lines*

applying second-order stationary perturbation theory [53]. We do not give the analytical expressions here. Instead, λ_0^s and λ_0^a for $k = 0$ are presented in Fig. 13.6[15] as functions of the amplitude of the electric field. The results with the two-level model are shown in dotted red lines, the results obtained from the perturbative model are shown in blue dashed lines. The results obtained by numerical diagonalization of the Floquet operator are shown in black full lines.

13.6 Control of the $\pi\pi^\star$ Excitation of Pyrazine by the Stark Effect *

As a last example, we present the control of the non-radiative decay after a $\pi\pi^\star$ excitation of pyrazine using a mechanism based on the dynamic Stark effect induced by a strong non-resonant laser pulse [54, 55]. The Stark effect corresponds to a non-resonant two-photon process where one photon is absorbed by the molecule and another photon is emitted by the molecule. Here, we use an effective Hamiltonian involving the electronic polarisability of the molecule. This effective Hamiltonian could be derived from the Floquet theory but this derivation is not given here. The molecule is depicted in Figs. 12.8 and 13.7.[16]

[15]Reprinted with permission from [53]. Copyright 2014, American Institute of Physics.

[16]Reprinted with permission from [54]. Copyright 2014, American Institute of Physics.

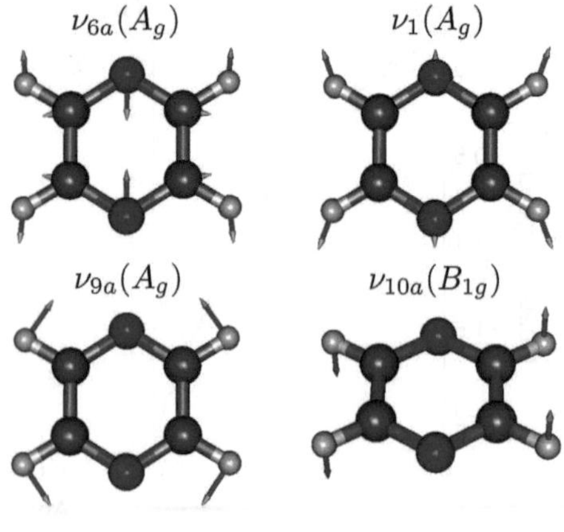

Fig. 13.7 Representation of the four normal modes included in the model for pyrazine: q_{6a}, q_1, q_{9a} and q_{10a}

For the BF frame, we use the principal axes of inertia with the following definition: at the ground-state equilibrium geometry the molecule lies in the yz-plane, with both nitrogen atoms on the z-axis. The molecule is planar at the equilibrium geometry in the electronic ground state and has a $\mathcal{D}_{2h}$ point group symmetry. Three electronic states are included in the present study: the ground state and the vibronically coupled diabatic $B_{3u}(n\pi^*)$ and $B_{2u}(\pi\pi^*)$ states, hereafter noted S_0, S_1 and S_2. The chemical meaning of the two excited states in terms of molecular orbitals can be seen on Fig. 4.11 of Sect. 4.4 in Chap. 4.

We adopt a description in terms of the normal coordinates of the electronic ground state. The molecule has $N = 10$ atoms and thus $3N - 6 = 24$ normal coordinates. For the sake of simplicity, we consider a model including the four most important vibrational normal modes only [56–60]: q_{6a}, q_1, q_{9a} and q_{10a}. They are depicted in Fig. 13.7. q_{6a}, q_1, q_{9a} are the totally symmetric normal modes and q_{10a} has B_{1g} symmetry.

As explained in Sect. 12.4, after a transition from the electronic ground state to the $B_{2u}(\pi\pi^*)$, there is a very fast transfer to the $B_{3u}(n\pi^*)$ state due to a conical intersection. The electronic transition to S_2 could be induced experimentally by a relatively weak resonant "pump pulse". We have seen in Sect. 12.5 that the physics induced by the presence of a conical intersection is linked to the topography around the latter. In order to modify the non-adiabatic transfer, it is thus natural to try to change this topography. A polar solvent could modify this topography but here we stay in the gas phase and use a second laser pulse that is a strong non-resonant laser pulse, hereafter referred to as "control pulse", that shapes the PESs via what is called the "Stark effect". In our simulations, the pump pulse is polarized along the y-direction while the control pulse is polarized along the z-direction.[17]

Thus, the term describing the interaction between the molecule and the resonant (pump) pulse simply reads in the diabatic representation

[17] We thus assume that the molecule is oriented in 3D space and thus does not rotate freely.

$$H^r_{\text{int}}(t) = -\begin{pmatrix} 0 & \mu^y_{01}(\boldsymbol{q}) & \mu^y_{02}(\boldsymbol{q}) \\ \mu^y_{01}(\boldsymbol{q}) & 0 & 0 \\ \mu^y_{02}(\boldsymbol{q}) & 0 & 0 \end{pmatrix} E_r(t)\,, \tag{13.61}$$

with $E_r(t)$ the resonant external field and μ^y_{nl} are y-components of the dipole matrix elements, as defined in Sect. 3.4, in the diabatic basis set.

For the term corresponding to the interaction between the molecule and the control pulse, we use the following effective operator in the diabatic representation:

$$H^c_{\text{int}}(t) = -\frac{1}{2}\begin{pmatrix} \alpha^{zz}_{00}(\boldsymbol{q}) & 0 & 0 \\ 0 & \alpha^{zz}_{11}(\boldsymbol{q}) & \alpha^{zz}_{12}(\boldsymbol{q}) \\ 0 & \alpha^{zz}_{12}(\boldsymbol{q}) & \alpha^{zz}_{22}(\boldsymbol{q}) \end{pmatrix} E^2_c(t)\,, \tag{13.62}$$

where α^{zz}_{nl} are the (z, z)-component of the electronic static polarisability matrix in the diabatic basis set and $E_c(t)$ the control external field. Several terms have been neglected, for instance α^{zz}_{01}. They could be added but their role in the dynamics would be very small.

Thus, in the the diabatic representation, the Hamiltonian of the system as a function of the four dimensionless normal modes reads

$$\begin{aligned} H(\boldsymbol{q}, t) = H^{gs}(\boldsymbol{q}) \begin{pmatrix} 1 & 0 & 0 \\ 0 & 1 & 0 \\ 0 & 0 & 1 \end{pmatrix} &+ \begin{pmatrix} 0 & 0 & 0 \\ 0 & V_{11}(\boldsymbol{q}) & V_{12}(\boldsymbol{q}) \\ 0 & V_{12}(\boldsymbol{q}) & V_{22}(\boldsymbol{q}) \end{pmatrix} \\ &- \begin{pmatrix} 0 & \mu^y_{01}(\boldsymbol{q}) & \mu^y_{02}(\boldsymbol{q}) \\ \mu^y_{01}(\boldsymbol{q}) & 0 & 0 \\ \mu^y_{02}(\boldsymbol{q}) & 0 & 0 \end{pmatrix} E_r(t) \\ &- \frac{1}{2}\begin{pmatrix} \alpha^{zz}_{00}(\boldsymbol{q}) & 0 & 0 \\ 0 & \alpha^{zz}_{11}(\boldsymbol{q}) & \alpha^{zz}_{12}(\boldsymbol{q}) \\ 0 & \alpha^{zz}_{12}(\boldsymbol{q}) & \alpha^{zz}_{22}(\boldsymbol{q}) \end{pmatrix} E^2_c(t)\,, \end{aligned} \tag{13.63}$$

where $H^{gs}(\boldsymbol{q}) = \sum_{i=1}^4 \frac{\omega_i}{2}\left(-\frac{\partial^2}{\partial q_i^2} + q_i^2\right)$ is the ground electronic state Hamiltonian in the harmonic approximation, $\boldsymbol{V}(\boldsymbol{q})$ is the potential matrix.

For the resonant laser pulse that drives the electronic transition and a non-resonant control pulse, we take the expression

$$E_{r,c}(t) = \epsilon_{r,c}(t)\cos(\omega_{r,c}t) = \epsilon^0_{r,c}\Lambda_{r,c}(t)\cos(\omega_{r,c}t)\,, \tag{13.64}$$

where the envelope $\Lambda_{r,c}(t)$ is chosen to be a sine-squared function $\Lambda_{r,c}(t) = \sin^2\left(\frac{\pi t}{t_p^{r,c}}\right)$, and $t_p^{r,c}$ is the pulse duration. The values $t_p^r = 100$ fs and $t_p^c = 500$ fs have been used. The control pulse is turned on at $t = -150$ fs. The pump pulse is turned on at $t = 0$ and ends when the control pulse reaches its peak intensity at $t = 100$ fs. The corresponding dynamics was obtained using MCTDH, the initial wavefunction being the vibrational ground state in the electronic ground state. The

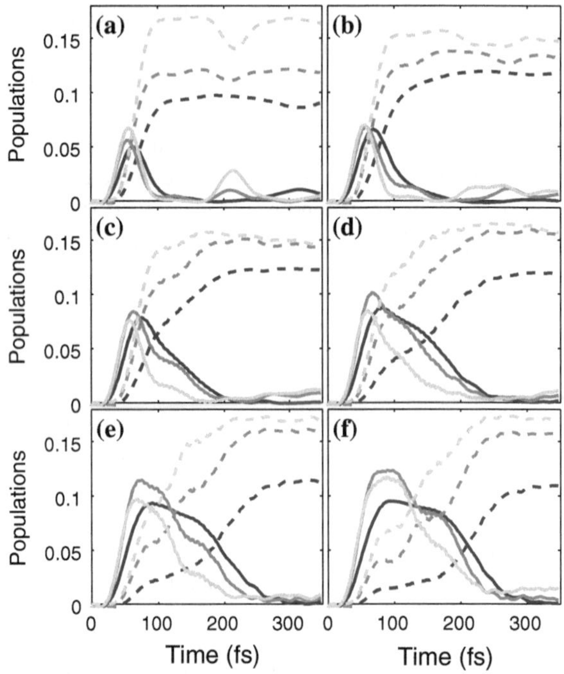

Fig. 13.8 Adiabatic populations of the S_1 (*dashed lines*) and S_2 (*full lines*) states for the pump pulse photon energies 4.6 eV (*blue*), 4.7 eV (*red*), and 4.8 eV (*green*) and for the control pulse intensities $I = 0$ (**a**), $I = 10$ TW/cm^2 (**b**), $I = 20$ TW/cm^2 (**c**), $I = 30$ TW/cm^2 (**d**), $I = 40$ TW/cm^2 (**e**), and $I = 50$ TW/cm^2 (**f**)

adiabatic populations of the S_1 and S_2 states are reported for various control pulse intensities and pump pulse photon energies in Fig. 13.8[18]: when there is no control pulse ($I = 0$) the S_2 state is populated by the pump pulse and we observe a very fast transfer to the S_1 state due to the presence of the conical intersection in agreement with the discussion of Sect. 12.4.

The dynamics of the molecule in the presence of a control pulse of 10 TW/cm^2 peak intensity (Fig. 13.8b) is hardly different from the dynamics without control pulse, when the intensity is further increased, a substantial fraction of the population remains in S_2 and decays at a slightly slower rate to S_1 between $t = 100$ fs and $t = 200$ fs. There is thus a trapping effect of the system in S_2 that is more and more pronounced (Fig. 13.8d–f). At the highest intensity ($I = 50$ TW/cm^2), the most part of the population excited to S_2 remains trapped up to $t = 200$ fs.

The impact of a non-resonant laser field on the topography of the PESs around the conical intersection that couples the S_1 and S_2 states can be analyzed in an effective way in terms of "dressed" PESs [61, 62]. The latter are obtained as the eigenvalues of the matrix:

[18]Reprinted with permission from [54]. Copyright 2014, American Institute of Physics.

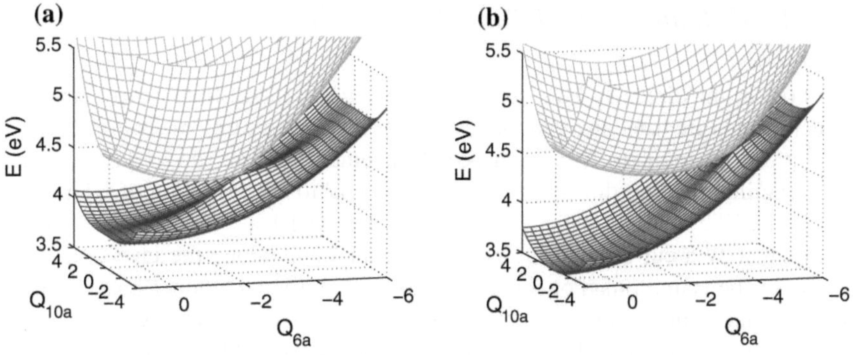

Fig. 13.9 S_1 (*green*) and S_2 (*red*) diabatic potentials plotted as functions of the q_{6a} and q_{10a} normal coordinates (**a**) in the field-free case and (**b**) dressed by a control field of intensity $I = 50$ TW/cm^2. $q_1 = 1.83$ and $q_{9a} = -0.15$ in the field-free case and $q_1 = 2.42$ and $q_{9a} = 0.13$ in the presence of a control field of intensity $I = 50$ TW/cm^2

$$\frac{1}{2}\sum_{i=1}^{4}\omega_i q_i^2 \begin{pmatrix} 1 & 0 \\ 0 & 1 \end{pmatrix} + \begin{pmatrix} V_{11}(\boldsymbol{q}) & V_{12}(\boldsymbol{q}) \\ V_{12}(\boldsymbol{q}) & V_{22}(\boldsymbol{q}) \end{pmatrix}$$
$$- \frac{1}{2}\begin{pmatrix} \alpha_{11}^{zz}(\boldsymbol{q}) & \alpha_{12}^{zz}(\boldsymbol{q}) \\ \alpha_{12}^{zz}(\boldsymbol{q}) & \alpha_{22}^{zz}(\boldsymbol{q}) \end{pmatrix}\epsilon_c^{0\,2}\,. \tag{13.65}$$

Two dimensional cuts of the dressed S_1 and S_2 PESs as a function of q_{6a} and q_{10a} are shown in Fig. 13.9[19]: (a) in the field-free case, here the dressed PESs are identical to the PESs of the isolated molecule, (b) in the presence of a control field of intensity $I = 50$ TW/cm^2. They show that the the non-resonant pulse induces a "Stark shift" of the potentials that moves the conical intersection away from the equilibrium geometry of the electronic ground state: in Fig. 13.9b, the CI lies higher in energy. Thus, one expects a trapping of the wavepacket on the S_2 PES for a much longer time than the natural S_2 lifetime since the CI is more difficult to reach.

13.7 Lab-Session V: Coherent Control by Laser Pulses

In the present lab-session, we follow Sects. 13.3–13.6 closely. Simplified Hamiltonian operators have been used to speed up the calculations: a one-dimensional analytical expression has been devised to describe the two-well potential energy of NHD_2 and two modes only have been used for pyrazine (instead of four in Sect. 13.6): ν_{10a} and ν_{6a} (see Fig. 13.7 for their definition and the molecule of pyrazine).

[19]Reprinted with permission from [54]. Copyright 2014, American Institute of Physics.

13.7.1 Enhancement of Tunneling in NHD_2

We work in the directory `lab-inputs/control/NHD2/enh_tun`. As usual, all commands are listed in a commands file, see `lab-inputs/control/NHD2/enh_tun/commands`. We can inspect the operator file `enh_field.op`: there is only one internal coordinate `qinv` corresponding to the inversion of NHD_2 (see Fig. 13.1). The analytical expression of the potential has been chosen to mimic approximately a cut through the potential of ammonia along the inversion mode. However, the Hamiltonian operator depends now on time (column `Time`): the system is in interaction with an external field of frequency `ome` = 1572.0 cm^{-1}, amplitude `ampl` = 0.01 a.u. and a pulse duration of `tp` = 5400.0 fs. The dipole moment is approximated as $0.714(1 - \exp(-1.655\,q))$. The parameters of the pulse that enhances the tunneling effect are close to those chosen in Sect. 13.3.

By typing "`mctdh84 -mnd gen_gss`", we converge the symmetric ground state (similar to the red state in Fig. 13.1). By typing "`mctdh84 -mnd gen_gsa`", we converge the antisymmetric ground state (similar to green red state in Fig. 13.1).

"`sumrst84 -mnd sum_l`" and "`sumrst84 -mnd sum_r`" make the difference and the sum of the two states creating wavepackets localized on the left side or on the right side of the potential, respectively.

"`mctdh84 -mnd gen_exc` generates an excited state corresponding to the state targeted by the laser pulse: see Fig. 13.3.

Let us assume that we start with a wavepacket localized on the right side of the double well. Without an external field, the system should tunnel to the left side in roughly 104 ps. But we have added a external field to enhance the tunneling process.

By typing "`mctdh84 -mnd prop_tun &`", we propagate the wavepacket in the presence of the external field. By typing "`mctdh84 -pes prop_tun`", we generate the potential surface. Already during the calculation we can enter the directory `prop_tun`: "`cd prop_tun`".

The command "`showsys84 -pes` allows us to plot the one-dimensional potential. The parameters can be optimized to zoom on the bottom of the potential by changing the `coordinate bounds`. It is also possible to change the unit of the potential (`change Y-axis units`).

"`showd1d84 -M -a -G -y 3.0 f1` gives the "movie" of the wavepacket: the transfer from the right to the left side is already completed after 5.4 ps. To speed up the movie one may additionally set the option `-n 4` by which only every fourth picture is displayed.

When the calculation is finished, we can see the evolution in time of the overlap of the wavepacket with three eigenstates: the symmetric and antisymmetric excited states and the excited state targeted by the laser pulse. This can be done by typing.

```
"crosscorr84 -r ../gs_s/restart"
"crosscorr84 -r ../gs_a/restart"
"crosscorr84 -r ../exc/restart"
"plgen -u 1:5 cross_gs_s cross_gs_a cross_exc".
```

The first three commands compute the overlap of the propagated wavefunction with the symmetric antisymmetric and excited states. The last command plots the absolute value squared of these overlaps. We see that we start from a linear combination of the symmetric and antisymmetric ground states. During the first half process the antisymmetric ground state is depopulated and the excited state is populated, while the symmetric ground state is only marginally affected. During the second half of the process the excited state is depopulated and the antisymmetric state is repopulated but with a minus sign (the sign cannot be seen on this figure).

We can also overlap the wavepacket with the wavefunctions localized on the left and right sides:

```
“crosscorr84 -r ../sup_r/restart”
“crosscorr84 -r ../sup_l/restart”
“plgen -u 1:5 cross_sup_r cross_sup_l”
```

We see that we start from a system localized on the right side that goes to the left through tunneling.

13.7.2 Destruction of Tunneling in NHD_2: Quasiresonant Regime

We work in the directory `lab-inputs/control/NHD2/cdt_qr/` We use essentially the same commands as above (they are listed in `lab-inputs/control/NHD2/cdt_qr/commands`). As only the operator file is different, the generated initial or reference states, (*i.e.* symmetric, antisymmetric, excited, right, left) are identical to the previous example.

We use parameters of a pulse similar to those of Sect. 13.5 (see the file `cdt_field.op`). The laser pulse is quasiresonant with an excited state. The dynamics is rather uninteresting since the system remains at its place and almost no tunneling occurs. The small oscillations observed in the population of the antisymmetric ground state are due to a slight population of excited states.

13.7.3 Destruction of Tunneling in NHD_2: Non-resonant Regime

We work in the directory `lab-inputs/control/NHD2/cdt_nr/` We use again essentially the same commands as above (they are listed in `lab-inputs/control/NHD2/cdt_nr/commands`).

We use parameters of a pulse similar to those of Sect. 13.6 (see the file `cdt_field.op`). The laser pulse is nonresonant with any excited state.

Again, the system remains at its place and almost no tunneling occurs. The destruction of tunneling is more efficient here but we need a higher intensity.

13.7.4 Control of the $\pi\pi^{\star}$ Excitation of Pyrazine by Stark Effect

We work in the directory

`lab-inputs/control/Stark_pyrazine/`

We have here three electronic states: the ground state, and two electronic states, $B_{2u}(\pi\pi^{*})$ and $B_{3u}(n\pi^{*})$, the latter are strongly coupled due to the presence of a conical intersection. We start from the system in the vibrational ground state of the electronic ground state. A resonant laser excites the system in the $B_{2u}(\pi\pi^{*})$ state, i.e. it induces a $\pi\pi^{\star}$ excitation of the system.

The corresponding dynamics (here with two degrees of freedom only) is performed by typing

"`mctdh84 -mnd no_Stark`"

The operator is in `no_Stark.op`: we use a model with two degrees of freedom `v10a` and `v6a`. the column `el` appears since we have three diabatic electronic states denoted "1" (the ground state), "2", and "3". The term followed by `S2&3` corresponds to the potential coupling between states "2" and "3". Because of the presence of an external field there is a term that depends on time. The laser couples the ground state (state "1") with the $\pi\pi^{*}$ excited state (state "3"), that is why the symbol `S1&3` appears for the electronic degree of freedom. The frequency of the pulse `ome` = 4.7 eV and its amplitude `Er` = 0.005 a.u.

The potential energy surfaces are generated by

"`mctdh84 -pes no_Stark`

The corresponding dynamics is in directory `no_Stark`. Type

"`cd no_Stark`"

and use `showsys84` to plot the potential surfaces and the reduced density. Use the menu point "60" (`change state selection`) to select the electronic potentials or nuclear densities to be displayed. Use the menu point "10" to select the plot task. In particular when the diabatic PES is to be plotted, it will be useful to switch to a 3D visualization (menu point 160). After you have viewed the PES and the reduced densities, close `showsys84` and type:

"`plstate`"

to study the diabatic populations. At the beginning only the ground state is populated, then the laser field populates state "3". After population of the latter, there is a very fast transfer to state "2" due to the presence of a conical intersection.

Now, we can perform the same dynamics but with a second non-resonant pulse that modifies the potential energy surfaces. We distinguish three cases here. They all correspond to a non-resonant laser pulse of frequency 1.8 eV but with laser-intensities of 10, 20, and 30 TW/cm^2, respectively. Return to the parent directory ("`cd ..`") and inspect `Stark.op`. Then run the propagations:

```
"mctdh84 -mnd Stark"
"mctdh84 -mnd -D Stark_2 -p Field 20.0 Stark"
"mctdh84 -mnd -D Stark_3 -p Field 30.0 Stark"
```

(In the first run we did not have to give `-D Stark_1 -p Field 10.0` because these are set as defaults in input and operator files). We can now compare carefully the different populations given by `plstate` and see the effect of the non-resonant pulse. The transfer from state "3" to state "2" becomes weaker and weaker with increasing intensity of the Stark laser. In other words, the non-resonant pulse traps the system in state "3" and slows down the non-adiabatic process. To see this, type

```
"plstate -i no_Stark"
"plstate -i Stark_1"
"plstate -i Stark_2"
"plstate -i Stark_3"
```

A clearer picture is provided by

```
"plstate -a 150 -G -y 0.35 -i no_Stark"
"plstate -a 150 -G -y 0.35 -i Stark_1"
"plstate -a 150 -G -y 0.35 -i Stark_2"
"plstate -a 150 -G -y 0.35 -i Stark_3"
```

To compare the populations of a selected state at different laser intensities, one has first to generate the state-population data files. Type:

```
"rdcheck84 -on no -oq no -oc st_0 -i no_Stark"
"rdcheck84 -on no -oq no -oc st_1 -i Stark_1"
"rdcheck84 -on no -oq no -oc st_2 -i Stark_2"
"rdcheck84 -on no -oq no -oc st_3 -i Stark_3"
```

and then

```
"plgen -u 1:3 -a 150 -G st_0.pl st_1.pl st_2.pl st_3.pl"
```

This plot shows the populations of the first excited state ($n\pi^*$) for the four different field strengths. To inspect the populations of the ground and second excited state, replace `-u 1:3` with `-u 1:2` and `-u 1:4`, respectively.

References

1. Judson RS, Rabitz H (1992) Teaching lasers to control molecules. Phys Rev Lett 68:1500
2. Asssion A, Baumert T, Bergt M, Brixner T, Kiefer B, Seyfried V, Strehle M, Gerber G (1998) Control of chemical reactions by feedback-optimized phase-shaped femtocecond laser pulses. Science 282:919
3. Rabitz H, de Vivie-Riedle R, Motkuz M, Kompa K (2000) Chemistry - whither the future of controlling quantum phenomena? Science 288:824
4. Shapiro M, Brumer P (2003) Principles of the quantum control of molecular processes. Wiley, New York
5. Chergui M (ed) (1996) Femtochemistry. World Scientific, Singapore
6. Zewail AH (1994) Femtochemistry—ultrafast dynamics of the chemical bond. World Scientific, Singapore
7. Ihee H, Lobastov V, Gomez U, Goodson B, Srinivasan R, Ruan C-Y, Zewail AH (2001) Direct imaging of transient molecular structures with ultrafast diffraction. Science 291:385
8. Shi S, Rabitz H (1990) Quantum mechanical optimal control of physical observables in microsystems. J Chem Phys 92:364
9. Zhu W, Rabitz H (1998) A rapid monotonically convergent iteration algorithm for quantum optimal control over the expectation value of a positive definite operator. J Chem Phys 109:385
10. Daniel C, Full J, González L, Lupulescu C, Manz J, Merli A, Vajda S, Wöste L (2003) Deciphering the reaction dynamics underlying optimal control laser fields. Science 299:536–539
11. Schröder M, Carreon-Macedo J-L, Brown A (2008) Implementation of an iterative algorithm for optimal control of molecular dynamics into MCTDH. Phys Chem Chem Phys 10:850
12. Schröder M, Brown A (2009) Realization of the cnot quantum gate operation in 6d ammonia using the oct-mctdh approach. J Chem Phys 131:034101
13. Schröder M, Brown A (2009) Generalized filtering of laser fields in optimal control theory: application to symmetry filtering of quantum gate operations. New J Phys 11:105031
14. Brumer P, Shapiro M (1986) Control of unimolecular reactions using coherent light. Chem Phys Lett 126:541
15. Drescher M, Hentschel M, Kienberger R, Uiberacker M, Scrinzi A, Westerwalbesloh T, Kleineberg U, Heinzmann U, Krausz F (2002) Time-resolved atomic inner-shell spectroscopy. Nature 419:803
16. Goulielmakis E, Loh Z-H, Wirth A, Santra R, Rohringer N, Yakovlev VS, Zherebtsov S, Pfeifero T, Azzeer AM, Kling MF, Leone SR, Krausz F (2010) Real-time observation of valence electron motion. Nature 466:739
17. Krausz F, Ivanov M (2009) Attosecond physics. Rev Mod Phys 81:163
18. Kuleff AI, Cederbaum LS (2012) Ultrafact correlation-driven electron dynamics. J Phys B 47:124002
19. Daems D, Guérin S, Hertz E, Jauslin HR, Lavorel B, Faucher O (2005) Field-free two-direction alignement alternation of linear molecules by elliptic laser pulses. Phys Rev Lett 95:063005
20. Madsen CB, Madsen LB, Viftrup SS, Johansson MP, Poulsen TB, Holmegaard L, Kumarappan V, Jorgensen KA, Stapelfeldt H (2009) Manipulating the torsion of molecules by strong laser pulses. Phys Rev Lett 102:073007
21. Holmegaard L, Hansen JL, Kalhøj L, Kragh SL, Stapelfeldt H, Filsinger F, Küpper J, Meijer G, Dimitrovski D, Martiny C, Madsen LB (2010) Photoelectron angular distributions from strong-field ionization of oriented molecules. Nat Phys 6:428
22. Bisgaard CZ, Clarkin OJ, Wu G, Lee AMD, Gessner O, Hayden CC, Stolow A (2009) Time-resolved molecular frame dynamics of fixed-in-space cs2 molecules. Science 323:1464
23. Sala M (2015) *Quantum Dynamics and Laser Control for Photochemistry*. PhD thesis, Université de Bourgogne
24. Sala M (2016) In quantum dynamics and laser control for photochemistry, Springer Theses. Recognizing Outstanding. PhD Research. Springer, Heidelberg
25. Shirley J (1965) Solution of the Schrödinger equation with a Hamiltonian periodic in time. Phys Rev 138:B979

26. Sambe H (1973) Steady states and quasi-energies of a quantum-mechanical system in an oscillating field. Phys Rev A 7:2203
27. Bayfield JE (1999) Quantum evolution. A Wiley-Interscience Publication, New York
28. Guérin S (1997) Complete dissociation by chirped laser pulses designed by adiabatic floquet analysis. Phys Rev A 56:1458
29. Guerin S, Jauslin HR (2003) Control of quantum dynamics by laser pulses: adiabatic floquet theory. Adv Chem Phys 125:147
30. Guérin S, Rouzée A, Hertz E (2008) Ultimate field-free molecular alignment by combined adiabatic-implusive field design. Phys Rev A 77:041404
31. Lapert M, Guérin S, Sugny D (2011) Field-free quantum cogwheel by shaping of rotational wave packets. Phys Rev A 83:013403
32. Cohen-Tannoudji C, Dupont-Roc J, Grynberg G (1998) Atom-photon interactions: basic processes and applications. Wiley Science Paperback Series
33. Guérin S, Monti F, Dupont J-M, Jauslin HR (1997) On the relation between cavity dressed states, Floquet states, RWA approximation and semiclassical models. J Phys A 30:7193
34. Yatsenko L, Guérin S, Jauslin HR (2002) Topology of adiabatic passage. Phys Rev A 65:233601
35. Guérin S, Hakobyan V, Jauslin HR (2011) Optimal adiabatic passage by shaped pulses: Efficiency and robustness. Phys. Rev. A 84:013423
36. Cohen-Tannoudji C, Diu B, Laloe F (1992) Quantum mechanics. Wiley-VCH
37. Rosen N, Zener C (1932) Double Stern-Gerlach experiment and related collision phenomena. Phys Rev 40:502
38. Marquardt R, Quack M, Thanopoulos I, Luckhaus D (2003) A global electric dipole function of ammonia and isotopomers in the electronic ground state. J Chem Phys 119:10724
39. Marquardt R, Sagui K, Klopper W, Quack M (2005) Global analytical potential enerrgy surface for large amplitude nuclear motions in ammonia. J Phys Chem B 109:8439
40. Sala M, Guérin S, Gatti F, Marquardt R, Meyer H-D (2012) Laser induced enhancement of tunneling in NHD_2. J Chem Phys 136:194308
41. Kartha SB, Singh K, Job VA, Karhta VB (1988) J Mol Struct (Theochem) 129:86
42. Snels M, Hollenstein H, Quack M (2006) Mode selectivity tunneling dynamics observed by high resolution spectroscopy of the bending fundamentals of $^{14}NH_2D$ and $^{14}ND_2H$. J Chem Phys 125:194319
43. Snels M, Hollenstein H, Quack M (2003) The NH and ND stretching fundamentals of $^{14}ND_2H$. J Chem Phys 119:7893
44. Hervé S, Le Quéré F, Marquardt R (2004) Effect of rotations on the generation of coherent internal molecular motion. Int J Quant Chem 99:439
45. Loesch HJ, Reimscheid A (1990) Brute force in molecular reaction dynamics: a novel technique for measuring steric effects. J Chem Phys 93:4779
46. Slenczka A, Friedrich B, Herschbach D (1994) Pendular alignment of paramagnetic molecules in uniform magnetic fields. Phys Rev Lett 72:1806
47. Kim W, Felker PM (1996) Spectroscopy of pendular states in optical-field-aligned species. J Chem Phys 104:1147
48. Cai L, Marango J, Friedrich B (2001) Time-dependent alignment and orientation of molecules in combined electrostatic and pulsed nonresonant laser fields. Phys Rev Lett 86:775
49. Seideman T (2001) On the dynamics of rotationally-broad, spatially-aligned wavepackets. J Chem Phys 115:5965
50. Marquardt R, Quack M, Thanopoulos I, Luckhaus D (2003) Tunneling dynamics of the NH chromophore in NHD_2 during and after coherent infrared excitation. J Chem Phys 118:643
51. Madsen CB, Madsen LB, Viftrup SS, Johansson MP, Poulsen TB, Holmegaard L, Kumarappan V, Jrgense KA, Stapelfeldt H (2009) A combined experimental and theoretical study on realizing and using laser controlled torsion of molecules. J Chem Phys 234310:234310
52. Thomas S, Guérin S, Jauslin HR (2005) State-selective chirped adiabatic passage on dynamically laser-aligned molecules. Phys Rev A 71:013402
53. Sala M, Gatti F, Guérin S (2014) Coherent destruction of tunneling in a six-dimensional model of NHD_2: a computational study using the multi-configuration time-dependent Hartree method. J Chem Phys 141:164326

54. Sala M, Saab M, Lasorne B, Gatti F, Guérin S (2014) Laser control of the radiationless decay in pyrazine using the dynamic stark effect. J Chem Phys 140:194308
55. Saab M, Sala M, Lasorne B, Guérin S, Gatti F (2014) Full-dimensional control of the radiationless decay in pyrazine using the dynamic Stark effect. J Chem Phys 141:134114
56. Woywod C, Domcke W, Sobolewski AL, Werner H-J (1994) Characterization of the S_1-S_2 conical intersection in pyrazine using ab initio multiconfiguration self-consistent-field and multireference configuration-interaction methods. J Chem Phys 100:1400
57. Worth GA, Meyer H-D, Cederbaum LS (1996) The effect of a model environment on the S_2 absorption spectrum of pyrazine: a wavepacket study treating all 24 vibrational modes. J Chem Phys 105:4412
58. Worth GA, Meyer H-D, Cederbaum LS (1998) Relaxation of a system with a conical intersection coupled to a bath: a benchmark 24-dimensional wavepacket study treating the environment explicitly. J Chem Phys 109:3518–3529
59. Worth GA, Meyer H-D, Cederbaum LS (1999) State filtering by a bath: up to 24 mode numerically exact wavepacket propagations. Chem Phys Lett 299:451
60. Raab A, Worth G, Meyer H-D, Cederbaum LS (1999) Molecular dynamics of pyrazine after excitation to the S_2 electronic state using a realistic 24-mode model Hamiltonian. J Chem Phys 110:936
61. Kinzel D, Marquetand P, González L (2012) Stark control of a chiral fluorethylene derivative. J Phys Chem A 116:2743
62. Ruiz-Barragan S, Blancafort L (2013) Photophysics of fulvene under the non-resonant stark effect. shaping the conical intersection seam. Faraday Discuss 163:497

Erratum to: Applications of Quantum Dynamics in Chemistry

Fabien Gatti, Benjamin Lasorne, Hans-Dieter Meyer and André Nauts

Erratum to:
F. Gatti et al., *Applications of Quantum Dynamics in Chemistry*, Lecture Notes in Chemistry 98, https://doi.org/10.1007/978-3-319-53923-2

In the original version of the book, the second affiliation of the author André Nauts was missing. His second affiliation "Laboratoire de Chimie Physique (UMR8000), Université de Paris-Sud, Orsay, Essonne, France" has now been inserted.

The updated online version of this book can be found at https://doi.org/10.1007/978-3-319-53923-2

F. Gatti et al., *Applications of Quantum Dynamics in Chemistry*, Lecture Notes in Chemistry 98, https://doi.org/10.1007/978-3-319-53923-2_14

Index

F. Gatti et al., *Applications of Quantum Dynamics in Chemistry*,
Lecture Notes in Chemistry 98, DOI 10.1007/978-3-319-53923-2

MIX
Papier aus verantwortungsvollen Quellen
Paper from responsible sources
FSC® C105338

If you have any concerns about our products,
you can contact us on
ProductSafety@springernature.com

In case Publisher is established outside the EU,
the EU authorized representative is:
Springer Nature Customer Service Center GmbH
Europaplatz 3, 69115 Heidelberg, Germany

Printed by Libri Plureos GmbH
in Hamburg, Germany